Histidine Kinases in Signal Transduction

Histidine Kinases in Signal Transduction

Edited by
Masayori Inouye
Rinku Dutta
Department of Biochemistry and Molecular Biology
Robert Wood Johnson Medical School
Piscataway, New Jersey

ACADEMIC PRESS
An imprint of Elsevier Science

Amsterdam Boston London New York Oxford Paris
San Diego San Francisco Singapore Sydney Tokyo

This book is printed on acid-free paper. ∞

Academic Press
An imprint of Elsevier Science
525 B Street, Suite 1900, San Diego, California 92101-4495, USA
http://www.academicpress.com

Academic Press
An imprint of Elsevier Science
84 Theobald's Road, London WC1X 8RR, UK
http://www.academicpress.com

Library of Congress Control Number: 2002107358

International Standard Book Number: 0-12-372484-8

PRINTED IN CHINA
02 03 04 05 06 CTP 9 8 7 6 5 4 3 2 1

CONTENTS

19 The Sln1-Ypd1-Ssk1 Multistep Phosphorelay System That Regulates an Osmosensing MAP Kinase Cascade in Yeast

Haruo Saito

20 Histidine Kinases of *Dictyostelium*

Christophe Anjard and William F. Loomis

21 Ethylene Perception in *Arabidopsis* by the ETR1 Receptor Family: Evaluating a Possible Role for Two-Component Signaling in Plant Ethylene Responses

Ronan C. O'Malley and Anthony B. Bleecker

PREFACE

During the last few years, a major achievement in the field of histidine kinase is the determination of their three-dimensional structures. Even if they are yet partial, the wealth brought from structural determination has provided undipustable new insights into our understanding of the function and regulatory mechanisms of histidine kinase. It is probably the most exciting time for those who study histidine kinase and their role in signal transduction. As I myself have been engaged in research on the signal transduction mediated by a histidine kinase for more than 20 years, I can certainly witness the recent excitement in the field. In addition to structural studies, sequencing of bacterial and plant genomes has revealed numerous histidine kinases that exist in bacteria and in plants, and the new roles of histidine kinases in a wide range of stresses.

It is highly fortunate that many major players in the field have agreed to contribute to this book, making it comprehensive and exciting first book on this subject. Indeed, this book covers topics from signal recognition at the receptor domain and its transduction through the membrane to regulation of the function of the cytoplasmic kinase/phosphotase domain and the response regulators executing the signal responses. It describes how widely histidine kinases respond toward numerous unique signals. In Chapter 1, I have written introductory remarks overviewing all of the chapters in this book, summarizing some of the major aspects and the uniqueness of histidine kinases, contrasting serine, threonine, and tyrosine protein kinases almost exclusively used in eukaryotes for stress responses and signal transduction. It is hoped that these remarks will provide a guideline to readers when reading this book.

Finally, I express my sincere gratitude to all of the contributors for their excellent chapters, to Dr. Rinku Dutta for her editorial assistance during the early stages of the book, to Janice Nappe for her secretarial and editorial assistance, to Yan Zhu for her editorial assistance, and to Aaron Johnson of Academic Press for the production of this book.

Masayori Inouye

CONTRIBUTORS

Numbers in parentheses indicate the pages on which the authors' contributions begin.

HIROFUMI AIBA (191), Laboratory of Molecular Microbiology, School of Agriculture, Nagoya University, Chikusa-ku, Nagoya 464-8601, Japan

CHRISTOPHE ANJARD (421), Center for Molecular Genetics, Division of Biology, University of California, San Diego, La Jolla, California 92093

MARIETTE R. ATKINSON (143), Department of Biological Chemistry, University of Michigan Medical School, Ann Arbor, Michigan 48109

BONNIE L. BASSLER (313), Department of Molecular Biology, Princeton University, Princeton, New Jersey 08544

ALEXANDRINE M. BILWES (47), Department of Chemistry and Chemical Biology, Cornell University, Ithaca, New York 14853

ANTHONY B. BLEECKER (439), Department of Botany, University of Wisconsin, Madison, Wisconsin 53706

MICHAEL J. BOWER (483), GlaxoSmithKline, Collegeville, Pennsylvania 19426

M. K. R. BURNHAM (459), Antimicrobials and Host Defence, GlaxoSmithKline, Collegeville, Pennsylvania 19426

BRIAN R. CRANE (47), Department of Chemistry and Chemical Biology, Cornell University, Ithaca, New York 14853

RINKU DUTTA (25), Department of Biochemistry and Molecular Biology, Robert Wood Johnson Medical School, Piscataway, New Jersey 08854

LEIV SIGVE HÅVARSTEIN (341), Department of Chemistry and Biotechnology, Agricultural University of Norway, N-1432 Ås, Norway

J. HUBBARD (459), Computational and Structural Sciences, GlaxoSmithKline, Harlow, United Kingdom

MITSUHIKO IKURA (11), Division of Molecular and Structural Biology, Ontario Cancer Institute and Department of Medical Biophysics, University of Toronto, Toronto, Ontario, Canada M5G 2M9

MASAYORI INOUYE (1, 25), Department of Biochemistry and Molecular Biology, Robert Wood Johnson Medical School, Piscataway, New Jersey 08854

HIDEO IWASAKI (297), Division of Biological Science, Graduate School of Science, Nagoya University and CREST, JST, Furo-cho, Chikusa, Aichi 464-8602, Japan

PENG JIANG (143), Department of Biological Chemistry, University of Michigan Medical School, Ann Arbor, Michigan 48109

SUNG-HOU KIM (123), Department of Chemistry, University of California, Berkeley, Berkeley, California 94720

TAKAO KONDO (297), Division of Biological Science, Graduate School of Science, Nagoya University and CREST, JST, Furo-cho, Chikusa, Aichi 464-8602, Japan

KRISTIN K. KORETKE (483), GlaxoSmithKline, Collegeville, Pennsylvania 19426

HIROFUMI KUROKAWA (11), Division of Molecular and Structural Biology, Ontario Cancer Institute and Department of Medical Biophysics, University of Toronto, Toronto, Ontario, Canada M5G 2M9

WILLIAM F. LOOMIS (421), Center for Molecular Genetics, Division of Biology, University of California, San Diego, La Jolla, California 92093

ANDREI N. LUPAS (483), GlaxoSmithKline, Collegeville, Pennsylvania 19426

MASAHIRO MATSUBARA (165), Laboratory of Molecular Microbiology, School of Agriculture, Nagoya University, Chikusa-ku, Nagoya 464-8601, Japan

TAKESHI MIZUNO (165, 191), Laboratory of Molecular Microbiology, School of Agriculture, Nagoya University, Chikusa-ku, Nagoya 464-8601, Japan

KENNY C. MOK (313), Department of Molecular Biology, Princeton University, Princeton, New Jersey 08544

HIROTADA MORI (191), CREST, Research and Education Center for Genetic Information, Nara Institute of Science and Technology, Ikoma 630-0101, Japan

H. JOCHEN MÜLLER-DIECKMANN (123), Department of Chemistry, University of California, Berkeley, Berkeley, California 94720

AUSTIN NEWTON (377), Department of Molecular Biology, Princeton University, Princeton, New Jersey 08544

ALEXANDER J. NINFA (143), Department of Biological Chemistry, University of Michigan Medical School, Ann Arbor, Michigan 48109

RODGER NOVAK (365), Institute of Microbiology and Genetics, Vienna Biocenter, Vienna A-1030, Austria

RONAN C. O'MALLEY (439), Department of Botany, University of Wisconsin, Madison, Wisconsin 53706

NORIKO OHTA (377), Department of Molecular Biology, Princeton University, Princeton, New Jersey 08544

TAKU OSHIMA (191), CREST, Research and Education Center for Genetic Information, Nara Institute of Science and Technology, Ikoma 630-0101, Japan

SANG-YOUN PARK (47), Department of Chemistry and Chemical Biology, Cornell University, Ithaca, New York 14853

JAMES A. PELISKA (143), Department of Biological Chemistry, University of Michigan Medical School, Ann Arbor, Michigan 48109

AUGEN PIOSZAK (143), Department of Biological Chemistry, University of Michigan Medical School, Ann Arbor, Michigan 48109

CINDY M. QUEZADA (47), Department of Biology, California Institute of Technology, Pasadena, California 91125

HARUO SAITO (397), Division of Molecular Cell Signaling, Institute of Medical Science, University of Tokyo, Tokyo 108-8639, Japan

MELVIN I. SIMON (47), Department of Biology, California Institute of Technology, Pasadena, California 91125

ANN M. STOCK (237), Center for Advanced Biotechnology and Medicine, Howard Hughes Medical Institute, Robert Wood Johnson Medical School, Piscataway, New Jersey 08854

JEFFRY B. STOCK (73), Departments of Molecular Biology and Chemistry, Princeton University, Princeton, New Jersey 08544

J. P. THROUP (459), Antimicrobials and Host Defence, GlaxoSmithKline, Collegeville, Pennsylvania 19426

CHIERI TOMOMORI (11), Cardiovascular Biology Department, Oklahoma Medical Research Foundation, Oklahoma City, Oklahoma 73104

ELAINE TUOMANEN (365), Department of Infectious Diseases, St. Jude Children's Research Hospital, Memphis, Tennessee 38105

KOTTAYIL I. VARUGHESE (203), Department of Molecular and Experimental Medicine, The Scripps Research Institute, La Jolla, California 92037

RICHARD DAVID VIERSTRA (273), Cellular and Molecular Biology Program and the Department of Horticulture, University of Wisconsin–Madison, Madison, Wisconsin 53706

CRAIG VOLKER (483), GlaxoSmithKline, Collegeville, Pennsylvania 19426

BARRY L. WANNER (191), Department of Biological Sciences, Purdue University, West Lafayette, Indiana 47907

ANN H. WEST (237), Department of Chemistry and Biochemistry, University of Oklahoma, Norman, Oklahoma 73019

PETER W. WOLANIN (73), Department of Molecular Biology, Princeton University, Princeton, New Jersey 08544

WEI YANG (219), Laboratory of Molecular Biology, National Institute of Diabetes and Digestive and Kidney Diseases, National Institutes of Health, Bethesda, Maryland 20892

YAN ZHU (25), Department of Biochemistry and Molecular Biology, Robert Wood Johnson Medical School, Piscataway, New Jersey 08854

CHAPTER 1

Histidine Kinases: Introductory Remarks

MASAYORI INOUYE
Department of Biochemistry and Molecular Biology, Robert Wood Johnson Medical School, Piscataway, New Jersey 08854

Histidine kinases (HKs) are the major players in signal transduction in prokaryotes as Ser/Thr/Tyr protein kinases are in the eukaryotes. Advances in research on HKs and signal transduction through these proteins have been remarkable, and now their structures and mechanisms of functions have begun to be unveiled. This short introductory chapter highlights a number of unique features of HKs and the systems regulated by their networks described in this book and attempts to cross-reference these features to each chapter in this book. © 2003, Elsevier Science (USA).

INTRODUCTION

Bacteria are always exposed to environmental changes. To survive under such conditions, bacteria adapt their signal transducing system, using histidine kinase

(HK) as sensors. Upon sensing the environmental signals, HKs unique to individual external signals transduce them to downstream factors called response regulators (RR; Chapter 12), which then promote necessary reactions for cells to acclimate to the new environmental condition. During this process, the RR is usually involved in the regulation of specific gene expression, interacting directly with the promoter regions to set new cellular physiology adapting to new environmental changes. RRs, however, can interact directly with proteins in some cases.

The signal transduction system using a HK and a RR is collectively called "two-component system" as it uses two distinctive components, HK and RR. However, more precisely reflecting its signal transducing mechanism, it is also called the "His-Asp phosphorelay signal transduction system (HAP system)." Indeed, these two names are used synonymously throughout the book.

The HAP system, believed to be unique to bacteria, is now found in lower eukaryotes, yeast (Chapter 19), slime mold (Chapter 20), and even in plants (Chapter 21), where it has been shown from genome analysis that the HAP system does not exist in *C. elegans*, Drosophila, mouse, and human (Chapter 23). The HAP system has been shown to also exist in some archaea (Chapter 23). These HAP systems respond to various external signals, including osmolarity (Chapters 2, 3, and 19), nutritional starvation (Chapters 7, 10, 15, and 20), specific chemicals causing taxis (Chapters 4, 5, and 6), oxygen deficiency (Chapter 8), light (Chapters 13 and 14), and chemical signals produced by its own or other systems (Chapters 15–17 and 21). Many other signals are used for the HAP system, although not described in this book, such as inorganic phosphate (e.g. PhoR/PhoB; [1]), metal ions (e.g. PhoP/PhoQ; [2]), temperature [3] and misfolded proteins in the bacterial envelope (e.g. CpxA/CpxR; [4]).

Indeed, *Escherichia coli* contains 29 HK genes and 32 RR genes, constituting at least 29 independent His-Asp phosphorelay systems, each of which is considered to be responsible for the response and adaptation to different stresses (Chapter 9). Deletion strains of every *E.coli* HK-RR operon and several RR genes have been constructed, and DNA microarray analysis of these deletion strains revealed that complex networks overlap a number of His-Asp phosphorelay signaling pathways to regulate *E.coli* physiology globally (Chapter 9). Interestingly, the multiple HAP systems are also used in a sophisticated regulatory network in cell cycle control in a gram–negative bacterium (Chapter 18), in developmental processes leading to spore formation (Chapter 10), and in fruiting body formation (Chapter 15). Needless to say, bacterial HKs are potentially good targets for the development of new antibiotics (Chapter 22).

It is interesting to note that all eukaryotic HKs found so far are hybrid HKs consisting of a HK domain and a response regulator domain (see Chapters

19–21). Although these eukaryotic HKs also sense environmental signals, all of them identified thus far have been shown to indirectly regulate specific gene expression via a MAP Ser/Thr kinase cascade. However, it has been reported that eukaryotic HKs involved in the cytokinin signal transduction in *Arabidopsis* directly activate specific genes through a phosphorelay pathway [5]. Although not firmly proven, the authors proposed that upon sensing the cytokinin signal, hybrid HKs in the cytoplasmic membrane transfer the high-energy phosphoryl group to a shuttle protein (on a His residue), which transmits the phosphoryl signal from the cytoplasm to the nucleus. Then the phosphoryl group is transferred to response regulators (on an Asp residue), which function as transcription activators upon phosphorylation for those genes involved in cell divisions, shoot formation, and senescence. This chapter discusses unique features of HKs and the HK-mediated signal transduction distinct from the eukaryotic Ser/Thr/Tyr protein kinase.

BASIC STRUCTURE OF HISTIDINE KINASES (HKS)

A typical HK such as EnvZ (Chapters 2 and 3) consists of four distinct major domains: the periplasmic sensor or receptor domain, the transmembrane domain, the linker domain (also called the HAMP domain; see Chapter 5) connecting the transmembrane domain, and the cytoplasmic kinase domain. HKs in this class (class I) sometimes lack both periplasmic and transmembrane domains. Some class I HKs contain other functional domains, such as the RR domain at the N- or C-terminal side of the kinase domain. In addition to the RR domain, some have yet another histidine-containing domain called the HPt domain to which the high-energy phosphate can be relayed (Chapters 5 and 8). These class I HKs are called hybrid HKs, which are also commonly found in eukaryotic systems (Chapters 19–21). Importantly, the kinase domain consisting of approximately 230 residues is composed of two subdomains: (1) the central dimerization domain containing the key histidine residue called the DHp domain and (2) the ATP-binding domain, called the CA domain (Chapters 2 and 3). The function of the CA domain is disputed as to whether it functions only for ATP binding to provide ATP for the kinase reaction or whether it also plays a role in the catalysis reaction (Chapters 3 and 7). Therefore, CA stands for either "<u>C</u>atalytic and <u>A</u>TP binding" or "<u>C</u>atalysis assisting and <u>A</u>TP binding" at present.

Another unique class of HKs (class II) functions in chemotaxis (Chapter 4). The function of class II HKs is regulated by interaction with an independent signal-transducing protein called MCP. Structural studies on the well-studied CheA, a typical class II protein kinase, revealed that the DHp domain of class I HKs is replaced with a simple four-helical bundle without the key histidine residue; this four-helical bundle serves only for dimerization. As a result, the CA domain of class II HKs is oriented in the opposite topology to the class I CA domain. Therefore, each of the CA domains in a dimer, facing outward, interacts with another domain consisting of a four-helical bundle called HPt (see Fig. 5 in Chapter 11). The HPt domain contains a single histidine residue and is used as the primary autophosphorylation domain for class II HKs (Chapters 3 and 4).

UNIQUENESS OF HKs

Of the two subdomains of the kinase domains just described, the CA domain shares three-dimensional structural similarities with the ATP-binding domains of gyrase B, heat shock protein Hsp90, and MutL, a DNA mismatch repair enzyme (Chapters 3 and 11). HK catalytic domain organization is therefore postulated to be a fusion of two independent evolved components: CA and DHp domains (Chapters 3 and 23).

The DHp domain consists of two helical hairpin structures, forming a four-helical bundle (Chapter 2). It is postulated that during the course of evolution, this primitive four-helical bundle acquired the key histidine residue on the center of helix I, which later became capable of phosphorylation by ATP. The physical connection of an ATP-binding domain at the C-terminal end of each hairpin α-helical structure of the DHp domain makes HKs highly efficient in the transfer of high-energy phosphoryl groups to the histidine residue. This phosphorylation step is the first key step in signal transduction as the γ-phosphoryl group of ATP is transferred to the N3 position of the histidine residue while maintaining its important high-energy state. When HKs act with their cognate RR molecules, the high-energy phosphoryl group at the histidine residue is transferred to a specific aspartate residue of the RR. As described later, HKs also function as phosphatases to dephosphorylate the RRs to produce inorganic phosphate. The DHp domain, by itself, has been shown to contain the phosphatase activity in which the conserved histidine residue plays a crucial role (Chapter 3).

Because the high-energy state of the phosphoryl group is maintained during the transfer reaction from HKs to RRs, the phosphoryl group on the aspartate residue can be further transferred to an additional histidine residue found in the HPt domain (Chapters 5 and 8). The HPt domain consists of a four-helical bundle similar to the DHp domain of HKs. However, the HPt domain consists of a single polypeptide having a single histidine residue. The phosphoryl group is then transferred to a second RR, but again maintains the high-energy state. In this fashion, the high-energy phosphate is transferred to the final RR, which in most cases acts as a transcription factor (activator and/or repressor) upon phosphorylation. This signal transduction using HKs is, therefore, called the His-Asp phosphorelay signal transduction system (HAP system). It should be noted that the His-Asp phosphorelay from HK through RR to HPt might happen within a single dimer complex of HK if the HK is a hybrid kinase. Interestingly, in some HAP systems, a DHp domain, together with a CA-like domain, without the ATP-binding site, is used instead of an HPt domain (Chapter 10).

DIFFERENCE BETWEEN HKs AND Ser/Thr/Tyr KINASES

The basic principle underlying the HAP system is utilization of the high-energy phosphoryl groups as a transferable means for signal transduction in the signal-transducing cascade. This creates clear distinctions from signaling systems using Ser/Thr/Tyr protein kinases used mainly in eukaryotes. First, the high-energy phosphoryl group on a HK, particularly on an aspartate residue, is unstable and therefore a high-energy phosphoryl group in any HAP system has an intrinsic instability with a half-life of a few seconds in some systems to several hours in some others. In addition, most of the class I HKs are bifunctional, having not only kinase (aspartate transphosphorylase) but also phosphatase (phosphoryl aspartate desphosphorylase) activities, although some HAP systems use separate phosphatases. Kinases themselves, therefore, play a major role in the dephosphorylation reaction of RRs. It has been proposed that the ratio of kinase-to-phosphatase activities of a HK is the key factor in exerting the response to the received signal (Chapters 3 and 7). In contrast to this rheostat model, the switch model proposes that HKs are only in one of two possible states, either kinase^{+} phosphatase^{-} (on–off) or kinase^{-} phosphatase^{+} (off–on). These two models are discussed in a later section.

However, in Ser/Thr/Tyr kinase systems, phosphorylated Ser/Thr/Tyr residues are extremely stable. This stability results in the requirement of another enzyme, a phosphatase, to remove the phosphoryl group to block the signaling cascade. Second, within a HAP cascade, there is no amplification of the initial signal; upon consuming one ATP molecule at the primary signal-transducing HK, the high-energy phosphate from this ATP is used as the signal all the way through to the final RR, without amplification. In contrast, in the Ser/Thr/Tyr kinase cascade system, an upstream kinase phosphorylates a large number of downstream kinases using ATP to effectively amplify the initial signal, which can be amplified exponentially at each step of the multiple kinase cascade.

Furthermore, because bacteria are, in general, able to grow much faster than eukaryotic cells, the intrinsic instablility of the HK system may not pose serious problems for bacteria or be even beneficial in some cases for a faster response to an external signal. As bacterial cells are much smaller than eukaryotic cells, the distance between the initial signal receptor and the site of action of the final RR is much shorter, making an extensive amplification of the input signal unnecessary. The significance of the multistep cascade seen in the HAP system is, therefore, quite different from the multiple Ser/Thr/Tyr kinase cascade. In the HAP system, multisteps appear to be used for fine-tuning of the signal transduction but not signal amplification. More specifically, the signal, in the form of the high-energy phosphate, can be pooled within the system before reaching the final RR or, alternatively, can also be quenched at each step of the phosphorelay system. It is interesting to note that some bacteria, such as *Mycobacterium tuberculosis* [6] and *Myxococcus xanthus* [7], have a large number of Ser/Thr kinases along with HKs. It is tempting to speculate that Ser/Thr kinases found in these bacteria are used for those cellular events that require long-lasting signaling systems and, therefore, cannot be sustained by HKs. The maintenance of an extremely long dormant state in *M. tuberculosis* and long developmental processes leading to fruiting body formation in *M. xanthus* are two such examples of long-lasting signaling in bacteria possibly requiring Ser/Thr kinases.

Some low eukaryotes and plants use HKs rather than Ser/Thr/Tyr kinases for sensing environmental signals. If a particular environment signal, such as osmolarity, ethylene, and cytokinin, is long lasting and stably maintaining, a cytoplasmic membrane HK can be activated constantly to maintain the active phosphorelay from the HK to the last component in the pathway without the signal being amplified. When the signal diminishes the phosphorelay, activity is simultaneously reduced. Such high coordinations between external signals and their outputs at the end of the pathway are unique in the HK–RR phosphorelay system, and the newly found cytokinin–phosphorelay signal transduction system in *Arabidopsis* [5], which directly regulates specific gene expression in the nucleus, may take advantage of the uniqueness of the His-Asp phosphorelay system.

SIGNAL TRANSDUCTION MECHANISM

Another important and unique aspect of HKs is their necessity to function as a dimer. HK autophosphorylation occurs through a trans mechanism between two molecules in the dimer. ATP bound to the CA domain of one molecule in the dimer is used to phosphorylate the histidine residue of the central dimerization domain (DHp) of the other molecule (Chapters 2 and 3). Interactions of a HK with its cognate RR are also carried out through its interaction with the dimerization domain of one molecule in the dimer and the ATP-binding domain of the other. In this manner, HKs are able to achieve the dephosphorylation reaction (phosphatase) for phosphorylated RRs as well as phosphotransfer reactions (kinase) for their cognate RRs. Notably, however, some HKs have no or very low phosphatase activities [CheA; see Chapter 5; NRII (NtrB); see Chapter 7] so that there is a separate phosphatase in the case of the CheA–CheY system or an extra accessory protein in the case of the NRII–NRI system.

Why do HKs function obligatorily as dimers? It seems that the transphosphorylation reaction and the dephosphorylation reaction carried out by HK dimers are coordinated with the signal recognition mechanism by the signal receptor domain. The receptor domain also forms a dimer, and it is reasonable to speculate from data obtained from Tar, a bacterial chemosensor for aspartate, and Taz1, a hybrid between Tar and EnvZ (HK), that external signals for HKs are recognized at the interface between two receptor domains in the dimer (Chapters 3, 5 and 6). Binding of a ligand at the interface then causes asymmetric movement of one molecule against the other in the dimer. This asymmetric molecular displacement of one molecule over the other is then likely to be physically transduced to the otherwise symmetrically arranged kinase dimer, causing topological rearrangement between the CA domain of one HK molecule in the dimer and the DHp domain of the other. Interestingly, ligand binding has been shown in the case of Taz1 to inhibit the phosphatase reaction, resulting in the stimulation of phosphorylation of the RR OmpR (Chapter 3). It should be noted that the significance of HK dimerization and transphosphorylation is disputed for CheA, a class II HK, as discussed in Chapter 1.

There is another important fact to support the necessity of dimer formation in Tar and other chemotaxis chemoreceptors called MCPs. In these MCPs, the binding of a ligand to one site results in an inhibitory effect on the binding of a second ligand to the other site of the same receptor dimer (Chapter 6). These negative cooperativities between two ligand-binding sites seem to be the basic principle to achieve the asymmetric mechanical movement between the two cytoplasmic domains in a dimer, which, in turn, indirectly regulates the function of CheA bound to MCPs in the case of chemotaxis and directly controls phosphatase/kinase activities in the case of class I HKs.

As discussed earlier, there is no signal amplification within the HAP system. However, in the chemotaxis system using CheA as the key HK, there seems to be a mechanism by which ligand binding to a small fraction of sensory membrane-bound MCP proteins is able to inactivate all the MCP-bound CheA molecules in the cells (Chapter 5). This signal propagation is speculated to be caused at the level of the sensory proteins, which are known to exist in clusters in the membranes. Therefore, a signal binding to even a minor sensory protein is able to propagate the signal through protein–protein interaction in the clusters so that even if an MCP molecule does not directly sense its ligand, the CheA bound to that MCP may be released to the cytoplasm to be inactivated. Under this circumstance, a successful cellular response may be achieved at a concentration of the ligand, which is much less than its actual K_d value to an MCP.

Receptor chimeras such as Taz1 have been used to provide insights into the mechanism of signal propagation in these pathways [8]. Such a chimeric receptor is considered to transduce the signal in the same manner as parent MCPs. However, there seems to be one important difference between Taz1 and Tar: the activation of Taz1 requires much higher concentrations (1 m*M*) of aspartate [8] as compared to Tar (5 μ*M*; [9]). This approximately 200-fold reduction in sensitivity to aspartate in Taz1 may be due to the structural impairment at the aspartate-binding sites in the hybrid protein. However, it could be due to the inability of Taz1 to form clusters in the membrane, as the Tar cytoplasmic domain is replaced with the EnvZ cytoplasmic domain in Taz1. In this respect, it is interesting to measure the K_d value for aspartate binding to Taz1.

REGULATION OF KINASE AND PHOSPHATASE ACTIVITIES: SWITCH MODEL AND RHEOSTAT MODEL

In considering how external signals through the periplasmic receptor domain regulate the two opposing functions, kinase and phosphatase, of the cytoplasmic domain, there are two alternative models. One is called the switch model, which proposes that the cytoplasmic domain functions either in the "on" mode (kinase$^+$ and phosphatase$^-$) or in the "off" mode (kinase$^-$ and phosphatase$^+$). The other model is called the rheostat model, in which the cytoplasmic domain always has both activities and the ratio of the opposing activities is controlled by the external signal. These two models are not necessarily conflicting and are discussed in detail in Chapter 3. However, it is important to clearly understand these two models in order to elucidate how individual HAP systems work. In terms of the final outcome of external signals, the extent of RR phosphorylation has been shown to be controlled by HK in a rheostat-like fashion (Chapter 7).

CONCLUDING REMARKS

It is now quite evident that HKs play a vital role in signal transduction in bacterial cells and are required for adaptation to environmental changes. During the course of evolution, HKs have diverged into a large number of systems sensing many different external signals. In addition, some HKs have acquired extra domains such as a RR domain and a four-helix bundle HPt domain (Chapter 23).

Elucidation of the molecular mechanism of HK function and its regulatory mechanisms is crucial for our understanding of cell growth and bacterial cell survival under various environmental conditions and the pathogenicity of disease-causing bacteria. Clearly, HKs are a novel target for developing new antibiotics, which may block either kinase and/or phosphatase activities, as they do not exist in higher eukaryotes. New antibiotics are urgently needed to treat emerging multidrug-resistant pathogens. Drug design may be carried out on the basis of structural information currently available, carefully avoiding structural similarities to eukaryotic proteins.

ACKNOWLEDGMENTS

I thank Dr. M. Ikura, Dr. A. Newton, and Dr. A. Khorchid and the members of my laboratory, Dr. R. Dutta, Dr. L. Qin, Y. Zhu, T. Yoshida, and S. Cai, for helpful discussions and careful reading of the manuscript

REFERENCES

1. Wanner, B. L. (1996). Phosphorus assimilation and control of the phosphate regulon. *In* "*Escherichia coli* and *Salmonella:* Cellular and Molecular Biology" (F. C. Neidhardt, R. Curtiss III, J. L. Ingraham, E. C. C. Lin, K. B. Low, B. Magasanik, W. S. Reznikoff, M. Riley, M. Schaechter, and H. E. Umbarger, eds.), pp. 1357–1381. ASM Press, Washington, DC.
2. Groisman, E. A. (2001). The pleiotropic two-component regulatory system PhoP-PhoQ. *J. Bacteriol.* **183**, 1835–1842.
3. Suzuki, I., Los, D. A., Kanesaki, Y., Mikami, K., and Murata, N. (2000). The pathway for perception and transduction of low-temperature signals in *Synechocystis*. *EMBO J.* **19**, 1327–1334.
4. Raivio, T. L., and Silhavy, T. J. (1999). The σ^E and Cpx regulatory pathways: Overlapping but distinct envelope stress responses. *Curr. Opin. Microbiol.* **2**, 159–165.
5. Hwang, I., and Sheen, J. (2001). Two-component circuitry in *Arabidopsis* cytokinin signal transduction. *Nature* **413**, 383–389.
6. Cole, S.T., Brosch, R., Parkhill, J., Garnier, T., Churcher, C., Harris, D., Gordon, S. V., Eiglmeier, K., Gas, S., Barry, C. E., III, Tekaia, F., Badcock, K., Basham, D., Brown, D., Chillingworth, T., Connor, R., Davies, R., Devlin, K., Feltwell, T., Gentles, S., Hamlin, N., Holroyd, S., Hornsby, T., Jagels, K., Krogh, A., McLean, J., Moule, S., Murphy, L., Oliver, S., Osborne, J., Quail, M. A., Rajandream, M. A., Rogers, J., Rutter, S., Seeger, K., Skelton, S., Squares, S., Sqares, R., Sulston, J. E., Taylor, K., Whitehead, S., and Barrell, B. G. (1998). Deciphering the biology of *Mycobacterium tuberculosis* from the complete genome sequence. *Nature* **393**, 537–544.
7. Inouye, S., Jain, R., Ueki, T., Nariya, H., Xu, C. Y., Hsu, M. Y., Fernandez-Luque, B. A.,

Munoz-Dorado, J., Farez-Vidal, E., and Inouye, M. (2000). A large family of eukaryotic-like protein Ser/Thr kinases of *Myxococcus xanthus*, a developmental bacterium. *Microb. Comp. Genom.* **5**, 103–120.

8. Utsumi, R., Brissette, R. E., Rampersaud, A., Forst, S. A., Oosawa, K., and Inouye, M. (1989). Activation of bacterial porin gene expression by a chimeric signal transducer in response to aspartate. *Science* **245**, 1246–1249.
9. Clarke, S., and Koshland, D. E. Jr. (1979). Membrane receptors for aspartate and serine in bacterial chemotaxis. *J. Biol. Chem.* **254**, 9695–9702.

CHAPTER 2

The Histidine Kinase Family: Structures of Essential Building Blocks

CHIERI TOMOMORI,* HIROFUMI KUROKAWA,† AND MITSUHIKO IKURA†

**Cardiovascular Biology Department, Oklahoma Medical Research Foundation, Oklahoma City, Oklahoma 73104 and †Division of Molecular and Structural Biology, Ontario Cancer Institute and Department of Medical Biophysics, University of Toronto, Toronto, Ontario, Canada M5G 2M9*

Protein phosphorylation, the covalent attachment of a phosphoryl group to a certain amino acid side chain in a protein, is an essential step in the signal transduction processes occurring in both prokaryotic and eukaryotic organisms. The histidine kinase protein, the principal component of protein phosphotransfer in bacteria, plays a central role in the signaling pathways required for the environmental adaptation of these organisms. Histidine kinases are composed of a central dimerization domain and an ATP-binding domain (which is highly conserved among members of this protein family, but distinct in primary sequence from Ser/Thr/Tyr protein kinases) and other structural modules such as the sensor domain and the phosphotransfer domain. Advances in three-dimensional structural studies of these domains, or the essential building blocks, from the various histidine kinases have deciphered both a functional and an evolutionary link between the members of this protein family as well as between these and other proteins. Most notably, the α/β sandwich fold identified in both EnvZ and CheA ATP-binding domains revealed a marked resemblance to the fold found in the GHL ATPase family containing Hsp90, DNA gyrase B, and MutL. This chapter discusses the structurally characterized building blocks that are essential for the activity and regulation of histidine kinases.

INTRODUCTION

The histidyl-aspartyl (His-Asp) phosphorelay system (also known as the two-component signal transduction system) is essential to the environmental adaptation of prokaryotes, as well as some eukaryotes, including *Saccharomyces cerevisiae*, *Dictyostelium discoideum*, *Neurospora crassa*, and *Arabidopsis thaliana* [1, 2]. In these organisms, a wide range of extracellular stimuli leads to the activation of a variety of intracellular adaptation pathways, many of which involve the His-Asp phosphorelay system.

Extensive biochemical studies on various His-Asp phosphorelay systems revealed a simple, but elegant molecular mechanism. For example, the osmosensing system of *Escherichia coli* consists of two protein components: EnvZ, a signal-sensing histidine kinase, and OmpR, a cognate response regulator. A change in osmolarity across the biological membrane activates the histidine kinase EnvZ, which autophosphorylates the conserved histidine residue within the protein in an ATP-dependent manner. This high-energy phosphoryl group attached to the active site histidine is then transferred to the conserved aspartate residue of OmpR. This response regulator functions as a gene transcription factor that controls the production of porin proteins OmpC and OmpF, both needed to adapt to the changing environment. The phosphorylation of OmpR results in a characteristic alteration in transcriptional activity.

Some other His-Asp phosphorelay systems consist of a number of signaling proteins and involve more complex phosphotransfer mechanisms. For chemotaxis and aerobic/anaerobic regulation, *E. coli* uses multicomponent His-Asp phosphorelay pathways previously characterized [3–5]. Complex multistep phosphotransfer pathways have been elucidated for osmoregulation in *S. cerevisiae* [6, 7] and for sporulation in *Bacillus subtilis* [8]. Despite the complexity in the phosphotransfer mechanism, each signaling system consists of proteins all comprising common building blocks or domains. Based on their domain organization, the histidine kinase family can be divided into two major classes: class I and class II (Fig. 1) [9]. The class I histidine kinase family, exemplified by EnvZ, comprises an N-terminal periplasmic sensor domain, a transmembrane domain, and a C-terminal cytoplasmic kinase domain. The kinase domain can be further divided into two portions: the dimerization domain containing histidine autophosphorylation and the ATP-binding domain. In class I histidine kinases, the active site histidine is located within the homodimerization domain [10, 11], immediately followed by the C-terminal ATP-binding domain. Contrary to this arrangement, the active site histidine in class II histidine kinases (such as CheA) is remote from the ATP-binding domain and resides in the phosphotransfer (HPt) domain atypically found at the N terminus of the kinase (Fig. 1). Interestingly, the adjacent arrangement of the dimerization domain and ATP-binding domain in class II kinases is similar to that of class I kinases (Fig. 1). Finally, in the CheA system, two response regulators (CheY and CheB) receive the phosphoryl group from the kinase.

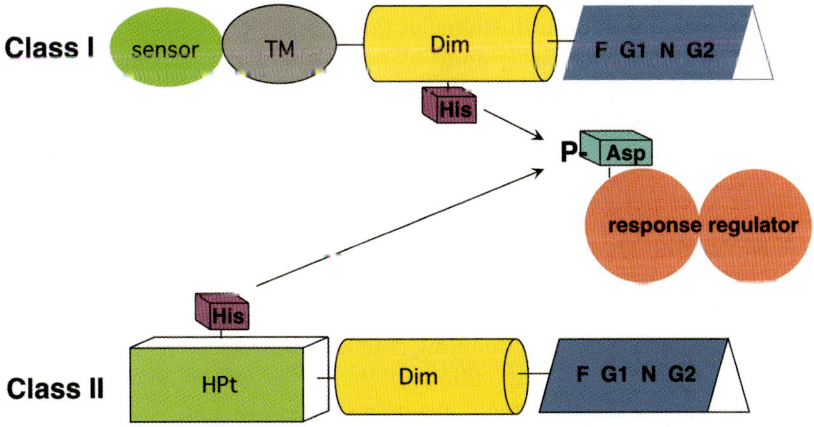

FIGURE 1 Schematic representation of the histidine kinase core domains: Sensor domain, transmembrane (TM) domain, dimerization (Dim) domain, histidine containing phosphotransfer (HPt) domain, and kinase ATP-binding domain (shown in blue triangular column). The response regulator contains two domains: the regulatory domain containing the asparatate residue that can be phosphorylated by the cognate histidine kinase. Histidine kinases are categorized into two classes according to the location of the conserved active site histidine (H box) with respect to the ATP-binding domain (N, G1, F, and G2 boxes). An adjacent positioning of the H box and the ATP-binding domain is found in type I histidine kinase. In type II histidine kinase, the H box is located a relatively larger distance from the ATP-binding domain.

Advances in the three-dimensional structure determination of histidine kinases significantly improved existing knowledge of the role of the His-Asp phosphorelay signal transduction system. Structural data available to date indicate that members of the histidine kinase family are made of structurally conserved building blocks. The organization of those building blocks in protein kinases is critical to the phosphorelay mechanism by which they work. Hence, structural information on each building block from various members of the kinase family cross-fertilizes our knowledge of individual signaling processes. As a number of excellent reviews have been published on the biological and biochemical aspects of the phosphorelay system [2, 10–14], this chapter focuses on the structural aspect of the histidine kinase family.

KINASE/PHOSPHATASE CORE DOMAIN

Osmosensor EnvZ

EnvZ is one of the best characterized class I histidine kinases. This transmembrane protein serves as an osmosensor in *E. coli* [15–22]. Like most histidine kinases, EnvZ is multifunctional in terms of phosphotransfer as it is able to (i) autophosphorylate the histidine residue (His-243) within a dimer; (ii) phosphorylate the aspartate residue (Asp-55) of OmpR; and (iii) remove the phosphoryl group from the phosphoaspartate of the response regulator.

EnvZ is a transmembrane protein consisting of 450 amino acid residues, in which all commonly conserved motifs within the histidine kinase family members are present. EnvZ is composed of an N-terminal short tail (residues 1–15) followed by the transmembrane domain (residues 16–46), a periplasmic putative sensor domain (residues 46–162), the second transmembrane domain (residues 163–179), the linker domain (residues 180–222), and the kinase/phosphatase core domain (residues 223–450).

The dimerization and ATP-binding core domain was dissected into its two functional fragments corresponding to the dimerization domain (residues 223–289) and the ATP-binding domain (residues 290–450) [23]. The dimerization domain contains the highly conserved histidine (His-243), whereas the ATP-binding domain features the rest of conserved motifs such as the N box (Asn-347), the F box (Phe-387), the G1 box (residues 373–377) and the G2 box (residues 403–405), and the recently recognized G3 box. To the first approximation, these two domains are structurally independent, but functionally complementary: the ATP-binding domain autophosphorylates His-243 of the dimerization domain, the prerequisite for phosphotransfer from EnvZ to Asp-55 of OmpR [23]. In the following section, structure properties of the ATP-binding domain and the dimerization domain are given in further detail.

ATP-Binding Domain

The structure of the *E. coli* EnvZ ATP-binding domain was determined by nuclear magnetic resonance (NMR) spectroscopy in the presence of a non-hydrolyzable ATP analog, AMP-PNP [24]. The ATP-binding domain assumes the α/β sandwich fold with left-handed αβα connectivity (Fig. 2a). The fold consists of an antiparallel five-stranded β sheet (strand B, residues 319–323; D, 356–362; E, 367–373; F, 420–423; G, 431–346) and three α helices (α1, residues 301–311; α2, 334–343; α4, 410–414) which are sealed within two β strands—A (residues 297–299) and C (residues 330–332) as well as a long flexible loop.

This first structure determination of the ATP-binding domain of histidine kinases revealed that the histidine kinase fold is distinct from the fold commonly observed in eukaryotic Ser/Thr/Tyr kinases. Instead, this fold was found to be similar to that of type II topoisomerases, DNA GyraseB [25], MutL DNA mismatch repair protein [26], and the eukaryotic molecular chaperon heat shock protein 90 (Hsp90) [27]. Interestingly, all four proteins

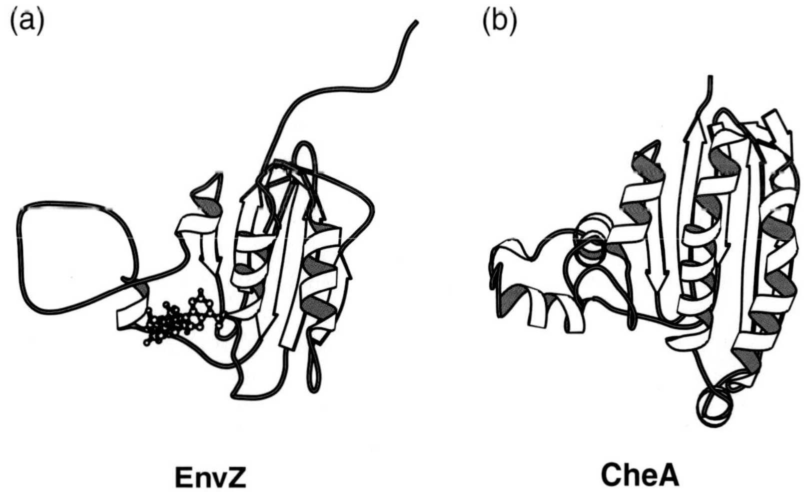

FIGURE 2 Ribbon representation of three-dimensional structures of the histidine kinase ATP-binding domain. (a) Class I histidine kinase EnvZ (residues 280–445, PDB: 1BXD). Bound AMP-PNP is drawn as a ball-and-stick model [24]. (b) Class II histidine kinase CheA (residues 354–539, PDB: 1B3Q) [9].

bind a nucleotide, whereas the α/β sandwich fold assumes nucleotide-binding activity. It should be noted that the structural similarity between EnvZ and the superfamily of GHL ATPases was suggested previously by Mushegian *et al.* [28] on the basis of a careful sequence analysis.

In the EnvZ structure, the AMP-PNP-binding site involves the N box at the edge of the α2 helix and the G1 box at edge β strand E and is surrounded by the F box and the G2 box located in the central flexible loop. The adenine ring of AMP-PNP is located in close spatial proximity to Asn-347, Asp-378, Leu-386, and Phe-387. These residues are highly conserved within the histidine kinase family. The triphosphate chain of AMP-PNP is fully exposed to the solvent in the structure. It should be noted, however, that the NMR-derived structure of EnvZ has low precision around the AMP-PNP-binding site, as only a dozen interatomic NOEs were observed between the protein and the nucleotide (despite the use of uniformly ^{13}C-labeled AMP-PNP). Also, a large portion of the binding site involves part of the central flexible loop. Nevertheless, it is interesting to note that the fold shared by EnvZ and the members of the GHL ATPase superfamily [26], Hsp90, DNA Gyrase B, and MutL, is responsible for nucleotide binding.

Dimerization Domain

The NMR-derived solution structure of the EnvZ dimerization domain (Fig. 4c) revealed an up-down-up-down, four-helix bundle in twofold symmetry along with the helix axis [29]. Each monomer is made of a pair of long antiparallel helices (helix I, residues 235–255; helix II, residues 265–286) with a connecting nine-residue turn. Helix I and II are similar in length, but different in surface and backbone characteristics as described by Tomomori *et al.* [29]. Furthermore, both helices also possess a significant difference in sequence conservation among members of the histidine kinase family: helix II is overall more variable than helix I. Helix I contains a highly conserved region of amino acid residues (H box) in which His-243 is located at the center of the helix. Not surprisingly, His-243 is exposed to the solvent, as it has to be accessible to the ATP-binding domain and the response regulator OmpR for enzymatic reactions. It should also be noted that His-243 is positioned at the edge of intra- and intermonomer surfaces: the former surface is formed by helix I (conserved helix) and helix II (variable helix) in the same monomer and the latter by adjacent helices from two monomers. An extensive surface around the active histidine may enable the ATP-binding domain of EnvZ and the regulatory domain of OmpR to simultaneously approach the active site, thereby effectively facilitating both autophosphorylation within EnvZ and phosphotransfer between EnvZ and OmpR.

The NMR-derived structure of the four-helix bundle dimerization domain has an interesting dynamic property. ^{15}N relaxation data studies [29] showed that helix I contains higher conformational flexibility than helix II in the region near the active histidine (residues 242–248). Many of the residues in this flexible region are highly conserved among members of the class I histidine kinase family. The significance of this high flexibility is unclear at present, but may help explain the molecular mechanisms underlying autophosphorylation and phosphotransfer to OmpR [30].

Chemotaxis Sensor CheA

The crystal structure of the class II histidine kinase ATP-binding core was first determined for CheA from *Thermotoga maritima* [9] (see Chapter 2). The structure of CheA (residues: 290–671) contains an enzymatically active ATP-binding domain, a dimerization domain, and a regulatory domain (Fig. 4a). The ATP-binding domain (residues: 355–540) assumes an α/β sandwich motif with a five-stranded β sheet and six α helices, which is essentially identical to that of EnvZ domain B [24]. More recently, Bilwes *et al.* [31] revealed the mode of nucleotide recognition from their studies on the crystal structures of the *T. maritima* CheA ATP-binding domain (residues: 350–540) in complex with various nucleotides (ADPNP, ADPCP, or ADP). Divalent metal ions are found in the CheA active center. In the ADPCP-Mg^{2+}-bound structure, three ADPCP phosphates, the Asn-409 carbonyl, and two water molecules coordinate Mg^{2+} in an octahedral geometry. The substitution of Mg^{2+} with Mn^{2+} revealed that Mn^{2+} has little effect on the ATP-binding cavity size but alters the position of His-405. It is also interesting to note that the mode of CheA–adenine base interaction is very similar in topology with the modes observed in structurally related proteins such as DNA GyraseB [25], Hsp90 [27], and MutL [26].

The homodimerization domain (residues: 290–354) consists of two antiparallel helices, a pair of which form a four-helix bundle, again similar to that found in EnvZ [29]. Alternatively, the regulatory domain (residues: 541–671) possesses two β barrels that are related to each other in a pseudo-twofold symmetry. Interestingly, each of these barrels is topologically related to the SH3 domain of human c-src tyrosine kinase [32, 33]. It is clear that class I kinase EnvZ and class II kinase CheA share similar building blocks (the ATP-binding domain and the dimerization domain), but exhibit different spatial arrangement of these building blocks, resulting in different phosphotransfer processes. Furthermore, class II kinases acquired an additional regulatory mechanism provided by the dual SH3 domains. More consideration on domain packing is given in the following section.

PHOSPHOTRANSFER DOMAIN

To date, three-dimensional structures of the phosphotransfer domain of various proteins have been determined [34–41]. These domains contain the active site histidine, which is phosphorylated by the ATP-binding domain of histidine kinase. However, because of the high-energy state of the histidine, the high-resolution structures thus far solved are all in the nonphosphorylated state. One attempt to study the phosphorylated form of a histidine kinase has been reported by Zhou and Dahlquist [35], who employed ^{1}H-^{15}N heteronuclear single quantum coherence (HSQC) spectroscopy to study phosphorylated CheA. It was observed that phosphorylation of the active histidine His-48 induced relatively small changes in backbone amide chemical shifts of only several amino acid residues, including His-48, Ser-49, Gly-52, Asn-71, and Asp74. This section summarizes the known three-dimensional structures of phosphotransfer domains.

Anoxic Redox Regulator ArcB

ArcB, a hybrid histidine kinase involved in a multistep His-Asp phosphorelay system with the HPt domain at the C terminus, functions as a transmitter of the phosphoryl group via the active histidine (His-717). The crystal structure of the anaerobic *E.coli* ArcB HPt domain (residues 654–778) [36] contains six α helices (helix A, residues 660–664; B, 667–676; C, 679–705; D, 709–726; E, 729–738; F, 746–775) (Fig. 3a). Active site His-717 is located on the surface of helix D that forms the four-helix bundle with helix E, C, and F. This bundle has an up-down-up-down topology with a left-handed twist [36, 37], similar to the one found in the CheA HPt domain.

Chemotaxis Sensor CheA

The NMR-derived structure of the *E. coli* CheA HPt domain (1–134, P1 domain) [34, 35] revealed a helix bundle structure. It consists of five α helices connected by turns (helix A, residues 10–26; B, 36–52; C, 60–77; D, 86–106; E, 112–131), and active site His-48 is located on the solvent-exposed surface of helix B. A high-resolution crystal structure of *Salmonella typhimurium* CheA [41] confirmed the topology of the five helix bundle and further demonstrated that this structural motif of the CheA P1 domain is identical to that of the HPt domain of ArcB [36] and Ypd1 [39, 40]. It is interesting to note that active site histidine is always found on a helix, which is surrounded by several other helices in a bundle architecture.

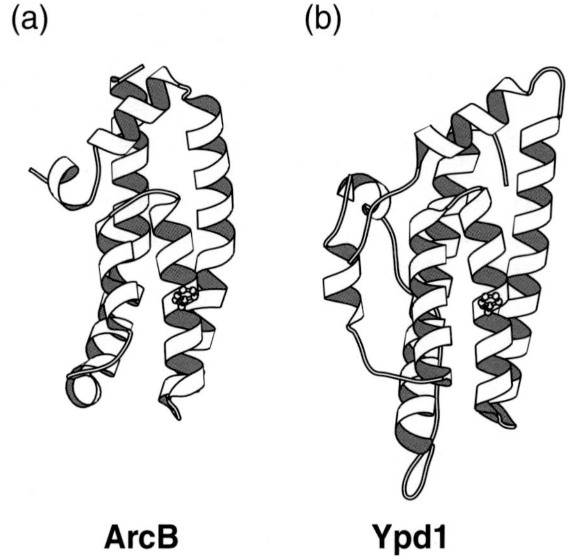

FIGURE 3 Ribbon representation of three-dimensional structures of histidine-containing phosphotransfer (HPt) domains of (a) ArcB (PDB: 2A0B) [37] and (b) Ypd1 (PDB: 1QSP Chain A) [39]. Active histidine residues are shown as a ball-and-stick model.

PHOSPHOTRANSFERASE SPO0B

In the multicomponent sporulation phosphorelay, Spo0B serves as a phosphotransferase, which transfers a phosphoryl group to Spo0F (a response regulator) (see Chapter 6). Subsequently, the phosphoryl group on Spo0F is transferred to Spo0A (a transcription factor). A four-helix bundle dimerization domain, similar to that found in EnvZ and CheA, was identified in the crystal structure of the Spo0B (residues 1–192) of *Bacillus subtilis* (Fig. 4b) [38]. The two monomers of Spo0B dimerize at the helical hairpin regions (α1, residues 10–45; α2, residues 48–71), forming a four-helix bundle. Phosphorylation site His-30 is located on the middle of helix α1, which is exposed to the solvent. A large number of hydrophobic residues found in the interior of the dimeric structure contribute largely to overall structure stabilization. A salt bridge can be seen at Arg-29 in helix α1 in one monomer and at Glu-65 in helix α2 in another monomer. The structure of the Spo0B HPt domain differs topologically from other HPt domain structures of ArcB [36, 37] or CheA [34, 35, 41], but shows close similarity to the homodimerization domain of *E. coli* EnvZ [29], as well as that of *T. maritima* CheA [9].

The C terminus of Spo0B also shows a remarkable resemblance to that of the ATP-binding domain of EnvZ and CheA (Fig. 4). This domain consists of five β strands (β1, residues 91–97; β2, residues 132–138; β3, residues 146–153; β4, residues 175–180; β5, residues 184–190) and two α helices (α3, residues 74–82; α4, residues 107–123). Analogous to EnvZ, this domain is located immediately at the C-terminal end of the homodimerization domain. However, this C-terminal domain lacks the conserved N, G1, G2, and G3 motifs and does not possess any ATP-binding activity. The function of this domain is still undetermined.

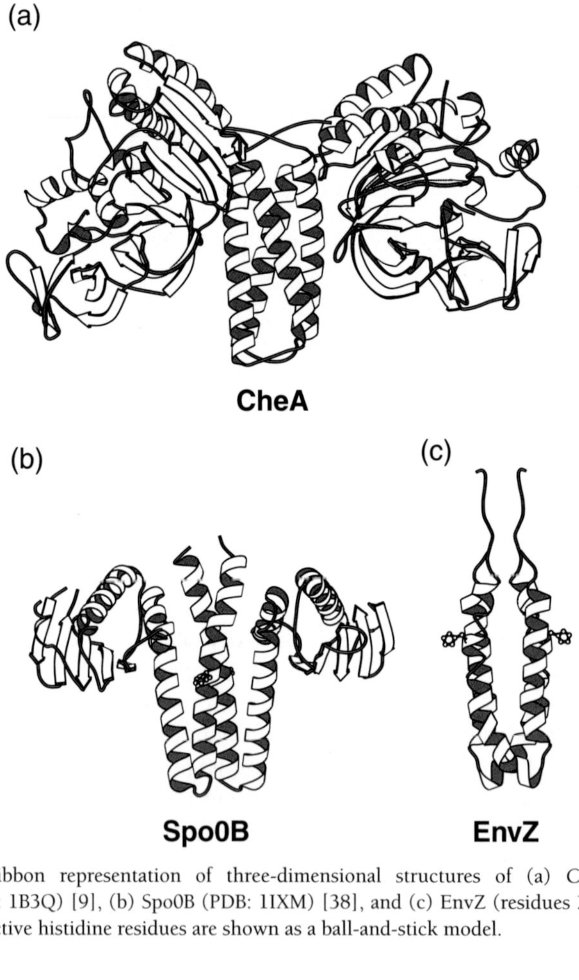

FIGURE 4 Ribbon representation of three-dimensional structures of (a) CheA (residues 290–671, PDB: 1B3Q) [9], (b) Spo0B (PDB: 1IXM) [38], and (c) EnvZ (residues 223–289, PDB: 1JOY) [29]. Active histidine residues are shown as a ball-and-stick model.

A report of the Spo0B–Spo0F (Y13S mutant) cocrystal structure [42] revealed that neither Spo0B nor Spo0F changes conformation upon forming a complex. The Spo0B dimer is associated with four monomers of Spo0F positioned asymmetrically in the cleft. A number of hydrophobic residues (Gln12m, Ile15, Leu18, and Glu21 in Spo0F) are identified at the intermolecular interface. The active site histidine (His-30) of Spo0B and the phosphoryl group receiver aspartic acid (Asp54) of Spo0F face each other within 4.3 Å distance. More details can be found in Chapter 6.

Yeast Ypd1

In *S. cerevisiae*, osmoregulation takes place via multistep phosphorelay signal transduction involving four partner proteins: Sln1, Ypd1, Ssk1, and Skn7. Sln1 is the sensor histidine kinase creating a primary signal, which is transmitted to the phosphotransfer protein Ypd1 and then to one of the independent response regulators, Ssk1 or Skn7. The crystal structures of Ypd1 were determined independently by Song *et al.* [39] and Xu *et al.* [40]. The HPt domain structure of Ypd1 consists of six α helices (helix A, residues 10–20; B, 26–52; C, 55–73; D, 75–90; E, 98–104; G, 134–164) and a short 3_{10} helix (residues 108–113) (Fig. 3b). Active site histidine His-64 is located at the middle of helix C and is fully solvent exposed. The central core of this molecule exhibits the up-down-up-down four-helix bundle (helices B, C, D, and G), similar to the fold of the other HPt domains of ArcB and CheA. The helix bundle is again a host of the active histidine, demonstrating the importance of this structural architecture for histidine-mediated phosphotransfer in the His-Asp phosphorelay system.

CONSIDERATIONS ON DOMAIN INTERACTIONS

The CheA structure [9] provided the first structural insight into the organization of functional domains within a kinase sensor protein. The four-helix bundle dimerization domain serves as a focal point, bringing other domains together to a close proximity and therefore enabling multiple reactions to occur almost simultaneously. Apparently, the dimerization domain is a common platform for both class I and II histidine kinase functions, but the mechanism of interaction between the ATP-binding domain and the active site histidine [which can be in the central dimerization domain (class I) or in the phosphotransfer domain in one extremity (class II)] differs.

The domain layout shown in Fig. 1 suggests that class I and class II histidine kinases possess different structural basis underlying the respective phosphotransfer reaction. In CheA, and most likely all other class II histidine kinases in which the active site histidine is located at the N-terminal HPt domain, the nucleotide-binding site within the ATP-binding domain should not face toward the dimerization domain, thereby allowing for interaction with the HPt domain, ultimately leading to transphosphorylation. However, the nucleotide-binding site of EnvZ and other class I histidine kinases should be close proximity to the dimerization domain, which contains the active histidine. Such structural diversity seems to have developed through the molecular evolution of these building blocks. Most remarkably, the C-terminal domain of Spo0B possesses the α/β sandwich fold common to the histidine kinase ATP-binding domain, yet it completely lacks enzymatic activity. In the multicomponent histidine kinase CheA, the HPt domain containing the active site histidine is positioned far from the ATP-binding domain in the primary sequence, but it must be recruited to the four-helix bundle platform where the catalysis should occur. This platform also recruits the regulatory domain of response regulators such that the phosphoryl group of the active site histidine can be transferred to the aspartate of the response regulators.

CONCLUDING REMARKS

Structural studies on histidine kinases have advanced the field of phosphorelay signal transduction into a “new millennium.” We now know the structure of many key building blocks of histidine kinase. A clear distinction, resulting from the differences in domain organization described earlier, can be seen between class I and II histidine kinases. A number of challenges for future structural studies include determining (1) how histidine kinase detects a specific stimulus and thus transfers that signal across the membrane; (2) the exact mechanism for the autophosphorylation reaction; and (3) how histidine kinase interacts with a response regulator to transfer a phosphoryl group. By elucidating high-resolution structures of protein–protein complexes such as EnvZ-OmpR and CheA-CheY, as well as larger protein constructs containing both ATP-binding and autophosphorylation domains, the goals outlined in this chapter can be met. We will soon see some of these structures, which will further enrich our understanding of this simple but elegant signal transduction system.

ACKNOWLEDGMENTS

We thank Dr. M. Inouye and members of his laboratory for discussions and Jane Gooding and Kit Tong for critical comments on the manuscript. This work was supported by a grant (to M.I.) from the Canadian Institute of Health Research (CIHR). M.I. is a Howard Hughes Medical Institute International Research Scholar and a CIHR Scientist.

REFERENCES

1. Loomis, W. F., Kuspa, A., and Shaulsky, G. (1998). *Curr. Opin. Microbiol.* **1**, 643–648.
2. Robinson, V. L., Buckler, D. R., and Stock, A. M. (2000). *Nature Struct. Biol.* 7, 626–633.
3. Hess, J., Bourret, R., and Simon, M. (1988). *Nature* **336**, 139–143.
4. Parkinson, J. (1976). *J. Bacteriol.* **126**, 758–770.
5. Ishige, K., Nagasawa, S., Tokishita, S., and Mizuno, T. (1994). *EMBO J.* **13**, 5195–5202.
6. Posas, F., Wurgler-Murphy, S. M., Maeda, T., Witten, E. A., Thai, T. C., and Saito, H. (1996). *Cell* **86**, 865–875.
7. Ota, I. M., and Varshavsky, A. (1993). *Science* **262**, 566–569.
8. Hoch, J. (1993). *Annu. Rev. Microbiol.* **47**, 441–465.
9. Bilwes, A. M., Alex, L. A., Crane, B. R., and Simon, M. I. (1999). *Cell* **96**, 131–141.
10. Swanson, R. V., Alex, L. A., and Simon, M. I. (1994). *Trends Biochem. Sci.* **19**, 485–490.
11. Egger, L. A., Park, H., and Inouye, M. (1997). *Genes Cells* **2**, 167–184.
12. Stock, J. (1999). *Curr. Biol.* **9**, R364–R367.
13. Hoch, J. (2000). *Curr. Opin. Microbiol.* **3**, 165–170.
14. Stock, A. M., Robinson, V. L., and Goudreau, P. N. (2000). *Annu. Rev. Biochem.* **69**, 183–215.
15. Forst, S., Delgado, J., Ramakrishnan, G., and Inouye, M. (1988). *J. Bacteriol.* **170**, 5080–5085.
16. Matsuyama, S., Mizuno, T., and Mizushima, S. (1986). *J. Bacteriol.* **168**, 1309–1314.
17. Forst, S., Delgado, J., and Inouye, M. (1989). *Proc. Natl. Acad. Sci. USA* **86**, 6052–6056.
18. Bernardini, M. L., Fontaine, A., and Sansonetti, P. J. (1990). *J. Bacteriol.* **172**, 6274–6281.
19. Yang, Y., and Inouye, M. (1991). *Proc. Natl. Acad. Sci. USA* **88**, 11057–11061.
20. Yang, Y., and Inouye, M. (1993). *J. Mol. Biol.* **231**, 335–342.
21. Skarphol, K., Waukau, J., and Forst, S. A. (1997). *J. Bacteriol.* **179**, 1413–1416.
22. Forst, S. A., and Roberts, D. L. (1994). *Res. Microbiol.* **145**, 363–373.
23. Park, H., Saha, S. K., and Inouye, M. (1998). *Proc. Natl. Acad. Sci. USA* **95**, 6728–32.
24. Tanaka, T., Saha, S. K., Tomomori, C., Ishima, R., Liu, D., Tong, K. I., Park, H., Dutta, R., Qin, L., Swindells, M. B., Yamazaki, T., Ono, A. M., Kainosho, M., Inouye, M., and Ikura, M. (1998). *Nature* **396**, 88–92.
25. Wigley, D. B., Davies, G. J., Dodson, E. J., Maxwell, A., and Dodson, G. (1991). *Nature* **351**, 624–429.
26. Ban, C., Junop, M., and Yang, W. (1999). *Cell* **97**, 85–97.
27. Prodromou, C., Roe, S. M., O'Brien, R., Ladbury, J. E., Piper, P. W., and Pearl, L. H. (1997). *Cell* **90**, 65–75.
28. Mushegian, A. R., Bassett, D. E. J., Boguski, M. S., Bork, P., and Koonin, E. V. (1997). *Proc. Natl. Acad. Sci. USA* **94**, 5831–5836.
29. Tomomori, C., Tanaka, T., Dutta, R., Park, H., Saha, S. K., Zhu, Y., Ishima, R., Liu, D., Tong, K. I., Kurokawa, H., Qian, H., Inouye, M., and Ikura, M. (1999). *Nature Struct. Biol.* **6**, 729–734.
30. Dutta, R., Yoshida, T., and Inouye, M. (2000). *J Biol Chem* **275**, 38645–38653.
31. Bilwes, A. M., Quezada, C. M., Croal, L. R., Crane, B. R., and Simon, M. I. (2001). *Nature Struct. Biol.* **8**, 353–360.
32. Xu, W., Harrison, S. C., and Eck, M. J. (1997). *Nature* **385**, 595–602.
33. Xu, W., Doshi, A., Lei, M., Eck, M. J., and Harrison, S. C. (1999). Mol. *Cell* **3**, 629–638.
34. Zhou, H., Lowry, D. F., Swanson, R. V., Simon, M. I., and Dahlquist, F. W. (1995). *Biochemistry* **34**, 13858–13870.
35. Zhou, H., and Dahlquist, F. W. (1997). *Biochemistry* **36**, 699–710.
36. Kato, M., Mizuno, T., Shimizu, T., and Hakoshima, T. (1997). *Cell* **88**, 717–723.
37. Kato, M., Mizuno, T., Shimizu, T., and Hakoshima, T. (1999). Acta. *Crystallogr. D. Biol.*

Crystallogr. **55**, 1842–1849.

38. Varughese, K. I., Madhusudan, Zhou, X. Z., Whiteley, J. M., and Hoch, J. A. (1998). *Mol. Cell* **2**, 485–493.
39. Song, H. K., Lee, J. Y., Lee, M. G., Moon, J., Min, K., Yang, J. K., and Suh, S. W. (1999). *J. Mol. Biol.* **293**, 753–761.
40. Xu, Q., and West, A. H. (1999) *J. Mol. Biol.* **292**, 1039–1050.
41. Mourey, L., Da Re, S., Pedelacq, J. D., Tolstykh, T., Faurie, C., Guillet, V., Stock, J. B., and Samama, J. P. (2001). *J. Biol. Chem.*
42. Zapf, J., Sen, U., Madhusudan, Hoch, J. A., and Varughese, K. I. (2000). *Struct. Fold. Des.* **8**, 851–862.

CHAPTER 3

Regulation of Porins in *Escherichia coli* by the Osmosensing Histidine Kinase/Phosphatase EnvZ

MASAYORI INOUYE, RINKU DUTTA, AND YAN ZHU
Department of Biochemistry and Molecular Biology, Robert Wood Johnson Medical School, Piscataway, New Jersey 08854

EnvZ of *Escherichia coli* is a transmembrane histidine kinase belonging to the family of His-Asp phosphorelay or two-component signal transducing systems prevalent in prokaryotes and also discovered recently in lower eukaryotes. In response to changes in medium osmolarity, EnvZ regulates the level of phosphorylated OmpR (OmpR-P), its cognate response-regulating transcription factor for *ompF* and *ompC* genes. EnvZ has dual-opposing enzymatic activities: OmpR-phosphorylase (kinase) and phospho-OmpR-dephosphorylase (phosphatase). The osmotic signal is proposed to regulate the ratio of the kinase to the phosphatase activity of EnvZ to modulate the level of OmpR phosphorylation. The C-terminal kinase domain of EnvZ has been dissected successfully into two independent domains: A and B. The structures of these domains have been solved by nuclear magnetic resonance spectroscopy, which provided the first insights into histidine kinase architecture. This chapter describes results that shed new light on various aspects of EnvZ function. We describe that domain A is not simply a structural scaffold for the EnvZ dimer formation by forming a central four-helix bundle, but plays an essential role in the phosphatase reaction, providing the invariant histidine residue at the active center. Indeed, in the absence of domain B, domain A by itself is able to dephosphorylate OmpR-P, and phosphorylated domain A is able to transfer the phosphoryl group to OmpR. However, domain B also plays an essential role in phosphorylation of the invariant histidine residue in domain A with ATP and significantly enhances domain A phosphatase activity when it is linked covalently to domain A. We also discuss how the opposing enzymatic activities of EnvZ are regulated on the basis of the biochemical characterization of individual domains A and B, mutagenesis analysis of these domains, and the experiments using Taz1, a Tar (aspartate chemoreceptor) and EnvZ hybrid. In addition, we demonstrate that EnvZ and OmpR form a stoichiometric complex and propose a model to comprehend how the cellular concentrations of OmpR-P are regulated, allowing the reciprocal expression of *ompF* and *ompC* genes under different osmolarities. © 2003, Elsevier Science (USA).

INTRODUCTION

EnvZ, the osmotic sensor for *Escherichia coli*, belongs to the largest class of histidine kinases (class I). Using ATP, EnvZ autophosphorylates its conserved His243 residue [1], which is subsequently transferred to the conserved Asp55 residue on its cognate response regulator OmpR. Like many other bifunctional histidine kinases, EnvZ can also function as a phosphatase to dephosphorylate phosphorylated OmpR (OmpR-P) [2–3]. ATP, ADP, and nonhydrolyzable ATP analogs such as AMPPNP can further stimulate this reaction. OmpR-P functions as a transcription factor for *ompF* and *ompC* genes, which are two major outer membrane porin proteins that allow passive access of solutes into the cell [4]. The kinase/phosphatase ratio of EnvZ determines the final levels of OmpR-P inside the cell and subsequently regulates the reciprocal expression of *ompF* and *ompC* according to environmental osmolarity changes [5–8].

EnvZ is a transmembrane protein consisting of 450 amino acid residues. The cytoplasmic domain of EnvZ consisting of 271 residues (residues 180–450), EnvZc, possesses both kinase and phosphatase activities similar to intact EnvZ. It contains all the highly conserved regions for histidine kinases: H, N, F, G1, G2, and G3 boxes [9–11]. Two distinct domains have been identified in EnvZc: domain A (residues 223–289) and domain B (residues 290–450) [12]. Both are functional enzymatically when mixed together. While nuclear magnetic resonance (NMR) structures of both domains have been solved [13, 14]; (see also Chapter 2), structural information on the topological linkage between domain A and domain B remains elusive. The junction region, termed the X region, that connects domains A and B is not well conserved [15] and has been shown to play an indirect role in controlling the enzymatic activities of EnvZ, presumably by adjusting the topology between domains A and B [16].

It is important to note that using EnvZ, it was demonstrated for the first time that the so-called autophosphorylation reaction occurs by the transphosphorylation mechanism between two EnvZ molecules in a dimer [1]. Since that time, the transphosphorylation mechanism is widely accepted in all histidine kinases tested to date, allowing one to safely conclude that histidine kinases function only by forming a homodimer and that they are autophosphorylated *in trans* in a dimer through this obligatory dimer formation. We propose that this is the key mechanism in regulating the two opposing functions of histidine kinases: kinase and phophatase. Experimentally, this notion has been confirmed by constructing a monomeric kinase consisting of two domain As followed by one domain B [17].

Earlier, we demonstrated that domain A, containing the autophosphorylation site His243, can be phosphorylated by domain B in the presence of ATP and that phosphorylated domain A is able to transfer the phosphoryl group to

OmpR [12]. Furthermore, domain A can, by itself, dephosphorylate OmpR-P in Mg^{2+}-containing buffer [16]. In contrast, domain B alone does not exhibit OmpR-P dephosphorylase activity. Mutational analysis of the His243 residue on the isolated domain A of EnvZ provided convincing evidence for the essential role of the His243 residue in the phosphatase activity of EnvZ [16]. Residues such as the highly conserved Thr247 residue present around the invariable His243 on domain A, have also been demonstrated to play an important role in influencing the catalytic activities of EnvZ [18]; (L. Qin and M. Inouye, unpublished results). On the basis of results obtained with the chimeric Tar-EnvZ receptor, Taz1, it has been proposed that the osmotic signal modulates the spatial arrangement between domains A and B, thereby altering the ratio of kinase/phosphatase activities, which in turn determines the OmpR-P output [16, 19].

This chapter first describes the structure and function of domain A and domain B and then discusses how these two domains are coordinated in regulating the two opposing functions of EnvZ. A model will be presented to understand how the cellular concentration of OmpR-P is modulated allowing the reciprocal expression of *ompF* and *ompC* genes. A comprehensive review on the historical perspective of the research on EnvZ and OmpR has been described previously by Pratt and Silhavy [7].

DOMAIN A IS THE CATALYTIC DOMAIN

CATALYTICALLY FUNCTIONAL DOMAIN

NMR studies demonstrated that domain A consists of a four-helix bundle serving as a dimerization and histidine phosphotransfer domain (DHp). In addition to these functions, the DHp domain in EnvZ also determines specificity by binding to the downstream cognate response regulator OmpR. NMR titration experiments performed with the DHp domain of EnvZ and the regulatory domain of OmpR strongly suggest that the OmpR-docking site lies toward the base of the four-helix bundle [14]. Moreover, molecular-docking studies indicate that the phosphate-accepting aspartate residue in the response regulator can be brought into close proximity of the donor histidine residue on the DHp domain without encountering steric hindrance (Spo0B in Varughese et al. [20]; EnvZ in C. Tomomori and M. Ikura, unpublished data). The four helices in EnvZ run nearly parallel to each other (the interhelical angles are $12 \pm 4^{\circ}$ for helix I–helix I and $5 \pm 3^{\circ}$ for helix II–helix II, respectively), which is very unusual in this topological class. Both EnvZ and Spo0B dimerization domains provide two symmetrically located histidine residues (His243 in EnvZ and His30 in Spo0B) that receive the phosphoryl group and subsequently donate it to downstream response regulator proteins.

We have demonstrated that domain A by itself has some default phosphatase activity both *in vitro* and *in vivo* [16]. This phosphatase activity is Mg^{2+} dependent and is not activated by ADP, ATP, and AMPPNP, which are known cofactors for the EnvZ phosphatase reaction. Interestingly, ADP-bound domain B modulates the phosphatase activity of domain A. Moreover, the covalently linked domain A and B protein exhibits a dramatic cofactor-dependent enhancement of the phosphatase activity. Interestingly, the N-terminally or C-terminally extended versions of domain A (domain A + 75 residues at the C-terminal end or + 44 residues at the N-terminal end) did not enhance its phosphatase activity. Experiments using substitution mutations at His243 strongly suggest that the autophosphorylating histidine residue plays an essential role in phosphatase activity. The X-region mutant L288P that is known to specifically abolish phosphatase activity in EnvZ [15] had no effect on the domain A phosphatase function. Accordingly, we have proposed that EnvZ phosphatase activity is regulated by the relative positioning of domains A and B, which is controlled by external signals [16]. In this topological arrangement, the junction X region may play an important role.

Role of the Invariant His243 Residue

The replacement of His243 with another residue (Ser, Asn, Lys, Tyr, and Val) completely abolished domain A phosphatase activity [16]. This demonstrates that EnvZ plays the major role in the OmpR-P phosphatase reaction and that EnvZ is not simply a cophosphatase that allosterically enhances the intrinsic phosphatase activity of OmpR-P. Furthermore, these substitution mutations at His243 indicate that His243 plays an essential role in the phophatase reaction, in addition to its essential role in the autophosphorylation and kinase reaction.

It has been disputed whether the phosphatase activity of EnvZ is caused by the reverse reaction of the kinase activity [21]. Indeed, the phosphoryl group of OmpR-P can be transferred back to His243 of EnvZc[N347D], an EnvZc kinase$^-$ phosphatase$^+$ mutant [21], and to His243 of domain A in an early period of the phosphatase reaction [16]. Although these results do not indicate that phosphorylated His243 is an intermediate of the OmpR-P dephosphorylation reaction, they clearly suggest that kinase and phophatase activities are not independent and that both His243 of EnvZ and Asp55 of OmpR are shared in the overlapping active centers for both reactions.

It should be noted that some substitution mutations at the His243 in EnvZ have been reported to still retain a low phosphatase activity [22, 23], whereas no such substitution mutations at the His243 in domain A displayed any detectable phosphatase activity [16]. In the EnvZ configuration, the role of the His243 residue in the domain A phosphatase reaction may be complemented by another residue(s) in the presence of the covalently linked ADP-bound domain B. It remains to be determined which residue(s) is capable of complementing the function of His243 in the substitution mutation.

Role of the Conserved Thr247 Residue

Threonine is the most preferred amino acid residue at the H+4th position in the conserved H box of histidine kinases [24]. The consensus sequence of the H box is h-HahbTPL (where h is hydrophobic, a is acidic, and b is basic amino acid residue). The highly conserved Thr247 residue is strategically positioned just one turn below the phospho-accepting His243 on helix I. In the NMR solution structure of domain A of EnvZ, the segment (residues 242–248) containing the invariant His243 and the conserved residues Thr247 and Pro248 are poorly defined [14]. Moreover, the backbone NH groups in this region exhibit a fast H/D rate, again indicating that this region is structurally dynamic, probably undergoing a conformational equilibrium between helical and unfolded states. It has been proposed that the structural dynamics observed in this segment in helix I might play a role in the catalytic function of EnvZ [14].

Substituting the Thr247 residue in EnvZc causes a range of outcomes for autokinase activity, from a negligible change (e.g., EnvZc[T247E]) to a 1.6-fold higher activity (e.g., EnvZc[T247R], EnvZc[T247Y]) than that of the wild-type EnvZc [18]. The effects of mutation of the Thr247 residue were more severe on the phosphotransferase activity of EnvZc. With the exception of EnvZc[T247S], EnvZc[T247A] and EnvZc[T247Q], all of the other EnvZc[T247X] mutant proteins were impaired in transferring their phosphoryl groups to OmpR. This suggests that neither hydrophobic (C/Y) nor charged (E/K/R) residues can functionally substitute threonine to execute the phosphotransfer process. Significantly, mutations at the Thr247 residue dramatically affect the phosphatase activity. Of the nine mutant proteins, only EnvZc[T247S] exhibited phosphatase activity comparable to that of the wild type. Additionally, replacement of the Thr247 residue by Arg in the isolated domain A abolished its intrinsic phosphatase function, strongly supporting the notion that the Thr247 residue plays a critical role in EnvZ function, and strengthens the emerging view that domain A is not only the dimerization and histidine phosphotransfer domain but also the phosphatase domain.

The observation that the Thr247 residue can be replaced only with the conservative Ser mutation to retain comparable levels of all activities of

EnvZc clearly indicated that Thr247 is a critical residue at the active center of EnvZ and is possibly involved directly in catalyzing the phosphatase reaction, while facilitating the autokinase and phosphotransferase reactions [18]. We hypothesize that imidazole of the proximal His243 residue could be functioning as an acid–base catalyst, enhancing the nucleophilicity of the hydroxyl group of Thr247, thereby enabling it to directly attack the phosphorus of the phosphoryl moiety on Asp55 of OmpR-P, forming a highly reactive ester acylenzyme, which is hydrolyzed rapidly. Alternatively, the hydroxyl group of Thr247 could provide an oxygen of a bound water molecule to make a nucleophilic attack on the phosphorus atom on Asp55. The structure of the N-terminal domain of OmpR has not been solved. However, it was found that five water molecules are in the active site of the Mg^{2+}-bound structure of the homologous response regulator CheY [25]. Use of the threonine hydroxyl group rather than direct attack of a water molecule on the substrate is considered to be more favorable, as alcohols are often better nucleophiles than water molecules [26]. It has been demonstrated that the His243 residue plays an important role in the phosphatase function of EnvZ [16, 22–23]. Therefore, the proposed model that involves both the invariant His243 and the highly conserved Thr247 residues in the phosphatase function of EnvZ seems to be quite plausible. Taken together, these results clearly indicate that EnvZ is not a passive partner in the dephosphorylation of OmpR-P and that conserved residues such as His243 and Thr247 on domain A of EnvZ are catalytically engaged in the hydrolysis of OmpR-P.

It has been demonstrated previously that the Taz1 chimeric receptor between Tar and EnvZ responds to aspartate in the medium by inducing the expression of *ompC-lacZ* in *E. coli* RU1012 cells [27]. In the absence of a known ligand for osmolarity, the Taz constructs have been employed successfully to study the regulation of EnvZ function *in vivo* by monitoring the production of β-galactosidase [1, 5, 19, 28]. The earlier proposal that binding of a ligand to the receptor increases the ratio of kinase to phosphatase activity (K/P) [19] was further supported by the analysis of the aspartate responsiveness of RU1012 cells carrying Thr247 mutations on Taz1 [18]. It has been demonstrated that if the K/P ratio of the wild-type EnvZc is considered to be 1, then the K/P ratio of EnvZc[T247S] was estimated to be 2.2 and that of EnvZc[T247N] to be 3.3. All of the mutants showing a K/P ratio equal to or higher than 3.3 were found to display OmpC constitutive phenotypes in Taz1.

Cysteine Scanning of Domain A

Twenty-four cysteine substitution mutants were created in domain A to examine their effect on phosphatase activity (L. Qin and M. Inouye, unpublished results). Results indicate that two regions in domain A affect its phosphatase activity the most. One region encompassing Ser242, His243, Arg246, and Thr247 is located in the middle part of helix I and the other region encompassing Asn271, Lys272, Glu275, Glu276, and Asn278 is located in the middle part of helix II. Cys mutations in these two regions significantly decrease the rate constant to below 0.02 min^{-1} as compared to 0.058 min^{-1} of wild-type domain A. In the three-dimensional structure, these residues are closely located on the external surface of the lower region of domain A, including the active site His243 residue (see Chapter 2), suggesting that these regions probably participate not only in the phosphatase active center, but also in OmpR-P binding. An NMR titration experiment has demonstrated that OmpR interacts with the middle and bottom regions of the four-helix bundle formed by domain A [14]. Through this interaction, Asp55 of OmpR is considered to be placed close to His243 of EnvZ.

Among all Cys mutations constructed, the mutation at Leu254, which is located at three helix turns downstream to His243, was found to affect the phosphatase activity most severely. This residue, even if it is a hydrophobic amino acid, has been shown to be fully exposed to the solvent as the His243 residue [14], suggesting that this residue may be directly involved in the OmpR binding to domain A.

Effect of Domain B on Domain A Phosphatase Activity

When domain A is linked covalently to domain B, the phosphatase activity increases significantly [16]. There are three possible mechanisms to explain the role of domain B in phosphatase activity. First, domain B may have an allosteric effect on the function of domain A, affecting the domain A conformation to stimulate its phosphatase activity. It is important to note that when detached from domain A, domain B can still stimulate the phosphatase activity only in the presence of ADP or AMPPNP, suggesting that the ADP or AMPPNP–domain B complex in the presence of Mg^{2+} is able to enhance the allosteric stimulatory effect. Second, domain B may facilitate the interaction between domain A and its substrate OmpR-P. Third, ADP-bound domain B interacts directly with the catalytic center on domain A as a cofactor to stimulate the phosphatase reaction.

The Leu288 residue exists at the assumed turn structure on the top of domain A (X region), linking to domain B. The L288P mutation in EnvZc has been shown to abolish phosphatase activity while retaining kinase activity [15]. However, this mutation has no effect on the phosphatase activity of domain A [16], suggesting that the L288P mutation alters the spatial arrangement between domains A and B rather than being involved directly in the phosphatase reaction. Compared with kinase activity, phosphatase activity is affected more easily by various mutations isolated so far, suggesting that the topological relationship between domains A and B plays a more crucial role in phosphatase activity than in kinase activity. It is interesting to note that EnvZ or EnvZc with the L288P mutation migrates abnormally in a sodium dodecyl sulfate–polyacrylamide gel electrophoresis [15] (T. Yoshida and M. Inouye, unpublished data) and that tryptic digestion of EnvZc L288P shows a different pattern from that of wild-type EnvZc (T. Yoshida and M. Inouye, unpublished data). These results further suggest that this mutation causes a distinct conformational change in EnvZ.

On the basis of the effect of domain B on the domain A phosphatase activity described earlier, and the proposed hypothesis that the external signal regulates the ratio of kinase to phosphatase activity of the EnvZ kinase domain mainly by inhibiting phosphatase activity [19], one may speculate that signals transduced across the membrane (osmolarity for EnvZ and aspartate for Taz1) alter the relative spatial arrangement between domains A and B to mainly modulate phosphatase activity. At a low osmolarity for EnvZ or in the absence of aspartate for Taz1, domains A and B are positioned in such a way that EnvZ or Taz1 exhibits both kinase and phosphatase activity. At high osmolarity for EnvZ or in the presence of a high concentration of aspartate for Taz1, the spatial arrangement between domains A and B is altered, resulting in negative regulation of the phosphatase function. Such a displacement of domain A of one subunit against domain B of the partner subunit within a dimer may be a consequence of a physical displacement of one helix in the four-helix bundle in the receptor domain upon ligand binding as proposed for chemotaxis chemosensors [29–31].

DOMAIN B IS THE CATALYSIS-ASSISTING AND ATP-BINDING DOMAIN

Compared with domain A consisting of 67 residues, which simply form an antiparallel α-helical hairpin, domain B consists of 161 residues and is composed of four α helices and seven β structures (see Chapter 2). Furthermore, domain B is responsible for ATP binding. Because of these facts, domain B was originally thought to play a key role in the catalytic reactions of EnvZ and was thus termed the CA domain for the Catalytic ATP-binding domain. However, as discussed in the previous section, domain A by itself is able to function as phosphatase [16], and phosphorylated domain A is capable of transferring the phosphoryl group to OmpR without the aid of domain B [12]. However, domain B by itself can neither dephosphorylate OmpR-P nor phosphorylate OmpR. Nevertheless, as described later, domain B plays important roles in both kinase and phosphatase reactions, and therefore we propose that CA should now stand for Catalysis-assisting ATP-binding domain.

Role of Domain B in Kinase and Phosphatase

Domain A cannot be autophosphorylated at His243 in the absence of domain B. Phosphorylation of domain A with ATP can be achieved when it is mixed with active-site mutant EnvZc[His243Val], which by itself is unable to autophosphorylate [12]. When domain B is added to domain A in the presence of ATP, domain A can be phosphorylated at a low level, suggesting that formation of a heterodimer between domain A and EnvZc[His243Val] significantly enhances the phosphorylation of domain A and that formation of the domain A–OmpR–domain B–ATP quadruple complex cannot be stably formed.

Interestingly, ATP affinity to EnvZc was found to be diminished significantly in EnvZc[His243Val] in comparison with wild-type EnvZc, indicating that the active-site His243 residue in domain A plays an important role in the stable binding of ATP to domain B in an EnvZc dimer (Y. Zhu and M. Inouye, unpublished results). It appears that the covalent linkage between domain A and domain B is essential for proper positioning of the phosphoryl group of ATP to the His243 residue for the autophosphorylation reaction.

It is important to note that ADP-bound domain B is able to stimulate domain A phosphatase activity, although the level of this enhancement is low. The result suggests formation of the domain A–OmpR-P–domain B–ADP quadruple complex. It remains to be determined if ADP-bound domain B functions as a cofactor for the phosphatase reaction by interacting directly with the OmpR molecule bound to domain A. The direct interaction between OmpR (OmpR-P) and domain B has not yet been demonstrated.

Conserved Motifs in Domain B

The α and β elements in domain B constitute the structural framework of the ATP-binding site, whereas the amino acid residues involved in making contact with the bound ATP mainly cluster in highly conserved surface loops. These residues are within five conserved motifs: N, G1, F, G2, and G3 boxes (Fig. 1). Among these motifs, the N box is the exception in that residues interacting with ATP lie on helix 2. The highly conserved Asn347 residue has been shown to be involved in ATP binding, as EnvZc[Asn347Asp] loses its ATP-binding ability completely [20]. The conserved D residue (Asp373) in the G1 box (DXGXGI) is considered to form a hydrogen bond with N6 amine

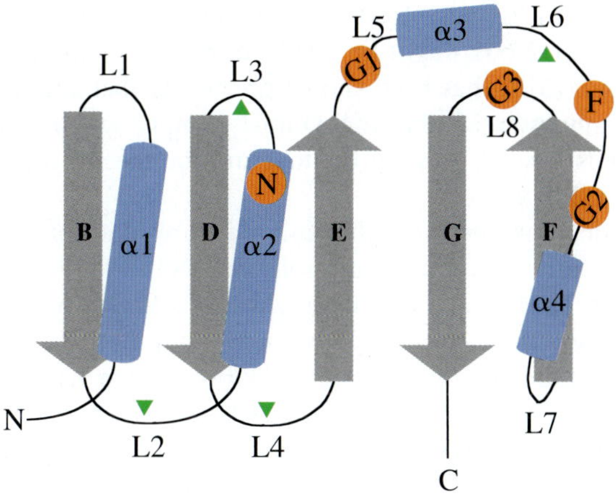

FIGURE 1 Schematic representation of the core elements of the ATP-binding fold of EnvZ domain B. β strands (gray arrow) are labeled alphabetically, whereas the first and third (β strands A and C) are omitted to simplify the diagram. In this α/β sandwich structure, α helices (blue cylinders) form a layer over the five β sheets. Orange circles represent all of the conserved regions in the CA domain, the N, G1, F, G2, and G3 boxes. Loop regions (L1 to L8) are also indicated by green triangles.

of the adenine ring [10]. This interaction possibly accounts for the specificity of ATP binding over GTP. The conserved glycine residues in the G1 and G2 boxes form its two hinges that confer flexibility of the movement of an intervening structure called the ATP lid [10]. This ATP lid consists of an unusually long disordered loop from Asp374 to Val409, containing a short helical structure (helix 3) and the F box. The Phe387 residue in the F box has been shown to be in close special proximity to the adenine ring of AMPPNP bound to domain B in its NMR structure [13].

The glycine residues in the G2 box of EnvZ are proposed to interact with the α and γ phosphates of ATP on the basis of the structural study on the mutL–ADPnP complex [10, 32]. There is another highly conserved glycine residue (Gly429) in loop 8 between β strands F and G. The region encompassing this glycine residue, together with Thr424 at the C-terminal end of strand F and Ser431 at the N-terminal end of strand G, is proposed to be another ATP-binding motif, termed the G3 box [10].

Mutational Analysis of Domain B Function

Biochemical and structural evidence has pointed out that the G2 box may play an important role in the phosphatase function of EnvZ. Comparison of the phosphatase activities of G1 box mutants and G1G2 double mutants indicated that the G2 box might be involved in both kinase and phophatase activities of EnvZ [5]. It has been shown previously that strains carrying the EnvZ T402K mutation exhibit the OmpF$^-$ OmpC^c phenotype [15]. The Thr402 residue is located between the first and the second conserved glycine residues of the G2 box.

It has been demonstrated that the amino acid residues of the G2 box and those in its vicinity play important roles in OmpR-P dephosphorylation. Residues Ser400, Thr402, and Gly403 might play different but related roles in this reaction. Of the three conserved glycine residues in the G2 box in EnvZc, mutations at Gly403 and Gly405 result in a loss of their ATP-binding ability, and therefore these residues are considered to be critical for the autokinase function (Y. Zhu and M. Inouye, unpublished results). The three other mutations studied, Ser400, Gly401, and Thr402, affected their autophosphorylation abilities at different levels. By intragenic suppressor screening, three independent second-site mutations were identified for the T402A mutation incorporated in a hybrid receptor, Taz1. All of them can suppress the Taz1-1 T402A OmpC$^-$ phenotype to restore wild-type-like phenotypes regulated by aspartate. Most interestingly, these suppressor mutations all fall in domain A, suggesting that domain A may directly interact with the G2 box region.

MONOMERIC HISTIDINE KINASE: TOPOLOGICAL ARRANGEMENT BETWEEN DOMAIN A AND DOMAIN B

It has not yet been determined how domain B is topologically arranged in the central domain A dimer in the three-dimensional structure. However, in order to explain the transautophosphorylation mechanism proposed earlier by a genetic approach [1], two EnvZ molecules are considered to be assembled as a dimer as shown in Fig. 2A. In this model, the ATP-bound domain B of the red molecule is presented to the His243 residue at domain A of the blue molecule. Therefore, an EnvZ dimer possesses two active centers, which are formed between domain B of an EnvZ subunit in a dimer and domain A of the other EnvZ subunit in the same dimer.

If this model is correct, one may construct a monomeric EnvZ histidine kinase by connecting the N-terminal end of the blue molecule in Fig. 2A to the C-terminal end of the second helix of domain A of the red molecule by removing domain B of the red molecule (shown by a dotted line). Indeed, such a construct, A–A–B, was made and shown to be active for all three reactions: autokinase, OmpR kinase, and OmpR-P phosphatase in a monomeric

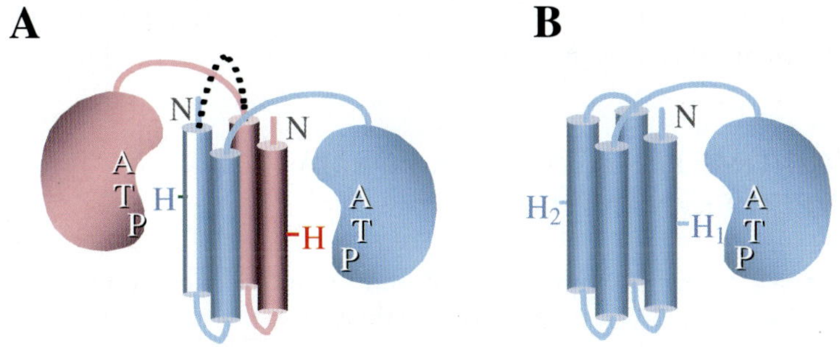

FIGURE 2 Molecular models for the EnvZc homodimer (A) and the EnvZc[AAB] monomer (B). (A) The four-helix bundle is formed by two domain As from two EnvZc subunits (colored red and blue). Domain B of one subunit is placed in close proximity to the His243 residue (H1 or H2) on helix I of domain A of the other subunit. The dotted line represents the linkage between two domain As in constructing EnvZc[AAB] (see text). (B) Molecular model of EnvZc [AAB] monomer. Only the H1 residue is phosphorylated.

form [17]. Although this monomeric kinase has two His residues, only the His243 residue of the amino proximal domain A (H1 in Fig. 2B) was found to be phosphorylated, proving the transphosphorylation mechanism. It is important to note that the A–A–B construct retains enzymatic function but loses its flexibility that regulates catalytic activity. The external signal is considered to cause asymmetric displacement of a subunit in a dimer against the other subunit in the same dimer to alter the three-dimensional configuration at the active center. This configuration change at the active center results in modulating the function of EnvZ. Thus, dimer formation is the obligatory requirement for signal transduction in the HAP system.

ROLE OF DNA IN EnvZ FUNCTION

An interesting aspect of EnvZ–OmpR-mediated osmoregulation is the role of promoter regions of target genes, *ompF* and *ompC*, in sequestering OmpR-P out of the reaction system. It has been reported that phosphorylation of OmpR was enhanced in the presence of DNA fragments containing OmpR-P-binding regions [33]. We reexamined the effect of DNA fragments on EnvZ function and found that the addition of DNA fragments has a negligible effect on OmpR phosphorylation by EnvZc, but dramatically reduces the dephosphorylation of OmpR-P by EnvZc [34]. The substantial stabilization of OmpR-P in the presence of DNA fragments occurs due to the sequestration of OmpR-P from the dephosphorylation reaction by its binding to DNA. This

OmpR-P sequestration is important in the reciprocal regulation of *ompF* and *ompC* genes in the cell, as discussed later.

STOICHIOMETRIC COMPLEX FORMATION BETWEEN EnvZ AND OmpR

The complex formation between EnvZ and OmpR has been observed previously by the Ni-NTA resin-binding method using His-tagged EnvZc and OmpR [12] or using His-tagged OmpR and EnvZc [17]. We have found that native polyacrylamide gel electrophoresis (PAGE) is very effective in identifying both the EnvZ/OmpR complex and the EnvZ/OmpR-P complex (T. Yoshida and M. Inouye, unpublished results).

When EnvZc and OmpR are mixed at the same concentration and the mixture is applied to native PAGE, individual EnvZc and OmpR bands (lanes 1 and 3, Fig. 3, respectively) disappear with the concomitant appearance of a new band near the top of the gel (lane 2). The new band was extracted from the gel to analyze its protein components by SDS–PAGE and it was confirmed that it indeed consisted of stoichiometric amounts of EnvZc and OmpR. Such a 1:1 EnvZc/OmpR complex can be found even in the presence of a large excess of EnvZc. It was also found that OmpR binding to EnvZc is cooperative, and its K_d value is estimated to be around 10^{-6} *M*. The complex formation is also observed between phosphorylated EnvZc and OmpR and between EnvZc and OmpR-P. Interestingly, Mg^{2+} is required for the former but not for the latter complex formation. Furthermore, OmpR-P bound to EnvZc is

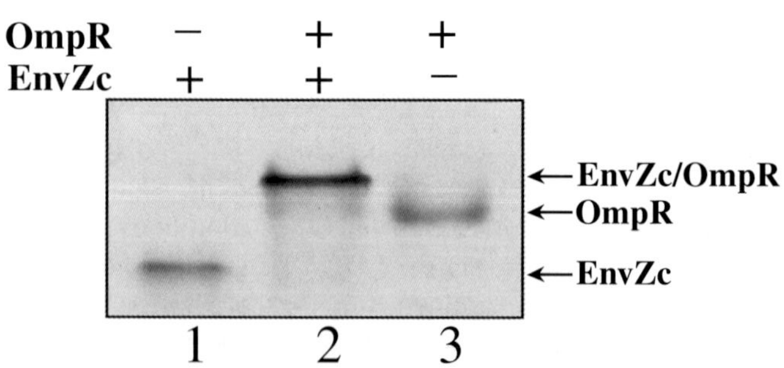

FIGURE 3 The EnvZc/OmpR complex. EnvZc 4 μM and OmpR 4 μM were mixed and incubated in the reaction buffer at room temperature for 5 min. After the addition of 2 × native loading solution, the samples were subjected to 10% native PAGE. The same amounts of EnvZc and OmpR as used for this experiment were applied in lanes 1 and 3, respectively.

released upon the addition of OmpR, suggesting that OmpR-P can be released easily in the presence of a large excess of nonphosphorylated OmpR into the cytoplasmic medium. The released OmpR-P is then trapped by the *ompF* and *ompC* promoter region (at a K_d value of approximately 1×10^{-8} *M*) to be sequestered from the EnvZ phosphatase reaction. This aspect is discussed in more detail later.

REGULATION OF KINASE AND PHOSPHATASE ACTIVITIES: SWITCH MODEL VERSUS RHEOSTAT MODEL

How do external signals through the periplasmic receptor domain regulate the function of the cytoplasmic kinase/phosphatase (K/P) domain? In one model, called the "on-off" model or "switch" model, the K/P domain is proposed to be only in one of the two possible states: either kinase$^+$ phosphatase$^-$ (on) or kinase$^-$ phosphatase$^+$ (off) [15]. In an alternative model, which we proposed to the "rheostat" model, the K/P domain always possesses both kinase and phosphatase activities, and the ratio of these two opposing activities is controlled by the external signal. Therefore, like an electrical rheostat, which is able to change the voltage continuously, a single histidine kinase dimer is able to determine a wide range of output of the downstream phosphorylated response regulator. Here we discuss these two models from a number of different perspectives.

Obligatory Dimerization

Histidine kinases function as dimers, and the active site is shared for both kinase and phosphatase activities. Two active sites are formed individually between two molecules in a dimer: a conserved His-containing DHp domain of one molecule in a dimer and a CA domain of the other molecule. Therefore, in the switch model, these two active centers in a dimer have to behave simultaneously in either "on" or "off" modes.

Asymmetric Signaling

Studies on Tar, the chemosensor for aspartate, have shown that the ligand (Asp) binds asymmetrically to the interface of two receptor domains in a Tar dimer [35]. This results in an asymmetric displacement of a helical structure with reference to the other domains in the dimer. Such asymmetric signal transduction has also been disputed in Tar within heterodimers of a full-length MCP and a truncated MCP [36, 37]. Asymmetric transmembrane signaling has also been demonstrated with Taz1 [28]. In this experiment, one of two aspartate-binding sites in the dimer interface was disrupted by mutations, and aspartate binding to only one of the ligand-binding pockets was shown to be sufficient for signal transduction. Furthermore, the subunit in a dimer, which plays a major role in ligand binding, is responsible for signal transduction.

These results indicate that the symmetrical arrangement of DHp and CA domains in the absence of signals is disturbed by asymmetric signaling through only one of the two histidine kinase molecules in a dimer. This likely results in different topological arrangements of DHp and CA domains between the two active sites in a histidine kinase dimer. Although it is not known how this asymmetric arrangement affects histidine kinase enzymatic activity, it is likely that enzymatic activities of the two active sites is different. Possibly at the site affected directly by the displacement of the signal-transducing helix, the ratio of kinase to phosphatase activity is altered, while at the other site, significant changes in the ratio of these activities may or may not occur. Thus, there may not be a distinct state of "on" and "off" configurations in the asymmetric signaling system through histidine kinase dimers.

Monomeric Histidine Kinase

As discussed earlier, by adding an extra domain A to the N-terminal end of an EnvZ kinase, the resulting A-A-B kinase is able to stay as a monomer and is fully functional as kinase and phosphatase [17]. It has been shown that the active site of the monomer is formed between the first domain A and the C-terminal domain B, and the conserved histidine residue in the second domain A is replaceable with another residue. These results demonstrate that a single active site functions simultaneously for both kinase and phosphatase.

Mutational Effects

A large number of mutations affect the enzymatic activity of EnvZ. A typical example is the series of substitutions at the highly conserved Thr247 residue on domain A of EnvZ [18], which alter phosphatase activity dramatically, and less pronouncedly affect autokinase and phosphotransferase activities. The overall result in these experiments is a wide range of final OmpR-P output, indicating that this Thr residue participates in both kinase and phosphatase reactions. However, all substitution mutations tested are less favorable to the phosphatase reaction in comparison with their effects on the kinase reaction. These results further indicate that EnvZ could be in any number of functional states having different ratios of kinase to phosphatase activities, which in turn determines the final cellular concentrations of OmpR-P.

In EnvZ, some mutations, such as those affecting ATP binding, diminish autokinase activity substantially, yet are still able to maintain *ompC* expression [15], indicating that the autokinase/kinase activity of EnvZ can be varied in a wide range if the phosphatase activity is affected simultaneously.

Signal Reception

External signals, such as medium osmolarity, are likely to change in continuous gradients. OmpF and OmpC can be produced in any ratios, dependent on the extraneous osmotic signal. However, receptors that recognize discrete chemical ligands exist in two forms, either ligand bound or ligand free; namely, either on or off. However, these "on" and "off" conformations are stably maintained only when the concentrations of the specific ligand are much higher ("on" conformation) or much lower ("off"conformation) than the K_d value of the ligand binding to the receptor. If the ligand concentration is at the K_d value, every dimeric receptor complex is constantly changing its conformation between "on" and "off," which may result in a kinetically intermediate conformation of the catalytic domain. Therefore, in a HAP system responding to a wide range of fluctuations in signal concentrations, it is working mostly in either "on" or "off" switch mode. This may be the case for the quorum-sensing histidine kinase system (Chapter 14). However, if a HAP system responds to a very narrow range of signal fluctuations centering around the K_d value for ligand binding, the outcome is controlled in a rheostat mode. This rheostat model is likely to be applied to the osmoregulation of *ompF* and *ompC* expression by EnvZ and OmpR.

Rheostatic Regulation of CheA

It is important to distinguish between class I and class II histidine kinases. Because CheA, a class II histidine kinase, lacks phosphatase activity, phospho-CheY is dephosphorylated by a separate enzyme, called CheZ. CheA kinase activity is enhanced 100-fold when it binds to MCPs such as Tar and Tsr (Chapter 5). As described in Chapter 5, if one-sixth of cellular CheA molecules are bound to MCPs in the absence of ligands, this results in an overall activation of the kinase activity by approximately 18 fold [=(100+5)/6] in comparison with conditions where no CheA binds to MCPs. Ligand binding to a particular kind of MCP (note that there are five different MCPs) apparently influences CheA binding not only to that particular kind of MCP, but also to other species of MCPs. This propagation of signals is proposed to occur through the clustering of cellular MCPs (Chapter 5). Interestingly, even if CheA itself functions in two distinct "on" (kinase-activated) and "off" (background kinase) states, the overall kinase activity is controlled by the extent of CheA kinase activity that is inactivated by ligand binding. For example, if 50 or 80% of MCP-based CheA is inactivated, the overall activation of CheA is reduced to 9-fold [=(50+5.5)/6] or 4-fold [(20+5.8)/6] over the background. Thus, the overall CheA kinase activity in the cell is regulated in the rheostat mode.

MECHANISM OF OSMOREGULATION

In considering HAP signal transduction, another important factor is the cellular concentrations of a histidine kinase and its cognate response regulator in individual HAP systems. In the osmoregulatory HAP systems in *E. coli*, EnZ is estimated to be approximately 100 molecules per cell, whereas OmpR exists at about 3500 molecules/cell, 35 times more than EnvZ molecules or 70 times more than EnvZ dimers [38]. We also estimated the K_d value of OmpR binding to EnvZ to be $1.2 \times 10^{-6}\,M$. Since the cellular OmpR concentration is calculated to be $6 \times 10^{-6}\,M$, OmpR molecules exist in the cell at a concentration about 5 times higher than the K_d value, indicating that 85% of EnvZ dimers are always occupied with OmpR.

One may wonder why OmpR molecules exist in such a large excess considering that there are probably at most 30 OmpR-binding sites per *E.coli* chromosome (8 for *ompF*, 6 for *ompC*, and assuming 16 more unknown OmpR-binding sites). This indicates that OmpR molecules exist more than 115 times in excess of their target sites. This is a rather interesting contrast to LacI, which exists only at the level of 10–20 molecules per cell to repress *lacZ* expression [39]. The critical difference between LacI and OmpR is due to the significant difference in their K_d values: $1 \times 10^{-13}\,M$ for LacI binding to the *lac* operator and approximately $1 \times 10^{-8}\,M$ for OmpR binding to the highest affinity sites such as F1 and C1 (0.68×10^{-8} and $0.77 \times 10^{-8}\,M$ for F1 and C1 sites, respectively) [40, 41].

Therefore, even if only 1/40 or 2.4% of the total OmpR molecules are phosphorylated by 50 EnvZ dimers in a cell (Omp-R-P concentration = $10^{-7}\,M$), 90% of F1 and C1 sites are calculated to be occupied with OmpR-P. However, under this condition, weaker OmpR-P-binding sites such as C2 and C3 are not able to serve as OmpR-P-binding sites, as several times higher concentrations of OmpR-P are needed for the binding of OmpR-P to these sites [42,

43]. Note that the binding of OmpR-P to F1 always results in cooperative OmpR-P binding to F2 and F3 sites, which is a few times tighter than OmpR-P binding to C2 and C3 sites [42]. Thus, the *ompF* gene is on while the *ompC* gene is off ($OmpF^+$ $OmpC^-$), a typical phenotype at low osmolarity. At higher osmolarity, when the cellular OmpR-P concentration increases several times than that at low osmolarity, the C2 and C3 sites become occupied with OmpR-P to induce *ompC* expression, whereas OmpR-P binding to F3 and F4 sites results in repression of the *ompF* expression.

Through these considerations, the HAP system regulated by EnvZ and OmpR appears to be designed for its ability to finely tune *ompF* and *ompC* reciprocal expression under different medium osmolarities. In summary, such fine-tuning can be achieved by the contribution of the following factors: (a) the presence of multiple OmpR-P-binding sites in both *ompF* and *ompC* promoters; (b) OmpR-P functions not only as a transcription activator, but also as a repressor; (c) OmpR molecules exist more than 115 times in excess over the OmpR-P-binding sites on the *E. coli* chromosome; (d) the K_d value for OmpR-P binding to the highest affinity DNA sites (K_{dDNA}) is designed to be about 1/170 of the K_d value of OmpR/OmpR-P binding to EnvZ (K_{dEnvZ}); and (e) the $K_{d_{DNA}}$ value is about 1/1000 of the total OmpR + OmpR-P concentration in the cell, which is five times higher than the $K_{d_{EnvZ}}$ value.

CONCLUDING REMARKS

The reciprocal expression of *ompF* and *ompC* genes regulated by medium osmolarity is designed in a most sophisticated fashion in bacteria. The fine-tuning of the expression of the two genes discussed in the last section is further achieved by using the antisense RNA against *ompF* mRNA [44]. The *micF* gene, which produces a small RNA molecule complementary to *ompF* mRNA, is located upstream of the *ompC* gene and its expression is induced at higher osmolarity so that *ompF* expression is controlled not only at the level of transcription, but also at the level of translation.

For a complete understanding of osmoregulation at the level of histidine kinases, several issues still remain to be addressed.

a. Defining the role of the periplasmic receptor domain of EnvZ. There may be specific ligands recognized by the receptor, which may controlled directly by medium osmolarity.

b. Establishing the notion that EnvZ, an osmosensor, and Tar, a chemosensor, share a common transmembrane signal-transducing mechanism. Clearly, the cytoplasmic region between the transmembrane domain and the kinase domain (domain A) called the linker region plays a crucial role in this mechanism. Characterization of the linker function is a key in determining the exact nature of the signal to be transmitted to the catalytic domain.

c. Establishing how the enzymatic function of EnvZ is regulated in the three-dimensional arrangement of domain A and domain B? It is certain that the junction region between domain A and domain B plays an important role in this aspect, and further characterization of this region may yield important insights into EnvZ function.

d. Establishing the mechanism by which EnvZ recognizes OmpR. In particular, it is of great interest how OmpR interacts with domain B.

e. Defining the function of ADP as a cofactor for the phosphatase reaction of EnvZ.

Addressing these issues will likely provide important clues in designing new antibiotics targeting histidine kinase.

ACKNOWLEDGMENTS

The authors are grateful to A. Newton, M. Ikura, U. Shinde, L. Qin, S. Phadtare, S. Cai, and T. Yoshida for their critical reading of this chapter.

REFERENCES

1. Yang, Y., and Inouye, M. (1991). Intermolecular complementation between two defective mutant signal-transducing receptors of *Escherichia coli*. *Proc. Natl. Acad. Sci. USA* **88**, 11057–11061.
2. Aiba, H., Nakasai, F., Mizushima, S., and Mizuno, T. (1989). Phosphorylation of a bacterial activator protein, OmpR, by a protein kinase, EnvZ, results in stimulation of its DNA-binding ability. *J. Biochem. (Tokyo)* **106**, 5–7.
3. Igo, M. M., Ninfa, A. J., Stock, J. B., and Silhavy, T. J. (1989). Phosphorylation and dephosphorylation of a bacterial transcriptional activator by a transmembrane receptor. *Genes Dev.* **3**, 1725–1734.
4. Nikaido, H., and Vaara, M. (1985). Molecular basis of bacterial outer membrane permeability. *Microbiol. Rev.* **49**, 1–32.
5. Yang, Y., and Inouye, M. (1993). Requirement of both kinase and phosphatase activities of an *Escherichia coli* receptor (Taz1) for ligand-dependent signal transduction. *J. Mol. Biol.* **231**, 335–342.
6. Forst, S. A., and Roberts, D. L. (1994). Signal transduction by the EnvZ-OmpR phosphotransfer system in bacteria. *Res. Microbiol.* **145**, 363–373.
7. Pratt, L., and Silhavy, T. J. (1995). Porin regulon of *Escherichia coli*. *In* "Two-Component Signal Transduction" (Silhavy, T. J. and Hoch, J. A., eds.), pp. 105–127. Am. Soc. Microbiol., Washington, DC.
8. Egger, L. A., Park, H., and Inouye, M. (1997). Signal transduction via the histidyl-aspartyl phosphorelay. *Genes Cells* **2**, 167–184.
9. Parkinson, J. S., and Kofoid, E. C. (1992). Communication modules in bacterial signaling proteins. *Annu. Rev. Genet.* **26**, 71–112.

10. Dutta, R., and Inouye, M. (2000). GHKL, an emergent ATPase/kinase superfamily. *Trends Biochem. Sci.* **25**, 24–28.
11. Swanson, R.V., Alex, L.A., and Simon, M.I. (1994). Histidine and aspartate phosphorylation: Two-component systems and the limits of homology. *Trends Biochem. Sci.* **19**, 485–490.
12. Park, H., Saha, S. K., and Inouye, M. (1998). Two-domain reconstitution of a functional protein histidine kinase. *Proc. Natl. Acad. Sci. USA* **95**, 6728–6732.
13. Tanaka, T., Saha, S. K., Tomomori, C., Ishima, R., Liu, D., Tong, K. I., Park, H., Dutta, R., Qin, L., Swindells, M. B., Yamazaki, T., Ono, A. M., Kainosho, M., Inouye, M., and Ikura, M. (1998). NMR structure of the histidine kinase domain of the *E. coli* osmosensor EnvZ. *Nature* **396**, 88–92.
14. Tomomori, C., Tanaka, T., Dutta, R., Park, H., Saha, S. K., Zhu, Y., Ishima, R., Liu, D., Tong, K. I., Kurokawa, H., Qian, H., Inouye, M., and Ikura, M. (1999). Solution structure of the homodimeric core domain of *Escherichia coli* histidine kinase EnvZ. *Nature Struct. Biol.* **6**, 729–734.
15. Hsing, W., Russo, F. D., Bernd, K. K., and Silhavy, T. J. (1998). Mutations that alter the kinase and phosphatase activities of the two-component sensor EnvZ. *J. Bacteriol.* **180**, 4538–4546.
16. Zhu, Y., Qin, L., Yoshida, T., and Inouye, M. (2000). Phosphatase activity of histidine kinase EnvZ without kinase catalytic domain. *Proc. Natl. Acad. Sci. USA* **97**, 7808–7813.
17. Qin, L., Dutta, R., Kurokawa, H., Ikura, M., and Inouye, M. (2000). A monomeric histidine kinase derived from EnvZ, an *Escherichia coli* osmosensor. *Mol. Microbiol.* **36**, 24–32.
18. Dutta, R., Yoshida, T., and Inouye, M. (2000). The critical role of the conserved Thr247 residue in the functioning of the osmosensor EnvZ, a histidine Kinase/Phosphatase, in *Escherichia coli*. *J. Biol. Chem.* **275**, 38645–38653.
19. Jin, T., and Inouye, M. (1993). Ligand binding to the receptor domain regulates the ratio of kinase to phosphatase activities of the signaling domain of the hybrid *Escherichia coli* transmembrane receptor, Taz1. *J. Mol. Biol.* **232**, 484–492.
20. Varughese, K. I., Madhusudan, Zhou, X. Z., Whiteley, J. M., and Hoch, J. A. (1998). Formation of a novel four-helix bundle and molecular recognition sites by dimerization of a response regulator phosphotransferase. *Mol. Cell.* **2**, 485–493.
21. Dutta, R., and Inouye, M. (1996). Reverse phosphotransfer from OmpR to EnvZ in a kinase–/phosphatase+ mutant of EnvZ (EnvZ.N347D), a bifunctional signal transducer of *Escherichia coli*. *J. Biol. Chem.* **271**, 1424–1429.
22. Hsing, W., and Silhavy, T. J. (1997). Function of conserved histidine-243 in phosphatase activity of EnvZ, the sensor for porin osmoregulation in *Escherichia coli*. *J. Bacteriol.* **179**, 3729–3735.
23. Skarphol, K., Waukau, J., and Forst, S. A. (1997). Role of His243 in the phosphatase activity of EnvZ in *Escherichia coli*. *J. Bacteriol.* **179**, 1413–6.
24. Grebe, T. W., and Stock, J. B. (1999). The histidine protein kinase superfamily. *Adv. Microb. Physiol.* **41**, 139–227.
25. Stock, A. M., Martinez-Hackert, E., Rasmussen, B. F., West, A. H., Stock, J. B., Ringe, D., and Petsko, G. A. (1993). Structure of the Mg(2+)-bound form of CheY and mechanism of phosphoryl transfer in bacterial chemotaxis. *Biochemistry* **32**, 13375–13380.
26. Fersht, A. (1999. Structure and Mechanism in Protein Science: A Guide to Enzyme Catalysis and Protein Folding, pp. 84–85. Freeman, New York.
27. Utsumi, R., Brissette, R. E., Rampersaud, A., Forst, S. A., Oosawa, K., and Inouye, M. (1989). Activation of bacterial porin gene expression by a chimeric signal transducer in response to aspartate. *Science* **245**, 1246–1249.
28. Yang, Y., Park, H., and Inouye, M. (1993). Ligand binding induces an asymmetrical transmembrane signal through a receptor dimer. *J. Mol. Biol.* **232**, 493–498.
29. Falke, J. J., Bass, R. B., Butler, S. L., Chervitz, S. A., and Danielson, M. A. (1997). The two-

component signaling pathway of bacterial chemotaxis: A molecular view of signal transduction by receptors, kinases, and adaptation enzymes. *Annu. Rev. Cell. Dev. Biol.* **13**, 457–512.
30. Gerstein, M., and Chothia, C. (1999). Perspectives: Signal transduction. Proteins in motion. *Science* **285**, 1682–1683.
31. Ottemann, K. M., Xiao, W., Shin, Y. K., and Koshland, D. E., Jr. (1999). A piston model for transmembrane signaling of the aspartate receptor. *Science* **285**, 1751–1754.
32. Ban, C., Junop, M., and Yang, W. (1999). Transformation of MutL by ATP binding and hydrolysis: A switch in DNA mismatch repair. *Cell* **97**, 85–97.
33. Ames, S. K., Frankema, N., and Kenney, L. J. (1999). C-terminal DNA binding stimulates N-terminal phosphorylation of the outer membrane protein regulator OmpR from *Escherichia coli*. *Proc. Natl. Acad. Sci. USA* **96**, 11792–11797.
34. Qin, L., Yoshida, T., and Inouye, M. (2001). The critical role of DNA in the equilibrium between OmpR and phosphorylated OmpR mediated by EnvZ in *Escherichia coli*. *Proc. Natl. Acad. Sci. USA* **98**, 908–913.
35. Yeh, J. I., Biemann, H. P., Pandit, J., Koshland, D. E., and Kim, S. H. (1993). The three-dimensional structure of the ligand-binding domain of a wild-type bacterial chemotaxis receptor. Structural comparison to the cross-linked mutant forms and conformational changes upon ligand binding. *J. Biol. Chem.* **268**, 9787–9792.
36. Tatsuno, I., Homma, M., Oosawa, K., and Kawagishi, I. (1996). Signaling by the *Escherichia coli* aspartate chemoreceptor Tar with a single cytoplasmic domain per dimer. *Science* **274**, 423–425.
37. Gardina, P. J., and Manson, M. D. (1996). Attractant signaling by an aspartate chemoreceptor dimer with a single cytoplasmic domain. *Science* **274**, 425–426.
38. Cai, S. J., and Inouye, M. (2002). EnvZ-OmpR interaction and osmoregulation in *Escherichia coli*. *J. Biol. Chem.*, in press.
39. Gibert, W., and Müller-Hill, B. (1970). The lactose repressor. *In* "The Lactose Operon." (D. Ziper, and J. Bechwith, eds.), pp. 93–109. Cold Spring Harbor Laboratory Press, Cold Spring Harbor, NY.
40. Head, C. G., Tardy, A., and Kenney, L. J. (1998). Relative binding affinities of OmpR and OmpR-phosphate at the ompF and ompC regulatory sites. *J. Mol. Biol.* **281**, 857–870.
41. Harlocker, S. L., Bergstrom, L., and Inouye, M. (1995). Tandem binding of six OmpR proteins to the ompF upstream regulatory sequence of *Escherichia coli*. *J. Biol. Chem*, **270**, 26849–26856.
42. Bergstrom, L. C., Qin, L., Harlocker, S. L., Egger, L. A., and Inouye, M. (1998). Hierarchical and co-operative binding of OmpR to a fusion construct containing the ompC and ompF upstream regulatory sequences of *Escherichia coli*. *Genes Cells* **3**, 777–788.
43. Forst, S., Delgado, J., Ramperand, A., and Inouye, M. (1990). *In vivo* phosphorylation of OmpR, the transcriptional activation of the *ompF* and *ompC* genes in *Escherichia coli*. *J. Bacterol.* **172**, 3473–3477.
44. Mizuno, T., Chou, M. Y., and Inouye, M. (1984). A unique mechanism regulating gene expression: Translational inhibition by a complementary RNA transcript (micRNA). *Proc. Natl. Acad. Sci. USA* **81**, 1966–1970.

CHAPTER 4

Structure and Function of CheA, the Histidine Kinase Central to Bacterial Chemotaxis

ALEXANDRINE M. BILWES,* SANG-YOUN PARK,* CINDY M. QUEZADA,† MELVIN I. SIMON,† AND BRIAN R. CRANE*

**Department of Chemistry and Chemical Biology, Cornell University, Ithaca, New York 14853 and †Department of Biology, California Institute of Technology, Pasadena, California 91125*

Most bacteria control their swimming by switching the direction of flagellar rotation in response to gradients of specific chemicals in their environment. The histidine kinase CheA couples changes in the ligand occupancy of transmembrane chemoreceptors to phosphorylation of the response regulator CheY; CheY directly modulates motion of the flagellar motor. Chemotaxis is the first signal transduction cascade where structures are known for all of the key protein components. The structure of CheA reveals a dimeric protein with each subunit composed of five domains. Each of the domains engenders a different functionality important for signaling: (1) histidine autophosphorylation, (2) CheY recognition, (3) dimerization, (4) ATP binding, and (5) receptor coupling. This chapter reviews how structural information has helped us understand the chemistries and interactions among the CheA domains and other proteins in the chemotactic signal transduction cascade. Interesting structural relationships among these proteins and those involved in functionally unrelated systems provide general insights into how ATP utilization can control molecular motions and associations.

INTRODUCTION

Protein histidine kinases (PHKs) regulate a wide variety of cellular responses in bacteria, fungi, and plants by initiating "phosphorelays" in response to environmental stimuli [1–4]. Bacterial chemotaxis, the movement of cells toward specific chemicals and away from others [5], uses a "two-component" signaling system with the PHK CheA as the central element (Fig. 1) [6–8]. Bacterial inner membranes contain membrane-spanning receptors, whose periplasmic domains bind attractants such as serine and aspartate. Receptor intracellular domains form stable complexes with dimeric CheA through the monomeric adaptor protein CheW [9]. In response to changes in receptor occupancy, CheA uses ATP to transphosphorylate a specific substrate histidine residue [10, 11] on the adjacent subunit within the CheA dimer [12, 13]. Immediately after autophosphorylation, the phosphoryl group is transferred from histidine to a specific aspartyl residue on the response regulator protein, CheY (Fig. 1) [14, 15]. Phosphorylated CheY interacts directly with the FliM protein of the flagellar motor and thereby affects swimming behavior [16]. When the flagella rotate counterclockwise, they form a coherent bundle and the bacterium swims smoothly. Clockwise rotation causes the filaments to fly apart and the bacterium to tumble and reorient. Phosphorylated CheY directly affects translocation by mediating changes in the direction of flagellar rotation.

Chemotaxis employs a fast excitation response (milliseconds) and a slow adaptation response (minutes). The fast response is generated by changes in the rate of CheA autophosphorylation and subsequent phosphotransfer to CheY when ligands bind or dissociate from the receptor. Three general activity

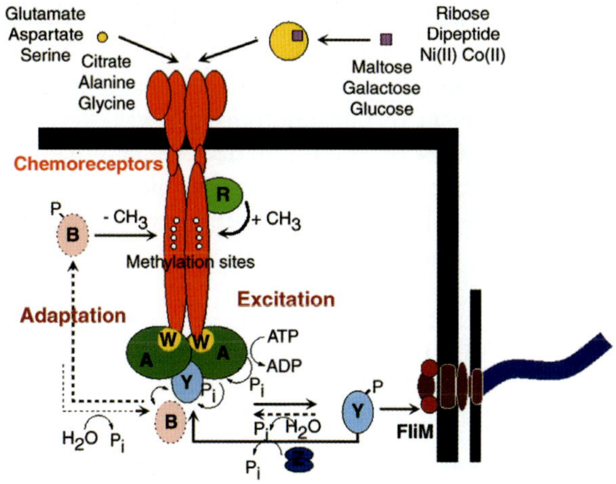

FIGURE 1 The signal transduction pathway that couples concentrations of external ligands to swimming behavior.

states have been characterized for CheA [11, 17]: (1) *in vitro* and in the absence of other proteins CheA autophosphorylates at a basal rate; (2) in the presence of activating receptor (no attractant bound) and the coupling protein CheW, CheA autokinase activity increases 10-fold or more relative to the basal rate; and (3) in the presence of inhibitory receptor (attractant-bound), CheA autophosphorylation inactivates completely. When the receptor is empty, the production of phospho-CheY biases the bacterium toward tumbling. CheY dephosphorylates spontaneously or by action of CheZ, which binds specifically to CheY and increases the rate of dephosphorylation by 100-fold [14]. Binding attractant inhibits CheA and favors smooth swimming.

Along with fast phosphorylation/dephosphorylation of CheY, a slower adaptation response also begins with CheA autophosphorylation (Fig. 1). In addition to CheY, CheA phosphorylates a methyl esterase CheB, which, in its activated form, removes methyl groups from glutamate residues on the C-terminal tail of the receptor. These methyl groups, which are added by the methyl transferase CheR, desensitize the receptor to external ligand [18] and also directly stimulate CheA kinase activity [19]. Thus, their removal by phosphorylated CheB downregulates the kinase and resets the system to respond to higher attractant concentrations. CheB competes with CheY for CheA [20] and CheB also becomes activated by aspartate phosphorylation. CheR and CheB together regulate the level of receptor modification, with demethylation (and hence kinase inhibition) controlled through the phosphorylation of CheB by CheA. Thus, through this feedback mechanism, CheA is essential to the slow adaptation response, as well as the fast excitation response in bacterial locomotion [21].

This chapter reviews the current structural information concerning PHK CheA and its associated proteins. We begin with an overview of CheA domains and their function. Then we discuss the kinase domain, its substrate domain, and the mechanism of phosphotransfer. Next we address response regulator coupling, the nature of CheA dimerization, and the CheA regulatory domain. We end with a discussion of interactions among the CheA domains, how the kinase may be regulated by chemoreceptors, and finally strategies for rational drug design targeting CheA.

MODULAR STRUCTURE OF CheA

Histidine kinases can be separated into two major classes by considering the position in the sequence of the substrate histidine and surrounding residues (H box) with respect to the ATP-binding domain (Fig. 2). Four regions of sequence similarity (N, G1, F, and G2 boxes) delineate the ATP-binding domain. The distinguishing feature of class I histidine kinases is that the H box is directly adjacent in sequence to the ATP-binding domain. In class II histidine kinases, which is exemplified by CheA, the histidine residue that becomes phosphorylated is located in the P1 domain, a distant separate domain at the N terminus of the protein.

The modular character of CheA was predicted by functional assays with isolated fragments [22–24]. These studies identified at least four different domains and associated them with five different functionalities: His-contain-

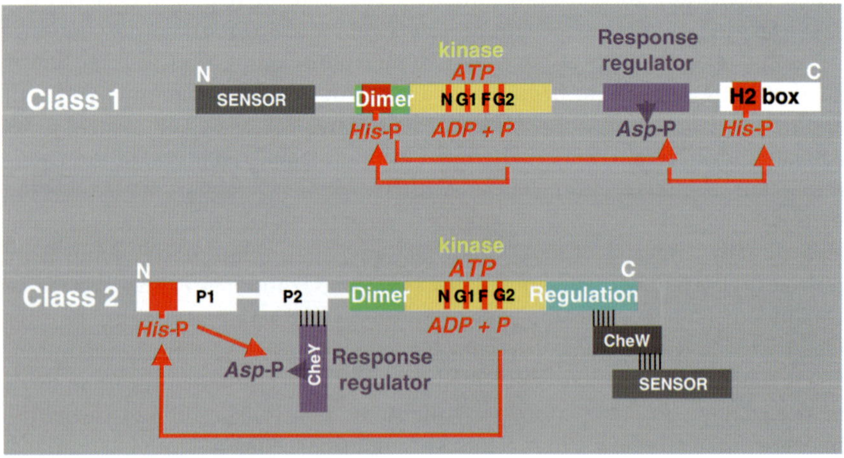

FIGURE 2 Two classes of protein histidine kinases can be distinguished by the position of the HPt/DHp domain relative to the ATP-binding (kinase) domain.

ing phosphotransfer (HPt), response regulator coupling (CheY binding), regulatory coupling (CheW binding), and dimerization with kinase activity. Crystallographic and nuclear magnetic resonance (NMR) structure determinations have further defined a molecular structure with five independent domains in which the region responsible for dimerization is separated from the one that provides kinase activity. Thus, CheA has five domains per monomer designated P1 to P5 from the N terminus to the C terminus. P1 constitutes the HPt domain, P2 docks CheY for phosphotransfer from P1 to CheY, P3 mediates dimerization, P4 binds ATP, and P5 regulates kinase activity in response to chemoreceptors (Fig. 3). P2 is separated from P1 and P3 by variable length linkers (typically 25 to 45 residues) presumed to be flexible [25]. In contrast, short hinges connect domain P4 to P3 and P5 [26] (at residues Arg354 and Thr540 for *Thermotoga maritima* CheA).

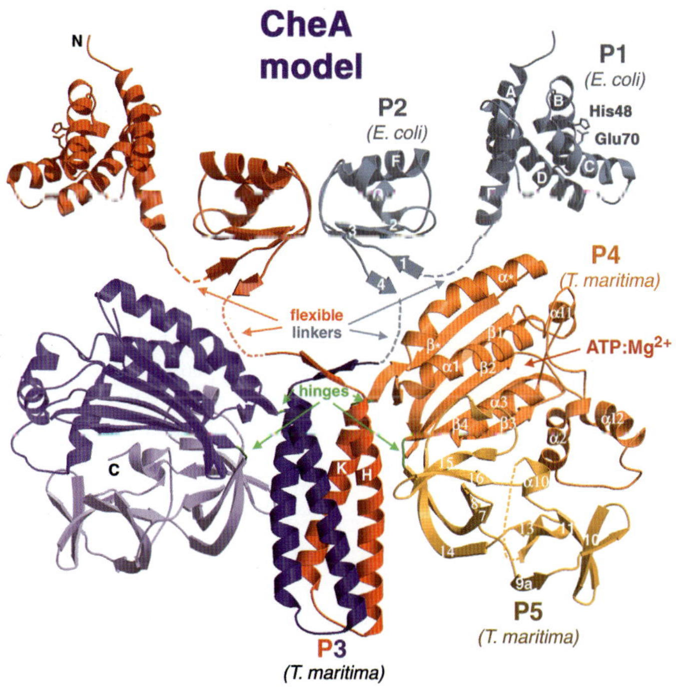

FIGURE 3 The PHK CheA in all its grandeur. The NMR structure of two *E. coli* CheA P1 [27], the crystal structure of two *E. coli* CheA P2 (pdb code 1eay [31]), and the crystal structure of one dimeric *T. maritima* CheA P3-P5 (pdb code 1b3q [26]) are shown. Dotted lines symbolize missing residues in the chain and putative linker regions between domains. One CheA subunit is represented in orange/yellow colors, while the other subunit is in purple/blue tones. Figures 3, 4, 7, 8, 9, 10, and 11 produced with Molscript [75] and Raster3d [76].

The "beads on a string" metaphor applies well to the structure of CheA in that the molecule contains separate folding units with distinct functions strung together over the length of the polypeptide (Fig. 3). The structure of P1 (from *Escherichia coli* and *Salmonella typhimurium*) consists of a four-helix bundle with the substrate histidine accessible to solvent on the outer surface of the second helix [27, 28]. A fifth helix connects the bundle to the P2 linker. The structure of *E. coli* P2, which was determined in isolation and in complex with *E. coli* CheY [29–32], reveals a small compact two-layer α/β sandwich (CATH classification) [33]. Finally, structures for the last three domains (dimerization, kinase, and regulatory modules, CheAΔ289) were determined together for the *T. maritima* enzyme [26]. In the dimeric structure of CheAΔ289, the two kinase (P4) and regulatory domains (P5) are arranged around a central four-helix bundle (P3) in a three-dimensional "X" pattern of dimension: $55 \times 120 \times 70$ Å^3. There are no contacts between the two regulatory domains or between the two kinase domains. This organization places the two ATP-binding pockets 90 Å apart and disfavors functional interaction between the two kinase domains. The P3 dimerization domain is composed of two antiparallel helices that pack against the analogous two helices of the second subunit to form the central four-helix bundle. The P4 kinase domain is a two-layered α-β sandwich made of a flat, mixed, five-stranded β sheet and seven α helices. This structure forms a deep cavity where Mg^{2+}-ATP binds. Three of the helices (α*, α1, α3) are amphipathic and pack parallel to the sheet, whereas the four shorter remaining helices point into the solvent; three of these helices (αI2, αI3, α2) border the ATP-binding site. Upon binding the nonhydrolyzable Mg^{2+}-ATP analog Mg^{2+}-ADPCP, the active site loop between α2 and α3 refolds into an additional short helix [34]. The P5 domain (541–671) regulates kinase activity by interacting with an activating receptor through the adaptor protein CheW [23, 35]. P5 displays two intertwined five stranded β barrels with an unusual topology that results in two adjacent strands being parallel in each barrel.

A SUPERFAMILY OF HISTIDINE KINASES AND ATPases

The CheA histidine kinase domain (P4) is structurally similar to the ATP-binding domain of a class of ATPases named the GHL family [36] after the three structurally defined members: the type II DNA topoisomerase GyraseB [37], the chaperone Hsp90 [38, 39], and the DNA repair enzyme MutL [40]. These functionally divergent ATPases are multidomain proteins whose other domains are unrelated to histidine kinases. The core structural elements in common among CheA, GyrB, Hsp90, and MutL consist primarily of the four β strands and three α helices that form a deep cavity for binding ATP (Fig. 4). The α/β hairpin made of α* and β* in CheA has also a structural equivalent in GHL ATPases but is transposed in sequence (Fig. 4). In summary, histidine

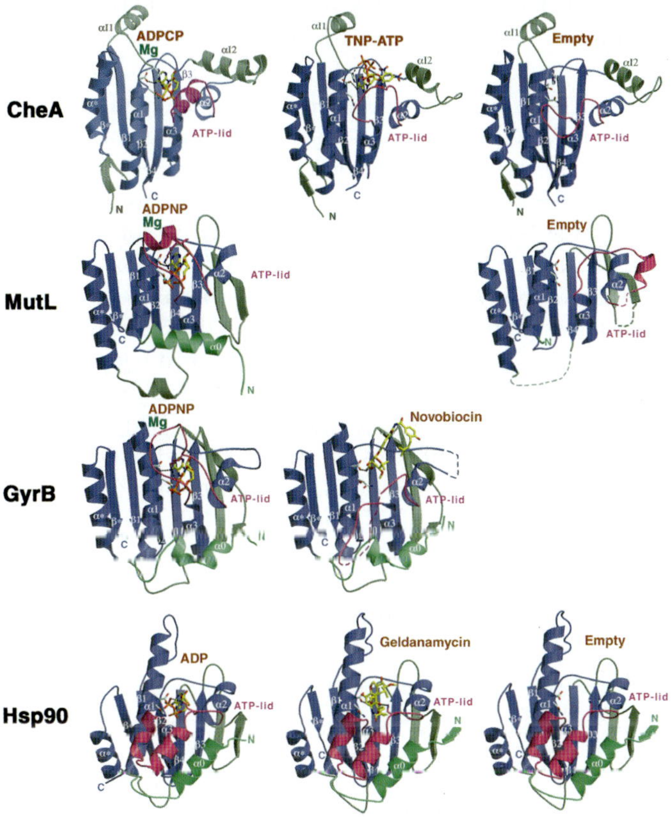

FIGURE 4 PHKs and GHL ATPases conserve a domain for binding ATP but vary in ATP lid conformation. Ribbon diagrams represent crystal structures of the ATP-binding domains of CheA with ADPCP:Mg^{2+} (pdb code 1i58), CheA with TNP-ATP (pdb code 1i5d), and CheA *"empty"* (pdb code 1b3q); MutL with ADPNP (pdb code 1b63 [36]) and MutL *"empty"* (pdb code 1bkn [40]); GyrB with ADPNP [37] and GyrB with the antibiotic novobiocin (pdb code 1aj6 [77]); and yeast Hsp90 with ATP (γ-phosphate is not visible; pdb code 1am1 [39]), human Hsp90 with the anti tumor drug geldanamycin (pdb code 1yet [38]), and human Hsp90 *"empty"* crystal form $P2_1$ (pdb code 1yes [38]). The secondary structure elements common between these proteins (in blue) are numbered in topological order α1 to α3 and β1 to β4; uncommon elements are gray or green (amino-terminal insertions of the GHL ATPases). One helix and one strand (α* and β*) are structurally similar but not topologically equivalent. The ATP lid (pink) is defined as the region between α2 and α3. In CheA, MutL, and GyrB, the ATP lid changes conformation depending on active site occupancy. For the structurally defined fragment of Hsp90, the ATP lid does not change conformation on geldanamycin or ATP binding.

kinases are unrelated to mammalian Ser/Thr or Tyr kinases and instead derive from an ancestor common to GyrB, Hsp90, and MutL.

NUCLEOTIDE BINDING BY CheA P4 AND THE GHL ATPases

The ATP-binding sites of PHKs and GHL ATPases are highly conserved. ATP analogs bind CheA in a deep cavity, whose back wall is formed by the P4 β sheet (Fig. 5). The cavity edges include four regions of sequence similarity that characterize the histidine kinase family (Fig. 2). These are: (1) the N box (helix α1), (2) the G1 box (the segment running in front of the sheet and forming a right angle turn after strand β2), (3) the F box (the end of helix α2), and (4) the G2 box (beginning of helix α3 with the end of the loop preceding it). Residues pointing into the cavity from the β strands form a mainly hydrophobic surface on which the adenine ring hydrogen bonds with the invariant Asp (449 in *T. maritima* CheA, Fig. 5). Four buried water molecules that bridge interactions between the nucleotide base and the cavity are also found in the nucleotide complexes of Hsp90 [39, 41] and MutL [36]. An invariant Asn (409 in *T. maritima* CheA) coordinates nucleotide -bound Mg^{2+} in CheA [34], MutL [36], GyrB [37], and Hsp90 [41].

Despite striking similarities in nucleotide binding by PHKs and GHL ATPases, there are some compelling differences. For example, an essential glutamate of GHL ATPases presumed to be the general base involved in water

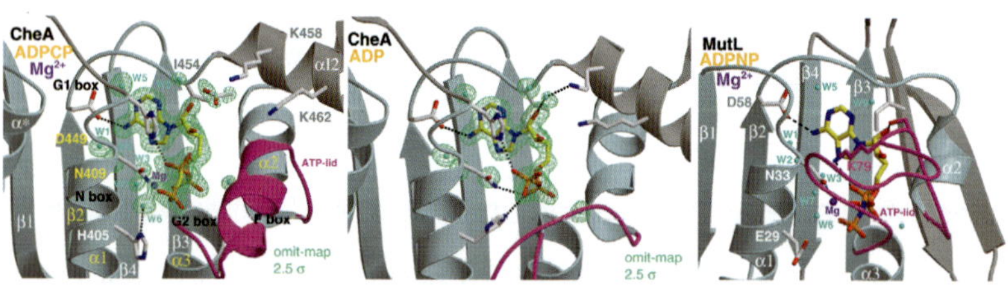

FIGURE 5 Recognition of nucleotides ADPCP and ADP and divalent cations by CheA and MutL. Representations of (F_o-F_c) omit electron density maps (in green) calculated following refinement by simulated annealing of the model in the absence of nucleotide, divalent cation, and active site solvent molecules. Ribbon representation of secondary structure elements in common between PHKs and GHL ATPases (blue) and unique to each class (gray). ATP lid (magenta) conformation varies with each complex. Side chains (white) involved in nucleotide binding and ordered solvent molecules (cyan) interacting with the nucleotide (yellow) are conserved in structure. The structures represented are: (left) the P4:ADPCP:Mg^{2+} complex (pdb codes 1i58), (center) the P4:ADP complex (pdb code 1i59), and (right) the MutL:ADPNP:Mg^{2+} complex (pdb code 1b63 [36]). Figure produced with Bobscript [78] and Raster3d [76].

activation for ATP hydrolysis [42] (Glu 29 for MutL) is replaced by His 405 in CheA. His 405 buttresses the G2 box when Mg^{2+} is bound. In contrast, the general base for histidine activation likely resides on the P1 domain (see below). Moreover, CheA and GHL ATPases appear to recognize the ATP phosphates in different ways. For example, the functional analog of CheA residue His 413, which hydrogen bonds to the ATP β-phosphate, comes from a different loop in the ATPases. Furthermore, interactions between nucleotide phosphates and main-chain nitrogens of the P loop (a glycine-rich segment found in many ATP-binding proteins that coordinates the α- and γ-phosphates of bound ATP) are not nearly as extensive in CheA as they are in GyrB and MutL. Perhaps P1 binding drives or stabilizes a more extensive interaction between the nucleotide and the CheA P loop that resembles structures observed for ATPases.

ATP HYDROLYSIS AND CONFORMATION OF P4

PHKs and GHL ATPases contain a region including the P loop that varies in conformation upon nucleotide binding: the ATP lid (Fig. 4). In CheA, the ATP lid (composed of the flexible loop between α2 and α3) changes conformation significantly among structures containing ATP analog(s), ADP, and no nucleotide. Only in the structure of separately expressed P4 with Mg^{2+}-ADPCP can the ATP lid be completely discerned [34]. The high mobility of the lid region is indicated by its poor order in all other P4 structures, the nucleotide-free structure of CheAΔ289 [26], and the NMR structure of the type I EnvZ PHK [43]. In the P4:ADPCP-Mg^{2+}structure, the ATP lid forms a helix that borders the nucleotide-binding cavity (Fig. 5). The resulting concave groove on the face of P4 surrounds the exposed γ-phosphate and has dimensions appropriate for binding P1 (Fig. 6). The shape of this groove, particularly its width nearest the bound nucleotide, depends on the presence of ATP analogs and Mg^{2+}. In P4 structures where the ATP pocket size is contracted due to molecular packing within the crystal lattice, Mg^{2+} does not bind and the γ-phosphate of nonhydrolyzable ATP analogs cannot be resolved due to disorder. This contracted conformation is also observed when ADP is bound by P4 (Fig. 5). Interestingly, CheA still binds ATP in the absence of Mg^{2+}, albeit six times weaker than in the presence of the divalent cation [44]. A change in cavity size and a loss of Mg^{2+} on ATP hydrolysis can be linked by the movement of His 405, which, in the absence of Mg^{2+}, swivels up from the position where it coordinates the metal ion and instead hydrogen bonds to the ADP β-phosphate (Fig. 5). In the ADP complex, G2 box residues change conformation because His 405 no longer buttresses the G2 box; this destabilizes the entire ATP lid structure (Fig. 5). If His 405 is forced to swivel by

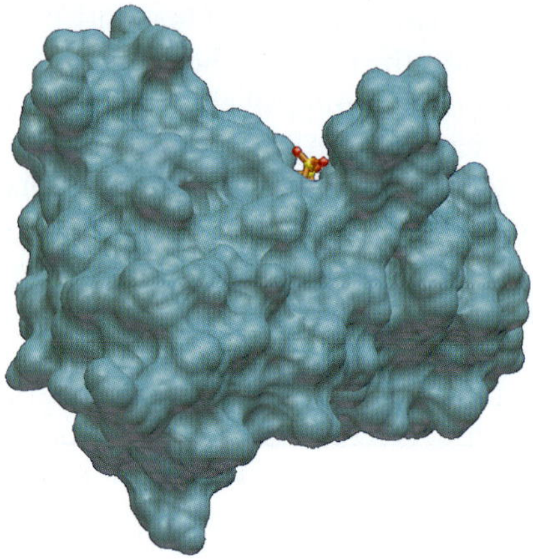

FIGURE 6 P4 has a shallow groove for binding P1. Solvent accessible surface of the P4 domain bound to ADPCP:Mg^{2+}. The ADPCP γ-phosphate (yellow bonds, red oxygen atoms) resides at the bottom of a large crevice of dimensions suitable for binding the α helix containing the substrate histidine of the P1 domain. Rendered with AVS (Advanced Visualization Systems Inc., Waltham, MA).

direct coordination to Mn^{2+} instead of water-mediated coordination to Mg^{2+}, the conformation of the ATP lid is similarly affected and interactions of the P loop with the γ-phosphate are weakened. Thus, conformational changes in regions that likely compose the P1-binding site on P4 (the ATP lid) are coupled to ATP hydrolysis and Mg^{2+} release by the movement of His 405.

HPt DOMAIN P1 AND PHOSPHORYL TRANSFER

P1 contains the substrate histidine that transfers phosphate from kinase bound ATP to the response regulators CheY and CheB. Although P1 and P4 must be contained within a dimeric CheA for physiological activity, some ATP-dependent histidine phosphorylation can be achieved *in vitro* by the two separated domains [34]. Thus, all the elements necessary for histidine phosphorylation are contained in the domains P1 and P4.

P1 is composed of an antiparallel four-helix bundle (helices A–D) plus a fifth helix (E) that connects to P2 *via* a 25–45 residue linker (Fig. 7A) [27, 28]. P1 helices are amphipathic with most hydrophobic residues buried in the core and most polar residues exposed to the surface. The five helices each display very different dynamic features. Residues from helices A, C, and D show strong protection from hydrogen exchange, indicative of local stability around

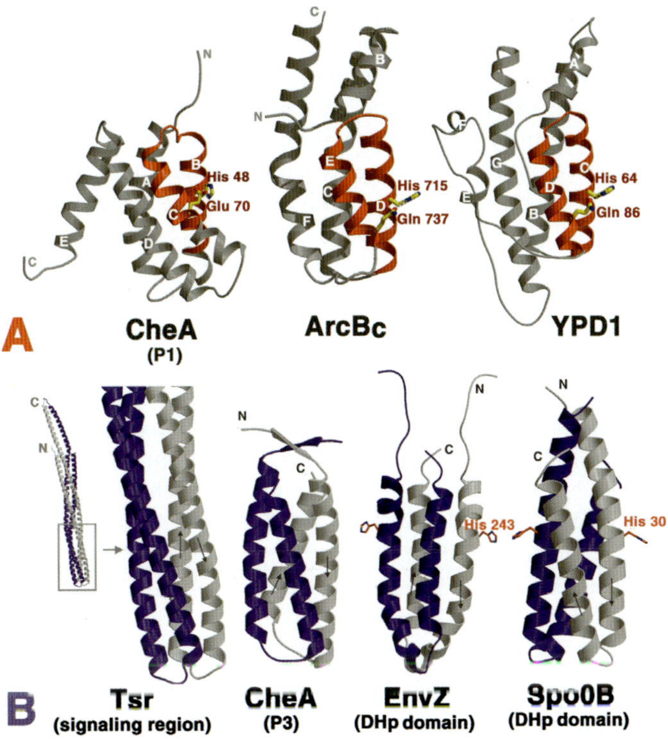

FIGURE 7 Comparisons among the four helix bundles that constitute monomeric HPt domains (A) and dimerization domains (B) among two component systems signaling pathways. (A) Sequence similarity among monomeric Hpt domains is concentrated in two helical segments (orange) around the active site histidine. In all cases, Glu or Gln is hydrogen bonded to the histidine. In HPt domains phosphorylated directly by ATP/kinase domains (*e.g.*, CheA P1), this Glu is believed to act as a general base to activate histidine for a nucleophilic attack on the ATP γ-phosphate. The NMR structure of *E. coli* CheA P1 [27], the crystal structure of the C-terminal domain of *E. coli* ArcB (pdb code1a0b [46]), and the crystal structure of yeast YPD1 (pdb code 1qsp [47]) are represented. (B) *E. coli* PHK EnvZ (pdb code 1joy [62]) and *B. subtilus* phosphotransferase Spo0B (pdb code 1ixm [63]) use a dimeric His-containing phosphotransfer domain (DHp) that presents two histidines for phosphorylation. DHp domains are structurally similar to the P3 domain of *T. maritima* CheA (pdb code 1b3q [26]) and the signaling region of *E. coli* chemoreceptor Tsr (pdb code 1qu7 [79]). Among this family, Spo0B is an outlier as its helix connectivity has a different handedness.

the amide hydrogen [27]. However, helix B, which contains the phosphoaccepting histidine, may be more variable in conformation, as its amide protons are not strongly protected from solvent exchange.

Sequence similarity among CheA P1 homologues is concentrated in helices B and C, where the active site residues are located. Given the high sequence conservation between *E. coli* and *T. maritima* CheA P1 in the region immediately surrounding the "substrate" histidine on helices B and C [27], yet the inability of *T. maritima* CheAΔ289 to phosphorylate an *E. coli* P1–P2 fragment [26], the interface between P1 and the kinase domain is likely to include residues on P1 not immediately surrounding the phospho-accepting histidine.

NMR studies of protein backbone dynamics indicate that P1 forms a rigid and compact helix bundle in both unphosphorylated and phosphorylated states [45]. Both these forms of P1 have very similar backbone conformation. Phosphorylation does not result in deprotonation of the histidine Nδ and results in only small chemical shift changes for residues on helices B and C surrounding the histidine. Alternations in the local electronic environment caused by phosphorylation are likely responsible for these changes. No interaction between P2 and phosphorylated P1 was detected by NMR [25].

The reactivity of the phospho-accepting histidine (His 48 in *E. coli*) located in the middle of helix B is tuned by its local environment. A hydrogen bond between His 48 Nδ and a neighboring glutamate side chain (Glu 70 in *E. coli*) may be responsible for the high pK_a (7.8) of the His 48 imidazole ring [45]. NMR studies indicate that His 48 Nε does not hydrogen bond with other P1 residues. Instead, His 48 Nδ is a hydrogen bond donor, which remains protonated at high pH and after phosphorylation [45]. The crystal structure of *Salmonella* P1 reveals that His 48 Nδ does indeed hydrogen bond to Glu 70 on helix C [28]. Furthermore, the hydrogen bond network His 48-Glu 70-Lys 51 likely stabilizes the otherwise unfavorable His 48 NδH tautomeric state and increases the nucleophilicity of Nε. Lys 51Ala and Glu 70Ala mutations reduce the ATP phosphotransfer rate drastically (*Salmonella*) [28]. However, these experiments could not distinguish whether the decreased transfer rate was due to loss of binding between the HPt domain and the ATP-binding domain or to a catalytic defect introduced by the mutation.

In contrast to GyrB, MutL, and Hsp90, which hydrolyze ATP, PHKs must transfer phosphate, and therefore the mechanism for nucleophilic attack on the ATP γ-phosphate must differ between the two enzyme types. In GyrB, mutagenesis studies [42] implicated the conserved Glu 42 residue (Glu 29 in MutL, Glu 47 in human Hsp90) as an essential general base for water activation. Despite a high conservation of active site residues between GyrB and histidine kinases, the latter do not contain a Glu at this position (His 405 for CheA proteins, Asn for other PHKs). Thus, the CheA P1 domain may provide not only the nucleophile for phosphate transfer (His 48), but also the activating glutamate (Glu 70), thereby completing the catalytic center observed in GyrB.

Despite minimal sequence similarity, a four-helix bundle motif is conserved among the phosphotransfer (HPt) domains of prokaryotes and eukaryotes. *E. coli* $ArcB_C$ [46] and *S.cerevisiae* Ypd1 [47, 48] are also four-helix bundles, although their loop regions and N and C termini differ significantly from P1

(Fig. 7A). In their respective phosphorelay pathways, $ArcB_C$ and Ypd1 accept phosphate from the aspartyl-phosphate of a response regulator, not ATP from a kinase domain. Interestingly, their phospho-accepting histidine residue hydrogen bonds to a Gln, not a Glu, as in P1. Consistent with the neutrality of Gln, the pK_a of ArcB's active site histidine (6.76) is unperturbed [49]; also, unlike P1 His 48, an equal ratio of the NδH and NεH tautomers is found for the ArcB His. When the conserved Gln was mutated to Ala in both Ypd1 [50] and ArcB [51], phosphotransfer activity to and from response regulator proteins was not curtailed. Furthermore, although the Glu 70Ala P1 mutation prevents phosphorylation by P4 it does not inhibit phosphotransfer to CheY [28]. Thus, activation of the phospho-accepting histidine by hydrogen bonding to a carboxylate may only be necessary for reaction with ATP bound in a kinase domain and not for reaction with the active site aspartate of a response regulator.

The type I PHK EnvZ and the phosphotransferase Spo0B have HPt domains that are structurally distinct from P1. These proteins still contain a four-helix bundle, but the bundle forms from a parallel dimerization of two helical hairpins, much like the CheA dimerization domain P3 (Fig. 7B). As a result, EnvZ and Spo0B have two active site histidines on opposite sides of the bundle. Hence, we name this functional and structural module DHp (dimeric his-containing phosphotransfer domain). The common helical architecture of DHp and HPt domains provides a rigid scaffold where the active site His can reside in stable secondary structure but also be highly exposed. Surrounding residues from adjacent helices can provide hydrogen bond networks to tune the histidine nucleophilicity and position the imidazole for reaction with phospho-acceptor and -donor domains.

P2 DOMAIN AND RESPONSE REGULATOR COUPLING

The P2 domain mediates interactions between CheA and both CheY and CheB. When CheA is activated by a receptor, the recruitment of CheY to CheA by P2 achieves a fast CheY phosphorylation rate of 750 s^{-1} [52]. Unphosphorylated CheY binds P2 with a dissociation constant (K_d = 2 μ*M* [20]) lower than its cellular concentration (8 μ*M* [53]). NMR and crystallographic studies characterize P2 as an open-face β sandwich with four antiparallel β strands packed against two antiparallel helices [29–32] (Fig. 8). CheY also folds as a small open-face β sandwich and the two proteins associate along their helical faces [31, 32]. In the complex, P2 helices run roughly perpendicular to the αD-β5-αE region of CheY [32]. Central hydrophobic contacts and peripheral hydrogen bonding interactions stabilize the interface.

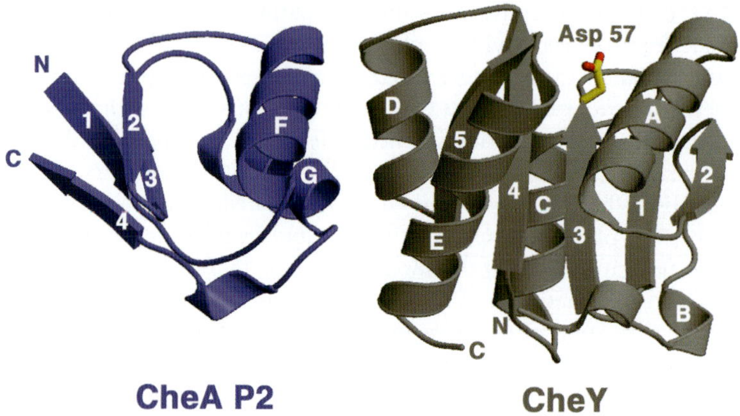

FIGURE 8 The complex formed by P2 and CheY. P2 (blue) orients CheY (gray) such that the phosphoryl-accepting Asp 57 is projected away from the interface of the complex for interaction with phosphoryl-donating P1 (pdb code 1eay [31]).

Two slightly different CheY-binding modes for nonequivalent molecules in one crystal structure of the complex indicate some plasticity in this interface [31]. Nevertheless, in both binding modes, P2 orients CheY so that its phosphate-accepting Asp 57 projects away from the P2 interface for interaction with phosphorylated P1. Moreover, P2 association causes CheY Phe 14 to change conformation and expose phospho-accepting Asp 57 to a greater degree than is observed in unbound CheY. P2 binding may also induce changes in the CheY active site that facilitate phosphotransfer from P1. Lengthening of the Asp 57-to-Lys 109 salt bridge in CheY by P2 binding has been proposed to prime CheY for accepting Mg^{2+} that is required to catalyze phosphotransfer [31].

After phosphorylation, the affinity of CheY for P2 drops sixfold [20]. Thus, structural changes resulting from phosphorylation must propagate to regions of CheY that interact with P2. The structure of phosphono-CheY, a stable analog of the phosphorylated, active form, shows that phosphorylation of CheY causes modest, localized, conformational changes that do affect at least one residue that interacts with P2 [54]. In the crystal structure of unphosphorylated CheY, Tyr 106 adopts two conformations: one where it is buried and one where it is exposed [55]. When CheY binds P2, Tyr 106 favors the exposed conformation and hydrogen bonds across the interface to P2 Glu 178. The exposure of Tyr 106 promotes the movement of Thr 87, the 91–97 loop, and helix D away from the CheY active center. Phosphorylation at Asp 57 repositions the 91–97 loop and helix D so that the internal position of Tyr 106 is favored. Thus, disruption of the Tyr 106 hydrogen bond to P2 may be a factor in promoting CheY release from CheA after phosphorylation [31]. However, other more subtle structural changes accompanying CheY phosphorylation may also be important. Although, a Thr 87Ile mutant that locks Tyr 106 in the exposed conformation increases the affinity of CheY for P2 by a factor of 2 [56], a Tyr 106Trp mutant that cannot supply the hydrogen bond to P2 and has the 106 aromatic side chain exclusively internalized is still phosphorylated readily by CheA [57].

Interestingly, many of the CheY residues that interact with P2 are not conserved by the response regulator domain of CheB [31]. Thus, either CheB recognizes a different site on P2 or CheB generates a chemically similar recognition surface with different residues. Given that CheB must also accept a phosphate from P1, it seems likely that its orientation, when bound to P2, is not drastically different from that of CheY.

Finally, it is important to note that if P2 is removed recombinantly from CheA, CheA can still catalyze CheY phosphorylation at rates that are many orders (~10^6) of magnitude faster than those for small molecule phosphodonors [52]. Thus, the most important function of P2 is not to facilitate the chemistry of phosphotransfer, but rather to increase the effective concentrations of response regulators near CheA.

A SEPARATE DIMERIZATION DOMAIN

Employment of a CheA dimerization domain that is separate from the catalytic machinery of P1 and P4 suggests that dimerization is not essential to the chemistry of histidine phosphorylation but may be so for signal transduction in the cell. Clearly, dimerization is required for CheA to undergo transautophosphorylation of the P1 domain [13, 58], but the significance of transphosphorylation remains to be determined. The dimerization domain of CheA contains an extensive, hydrophobic interface that generates a large energy barrier for dissociation of the subunits. Over 97% of the 1600 $Å^2$ of surface area buried on each subunit in the dimer interface lies within P3. The four-helix bundle formed by dimerization has both amino termini on the same end of the bundle and a left-handed twist (Fig. 7B). This symmetry allows each helix to be antiparallel to the two adjacent helices, which is the most stable arrangement for a four-helix bundle [59]. To complete the dimerization domain of CheA, a six-residue-long amino-terminal strand interacts with the equivalent region on the adjacent subunit to form an antiparallel, two-stranded β sheet, capping the exclusively hydrophobic interface between the four helices. As a result, the N terminus of the dimerization domain is likely to direct P2 and P1 toward the symmetry-related kinase domain, consistent with the transphosphorylation of P1 [13, 58].

The helical hairpins from each subunit associate in an arrangement that is uncommon for four-helix bundles. Most dimeric four-helix bundles fold so that the sequential hairpins are antiparallel, *i.e.*, their termini are at opposite ends of the bundle. The P3 domain generates a topology where the hairpins are parallel, with both sets of termini at the same end of the barrel. With the exceptions of DNA polymerase γB [60] and hepatitis B virus capsid protein [61], this arrangement has only been found in proteins involved in two-component signaling pathways and, most interestingly, in the signaling region of the helical chemoreceptors. Because projection of the dimeric receptor from the membrane dictates a parallel bundle, the same arrangement in CheA may have derived evolutionarily from the chemoreceptors. Alternatively, having the twofold symmetry axis directed along the P3 barrel may satisfy currently unknown spatial constraints imposed by receptor association.

Class I histidine kinases also share a similar dimerization motif with CheA even though there is only moderate sequence similarity in this region between the two classes. Unlike CheA, this domain contains the target histidine for phosphorylation. Homodimerization of EnvZ subdomain A produces a four-helix bundle [62] similar in overall structure to the monomeric HPt domains (see earlier discussion).

Finally, this dimeric arrangement is also found in the phosphotransferase Spo0B [63]. However, Spo0B is a structural outlier in this group because the connectivity of its helical hairpins has a different handedness than the other members (Fig. 7B). For further discussion, see Chapter 23.

Motions of PHK domains around their respective dimerization domains are suggested by comparing the architectures of CheA, EnvZ, and Spo0B. Attached to its dimerization domain, Spo0B has a kinase-like domain that does not bind ATP. Orientation of this domain relative to the dimerization domain is very different than the juxtaposition of P3 and P4 in CheA (Fig. 9). However, in CheA, some movement at the P3–P4 hinge is indicated by non-equivalent subunit conformations in the dimer structure. A similar hinge in EnvZ-related kinases (type I) may allow different spatial arrangements of the DHp and kinase domains to mediate transphosphorylation and subsequent response regulator activation [64].

RECEPTOR COUPLING BY THE P5 REGULATORY DOMAIN

The P5 regulatory domain mediates the interaction between CheA and CheW [23, 35]. P5 consists of two small β barrels that are related to each other by pseudosymmetry and to mammalian SH3 domains by topology (Fig. 10). SH3 domains regulate kinase activity in higher organisms by mediating transient

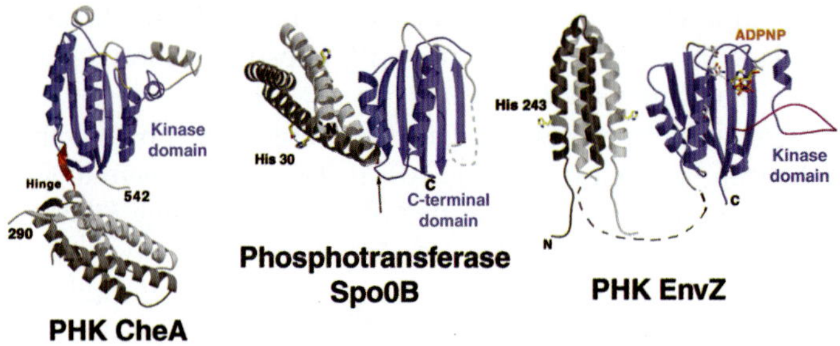

FIGURE 9 A flexible hinge may allow reorientation of dimerization and kinase domains in CheA and EnvZ (pdb code 1bxd for the kinase domain [43]). In Spo0B, orientation of the helical dimerization domain (gray) relative to the kinase-like domain (blue) differs from that of the topologically analogous domains in CheA [for all the dimeric proteins, only one kinase(-like) domain is represented]. Dotted lines symbolize missing residues in the chain and putative linker regions between domains (not to scale). Although the Spo0B kinase-like domain has no catalytic activity, the dimerization domain contains a substrate histidine, like type I kinases, such as EnvZ. In EnvZ, the two domains must associate for autophosphorylation. Thus, the structures of CheA and Spo0B may represent two conformational extremes that are accessible to this common structural unit in phosphorelay proteins.

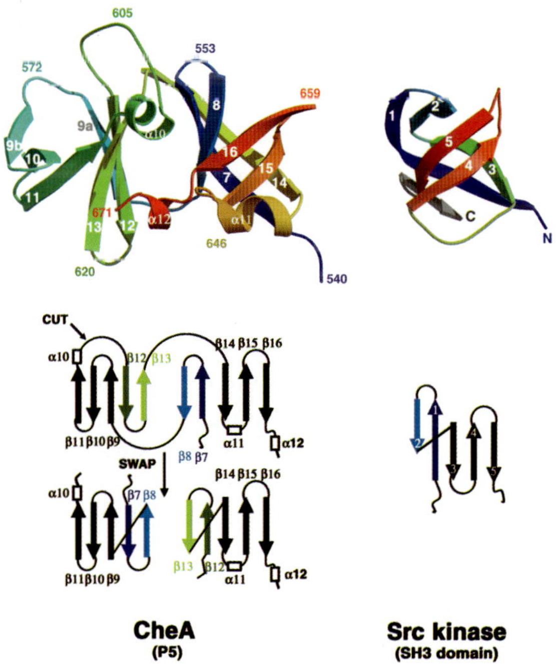

FIGURE 10 The P5 regulatory domain has a topology similar to SH3 domains. CheA domain P5 forms two β barrels that are related by pseudo-two-fold symmetry (left). Each barrel is related topologically to the SH3 domain of human c-Src kinase (right).

protein–protein interactions [65]. Similarly, the role of P5 in chemotaxis is to couple changes in receptor structure to P1 phosphorylation — an activity dependent on interactions with CheW.

The P5 structure can be described as two consecutive β barrels that swap a central β hairpin between them (Fig. 10). If the hairpins were to swap back, each barrel has a topology similar to that of an Src homology 3 (SH3) domain [66]. The exclusively hydrophobic interface between the two P5 β barrels partially contains the molecular surface that recognizes polypeptides in typical SH3 domains. Thus, the CheA regulatory domain and typical SH3 domains recognize different protein targets with different surfaces. Nonetheless, it appears that SH3-like domains have been employed for coupling protein or peptide recognition surfaces in bacteria as in other organisms.

Surprisingly, P5 is structurally related to CheW. This relationship was first identified by the conservation of essential P5 structural residues by CheW sequences [26] and later confirmed by the NMR structure of CheW [67]. The most extensive region of sequence similarity between P5 and CheW contains P5 β10 and β11 and forms the exposed hydrophobic face of the regulatory domain that is most peripheral to the dimerization domain. This surface may participate in a functional interface between CheA and CheW or the chemoreceptor (Fig. 11A). Thus, the protein module represented by P5 and CheW is an adapter for associating proteins, with each β barrel generating hydrophobic surfaces at both ends that may bind specific targets.

IS FLEXIBILITY BETWEEN DOMAINS IMPORTANT FOR SIGNALING?

The modular structure of CheA allows functional elements to reposition relative to each other. For example, P1 must move between P4 and P2 to transfer phosphate from ATP to CheY. Relative movement of the P3, P4, and P5 domains is also indicated by their limited interactions with each other, differences in their relative positioning within each subunit of the CheAΔ289 dimer, and conserved hinge regions. Such motions are likely essential features of signaling by multidomain histidine kinases.

The CheAΔ289 dimer observed crystallographically is not symmetric. Different conformations of the conserved hinge residues that link P3 to P4 and P4 to P5 generate different interdomain contacts and relative domain arrangements within the two subunits. In the case of class I histidine kinases, large-amplitude motion about a hinge between dimerization and kinase domains would be necessary for the kinase domain to reach the substrate domain on the dimer interface and then release it for subsequent transfer of the phosphate from the DHp to a response regulator domain (Fig. 9).

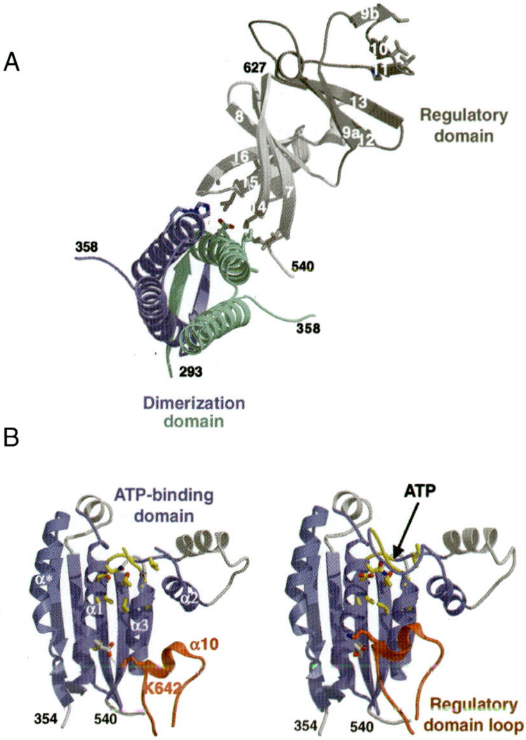

FIGURE 11 Interactions of the regulatory domain with the rest of CheA and other proteins. (A) P5 likely acts to couple another protein to CheA via interfaces at its barrels ends. One barrel (light gray) interacts with the dimerization domain, whereas the other barrel (dark gray) exposes a conserved hydrophobic surface for protein recognition. Both surfaces are at opposite ends of P5, following the pseudosymmetry of the domain. (B) Movement of P5 may influence P4 activity. In only one subunit of the asymmetric CheA dimer, a P5 loop (orange) interacts with P4 (blue) and may compete for P1 binding.

Domain motion about hinges may also allow regulation of kinase activity by the chemoreceptor and CheW. In the CheAΔ289 structure, the position of P5 may interfere with P1 binding to the kinase in only one subunit (Fig. 11B). In this subunit, α10 of P5 resides between α1 and α3 in the kinase domain, near the shallow groove where P1 is presumed to access the ATP γ-phosphate. In the other subunit, a rigid body rotation of P5 about a hinge at Thr 540 produces a more open conformation with α10 along side the C-terminal end of α3. P3 and P5 pivot about a conserved interface involving invariant hydrophobic residues. Movement within the receptor due to changes in ligand occupancy may be propagated through CheW to P5. Thus,

P5 may mediate P1 access to the ATP-binding site by changing its orientation relative to P3, P4, and CheW.

CONTROLLING PROTEIN–PROTEIN INTERACTIONS WITH ATP

The architecture of CheA is then well designed to meet a key requirement of a signaling system: the dynamic association and dissociation of protein domains. However, to propagate signals effectively over large time scales, such transient interactions must be linked to chemistry. The structural similarity among histidine kinases, type II topoisomerases, and Hsp90 protein chaperones suggests a common mechanism linking ATP hydrolysis to relative subunit motion. Histidine kinases use conformational motion for coupling extracellular signals to histidine phosphorylation; GyrB couples ATP hydrolysis to subunit dissociation and the release of relaxed DNA [68, 69]. The molecular motions common to these events likely involve movements of the kinase or ATPase domain relative to other domains or subunits in these molecules. CheA has a molecular structure that allows separate catalytic, substrate, organizing, and regulatory modules to influence and respond to kinase activity through conserved hinge regions and adaptable interfaces.

The coupling of mobile protein elements to ATP hydrolysis and protein–protein interactions may be a common feature of the PHK/GHL family [26, 70]. A comparison among the different structures of PHKs and GHL ATPases with and without nucleotides and inhibitors shows high conservation in the mode of ATP binding but divergence in the length, secondary structure, and conformation of the ATP lid (Fig. 4). Although in many cases throughout the superfamily, the ATP lid changes conformation on nucleotide binding. This action alters each domain and produces a new recognition surface only available when the nucleotide is bound. Although PHKs, GyrB, MutL, and Hsp90 have very different functions, they all couple ATP hydrolysis to modulating interdomain or intersubunit interactions using the same ATP binding cavity. Thus, PHKs and GHL ATPases adapt a common mode of ATP binding to different protein–protein associations *via* a variable, flexible ATP lid.

PROSPECTS FOR THE DESIGN OF ANTIBIOTICS DIRECTED AT CheA

Histidine kinases have thus far been found only in plants, fungi, and bacteria, but not in mammalian organisms; thus, they represent excellent targets for antibiotics, herbicides, and fungicides. In fact, a functional CheA is necessary for the tumoregenic pathogen *Helicobacter pylori* to colonize the gastric mucosa [71]. The structure of CheA P4 with the nanomolar inhibitor TNP-ATP suggests a strategy for drug design. TNP-ATP binds unlike other ATP analogs due to interactions of its trinitrophenyl ring with a hydrophobic pocket adjacent to the ATP-binding cavity. This CheA-specific hydrophobic pocket could be targeted by rigid CheA-specific inhibitors designed to exploit the binding mode of TNP-ATP. The hydrophobic pocket recognizing the TNP-ATP trinitrophenyl ring is not conserved among GHL ATPases and PHKs. Indeed, in GHL ATPases, the pocket is partially altered by two additional strands inserted between β5 and β6. Moreover, GHL ATPases do not have the F box, a region of sequence similarity among PHKs on helix α2 that forms a hydrophobic face of the binding pocket. This hydrophobic pocket of CheA is not completely conserved among PHKs either; most PHKs lack helices α6 and α7, which in CheA are inserted between strand β4 and helix α8. Thus, rigid bifunctional inhibitors could be designed to interact with both the conserved adenine-binding cavity and the adjacent variable site to achieve high-affinity binding and selectivity for a given PHK. For example, inhibitors designed to bind the conserved and variable regions of the *H. pylori* CheA nucleotide pocket, which is very similar in residue composition to that of *T. maritima* CheA, may lead to antibiotics that would not impair the function of essential mammalian ATPases.

WHAT IS NEXT?

Bacterial chemotaxis is one of the few biological systems where we can endeavor to comprehend behavior at the level of molecular interactions. Currently we believe that we have defined all of the signaling components, the sequence of their actions, the nature of their chemistries, and now their detailed structures; what remains is to fully understand the mechanisms of their interactions. A central unanswered question concerns how receptor occupancy influences kinase activity. We envisage three general mechanisms for how changes in receptor structure could affect CheA: (1) by direct perturbation of the catalytic machinery responsible for autophosphorylation, (2) by modulation of binding interfaces among CheA domains and response regulators, and/or (3) by control of motions between CheA domains. The asymmetry of the CheAΔ289 structure does suggest that movement of P5 relative to P4 could inhibit kinase activity by excluding P1 from the ATP-binding site. However, further details of possible mechanisms await a greater understanding of the dynamic associations among chemotactic proteins and their domains. CheW could play a key role in any or all of the aforementioned scenarios. With application of the current structures, modeling studies have begun to place constraints on how chemoreceptors, CheA and CheW may assemble [72]. As it is clear that receptor clustering is important for function [9], networks of proteins are likely key to understanding the dynamic range and amplification in bacterial signal transduction [73, 74]. The future for structural work in chemotaxis lies in the characterization of active signaling complexes.

REFERENCES

1. Parkinson, J. S., and Kofoid, E. C. (1992). Communication modules in bacterial signaling proteins. *Annu. Rev. Genet.* **26**, 71–112.
2. Swanson, R. V., Alex, L. A., and Simon, M. I. (1994). Histidine and aspartate phosphorylation: Two-component systems and the limits of homology. *Trends Biochem. Sci.* **19**, 485–490.
3. Appleby, J. L., Parkinson, J. S., and Bourret, R. B. (1996). Signal transduction via the multi-step phosphorelay: Not necessarily a road less traveled. *Cell* **86**, 845–848.
4. Pirrung, M. C. (1999). Histidine kinases and two-component signal transduction systems. *Chem. Biol.* **6**, R167–R175.
5. Berg, H. C. (1975). Bacterial behaviour. *Nature* **254**, 389–392.
6. Blair, D. F. (1995). How bacteria sense and swim. *Annu. Rev. Microbiol.* **49**, 489–522.
7. Falke, J. J., Bass, R. B., Butler, S. L., Chervitz, S. A. and Danielson, M. A. (1997). The two-component signaling pathway of bacterial chemotaxis: A molecular view of signal transduction by receptors, kinases, and adapation enzymes. *Annu. Rev. Cell Dev. Biol.* **13**, 457–512.
8. Aizawa, S., Harwood, C. S., and Kadner, R. J. (2000). Signaling components in bacterial locomotion and sensory reception. *J. Bacteriol.* **182**, 1459–1471.
9. Maddock, J. R., and Shapiro, L. (1993). Polar location of the chemoreceptor complex in the *Escherichia coli* cell. *Science* **259**, 1717–1723.
10. Hess, J. F., Bourret, R. B., and Simon, M. I. (1988). Histidine phosphorylation and phosphoryl group transfer in bacterial chemotaxis. *Nature* **336**, 139–143.
11. Borkovich, K. A., and Simon, M. I. (1990). The dynamics of protein phosphorylation in bacterial chemotaxis. *Cell* **63**, 1339–1348.
12. Wolfe, A. J., and Stewart, R., 1993). The short form of the CheA protein restores kinase activity and chemotactic ability to kinase-deficient mutants. *Proc. Natl. Acad. Sci. USA* **90**, 1518–1522.
13. Swanson, R. V., Bourret, R. B., and Simon, M. I. (1993). Intermolecular complementation of the kinase activity of CheA. *Mol. Microbiol.* **8**, 435–441.
14. Hess, J. F., Oosawa, K., Kaplan, N., and Simon, M. I. (1988). Phosphorylation of three proteins in the signaling pathway of bacterial chemotaxis. *Cell* **53**, 79–87.
15. Sanders, D. A., Gillece-Castro, B. L., Stock, A. M., Burlingame, A. L., and Koshland, D. E., Jr. (1989). Identification of the site of phosphorylation of the chemotaxis response regulator protein, CheY. *J. Biol. Chem.* **264**, 21770–21778.
16. Welch, M., Oosawa, K., Aizawa, S., and Eisenbach M. (1993). Phosphorylation-dependent binding of a signal molecule to the flagellar switch of bacteria. *Proc. Natl. Acad. Sci. USA* **90**, 8787–8791.
17. Ninfa, E. G., Stock, A., Mowbray, S., and Stock, J. (1991). Reconstitution of the bacterial chemotaxis signal transduction system from purified components. *J. Biol. Chem* **266**, 9764–9770.
18. Li, G., and Weis, R. M. (2000). Covalent modification regulates ligand binding to receptor complexes in the chemosensory system of *Escherichia coli*. *Cell* **100**, 357–365.

19. Borkovich, K. A., Alex, L. A., and Simon, M. I. (1992). Attenuation of sensory receptor signaling by covalent modification. *Proc. Natl. Acad. Sci. USA* **89**, 6756–6760.
20. Li, J., Swanson, R. V., Simon, M. I., and Weis, R. M. (1995). The response regulators CheB and CheY exhibit competitive binding to the kinase CheA. *Biochemistry* **34**, 14626–14636.
21. Jurica, M. S., and Stoddard, B. L. (1998). Mind your B's and R's: Bacterial chemotaxis, signal transduction and protein recognition. *Structure* **6**, 809–813.
22. Swanson, R. V., Schuster, S. C., and Simon, M. I. (1993). Expression of CheA fragments which define domains encoding kinase, phosphotransfer and CheY binding activities. *Biochemistry* **32**, 7623–7629.
23. Bourret, R. B., Davagnino, J., and Simon, M. I. (1993). The carboxy-terminal portion of the CheA kinase mediates regulation of autophosphorylation by transducer and CheW. *J. Bacteriol.* **175**, 2097–2101.
24. Morrison, T. B., and Parkinson, J. S. (1994). Liberation of an interaction domain from the phosphotransfer region of CheA, a signaling kinase of *Escherichia coli*. *Proc. Natl. Acad. Sci. USA* **91**, 5485–5489.
25. Zhou, H., McEvoy, M. M., Lowry, D. F., Swanson, R. V., Simon, M. I., and Dahlquist, F. W. (1996). Phosphotransfer and CheY-binding domains of the histidine autokinase CheA are joined by a flexible linker. *Biochemistry* **35**, 433–443.
26. Bilwes, A. M., Alex, L. A., Crane, B. R., and Simon, M. I. (1999). Structure of CheA, a signal-transducing histidine kinase. *Cell* **96**, 131–141.
27. Zhou, H., Lowry, D. F., Swanson, R. V., Simon, M. I., and Dahlquist, F. W. (1995). NMR studies of the phosphotransfer domain of the histidine kinase CheA from *Escherichia coli*: Assignments, secondary structure, general fold, and backbone dynamics. *Biochemistry* **34**, 13858–13870.
28. Mourey, L., Da Re, S., Pedelacq, J. D., Tolstykh, T., Faurie, C., Guillet, V., Stock, J. B., and Samama, J. P. (2001). Crystal structure of the CheA histidine phosphotransfer domain that mediates response regulator phosphorylation in bacterial chemotaxis. *J. Biol. Chem* **276**, 31074–31082.
29. McEvoy, M. M., Zhou, H., Roth, A. F., Lowry, D. F., Morrison, T. B., Kay, L. E., and Dahlquist, F. W. (1995). Nuclear magnetic resonance assignments and global fold of a CheY-binding domain in CheA, the chemotaxis-specific kinase of *Escherichia coli*. *Biochemistry* **34**, 13871–13880.
30. McEvoy, M. M., Muhandiram, D. R., Kay, L. E., and Dahlquist, F. W. (1996). Structure and dynamics of a CheY-binding domain of the chemotaxis kinase CheA determined by nuclear magnetic resonance spectroscopy. *Biochemistry* **35**, 5633–5640.
31. McEvoy, M. M., Hausrath, A. C., Randolph, G. B., Remington, S. J., and Dahlquist, F. W. (1998). Two binding modes reveal flexibility in kinase/response regulator interactions in the bacterial chemotaxis pathway. *Proc. Natl. Acad. Sci. USA* **95**, 7333–7338.
32. Welch, M., Chinardet, N., Mourey, L., Birck, C., and Samama, J. P. (1998). Structure of the CheY-binding domain of histidine kinase CheA in complex with CheY. *Nature Struct. Biol.* **5**, 25–29.
33. Orengo, C. A., Michie, A. D., Jones, S., Jones, D. T., Swindells, M. B., and Thornton, J. M. (1997). A hierarchic classification of protein domain structure. *Structure* **5**, 1093–1108.
34. Bilwes, A. M., Quezada, C. M., Croal, L. R., Crane, B. R., and Simon, M. I. (2001). Nucleotide binding by the histidine kinase CheA. *Nature Struct. Biol.* **8**, 353–360.
35. Morrison, T. N., and Parkinson, J. S. (1997). A fragment liberated from the *E. coli* kinase that blocks stimulatory, but not inhibitory, chemoreceptor signaling. *J. Bacteriol.* **179**, 5543–5550.
36. Ban, C., Junop, M., and Yang, W. (1997). Transformation of MutL by ATP binding and hydrolysis: A switch in DNA mismatch repair. *Cell* **97**, 85–97.
37. Wigley, D. B., Davies, G. J., Dodson E. J., Maxwell, A., and Dodson, G. (1991). Crystal structure of an N-terminal fragment of the DNA gyrase B protein. *Nature* **351**, 624–629.

38. Stebbins, C. E., Russo, A. A., Schneider, C., Rosen, N., Hartl, F. U., and Pavletich, N. P. (1997). Crystal structure of an Hsp90-geldamycin complex: Targeting of a protein chaperone by an antitumor agent. *Cell* **89**, 239–250.
39. Prodromou, C., Roe, S. M., O'Brien, R., Ladbury, J. E., Piper, P. W., and Pearl, L.H. (1997). A molecular clamp in the crystal structure of the N-terminal domain of the yeast Hsp90 chaperone. *Cell* **90**, 65–75.
40. Ban, C., and Yang, W. (1998). Crystal structure and ATPase activity of MutL: Implications for DNA repair and mutagenesis. *Cell* **95**, 541–552.
41. Obermann, W. M. J., Sondermann, H., Russo, A. A., Pavletich, N. P., and Hartl, F. U. (1998). *In vivo* function of Hsp90 is dependent on ATP binding and ATP hydrolysis. *J. Cell Biol.* **143**, 901–910.
42. Jackson, A. P., and Maxwell, A. (1993). Identifying the catalytic residue of the ATPase reaction of DNA gyrase. *Proc. Natl. Acad. Sci. USA* **90**, 11232–11236.
43. Tanaka, T., Saha, S. K., Tomomori, C., Ishima, R., Liu, D., Tong, K. I., Park, H., Dutta, R., Qin, L., Swindells, M. B., Yamazaki, T., Ono, A. M., Kainosho, M., Inouye, M., and Ikura, M. (1998). NMR structure of the histidine kinase domain of the *E. coli* osmosensor EnvZ. *Nature* **396**, 88–92.
44. Hirschman, A., Boukhvalova, M., VanBruggen, R., Wolfe, A. J., and Stewart, R. C. (2001). Active site mutations in CheA, the signal-transducing protein kinase of the chemotaxis system in *Escherichia coli*. *Biochemistry* **40**, 13876–13887.
45. Zhou, H. D. (1997). Phosphotransfer site of the chemotaxis-specific protein kinase CheA as revealed by NMR. *Biochemistry* **36**, 699–710.
46. Kato, M., Mizuno, T., Shimizu, T., and Hakoshima, T. (1997). Insights into multistep phosphorelay from the crystal structure of the C-terminal HPt domain of ArcB. *Cell* **88**, 717–723.
47. Xu, Q., and West, A. H. (1999). Conservation of structure and function among histidine-containing phosphotransfer (HPt) domains as revealed by the crystal structure of YPD1. *J. Mol. Biol.* **292**, 1039–1050.
48. Song, H. K., Lee, J. Y., Lee, M. G., Moon,, J., Min, K., Yang, J. K., and Suh, S. W. (1999). Insights into eukaryotic multistep phosphorelay signal transduction revealed by the crystal structure of Ypd1 from *Saccharomyces cerevisiae*. *J. Mol. Biol.* **293**, 753–761.
49. Ikegami, T., Okada, T., Ohki, I, Hirayama, J., Mizuno, T., and Shirakawa, M. (2001). Solution structure and dynamic character of the histidine-containing phosphotransfer domain of anaerobic sensor kinase ArcB form *Escherichia coli*. *Biochemistry* **40**, 375–386.
50. Janiak-Spens, F., and West, A. H. (2000). Functional roles of conserved amino acid residues surrounding the phosphorylatable histidine of the yeast phosphorelay protein YPD1. *Mol. Microbiol.* **37**, 136–144.
51. Matsushika, A., and Mizuno, T. (1998). Mutational analysis of the histidine-containing phosphotransfer (HPt) signaling domain of the ArcB sensor in *Escherichia coli*. *Biosci. Biotechnol. Biochem.* **62** 2236–2238.
52. Stewart, R. C., Jahreis, K., and Parkinson, J. S. (2000). Rapid phosphotransfer to CheY from a CheA protein lacking the CheY-binding domain. *Biochemistry* **39**, 13157–13165.
53. Kuo, S.C., and Koshland, D.E., Jr. (1987). Roles of cheY and cheZ gene products in controlling flagellar rotation in bacterial chemotaxis of *Escherichia coli*. *J. Bacteriol.* **169**, 1307–1314.
54. Halkides, C.J., McEvoy, M. M., Casper E., Matsumura P., Volz, K., and Dahlquist, F. W. (2000). The 1.9 Å resolution crystal structure of phosphono-CheY, an analogue of the active form of the response regulator, CheY. *Biochemistry* **39**, 5280–5286.
55. Volz, K., and Matsumura, P. (1991). Crystal structure of *Escherichia coli* CheY refined at 1.7-Å resolution. *J. Biol. Chem.* **266**, 15511–15519.
56. Shukla, D., and Matsumura, P. (1995). Mutations leading to altered CheA binding cluster on a face of CheY. *J. Biol. Chem.* **270**, 24414–24419.

57. Zhu, X., Rebello, J., Matsumura, P., and Volz, K. (1997). Crystal structures of CheY mutants Y106W and T87I/Y106W. *J. Biol. Chem.* **272**, 5000–5006.
58. Ellefson, D.D., Weber, U., and Wolfe, A.J. (1997). Genetic analysis of the catalytic domain of the chemotaxis-associated histidine kinase CheA. *J. Bacteriol.* **179**, 825–830.
59. Chou, K.-C., Maggiora, G.M., Nemethy, G., and Scheraga, H.A. (1988). Energetics of the structure of the four-a-helix bundle in proteins. *Proc. Natl. Acad. Sci. USA* **85**, 4295–4299.
60. Carrodeguas, J. A., Theis, K., Bogenhagen, D. F., and Kisker, C. (2001). Crystal structure and deletion analysis show that the accessory subunit of mammalian DNA polymerase γ, PolγB, functions as a homodimer. *Mol. Cell* **7**, 43–54.
61. Wynne, S. A., Crowther, R. A., and Leslie, A. G. W. (1999). The crystal structure of the human hepatitis B virus capsid. *Mol. Cell* **3**, 771–780.
62. Tomomori, C., Tanaka, T., Dutta, R., Park, H., Saha, S. K., Zhu, Y., Ishima, R., Liu, D., Tong, K. I., Kurokawa, H., Qian, H., Inouye, M., and Ikura, M. (1999). Solution structure of the homodimeric core domain of *Escherichia coli* histidine kinase EnvZ. *Nature Struct. Biol.* **6**, 729–734.
63. Varughese, K. I., Madhusudan, Zhou, X. Z., Whiteley, J. M., and Hoch, J. A. (1998). Formation of a novel four-helix bundle and molecular recognition sites by dimerization of a response regulator phosphotransferase. *Mol. Cell* **2**, 485–493.
64. Park, H., Saha, S. K., and Inouye, M. (1998). Two-domain reconstitution of a functional protein histidine kinase. *Proc. Natl. Acad. Sci. USA* **95**, 6728–6732.
65. Schlessinger, J. (1994). SH2/SH3 signaling proteins. *Curr. Opin. Gen. Dev.* **4**, 25–30.
66. Xu, W., Harrison, S. C., and Eck, M. J. (1997). Three-dimensional structure of the tyrosine kinase c-Src. *Nature* **385**, 595–601.
67. Grisold, I. S., Zhou, H., Matison, M., Swanson, R. V., McIntosh, L. P., Simon, M. I., and Dahlquist, F. W. (2002). The solution structure and interactions of CheW from *Thermotoga maritima*. *Nature Struct. Biol.* **9**, 121–125.
68. Bates, A. D., and Maxwell, A. (1997). DNA topology: Topoisomerases keep it simple. *Curr. Biol.* **7**, R778–R781.
69. Champoux, J. J. (2001). DNA topoisomerases: Structure, function and mechanism. *Annu. Rev. Biochem.* **70**, 369–413.
70. Dutta, R., and Inouye, M. (2000). GHKL, an emergent ATPase/kinase superfamily. *Trends Biochem. Sci.* **25**, 24–28.
71. Foynes, S., Dorrell, N., Ward, S. J., Stabler, R. A., McColm, A. A., Rycroft, A. N., and Wren, B. W. (2000). *Helicobacter pylori* possesses two CheY response regulators and a histidie kinase sensor, CheA, which are essential for chemotaxis and colonization of the gastric mucosa. *Infect. Immun.* **68** 2016–2023.
72. Shimizu, T. S., Le Novere, N., Levin, M. D., Beavil, A. J., Sutton, B. J., and Bray, D. (2000). Molecular model of a lattice of signaling proteins involved in bacterial chemotaxis. *Nature Cell Bio.* **2**, 792–796.
73. Alon, U., Surette, M. G., Barkai, N., and Liebler, S. (1999). Robustness in bacterial chemotaxis. *Nature* **397**, 168–171.
74. Yi, T. M., Huang, Y, Simon, M. I., and Doyle, J. (2000). Robust perfect adaptation in bacterial chemotaxis through integral feedback control. *Proc. Natl. Acad. Sci. USA* **97**, 4649–4653.
75. Kraulis, P. J. (1991). Molscript: A program to produce both detailed and schematic plots of protein structures. *J. Appl. Crystallogr.* **24**, 946–950.
76. Merritt, E. A., and Murphy, M. E. P. (1994). Raster3D Version 2.0: A program for photorealistic molecular graphics. *Acta Crystallogr.* **D50**, 869–873.
77. Holdgate, G. A., Tunnicliffe, A., Ward, W. H., Weston, S. A., Rosenbrock, G., Barth, P. T., Taylor, I. W., Pauptit, R. A., and Timms, D. (1997). The entropic penalty of ordered water accounts for weaker binding of the antibiotic novobiocin to a resistant mutant of DNA

gyrase: A thermodynamic and crystallographic study. *Biochemistry* **36**, 9663–9673.
78. Esnouf, R. M. (1997). An extensively modified version of Molscript that includes greatly enhanced coloring capabilities. *J. Mol. Graph.* **15**, 133–138.
79. Kim, K. K., Yokota, H., and Kim, S. H. (1999). Four-helical-bundle structure of the cytoplasmic domain of a serine chemotaxis receptor. *Nature* **400**, 787–792.

CHAPTER 5

Transmembrane Signaling and the Regulation of Histidine Kinase Activity

PETER M. WOLANIN* AND JEFFRY B. STOCK*,†

*Departments of *Molecular Biology and †Chemistry, Princeton University, Princeton, New Jersey 08544*

Histidine Kinases in Signal Transduction

Transmembrane signal transduction plays a central role in biology, allowing cells to transport information from the environment into the cytoplasm. In prokaryotes, histidine protein kinases (HPKs) transduce sensory inputs into protein phosphorylation outputs. Although all HPKs share homologous kinase catalytic domains, their activities are generally regulated by external stimuli *via* a wide variety of sensory input domains. Like tyrosine protein kinases, dimerization is important for HPK function. A possible insight into the mechanism of signaling by HPKs comes from observations of tight clustering by the chemotaxis receptors in *Escherichia coli* and other bacteria. Through examination of the *E. coli* chemotaxis system, it seems that changes in lateral interaction among hundreds or even thousands of receptors in large clusters play a key role in chemotaxis signal transduction.

INTRODUCTION

Transmembrane signal transduction plays a central role in biology. All cells transport information from surface receptors into the cytoplasm where it is processed and used to regulate virtually every aspect of biological activity. This is analogous to the uptake of nutrient molecules from the environment. In the case of membrane transporters and channels, the molecular basis for their function has become clear through high-resolution structural studies [1, 2]. For transmembrane signal transduction proteins, however, a detailed picture of molecular function is still in development. This chapter focuses on what is known about the function of histidine protein kinase (HPK) and HPK-linked receptors in well-characterized prokaryotic systems that regulate gene expression and motility.

MEMBRANE RECEPTOR KINASES

Sequence Relationships between Membrane Associated Histidine Protein Kinases

The majority of HPKs are integral membrane proteins with hydrophobic membrane-spanning sequences that are usually N-terminal to the conserved histidine kinase core. In *Escherichia coli* K-12, of the 27 genes [3] identified as encoding a protein with a histidine kinase domain (using the Pfam database definition [4]), 25 are likely transmembrane receptors [5]. Similarly, there are 25 histidine kinases identified in *Bacillus subtilis* [6], of which 21 are likely to be associated with the membrane [5]. The N-terminal membrane-associated portions of these receptors show very little sequence similarity. Even the membrane topology tends to vary dramatically. Most frequently the proteins have a simple type I transmembrane topology with an uncleaved signal sequence near the N terminus leading to extracytoplasmic domains followed by a second transmembrane sequence that brings the polypeptide chain back into the cytoplasm (e.g., EnvZ in Fig. 1). In some cases, however, there is no distinct extracytoplasmic domain, and the sensing of signals is believed to be accomplished by a cytoplasmic domain and/or auxiliary protein components (e.g., ArcB in Fig. 1). There are also several examples where the N-terminal region contains several transmembrane sequences that would be expected to form a distinct hydrophobic domain embedded within the plane of the membrane bilayer (e.g., ComD in Fig. 1). Conversely, a number of histidine kinases are not integral membrane proteins, but are associated with membrane receptors (e.g., CheA in Fig. 1). Other HPKs, such as NtrB [7, 8], are soluble enzymes whose regulation does not appear to involve interactions with the membrane. The divergence of N-terminal regulatory sequences within a given bacterial genome presumably reflects the fact that each paralogous HPK has evolved to respond to a unique set of stimuli. Orthologous HPKs with similar regulatory inputs tend to have highly conserved N-terminal regulatory domains. For instance, EnvZ sensor kinases from *E. coli* and *Vibrio cholera* [3, 9] are conserved over their entire lengths, as are CheA proteins from *E. coli* [3], *B. subtilis* [6], and even archael species such as *Halobacterium salinarum* [10].

In contrast to their varied sensory input domains, HPKs all share a common output mechanism: the ATP-dependent phosphorylation of a specific histidine and the subsequent transfer of this phosphoryl group to an aspartate residue in the receiver domain of a cognate response regulator protein [11]. This shared function is reflected in homologies between the histidine kinase cores of different HPKs [12].

The HPK catalytic core shows no apparent homology to the eukaryotic protein kinases such as tyrosine protein kinases (TPKs) insofar as the two types of proteins do not have substantial sequence similarity. However, a possible structural similarity has been identified. The ATP-binding small lobe of protein kinase C has a similar fold to the histidine kinase-like ATP-binding domain found in HPKs, type II topoisomerases, Hsp90, and MutL [13]. While the similarity is weak, this, together with the possible homology between the

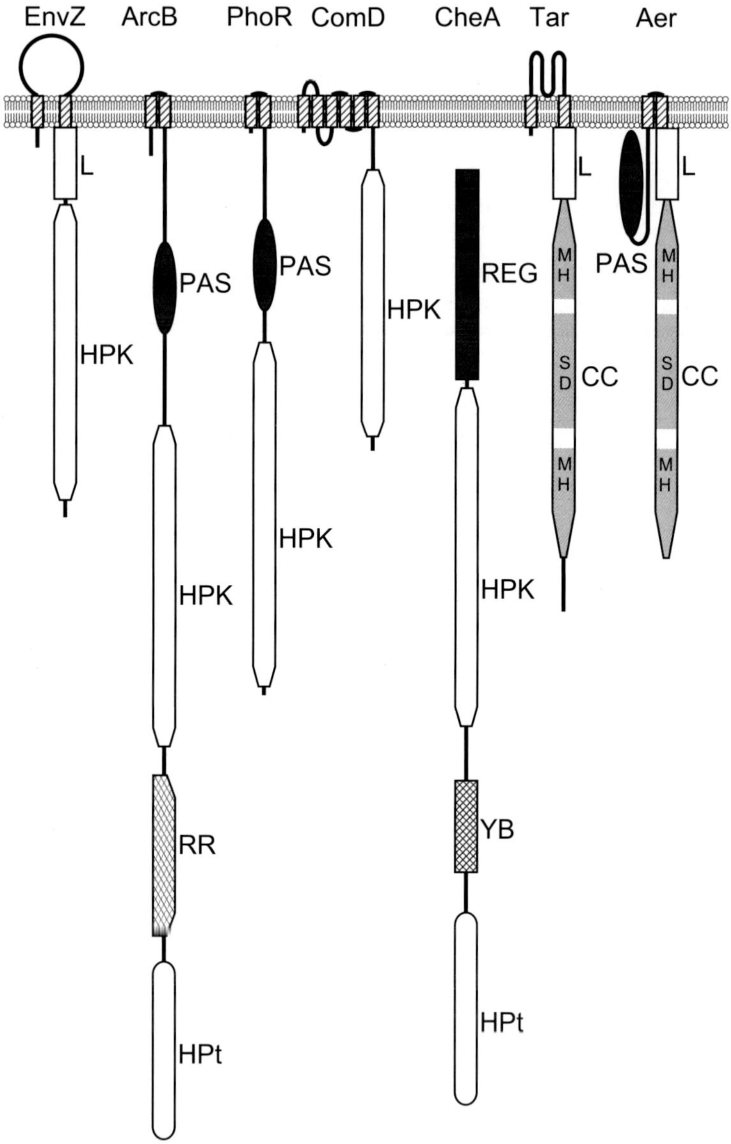

FIGURE 1 Schematic diagram showing the domain organization of examples of several different families of histidine kinase receptors. The length of each domain is proportional to the length of its amino acid sequence. L, linker region (HAMP domain); HPK, histidine protein kinase core domain (see Fig. 2); PAS, PAS domain; HPt, histidine phosphotransfer; RR, response regulator; YB, CheY binding; REG, regulatory domain; CC, coiled-coil region; MH, methylated helix; SD, signaling domain (a.k.a. CheA/CheW-binding region).

Ras family and the response regulators [14, 15], suggests that a common ancestral phosphorylation system may have existed in primordial cells [13].

Role of Dimerization in Receptor Regulation

The enormous diversity of sensory input domains suggests that the mechanism of signaling across the membrane involves an underlying principle that can accommodate numerous different types of protein structures. One possibility is that stimulatory ligands bind to sites at the interface between receptor monomers, thereby favoring the formation of dimers. Ligand-induced dimerization would be expected to cause kinase activation by shifting an equilibrium between inactive histidine kinase monomers and active dimers. Dimerization has been advanced as the primary mechanism for stimulus–response coupling by type I TPK receptors in vertebrate cells [16, 17]. For instance, hormones such as vascular endothelial growth factor and human growth hormone have been shown to bind to their respective TPK receptors at sites that bridge the receptor dimer interface, thereby favoring the formation of receptor dimers in the membrane [17]. In several other cases, including the epidermal growth factor (EGF) receptor, it has been demonstrated that agonist binding induces receptor dimerization [18]. Insulin also binds between receptor subunits, but the insulin receptor is permanently locked in a dimeric state by disulfide cross-links between sensory domains at the outside surface of the membrane [19]. Insulin binding is thought to cause a conformational change in the receptor that leads to tyrosine phosphorylation in one kinase domain by the opposing subunit [19]. Tyrosine phosphorylation activates the kinase signaling domains to phosphorylate other substrates to produce an insulin response. Thus, at least in the case of insulin responses, receptor dimerization is necessary but not sufficient for receptor signaling.

It is clear that dimerization also plays an important role in HPK receptor function. The conserved histidine kinase core is composed of an antiparallel coiled-coil dimerization domain that connects to an ATP-binding catalytic domain (Fig. 2). Histidine kinase activity depends on homodimer formation with the two-stranded coiled-coils coming together to form a four-helix bundle [20–23]. As in the case of the insulin receptor, HPK-mediated phosphorylation generally occurs *in trans* with the kinase catalytic domain of one subunit in a dimer phosphorylating a specific histidine in the dimerization domain of the other subunit [11, 24, 25].

The highly variable membrane topology of HPK N-terminal regulatory domains parallels what occurs with TPKs in metazoans. EnvZ is like typical type I TPK receptors such as the EGF and insulin receptors, whereas CheA is like soluble TPKs such as Src and Jak that bind to and are regulated by auxil-

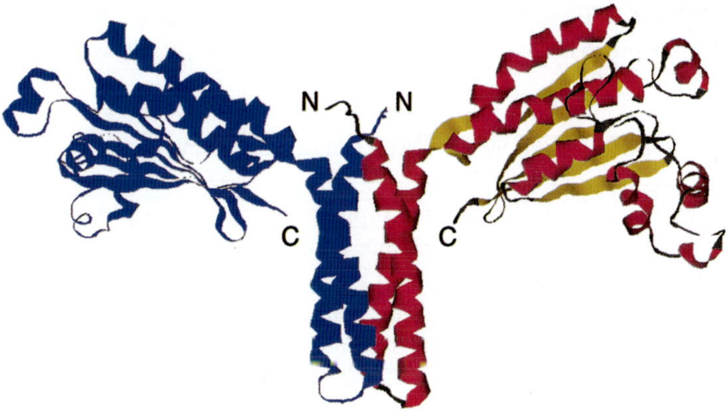

FIGURE 2 The homodimeric histidine kinase core of *T. maritima* CheA [21]. Figure produced from PDB entry 1B3Q with RasMac.

iary type I receptor proteins [17, 19, 26, 27]. The chemotaxis receptors that function together with CheA have homodimeric extracytoplasmic-sensing domains with ligand binding sites bridging the dimer interface [28–33]. The chemotaxis receptors are commonly called methyl-accepting chemotaxis proteins and are referred to as MCPs. The binding of attractants such as serine or aspartate stabilizes the dimeric form of their corresponding MCPs, Tsr or Tar, respectively, but the receptor signaling mechanism does not involve a transition from a monomeric to a dimeric state [34, 35]. Although CheA dimerization is essential for kinase activity, and MCP dimerization is essential for ligand binding, ligands such as aspartate and serine that favor dimerization actually cause a dramatic inhibition of CheA kinase activity. Moreover, if the MCP Tar is locked in a dimeric state by engineered disulfide cross-links, it retains its ability to respond to aspartate [34]. Thus, although dimerization is generally a necessary first step toward the assembly of a signal transduction apparatus, it is not sufficient for signaling.

Alteration of Protein–Protein Interactions Associated with Receptor Signaling

Both sensory inputs and signaling outputs of HPK receptors involve alterations of interactions between protein surfaces. For instance, in the homodimeric MCPs Tsr and Tar, Tsr binds serine and Tar binds aspartate at sites composed of residues from both subunits [29, 36, 37]. The energy of ligand binding acts to fix the relationship between monomers in a particular orienta-

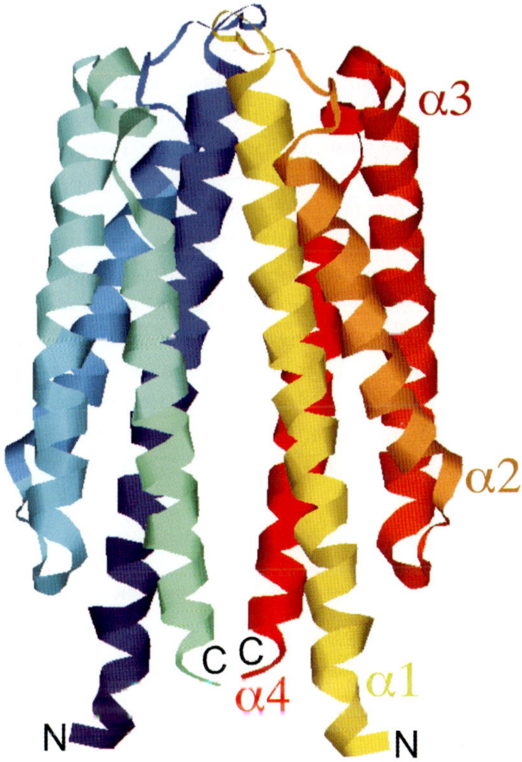

FIGURE 3 Structure of the Tar periplasmic domain [38] with the dimer generated from a monomer by symmetry. Figure produced from PDB file 1VLS with RasMac.

tion (see Fig. 3) [28, 38]. For several MCPs, the small stimulatory ligand binds to a soluble periplasmic-binding protein, which then binds to the homodimeric receptor. Binding proteins are composed of two domains connected by a hinge region. Ligand binding between these domains stabilizes a closed conformation that can then bind to the dimeric sensory domain of a MCP. For example, the *E. coli* Tar protein is activated directly by aspartate binding and indirectly by maltose through the periplasmic maltose-binding protein [39–42].

Alteration of protein–protein interactions in receptor-mediated signal transduction does not stop at the membrane. There are several examples where stimulatory ligands regulate HPK activity by modulating interactions with auxiliary proteins in the cytoplasm. It is apparent that the membrane-spanning sequences at the N terminus of HPK receptors often function to position the protein with respect to other auxiliary receptor proteins in the membrane. For example, the *E. coli* HPK receptor PhoR controls phosphorylation of the response regulator transcription factor PhoB, which turns on genes such as alkaline phosphatase (*phoA*) in response to phosphate limitation [43, 44]. PhoR has the same transmembrane topology as a type I histidine kinase receptor, but lacks a periplasmic-sensing domain between the two transmembrane regions (Fig. 1) [45]. Genetic evidence indicates that PhoR activity is regulated by an auxiliary cytoplasmic protein, PhoU, which interacts with an ABC transporter for phosphate, the PST system [44]. The PhoR kinase is activated when the rate of phosphate transport is low due to depletion of the exogenous phosphate [43, 44, 46].

HPK activity requires a cycle of changes in protein conformation, with concomitant alterations in protein domain interactions (see Fig. 4). The histidine kinase catalytic domain is homologous to the ATPase domains of Hsp90, type II topoisomerases, and MutL [21, 47–49]. This conserved structure con-

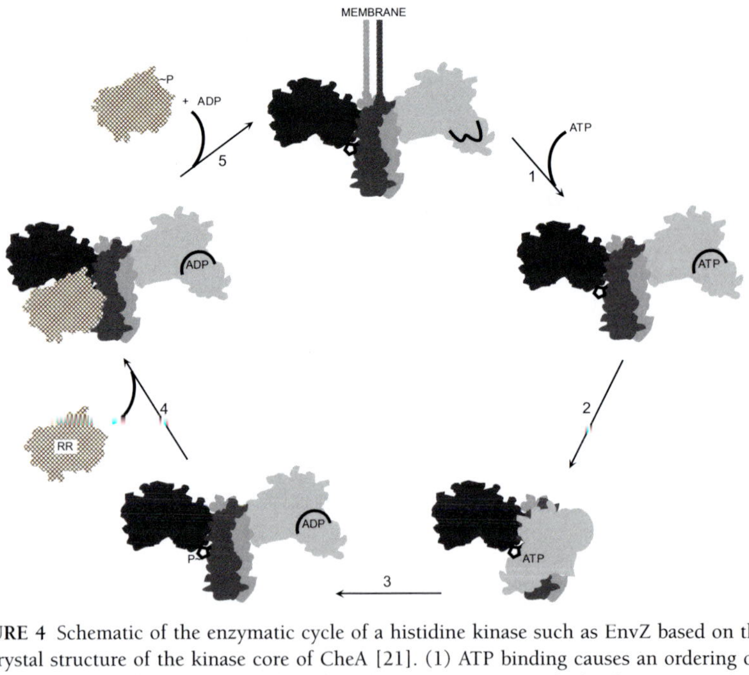

FIGURE 4 Schematic of the enzymatic cycle of a histidine kinase such as EnvZ based on the X-ray crystal structure of the kinase core of CheA [21]. (1) ATP binding causes an ordering of the ATP binding loop; (2) the ATP-bound catalytic domain associates with the dimerization domain; (3) the dimerization domain is phosphorylated *in trans*; (4) the response regulator binds to the kinase and transfers the phosphoryl group to an aspartate; and (5) the phospho-response regulator and ADP are released. The illustration of response regulator binding is based on the interaction of Spo0B and Spo0F [216].

sists of several α helices packed over one face of a large, mostly antiparallel, β sheet. The catalytic domain contains a disordered ATP-binding loop, and the entire ATP-binding site is poorly organized in the absence of nucleotide [21, 48–52]. In MutL, it has been shown in the X-ray crystal structure that ATP binding causes a large rearrangement, with the ATP-binding loop closing down over the nucleotide [51]. This ATP-dependent conformational change creates a new surface that serves as a binding site for another protein domain. The subsequent conversion of bound ATP to ADP facilitates dissociation of the other protein domain with concomitant release of the ADP and restoration of the binding loop to its original, relatively disordered, state. Similar conformational changes in an ATP-binding loop can be seen in the X-ray crystal structures of nucleotide-bound forms of the type II toposiomerase GyrB [47], as well as the HPK CheA [53].

In ATPases, the cycle of protein domain binding and dissociation is used to manipulate the structures of macromolecular assemblies: protein complexes in the case of Hsp90 and DNA in the case of topoisomerases and MutL [51, 54–56]. HPKs such as CheA appear to undergo a similar cycle (Fig. 4) [11, 53, 57]. In HPKs, the ATP-bound kinase active site associates with a domain having a phospho-accepting histidine, usually the dimerization domain of a second subunit. The portion of the dimerization domain where the phosphohistidine is generated must then dissociate from the kinase active site so that it is free to pass the phosphoryl group to the aspartate side chain in a cognate response regulator protein. The response regulator output from a given receptor could be affected by receptor-induced changes in rates of transition through any stage of this cycle of alternating domain interactions.

Figure 4 illustrates the HPK cycle based on our hypothesis that the histidine kinase core of all HPKs has a topology like that of CheA. This implies that a large conformational change is required in step 2 in order to achieve transphosphorylation. It has also been suggested that other HPKs have a different topology than CheA, with the difference consisting primarily of a crossing of their catalytic domains [23, 58]. This would facilitate transphosphorylation by placing each catalytic domain adjacent to the H box in the dimerization domain of the opposing subunit.

Phosphatase Activities Associated with Histidine Kinase Receptors

Many histidine kinase receptors mediate both phosphorylation and dephosphorylation of their cognate response regulators. A well-studied example of this is EnvZ. The phosphatase activity of EnvZ is associated with its dimerization domain [59] and appears to be mediated by protein–protein interactions that are similar to those involved in phosphotransfer from the dimerization domain phosphohistidine to the response regulator aspartate. In many HPKs, it is the phosphatase activity whose regulatory effects predominate [44, 59, 60]. Alternative mechanisms for response regulator phosphorylation do not require the activity of a cognate HPK, including phosphorylation by small molecule phosphodonors such as acetylphosphate or carbamoylphosphate, as well as a relatively nonspecific phosphotransfer from phosphohistidines in noncognate histidine kinases [43, 61, 62]. In contrast, the phosphatase activities of HPKs appear to be relatively specific to the cognate response regulator. Thus, the dominant phenotype of a mutant strain lacking a particular HPK receptor is often a low-level constitutive response regulator output under conditions where, in wild-type cells, receptor phosphatase activity leads to inactivation. Regulation of these dual-function HPK receptors appears to involve modulation of a balance between two distinct states: kinase on/phosphatase off and kinase off/phosphatase on [63].

TYPE I HISTIDINE KINASE RECEPTORS

Kinase Classifications

The majority of HPKs are type I receptors with an uncleaved signal sequence near the N terminus leading to a sensory input domain outside the cytoplasmic membrane. The sensory domain is connected *via* a single transmembrane sequence to a histidine kinase catalytic domain in the cytoplasm. These receptors have been characterized as either "orthodox" or "hybrid" HPKs depending on whether they have a response regulator domain linked C-terminal to the histidine kinase core. The so-called hybrid kinases have an attached response regulator domain (see Fig. 1; EnvZ is an orthodox HPK and ArcB is a hybrid HPK). They constitute a distinct subfamily that includes all eukaryotic HPK receptors, as well as many receptors in prokaryotes [12, 64]. In addition to the histidine kinase core and response regulator domains, hybrid kinases often interact with a third phosphorylated domain, termed a histidine phosphotransfer (HPt) domain. HPt domains may be attached to a hybrid histidine kinase (e.g., ArcB in Fig. 1) or may exist as distinct soluble proteins.

Orthodox Kinases — EnvZ

EnvZ is a membrane kinase thought to mediate changes in porin expression in response to changes in osmolarity in the medium around the cell [65, 66].

It is a type I membrane receptor with an uncleaved signal sequence leading to a periplasmic domain followed by a second transmembrane sequence that leads to a HAMP domain just inside the cytoplasmic membrane (see Fig. 1 and the section "Transmembrane Signalling in Bacterial Chemotaxis") [67, 68]. The HAMP domain leads directly into the dimerization domain of an archetypal histidine kinase core. The site of histidine phosphorylation is located within the first helix of the dimerization domain, near to where it merges with the HAMP domain [69]. The phosphoryl group is specifically transferred to an aspartate residue in the cognate response regulator-transcription factor, OmpR [70, 71]. In addition to its histidine kinase activity, EnvZ also acts as a phosphatase to catalyze the dephosphorylation of phospho-OmpR [72–74]. The phospho-accepting histidine residue in the dimerization domain plays an important role in the phosphatase reaction, and the phosphatase active site appears to be localized to the dimerization domain [59, 75, 76]. The phosphatase reaction does not generally proceed *via* a phosphohistidine intermediate, as mutation of the conserved histidine does not completely eliminate phosphatase activity [76]. A fragment of EnvZ consisting of only the dimerization domain retains significant phosphatase activity [59]. However, the ATP-binding catalytic domain also plays an important role, as nucleotide binding enhances phosphatase activity greatly [59, 77].

The ratio of EnvZ kinase to phosphatase activities is thought to increase with increasing extracellular osmotic strength to give increasing levels of phospho-OmpR [72, 74, 78]. It has generally been assumed that this effect derives from a change in the periplasmic part of the receptor. Several different missense mutations in this region cause constitutive extremes of receptor signaling ranging from low to high ratios of kinase to phosphatase activity [79]. These findings indicate that kinase/phosphatase activities are sensitive to changes in the structure of the periplasmic domain. However, large deletions within the periplasmic domain do not block OmpR-mediated responses to changes in osmotic pressure [80]. It should be noted, however, that strains that completely lack EnvZ can still show OmpR-dependent transcriptional regulation in response to changes in osmotic pressure [81]. OmpR is subject to phosphorylation by acetylphosphate as well as by other HPK receptors such as ArcB [82]. In addition, osmotic and anaerobic stress can act to directly affect the transcription of OmpR-regulated genes by inducing changes in DNA supercoiling [83, 84]. These alternative modes of regulation presumably act to control OmpR signaling in ways that are more or less coordinated with EnvZ-mediated regulation, thereby providing backup regulatory mechanisms in strains with deficiencies in EnvZ function.

Numerous missense mutations in cytoplasmic portions of EnvZ cause dramatic shifts in the kinase/phosphatase equilibrium. In this case, one must distinguish between mutations in residues that participate directly in kinase or phosphatase catalysis and mutations that alter the balance between kinase and phosphatase receptor states. There are, for example, several well-characterized missense mutations in residues adjacent to the phospho-accepting histidine that cause a dramatic reduction in phosphatase activity [63, 85]. The effect is probably related to a direct effect on phosphatase catalysis rather than a shift away from a hypothetical phosphatase receptor conformation. Conversely, mutations in the ATP-binding catalytic domain that differentially reduce kinase activity compared to phosphatase activity probably result from problems associated with the kinase catalytic mechanism [63, 86].

Hybrid Kinases — ArcB

The best characterized example of a hybrid kinase is the *E. coli* anaerobic sensor ArcB (Fig. 1). This protein has a rather typical type I HPK organization similar to EnvZ and PhoR except that instead of ending at the histidine kinase core, the sequence continues with a C-terminal response regulator domain followed by an HPt domain [87]. Phosphoryl groups can be passed between any of the three phosphorylation sites: the kinase dimerization domain histidine, the HPt histidine, and the response regulator aspartate [87]. The phosphorylated HPt domain acts as a phosphodonor for at least two response regulators, ArcA and OmpR, both of which are phosphorylation-activated transcription factors [82, 88]. The ArcB receptor kinase is activated under anaerobic conditions, which leads *via* HPt phosphorylation to increased levels of phospho-ArcA and phospho-OmpR [82, 88]. ArcA regulates the transcription of several genes related to aerobic metabolism [89], whereas OmpR, described earlier, regulates porin expression [90, 91].

There are two-component systems in several organisms, including most found in eukaryotes, where there are two or more hybrid kinases with HPK and response regulator domains, but no attached HPt domain [92–94]. Phosphoryl groups from these hybrid kinases are funneled into a common HPt that is produced as an independent protein. The phosphorylated HPt then acts as a phosphodonor for one or more common response regulator targets. For example, in *Dictyostelium discoideum*, the HPt protein RdeA interacts with the histidine kinase DokA [95], and possibly several other histidine kinases. RdeA transfers a phosphoryl group to a response regulator with phosphodiesterase activity, RegA [93]. A similar phosphorylation network occurs in the yeast *Schizosaccharomyces pombe* [96].

In these hybrid kinase systems, the primary regulatory input from a given receptor seems to be activation of a phosphatase activity that is associated with its response regulator domain [93, 95, 97]. This leads to a reverse flow of phosphoryl groups that dephosphorylate the HPt, and its response regulator targets. Thus, in hybrid histidine kinase receptors, the signal transduction process acts primarily to regulate a phosphatase activity associated with the receptor response regulator domains. The involvement of the histidine kinase core in phosphatase regulation is not clear. Genetic results indicate that, unlike with EnvZ, response regulator phosphatase activities in hybrid HPKs do not depend on the region around the phospho-accepting histidine in the kinase dimerization domain [97].

RECEPTORS WITH SEVERAL MEMBRANE-SPANNING SEGMENTS

SIX-TRANSMEMBRANE HPK RECEPTORS

A distinct and widely distributed family of HPK receptors have hydrophobic N-terminal domains that are predicted to be composed of between five and seven transmembrane helices. For many of these kinases, the number of predicted transmembrane segments depends on the algorithm used. Even in cases where the membrane topology of the histidine kinase has been assessed experimentally, the results only confirmed that there were between five and seven transmembrane segments [98–102]. For simplicity, we will refer to these as six-transmembrane proteins. These receptors appear to generally function in cell–cell communication. In the few cases where they have been identified, the stimulatory ligands are small peptides or modified peptides that are secreted by the same organism. For example, the AgrC system in *Staphylococcus aureus* is a quorum-sensing system that exports and then senses modified peptides [103, 104]. Another good example is provided by the ComD receptors in *Streptococcus* spp. that function to turn on transformation competence in response to small peptides termed competence factors [105–107]. Each species has a different competence factor and a ComD with a divergent six-transmembrane-sensing domain [108]. Genes that encode the competence factor and ComD are adjacent to one another [107]. Different peptide – receptor gene pairs from different *Streptoccus* species can recombine into this locus so that competence in one species can be turned on by the presence of another [108]. One can see how this type of mechanism could result in the rapid evolution of cellular communication networks. It is clear from the examination of various cognate competence factor – ComD pairs that the six-transmembrane domain offers a very flexible framework for the design of a ligand binding signal transduction mechanism.

The six-transmembrane HPK receptors appear to be the prokaryotic correlate of the seven-transmembrane G-protein-coupled receptors that mediate a wide range of different sensory responses in eukaryotic cells. It has been estimated that in humans there are over 1000 different members of the seven-transmembrane receptor family [109]. These proteins are responsible for taste, smell, and vision, as well as responses to numerous hormones and neurotransmitters. Seven-transmembrane receptors generally bind stimulatory ligands within the plane of the membrane in a central pocket surrounded by the seven helices [110–113]. Ligand binding appears to affect a response by changing the positioning of the helices so as to alter the surface formed by connecting loops at the membrane – cytoplasm interface [114].

Because other classes of HPKs function as homodimers, it could be assumed that the six-transmembrane HPKs do as well. There is no direct evidence, however, to demonstrate this contention. Differences in the histidine kinase core domain distinguish six-transmembrane HPKs from other HPKs. They fall into HPK family 10, which is distinguished by the absence of a D box [12]. In addition, there is no X box and the region near the H box does not show a high similarity to other HPK dimerization domains [12]. In CheA and other HPKs, the D box forms part of the nucleotide-binding pocket, with the aspartate hydrogen bonding to the adenine ring [53]. This difference between six-transmembrane proteins and other HPKs may be indicative of significant differences in the active site geometry. Signal transduction mechanisms for these proteins may proceed more in analogy to the seven-transmembrane receptors than to the other classes of HPK receptors.

UhpB and UhpC

Expression of genes that encode the sugar phosphate transport system in *E. coli* and *S. typhimurium* is regulated by the activity of a HPK, UhpB, through the response regulator-transcription factor UhpA [98, 115]. Based on *phoA* fusion experiments, UhpB has from 6 to 10 transmembrane segments [98]. Transmembrane prediction algorithms indicate 7 to 9 transmembrane segments [116, 117]. In addition, there is another protein, UhpC, that is homologous to the hexose phosphate transporter, UhpT [98, 118]. UhpC is necessary for UhpB to sense sugar phosphates, and many constitutive signaling mutants of UhpB are inactive without UhpC [119].

The dependence of UhpB on UhpC suggests a signal transduction mechanism that depends on protein–protein interactions between them. This is reminiscent of PhoR, which is also regulated in interaction with a transporter [44]. In addition, the homology between UhpC and UhpT suggests that essentially the same protein architecture can be adapted to either sense or transport a substrate ligand.

TRANSMEMBRANE SIGNALING IN BACTERIAL CHEMOTAXIS

OVERVIEW OF THE CHEMOTAXIS SYSTEM

The *E. coli* chemotaxis system is the best characterized HPK receptor-mediated signal transduction network. There are five membrane-bound MCPs in *E. coli*: Tsr, Tar, Trg, Tap, and Aer (see Fig. 1; Tsr, Trg, and Tap have essentially the same structure as Tar). All except Aer are type I receptors that sense the presence of attractants in the periplasm either through direct binding or through interaction with periplasmic-binding proteins [120, 121]. Aer senses the cellular redox potential using a cytoplasmic-sensing domain with a noncovalently associated flavin [122, 123]. All five MCPs have homologous, highly conserved cytoplasmic-signaling domains. The periplasmic-sensing domains of the MCPs are essentially homodimeric [35], but the cytoplasmic-signaling domains at the other side of the membrane interact to form a higher ordered structure together with an adapter protein, CheW, and the chemotaxis HPK, CheA [124–127]. MCPs, together with CheW and CheA, have been found to cluster into a single patch at one pole of the cell [128–131]. CheA and CheW interact with the highly conserved central portion of the MCP coiled-coil region in the cytoplasm (the CheA/CheW-binding domain). This portion of the receptor, together with CheW, is required for activation of CheA [132–134]. CheA has the same histidine kinase catalytic core structure as membrane HPKs like EnvZ [12]; however, its target site of phosphorylation is a histidine residue in a separate HPt domain rather than a histidine within the dimerization domain [135, 136].

The chemotaxis response regulator, CheY, accepts a phosphoryl group from the phosphorylated CheA HPt domain [137–140]. CheY diffuses freely through the cytoplasm and, when phosphorylated, binds to the flagellar motor to cause a change in the direction of cell swimming [141]. *In vitro*, phospho-CheY hydrolyzes spontaneously, with a pseudo first order rate constant of about 0.04 s^{-1} under physiological conditions of temperature and pH [15, 142–144]. Phospho-CheY hydrolysis is accelerated by the presence of CheZ [137]. For reviews on chemotaxis, see Stock *et al.* [94] and Falke and Kim [145] regarding structural studies of the chemotaxis proteins and Armitage [146], Stock and Levit [147], and Berg [148] for more general reviews of the entire chemotaxis system.

MCP CLUSTERING IN CELLS

Based on the number of serine and aspartate binding sites, the number of MCPs has been estimated at 5800 monomers of Tsr and 1200 of Tar [149] per

E. coli cell. Tsr and Tar are called the major receptors. Other measurements, using Western analysis, have found a similar number of 6000 MCPs total per cell [124]. Each of the so-called minor receptors, Trg and Tap, is present in substantially lower numbers than Tsr or Tar, probably less than 10% of the level of Tar [150]. The recently discovered aerotaxis receptor Aer [123] has not been quantitated, but is assumed to be present at levels similar to that of Trg and Tap. Based on cell-swarming assays, all five MCPs have an apparently equal ability to mediate chemotaxis [123, 151, 152]. However, it is not known whether the minor receptors will fully suppress CheA activity in the presence of the major receptors. Using electron microscopy (EM) with cytological immunolocalization techniques, Maddock and Shapiro [128] discovered that an *E. coli* cell typically has almost all of its MCPs clustered together with CheW and CheA in a dense patch at one pole of the cell. These clusters are tightly packed and are not simply localized to the entire area of the cell pole. Subsequent EM and fluorescence microscopy studies have confirmed these findings [129–131].

One piece of evidence for the importance of clustering for MCP function is the fact that the minor receptors do not function in the absence of the major receptors [123, 153–155]. Immunolabeling EM studies show that in cells lacking both the major receptors, Tap or Trg are still generally localized to the end of the cell and form ternary complexes with CheA and CheW. However, Tap or Trg alone does not appear capable of forming a single tight cluster [156]. Thus, the presence of one of the major receptors seems to be required for the proper clustering of Tap and Trg, and they may only function when in an MCP cluster.

It has been established that cross talk between major and minor MCPs is crucial for methylation-dependent adaptation. The minor receptors do not adapt by methylation in the absence of the major receptors [123, 153–155]. The C-terminal tails of Tsr and Tar serve as binding sites for the methylating and demethylating enzymes[122, 157], and the last five residues are essential for this interaction [158, 159]. Removing this portion of Tar eliminates adaptation to aspartate by methylation [160]. The minor receptors are at least 18 residues shorter [161, 162], and even when a minor receptor alone is overproduced to the level of a major receptor, adaptation by methylation does not occur [152, 154]. Adding the C-terminal tail to Tap does not allow Tap to function in the absence of the major receptors [154], but attaching the C-terminal tail to Trg greatly improves both its ability to undergo methylation-dependent adaptation and to mediate chemotaxis [152, 155, 157, 163]. This suggests that the structural differences between major and minor receptors are more complex than just the presence or absence of the C-terminal tail.

The improved functioning of Trg associated with the addition of the C-terminal tail suggests that receptor methylation could play a role in receptor clustering. However, other studies have shown that receptors that are either highly methylated or highly demethylated cluster equally well [129]. In studies of various soluble constructs containing a leucine zipper-linked Tar cytoplasmic domain (LZ-Tar_C), the one with the highest propensity to form higher-order complexes with CheA and CheW mimics the methylated state. This construct has the two methylatable glutamates changed to glutamine [125, 126], which has an effect similar to methylation [133, 164, 165]. Deamidation of LZ-Tar_C (i.e., converting specific glutamines to glutamates) results in dissociation of the complexes [125, 126]. Perhaps the neutralization of the charge on the glutamates is more critical for clustering in this system using LZ-Tar_C than for membrane receptors.

Transmembrane MCPs

The *E. coli* chemotaxis receptors are comprised almost entirely of alpha-helical coiled-coils [28, 125, 127, 166–168]. Figure 1 shows a schematic view of a Tar receptor: a short N-terminal cytoplasmic region is followed by the first transmembrane helix (TM1), the aspartate-binding sensing domain in the periplasm, the second trans-membrane helix (TM2), the linker region (or HAMP domain), and a long coiled-coil region. The coiled-coil contains functionally distinct regions termed the first methylated helix (MH1), the CheA/CheW-binding region (or signaling domain), and the second methylated helix (MH2).

Ligand-Binding Domain

The X-ray crystal structure of the aspartate binding domain of Tar consists of a homodimeric four helix bundle (see Fig. 3) [28, 38]. Sequence alignments, modeling, and structural studies suggest that the periplasmic domains of Tsr, Trg, and Tap are homologous to Tar, despite their relatively divergent sequences [29, 169]. The fifth *E. coli* MCP, Aer, has a completely different sensory domain: a cytoplasmic redox potential-sensing PAS domain that contains a noncovalently associated flavin [122, 123, 170].

Completely different extracytoplasmic-sensing domains can couple to homologous cytoplasmic chemotaxis machinery in different species. In general, the extracytoplasmic domains of MCPs in a given species tend to be structurally similar, but divergent from those in other species. Except for a few exceptions from closely related species such as *S. typhimurium* and *Enterobacter aerogenes*, MCP-sensing domains from other prokaryotes are structurally distinct from the periplasmic domain of *E. coli* Tar [4, 171].

X-ray crystallographic studies indicate that aspartate binds to Tar in a cleft between the two subunits of isolated sensory domain homodimers [28, 38].

Three arginine residues in Tar (R69, R73 in one subunit, and R64′ in the other subunit) form an amino-acid binding site where aspartate binds [172]. Mutation of these residues reduces the ligand-binding affinity of Tar but not the ability of Tar to regulate CheA in response to aspartate [173]. The cluster of three arginines seems to be part of a general amino acid binding motif [172], and they are conserved in Tsr where they are thought to participate in serine binding [174]. Evidence from sequence similarity, genetic studies, NMR studies, and modeling suggests that ligands or periplasmic binding proteins bind between the two dimer subunits in all *E. coli* and *S. typhimurium* MCPs with ligand binding domains that are homologous to Tar [30–33].

Transmembrane Helices

The transmembrane helices from the two dimer subunits appear to pack together in a four-helix bundle, with TM1 and TM1′ forming the interface between the dimers [169, 175–177]. The primary sequences of the transmembrane helices are not well conserved among the receptors, but these regions do play a role in signaling. The overall phasing and symmetry of the residues may be more important than their exact identities [178]. The importance of transmembrane helices is indicated by the results of mutagenesis experiments where the introduction of cysteine residues, or even a conservative mutation of one hydrophobic residue for another, may disrupt receptor function [168, 179, 180]. If the introduced cysteines are allowed to form disulfide cross-links between neighboring helices, additional effects are observed, including cross-links that lock the receptor into a CheA-activating or CheA-inactivating state [180]. There are also structural requirements beyond those merely provided by a transmembrane domain from another protein. For instance, when Tar TM2 is replaced by the transmembrane helix from the insulin receptor or replaced by a random hydrophobic sequence, *E. coli* swarming to aspartate is obliterated [181]. However, the central seven residues of TM2 can be mutated without causing a loss of swarming [181].

The transmembrane helices of *E. coli* Tar, Tsr, Trg, and Tap are predicted to be 23 residues long using the TMHMM algorithm [117] and have residues of a hydrophobic, noncharged character (see Fig. 5). The higher degree of sequence conservation in TM1 may be due to the fact that TM1 and TM1′ pack together across the dimer interface and because TM1 is part of a signal sequence that directs the protein to the membrane.

The Linker Region

Perhaps the least understood portion of the receptor is the linker region consisting of about 46 residues just after TM2 (Fig. 1) [182]. The linker region

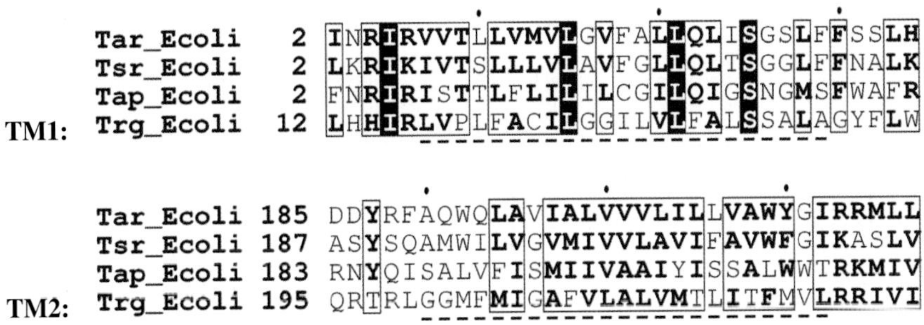

FIGURE 5 Alignments of the TM1 and TM2 regions of *E. coli* MCPs. The 23 transmembrane residues plus 5 flanking residues on each end are shown. Alignments of the entire MCP sequences were made using ClustalW [185], and shading according to 75% equivalent physicochemical properties was done with ESPript 1.9. Identification of the transmembrane segments was based on the TMHMM algorithm [117].

may play a critical role in signal transduction, as it lies between the sensing domain and the region where CheA and CheW interact with the receptor. There is a pattern of sequence conservation in this region between HPK receptors [182] as well as between MCPs [167, 103]. Cysteine scanning studies of the Tar linker region indicate a structure consisting of two helical regions, with a 14 residue stretch in between that forms a compact subdomain [184]. These descriptions of the linker region fit within the broader proposal by Aravind and Ponting [183] that this region is part of a ~50 residue, two-helix "HAMP domain" that is a widely distributed regulatory element in transmembrane and other signaling proteins. The term HAMP is due to the presence of this domain in histidine kinases, adenylyl cyclases, methyl accepting proteins, and phosphatases [183]. Figure 6 shows an alignment of this domain for the five *E. coli* and eight *B. subtilis* MCPs predicted to contain it.

The Pfam 6.0 database has 311 examples of the HAMP domain and it is found in most known or putative MCPs that have a CheA/CheW-binding domain, as well as in many HPK receptors (see Fig. 1) [4]. A HAMP domain phylogenetic tree calculated using ClustalW [185] indicates that MCPs from *E. coli*, *S. typhimurium*, and *E. aerogene* group together relative to other HAMP domains, although there is substantial sequence divergence even within this closely related group (unpublished results). Despite its anticipated importance for signaling, the HAMP domain is not critical to the interaction of MCPs with CheA and CheW, as soluble proteins consisting of only the CheA/CheW-binding region of the Tar or Tsr cytoplasmic domain retain an ability to activate or inhibit CheA [125, 126, 134, 186].

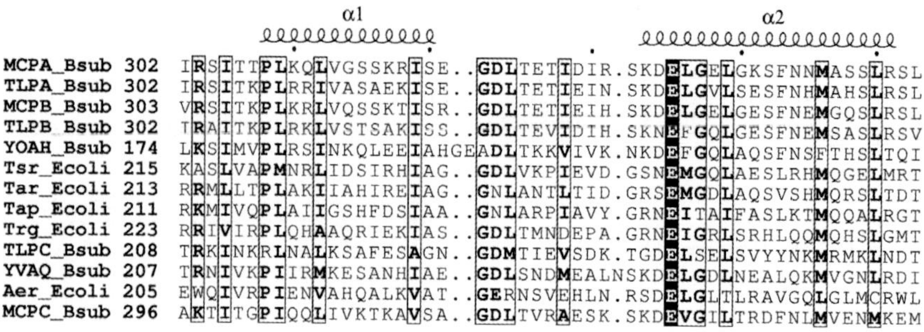

FIGURE 6 ClustalW [185] alignment of the HAMP domains [183] from *B. subtilis* [6] and *E. coli* [3] MCPs. Secondary structure prediction from Aravind and Ponting [183]. Shading according to 80% equivalent physicochemical properties was done using ESPript 1.9.

The secondary structure prediction for the HAMP domain indicates that, for MCPs, the conserved proline is followed by 12 residues that form the first helix, next comes a 12 residue nonhelical stretch, and then a 19 residue second helix (see Fig. 6) [183]. One face of the first helix is protected from the solvent, as shown by chemical reactivity studies of single cysteine mutants [184]. This protected face of the helix is primarily hydrophobic in character and would face away from the central axis of the bundle formed by the transmembrane helices, assuming a continuous helical phasing from TM2 to this helix [184]. These structural studies of the HAMP domain suggest a topological similarity to the helix-loop-helix regions of leucine-zipper transcription factors such as Max and Myc [187, 188] (see Fig. 7).

Cytoplasmic Domain Helices

The cytoplasmic regions of MCPs are long α-helical coiled-coil structures, as shown by a partial X-ray crystal structure (see Fig. 8) [127], by numerous other structural studies [127, 168, 189–191], and by structure prediction algorithms [166, 167]. Within this coiled-coil region, there are functionally discrete portions termed the methylated helices and the CheA/CheW-binding region. Coiled-coil domains are much more highly conserved than periplasmic domains, the transmembrane sequences, or linker regions. The degree of sequence identity in this region over vast evolutionary distances is truly remarkable. For example, MCPs in *E. coli* and the archaebacterium *H. salinarum* are over 30% identical in the cytoplasmic domain helices [192].

The first methylated helix (MH1) follows the linker region at the N-terminal end of the coiled-coil region and contains three or four glutamate residues that can be covalently modified by enzymatic addition of a methyl group

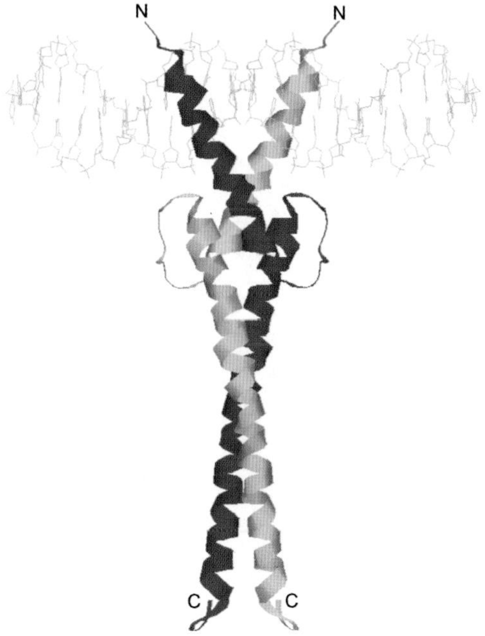

FIGURE 7 A homodimer of the helix-loop-helix transcription factor Max bound to DNA [188]. Figure produced from PDB file 1AN2 using RasMac.

[193–196]. The second methylated helix (MH2) is at the C-terminal end of the coiled-coil region and contains one or two methylation sites, which are generally found to be methylated less efficiently than those on MH1 [193 199]. Some of the modified residues in MH1 and MH2 are glutamines that are deamidated prior to methylation [194, 197, 198]. The deamidase that catalyzes this reaction, CheB, also catalyzes hydrolysis of the glutamyl methyl esters, effectively reversing the MCP methylation reaction [199, 200]. Elevated levels of methylation (or amidation) decrease the sensitivity of the receptors to attractant ligands [133, 165, 201, 202]; however, the K_D for ligand binding is not substantially altered [247–249].

Most evidence supports the idea that helices of the coiled-coil region in the cell come together to form an antiparallel coiled-coil. In particular, the X-ray crystal structure of the coiled-coil region of Tsr [127], which includes parts of MH1 and MH2, shows an antiparallel coiled-coil conformation (Fig. 8). In addition, across species, four 14 residue insertions have occurred in MCPs [167]. These occur in two sets of two insertions. The first set consists of one insertion at the beginning of MH1 and another at the end of MH2. The other set consists of one insertion at the beginning of the CheA/CheW-binding

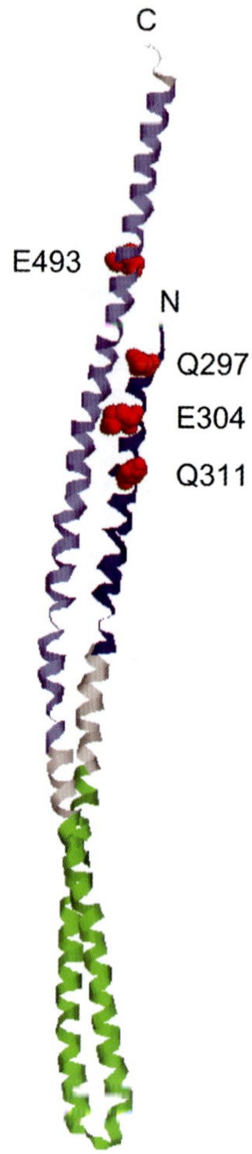

FIGURE 8 One subunit from the X-ray crystal structure of a fragment of the cytoplasmic domain of Tsr [127]. The partial MH1 is in blue, the CheA/CheW-binding region is in green, and MH2 is in violet. Methylation sites are labeled by the residue number and are shown in red as space-filling models of the amino acids. Figure produced from PDB file 1QU7 using RasMac.

region and another insertion at the end. Based on the X-ray crystal structure of Tsr [127], the two inserted 14 residue segments in the CheA/CheW-binding region lie on the opposite side of the hairpin turn and form an antiparallel coiled-coil. If the same coiled-coil structure is extended, the first set of insertions would also be expected to form an antiparallel coiled-coil. Thus, these insertions would each be four turns of the coiled-coil that extend the length of the MCP while maintaining phasing of the remainder of the antiparallel coiled-coil [167]

It is also possible, however, that MH1 and MH1′ from the other half of the dimer form a parallel coiled-coil as part of the signal transduction process. Evidence for this comes from the fact that soluble fragments of the Tar cytoplasmic domain do not activate CheA unless fused to a leucine zipper dimerization domain, which would strongly promote a parallel coiled-coiled interaction [125, 186].

The region between MH1 and MH2 is denoted the CheA/CheW-binding region. Based on the X-ray structure [127], this region consists of two helices separated by a hairpin or a flexible loop (Fig. 8) and exhibits substantial flexibility. Despite its large helical content [125, 127, 189–191], residues of this region in MCP fragments are as solvent exposed as would be expected for a molten globule [191]. In addition, disulfide-trapping experiments in the CheA/CheW-binding region show that movements of up to 19 Å and rotation of up to 180° can occur in the hairpin, based on the ability of pairs of cysteine residues to form disulfide cross-links [168].

The location for binding of CheA and CheW to MCPs is close to the hairpin. MCP fragments containing this region can activate CheA [125, 134, 203, 204], and mutations in this region have substantial effects on the MCP interaction with CheW and CheA [205, 206]. Studies with mutant receptor cytoplasmic domain fragments [134] indicate that CheA can bind directly to the hairpin region rather than requiring CheW as an intermediate adapter.

The MCP-Linked Kinase, CheA

CheA is composed of five distinct domains (Fig. 9), and X-ray crystal structures of individual domains from *T. maritima, E. coli,* and *S. typhimurium* CheA give a complete picture of the three-dimensional structure.

The HPt Domain (P1)

The N-terminal HPt (or P1) domain is a helical bundle that is structurally and functionally homologous to the phospho-accepting HPt domains associated with hybrid HPKs such as ArcB [136]. The X-ray crystal structure of the HPt domain from *S. typhimurium* shows a four helix bundle structure, with the phospho-accepting histidine on an exposed face of the bundle in a hydrogen bond network with the conserved glutamate and lysine residues [136]. The CheA HPt domain is connected *via* a flexible linker to a domain termed P2 [207, 208].

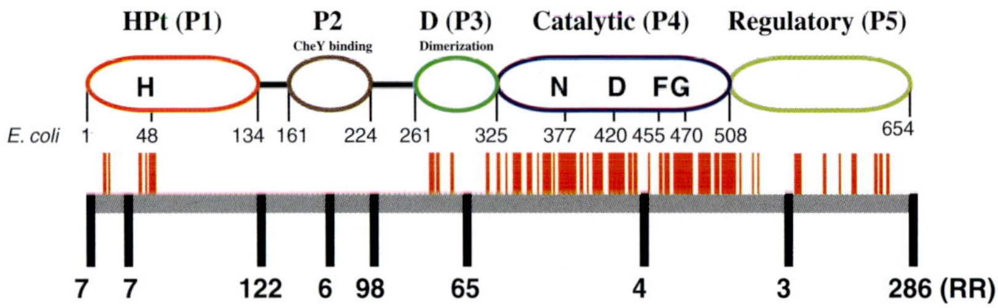

FIGURE 9 The domain structure of CheA from *E. coli*. The positions of conserved histidine kinase motifs, the H, N, D, F, and G boxes, are indicated [12, 246]. Below the domains, amino acid positions from the *E. coli* sequence [247] are listed at the beginning and end of each domain, as well as at each conserved motif. Alignments of 36 unique CheA sequences were made using ClustalW [185]. The red bars below indicate positions where 95% of the residues have equivalent physicochemical properties. Positions of insertions not present in the *E. coli* sequence are indicated by black bars with the approximate maximum size of the insertion. The 36 unique CheA proteins were identified as all those in the SMART database containing a single HPt and a CheW-like domain [248]. Proteins with greater than 95% identity to another in this database were excluded. The 29 species represented are *Agrobacterium radiobacter, Archaeoglobus fulgidus, B. subtilis, Bacillus cereus* (2), *Bacillus halodurans, Borrelia burgdorferi* (2), *Campylobacter jejuni, Caulobacter crescentus, E. coli, Halobacterium salinarum, Helicobacter pylori, Helicobacter pylori J99, Listeria monocytogenes, Mesorhizobium loti, Myxococcus xanthus, Pseudomonas aeruginosa* (3), *Pseudomonas putida, Pyrococcus abyssi, Pyrococcus horikoshii, Rhizobium meliloti, Rhodobacter sphaeroides* (2), *Rhodospirillum centenum, S. typhimurium, Synechocystis sp., T. maritima, Treponema denticola, Treponema pallidum, Vibrio cholerae* (3), *and Vibrio parahaemolyticus*. At the C terminus, a response regulator domain is found in the CheA proteins from *C. jejuni, H. pylori, M. xanthus, P. aeruginosa, R. centenum*, and *Synechocystis sp.*

The CheY-Binding Domain (P2)

The X-ray crystal structure of P2 from *E. coli* in complex with CheY has been determined [209, 210]. The P2 domain is an α/β structure that binds the chemotaxis response regulator CheY and appears to be the primary CheY recognition component in CheA [207, 209, 210]. Binding of CheY to P2 facilitates phosphotransfer from phospho-HPt to CheY by increasing the local concentration of CheY [211, 212]. P2 is connected *via* a second flexible linker to the dimerization domain [207].

The Dimerization Domain (P3)

The X-ray crystal structure of the core of CheA from *T. maritima* consisting of dimerization, catalytic, and regulatory domains has been determined [21]. The dimerization domain is a four-helix bundle formed from a helix-turn-helix in each of the two subunits and is connected directly to the ATP-binding catalytic domain [21]. This four-helix dimerization structure closely overlays that of the dimerization domain determined for the orthodox HPK EnvZ [22](J.B. Stock, unpublished results).

The Catalytic Domain (P4)

Figure 2 shows the X-ray structure of the dimerization and catalytic domains of CheA. Together these comprise the highly conserved histidine kinase catalytic core that defines the protein histidine kinase superfamily [12]. However, in CheA the HPt domain, rather than the dimerization domain, contains the site of histidine phosphorylation [135]. CheA shows other features in its sequence that places it into a distinct subclass compared to HPKs such as EnvZ. In the CheA N box, the first asparagine has been replaced by a histidine, and the "KFT" motif three residues beyond the second asparagine has been replaced by "DHG" [12]. The X-ray crystal structure of the CheA catalytic domain in complex with ADP and three ATP analogs has contributed further to our understanding of the kinase mechanism [53]. In particular, one of the ATP analogs induces an ordering of the ATP lid and formation of a complete ATP-binding pocket [53].

The Regulatory Domain (P5)

The kinase catalytic domain is followed by the so-called regulatory domain, which is composed of a pair of SH3-like subdomains [21]. This domain is required for binding to MCPs [213]. It has sequence similarity over its entire length to CheW and has essentially the same three-dimensional structure [21, 214]. The regulatory domain may have a role in controlling CheA activity based on the observation that a truncated version of CheA lacking this domain shows about a twofold higher specific activity in solution [57]. Given the very high structural similarity between CheW and the regulatory domain of CheA, it seems likely that this region of CheA can bind directly to the hairpin region of the MCPs. No good evidence supports the idea that CheW functions as an adapter to mediate the binding of CheA to MCPs.

Relationship of CheA to Other HPKs

In the first half of the CheA dimerization domain is a region of high sequence conservation across CheAs from many species (Fig. 9). Comparison of the CheA X-ray crystal structure [21] with the EnvZ solution NMR structure [22] suggests that this region of sequence conservation occurs around the position corresponding to the H box in the EnvZ dimerization domain. In CheA, this region usually contains a glycine in place of a histidine. We term this region the "pseudo-H box" and wonder whether it plays some role in the CheA activity.

Comparison of the CheA sequence with that of EnvZ and other orthodox histidine kinases suggests that they have the same spatial relationship between their dimerization and catalytic domains. For example, between the *E. coli* EnvZ H box histidine and the N box (the site of ATP binding) are 100 residues with a proline almost exactly halfway between at the boundary between dimerization and kinase domains [22, 49]. Aligning the 21 *E. coli* HPK receptors with clear H and N boxes[1] gives an average H to N box distance of 109.8 residues with a standard deviation of 3.5. In almost all of these *E. coli* HPK receptors, there is a proline about midway between H and N boxes. By comparison, in 33 CheA proteins,[2] there is an average of 95.4 residues with a standard deviation of 3.4 between the pseudo-H box and the N box. Almost all CheAs have a proline halfway between at the boundary between dimerization and catalytic domains [21].

Viewing the CheA crystal structure [21], and assuming a similar three-dimensional relationship between EnvZ dimerization and catalytic domains, a large conformational change would seem to be required to achieve an activated form where the ATP binding site of one catalytic domain contacts the dimerization domain of the opposing subunit. However, the relationship of HPKs to type II topoisomerases suggests that such a large movement is possible, and perhaps even to be expected. One possibility for this conformational change is that the activated form of HPKs has a dimerization domain conformation that resembles the structure of Spo0B [215, 216], whereas the observed X-ray structure of CheA [21] represents an inactive conformation.

CheA Sequence and Function

Figure 9 shows regions of CheA that are conserved between different species, suggesting that they are critical for CheA function. The flexible linker regions on both sides of P2 show a particular lack of conservation [207]. These flexible linkers are thought to function solely as tethers to keep the HPt, P2, and kinase core in close proximity to one another. The enormous divergence in these linkers, even between *E. coli* and *S. typhimurium* CheA sequences, suggests that other regions that have high sequence conservation are playing critical functional roles. Across species, the P2 domain also shows a low level

[1]These 21 HPK receptors are YgiY, BasS, YbcZ, YedV, KpiD, RstB, EnvZ, BaeS, PhoR, CpxA, TorR, NtrB, YfhK, PhoQ, RcsC, EvgS, AtoS, HydH, BarA, ArcB, and CreC.

[2]Average based on the alignment of 33 CheA proteins out of the 39 used in Fig. 9. One CheA from *Borrelia burgdorferi* and those from *Treponema denticola* and *Treponema pallidum* have insertions in the middle of the dimerization domain and were excluded from this analysis.

of sequence conservation. This suggests a coevolution of the P2 domain of CheA and its cognate response regulators.

Figure 10 shows a sequence alignment between CheA regulatory domains and CheWs. Based on this alignment, Fig. 11 shows a mapping of conserved residues onto the X-ray crystal structure of the regulatory domain from *T.*

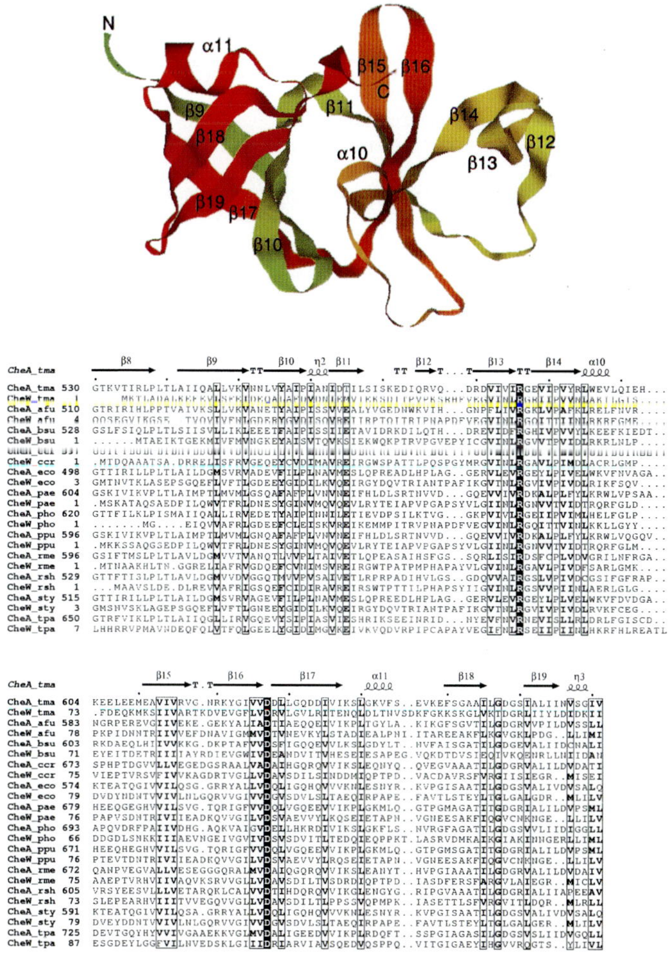

FIGURE 10 Alignment of CheA and CheW from 11 organisms where both sequences are available using ClustalW [185]. The organisms represented are *T. maritima, A. fulgidus, B. subtilis, C. crescentus, E. coli, P. aeruginosa, P. horikoshii, P. putida, R. meliloti, R. sphaeroides, S. typhimurium,* and *T. pallidum.* The secondary structure is from the *T. maritima* CheA crystal structure [21]. Shading according to 80% equivalent physicochemical properties was done using ESPript 1.9. The ribbon diagram was produced with RasMac.

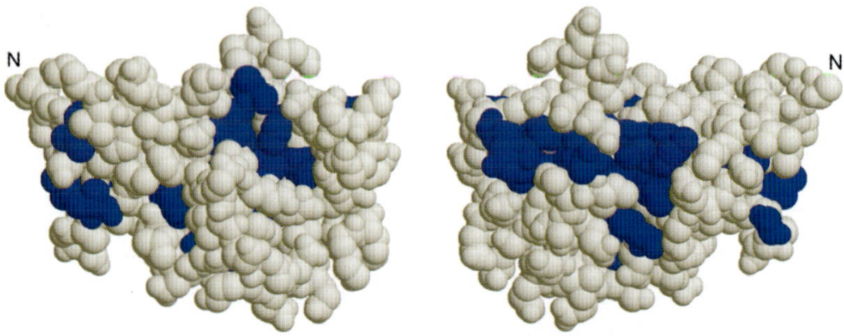

FIGURE 11 The positions of equivalent residues from Fig. 10 are shown in blue mapped onto the regulator domain of the X-ray crystal structure of the *T. maritima* CheA regulator domain [21]. The left view is in the same orientation as the ribbon diagram in Fig. 10; the right view is rotated by 180° about the vertical axis. Figure produced with RasMac.

maritima CheA. The region of conservation near the N terminus is, for the CheA regulatory domain, located close to the dimerization domain and hence may be a site of interaction for both the CheA regulatory domain and for CheW with the kinase core of CheA. The regions of conservation at the opposite end and along the side may be sites of interaction between CheA/CheW and MCPs.

The CheA Activator, CheW

CheW is composed of two SH3-like subdomains and has the same overall fold as the CheA regulatory domain [21, 214]. CheW serves to activate CheA in association with MCPs, and CheW is necessary for CheA activation, but not CheA inhibition [134]. Higher levels of CheW inhibit CheA binding in a manner consistent with the notion that CheA and CheW are competing for overlapping binding sites on MCPs [134, 217]. Direct binding between CheA and CheW has also been observed [218], but in the absence of MCPs or a suitable MCP cytoplasmic fragment, the binding of CheW to CheA does not serve to activate CheA.

Formation of CheA–CheW–MCP Signaling Complexes

CheA can be activated over 100-fold in ternary complexes with CheW and the MCPs. A detailed kinetic analysis of this process indicates that CheA activation is achieved by facilitating the formation of a tight, activated complex between the HPt domain and the ATP-bound catalytic core [219]. These results indicate that activation involves a shift in the equilibrium between active and inactive forms. Conversely, the catalytic rate constant, affinity of the catalytic core for ATP, and the affinity of the catalytic core for the HPt domain are not altered by formation of the ternary complex with the receptors [219]. Figure 12 shows the details of this kinetic model.

The colocalization of various chemotaxis proteins in the signaling complexes has been studied using yellow fluorescent protein (YFP) fusions of CheY, CheA, and CheZ, as well as anti-Tsr antibodies [131]. The effect of single knockouts of *tsr*, *tar*, and *trg* was examined, as well as a combined knockout of *tsr*, *tar*, *trg*, and *tap*. All three of the single knockouts had some co-localization of CheA, CheY, and CheZ with the receptor patch, although this colocalization in both *tsr* and *tar* mutants was less than in wild-type cells. The *trg* knockout had localizations closer to wild type, whereas the quadruple knockout had no localization, as expected.

MCP dimers containing a single CheA/CheW-binding region can regulate CheA [220, 221]. This suggests that CheA binds between two receptor dimers. Binding between dimers would also explain how CheA might act to facilitate receptor clustering at the cell pole [130].

Binding of attractant ligands such as serine and aspartate to MCPs causes inactivation of CheA [132, 133], such that at saturating concentrations of

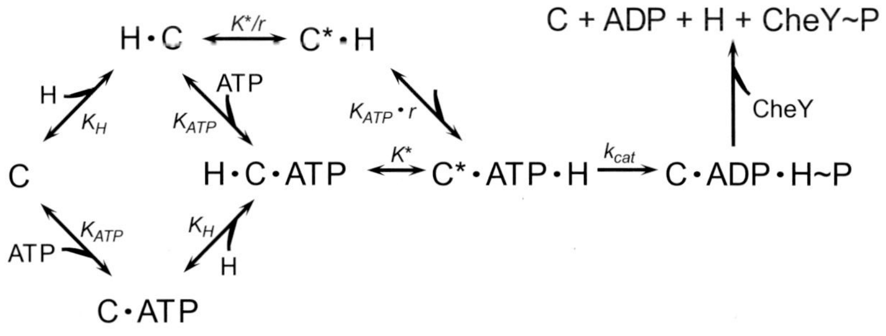

FIGURE 12 Diagram of the kinetic model for CheA autophosphorylation. H, HPt domain; C, catalytic kinase core; K_{ATP}, dissociation constant for C•ATP; K_H, dissociation constant for H•C; C*, activated form of the kinase; K*, equilibrium constant between inactive and active forms of the enzyme–substrate complex; *r*, coefficient representing the effect of activation on ATP binding; and k_{cat}, catalytic rate in the active enzyme–substrate complex. Experimental values for these kinetic constants are K_H =26 μ*M*; K_{ATP} = 370 μ*M*; r = 0.36; k_{cat} = 37 s^{-1}; and K* = 77 [219]. Figure adapted from Levit *et al.* [219].

attractant CheA activity is reduced to a very low level. When bound in a complex with MCPs, phospho-CheA rapidly releases its phosphoryl group in the presence of excess ADP to produce ATP. However, in the presence of high concentrations of aspartate, phospho-CheA in CheA–CheW–Tar complexes does not transfer its phosphoryl group to ADP [222]. With pure CheA, the fraction of monomeric CheA can be calculated based on the known dimer disassociation constant (K_d = 0.2 μ*M*) [57]. Purified CheA is only active as a dimer [20], and the receptor-inactivated form may be the inactive monomer. If pure CheA is labeled with radioactive phosphate and then cold ADP is added, the fraction retaining the label is equal to the fraction that is monomeric, and the remaining phospho-CheA disappears at a rate corresponding to the rate of monomer–dimer exchange [20]. The mechanism of receptor-mediated CheA inactivation could therefore involve dimer dissociation. There are numerous other possibilities, however. For example, MCPs could interact with the HPt domain to prevent its phosphorylation [223] or impose a conformational change on the catalytic domain to inhibit its function. It may not be necessary to inhibit CheA activity to below the that of the isolated dimer, as there is a substantial pool of free CheA in the cytoplasm [131] that should give a constant background activity. The critical step in regulating CheA is the greater than 100-fold activation achieved by formation of the CheA–CheW–MCP complexes in the absence of attractants.

The stoichiometry of CheA, CheW, and MCPs in active signaling complexes is important for further considerations of the mechanism of signal transduction. To date, there has been only one report quantifying CheA and CheW binding to a MCP (Tsr) in membranes [124]. Results indicated that the binding of CheW and CheA to Tsr approached a saturating stoichiometry of one CheW and one CheA subunit per Tsr subunit. Because MCPs and CheA are both dimeric, it has generally been assumed that the essential receptor signaling unit is a 2:2:2 complex, with CheW acting as a bridge to mediate the MCP–CheA interaction. Before the discovery of MCP clusters, it was assumed that these 2:2:2 complexes were distributed randomly in the cytoplasmic membrane where they functioned as independent signaling units. Signal integration was thought to occur through the sum of effects of the phospho-CheY produced by ~5000 independent CheA dimers in a one-to-one association with ~5000 independent MCP dimers with CheW sandwiched in between.

Support for a more complex view of interactions within the receptor clusters has come from investigations of soluble receptor signaling complexes formed between LZ-Tar_C, CheW, and CheA [125, 186]. These complexes are very large (radius of gyration, 20 nm; molecular mass, 1,400,000) stable structures [126]. Their composition of 28 receptor: 6 CheW: 4 CheA subunits does not reflect the expected 1:1:1 stoichiometry of MCP:CheW:CheA expected from previous studies of Tsr:CheW:CheA complexes in membranes [126]. More recently, the stoichiometry of complexes formed among CheA, CheW, and Tsr in membranes has been examined [217]. Results from this work show that CheW is not required for binding of CheA to receptors. Instead, CheA and CheW compete for binding, as expected from the similarity between CheW and the receptor-binding regulatory domain of CheA. Furthermore, the maximum ratio of CheA:MCP appears to be only about 1:6, whereas the maximum CheW:MCP ratio is 1:1. In contrast to the model with 2:2:2 complexes, these results suggest that only a minority of MCPs are interacting directly with CheA at a given time and provide further evidence for the importance of lateral interactions within the MCP cluster.

The Mechanism of Transmembrane Signaling in Chemotaxis

As mentioned earlier, all five *E. coli* MCPs have a very high degree of sequence conservation in the coiled-coil region of their cytoplasmic domains. This region is conserved in virtually all motile prokaryotes. The reason for this is assumed to be related to the specific, unknown, mechanism by which MCPs regulate CheA activity. Thus, the method of signal transduction must be simple and robust enough to couple to the wide variety of sensor domains found in MCPs in these different species.

Evidence for two distinct states of MCPs comes from missense mutations that have been found along the entire length of the MCPs that lead to constitutive activation or repression of CheA [205, 224]. For example, mutation of residue 19 in TM1 of *E. coli* Tar from alanine to arginine results in smooth-swimming cells, i.e., a continuous repression of CheA activity as if an attractant ligand was bound [224]. This mutant can still respond weakly to the removal of aspartate, suggesting that the structure of the receptor is largely intact. Second-site mutations in the region of residues 263 to 301, between the linker region and MH1 (see Fig. 1), were found that restore chemotaxis to aspartate [224]. Because MH1 is far from the transmembrane region [127], the two sites of mutation are probably not interacting directly. Instead, this restoration of function suggests that these mutations are able to shift the signaling state of the receptor in an opposing manner, and this type of mutational work suggests that every region of the MCP structure can play a role in modulating the signal transduction process.

The mechanism by which CheA is regulated during transmembrane signaling does not require a membrane. For example, in *E. coli*, signal transduction between Aer and CheA appears to occur entirely within the cytoplasm [122, 123, 170]. In addition, soluble cytoplasmic MCPs are found in species such as *H. salinarum* [225] and *Rhodobacter sphaeroides* [226].

Conformational Changes in the MCP Ligand-Binding Domain

Any detailed conformational changes that are associated with the ligand-binding domains of the *E. coli* and *S. typhimurium* MCPs are not generally applicable to MCPs in other prokaryotic cell types, as MCP-sensing domains from other prokaryotes do not share substantial similarity to the periplasmic domain of Tar [4, 171]. In addition, it seems likely that the physiologically relevant structural changes associated with ligand binding may only occur in the context of a tightly packed array of MCPs. Despite this, efforts to understand the general mechanism of signal transduction in the chemotaxis system have focused on small motions of the periplasmic domain of Tar. A new analysis of four X-ray crystal structures of the periplasmic ligand-binding domain from *S. typhimurium* Tar shows a variety of inter- and intrasubunit motions on ligand binding [173]. One pair of structures has the two subunits cross-linked by a cysteine at position 36 [28], whereas the other two structures consist of the uncross-linked wild-type sequence [38]. One of the cross-linked and one of the wild-type structures had aspartate bound. These had different crystal forms depending on whether they had aspartate bound, and the crystal packing contacts were sufficiently different to account for all of the small conformational differences between the different forms [173]. In the first set of structures, a piston-like motion of TM2 is observed in the aspartate bound versus the unbound structure [173, 227], whereas in the second set there is a relative rotation of the helices [173].

Signaling by MCP Chimeras

Another approach to understanding the mechanism of signal transduction has involved the construction of MCP chimeras. The most interesting ones in terms of the mechanism of transmembrane signaling involve Aer. Despite the very different structure and localization of the sensing domains of Aer and Tsr, chimeras of the Aer-sensing domain and the Tsr cytoplasmic domain restore *E. coli* aerotaxis [228, 229]. This provides strong evidence for a common mechanism of signaling among all the *E. coli* MCPs, independent of the sensing domain structure.

Receptor chimeras have also been formed between the N-terminal domains of *E. coli* MCPs and the histidine kinase core of EnvZ [230–233]. One of these is denoted Taz1, a fusion of the sensing, transmembrane, and HAMP domains of *E. coli* Tar to the EnvZ core [230]. Taz1 responds to aspartate by causing increased levels of phospho-OmpR. This response appears to be due primarily to an attractant-induced decrease in phosphatase activity of the EnvZ core [231]. Taz1 activation requires high (millimolar) concentrations of aspartate [230], suggesting that the aspartate-binding domain of the chimera has a much lower affinity for aspartate than Tar itself. However, this effect could also involve the relative clustering of Taz1 versus Tar. The clustering of MCPs may be critical to their sensitivity (see later), however, there is no evidence as to whether Taz1 forms similar tight clusters. Thus, the high levels of aspartate required to induce a Taz1 signal may reflect a loss of signal amplification due to the absence of the lateral interactions present in the MCP cluster [234].

Analysis of the effects of aspartate binding to, and of mutations, in MCPs and Taz1 implies a common mechanism of transmembrane signaling for MCPs and EnvZ. Aspartate binding to Taz1 leads to an increased level of phospho-OmpR [230], but aspartate binding to Tar causes a reduction in phospho-CheY [132, 133]. In both cases, the HAMP domain apparently plays a significant role in transmembrane signaling. Mutations in this domain that cause a locked signaling state in Tsr also affect Taz1 (whose HAMP domain is from Tar) [230, 231, 233]. As with aspartate binding, a mutation that causes a low level of phospho-CheY in the chemotaxis system causes a high level of phospho-OmpR with Taz1, and *vice versa*. Despite these results, it is not clear whether activation of the chimera is related to the physiologically relevant mechanism by which EnvZ normally affects a shift in the balance between phosphatase and kinase activities in response to changes in osmotic pressure.

Role of the MCP Linker Region

HAMP domains may be critically involved in the regulation of coiled-coil interactions [183], and the MCP linker region may be essential for transmembrane signal transduction [220, 221]. This was shown in studies of heterodimeric MCPs, two derivatives of *E. coli* Tar with different mutations expressed in the same cell. Heterodimers composed of a full-length Tar and a truncated Tar could not regulate CheA activity in response to aspartate if the truncated subunit lacked the linker region [220, 221]. In order to restrict the response to heterodimers, Tar proteins included mutations in the sensing domain that affected two different portions of the aspartate-binding pocket. The different Tar receptors were expressed together in combinations where a heterodimer could bind aspartate, but a homodimer could not bind aspartate [220, 221].

In an additional role, the HAMP domain of Aer appears to interact with the PAS domain to bind FAD [228]. In analogy with other MCPs, the HAMP domain may play an essential role in signal transduction, and the signal transduction mechanism may involve a change in the interactions between PAS domains and HAMP domains [228, 229] or possibly a change in interaction between the PAS domains of the two subunits in a homodimer.

Sensitivity and Gain

The ability to amplify a faint signal is critical for biological systems to respond to their environment [235]. In general, the amplification of a signal has been categorized as either magnitude amplification or sensitivity amplification. Magnitude amplification (or gain) represents the production of a larger number of output molecules than stimulus molecules [235]. In contrast, sensitivity amplification occurs when the fractional change in the output is greater than the fractional change in the input [235]. In the chemotaxis system, the cell needs to efficiently use information from its environment to modulate the activity of the flagellar motor *via* the response regulator CheY. The degree of CheA inactivation in the chemotaxis system is much greater than can be explained by a stoichiometric inactivation of one CheA by one ligand-bound MCP [236]. It has been estimated that a change in occupancy of only one MCP can be enough to produce an observable chemotaxis response from the entire system [237]. This is an example of sensitivity amplification, as the percentage change in CheA inhibition is much greater than the percentage change in receptor occupancy [235]. Several mechanisms have been proposed to explain this amplification. For example, MCP clustering and cooperativity among MCPs may be important as suggested by simulations of receptor arrays in which the conformational state of a receptor influenced that of its neighbors [238, 239]. Another possible mechanism for this amplification could be a very high cooperativity of phospho-CheY binding to flagellar motors. A high cooperativity has been observed in some experiments, but not in others [240, 241].

One alternative to these cooperative models suggests that the proteins of the methylation–adaptation system are responsible for the amplification. Preferential binding by the methylating enzyme to the CheA-activating state of the MCPs and preferential binding by the demethylating enzyme, CheB, to the CheA-inactivating state may prolong the lifetimes of these conformations and amplify the signal due to ligand binding [242].

Another alternative mechanism for amplification may involve CheB functioning as a phosphatase. In the chemotaxis system, there is a balance between the flow of phosphoryl groups to CheY and CheB, which suggests a possible mechanism for zero-order ultrasensitivity in this system [235]. This flow of phosphoryl groups to CheB could either proceed directly from the CheA HPt domain to CheB or from CheY to CheB *via* a CheA phospho-HPt intermediate. Phosphotransfer from CheY to the CheA HPt domain has been demonstrated *in vitro* [136]. Whenever CheA activity is inhibited, CheB may serve as a sink for phosphoryl groups to reduce the level of phospho-CheY more rapidly. This mechanism may explain the observation that CheB is essential for high sensitivity in the *E. coli* chemotaxis system [242]. In this respect, CheB may function like CheY1 in *R. meliloti*, which acts as a CheY2 phosphatase by participating in a phosphotransfer from CheY2 *via* a phospho-CheA intermediate [243].

MCP Methylation

The two methylated helices are central to an adaptation mechanism that allows the chemotaxis system to maintain its sensitivity over a wide range of concentrations of attractant or repellant ligands. Methylation appears to change the sensitivity of MCPs to ligand-binding [202, 249], as well as affecting their propensity to activate CheA [133, 165]. The effect of methylation on the regulation of CheA activity is not entirely understood. The most obvious effect is the neutralization of negative charges, which seems likely to lead to changes in the packing of methylated helices. Deamidation or demethylation of soluble receptor constructs tends to reduce their oligimerization and reduce the formation of active signaling complexes [125, 126, 219]. In addition, it has been shown that methylation of Tsr reduces the concentration of serine needed to inhibit CheA activity [202, 249] without substantially changing the affinity of Tsr for serine [247–249]. Thus, the system cannot be modeled simply in terms of a two-state system, with MCPs in either a ligand-free CheA-activating state or a ligand-bound CheA-inactivating state.

In many *in vivo* studies, the compensatory effects of methylation may mask the effect of mutations or other perturbations to the chemotaxis system. Thus, while the methylation system is critical for chemotaxis, it may confound attempts to understand the mechanism of signaling.

Symmetry Breaking in the MCP Cluster

A further consideration of the effect of ligand binding concerns symmetry breaking. The binding of an attractant ligand across a MCP dimer causes the two halves of the dimer to become asymmetric, and negative cooperativity prevents occupation of the second ligand-binding site [28, 31, 38]. While signal transduction has generally been believed to involve receptor homodimers [35, 244], this symmetry breaking effect may change the higher order interactions between the receptor periplasmic domains of dozens of MCPs in the cluster at the cell pole. This effect may be more important than any change directly propagated through the membrane.

Thus, signal transduction seems likely to occur *via* the higher order MCP assemblies that occur at the cell pole. Ligand binding and lateral interactions in the periplasm could shift the whole MCP cluster into a state incompatible with CheA activation. As described earlier, the cytoplasmic domain of a single MCP monomer forms an antiparallel coiled-coil formed within one subunit, but the CheA-activating conformation may be a parallel coiled-coil formed between the two subunits of a dimer. The latter interaction is suggested by the fact that soluble fragments of the Tar cytoplasmic domain do not activate CheA unless fused to a leucine zipper dimerization domain that strongly promotes a parallel coiled-coil interaction [125, 186]. Thus, changes in interactions among MCP periplasmic domains may modulate a shift between CheA-activating (parallel) and CheA-inactivating (antiparallel) coiled-coil states in MCP cytoplasmic domains.

Dynamic interactions between thousands of receptors open the possibility for a much more complex information processing mechanism than has previously been anticipated. Thus, an appropriate view of the MCP cytoplasmic domain may not be as a four-helix bundle, but rather as a thousand-helix bundle involving many interactions between the cytoplasmic domains of one MCP and all its neighbors in the polar cluster. Within the cluster of MCPs, distinct islands consisting of MCPs with a higher coaffinity may form. Aggregation according to the type of receptor, as well as methylation state, and CheA or CheW binding are likely to occur. The position of a MCP within these islands and with the overall cluster may affect the response of the chemotaxis system to ligand binding at that receptor.

CONCLUSIONS

An emerging paradigm for the chemotaxis system is that the mechanism of transmembrane signaling requires a consideration of the many lateral interactions occurring within the MCP cluster. The scheme by which CheA is regulated may depend in a fundamental way on this clustering, and this must be explored thoroughly before any firm conclusions can be drawn. The association of CheA, CheW, and MCPs into an active signaling complex involves a process of regulated self-assembly that may extend to other HPK signal transduction systems. In *Caulobacter crescentus*, the polar clustering of the HPK receptor CckA may serve to activate the kinase at a specific phase of the cell cycle [243]. During the remainder of the cell cycle, CckA is dispersed and apparently inactive. Further work is needed to investigate whether, in general, HPK receptors require clustering for transmembrane signaling in a fashion similar to *E. coli* MCPs.

ACKNOWLEDGMENTS

We thank Mikhail Levit, Peter Thomason, Reem Hussein, and Sandra Da Re for their assistance and for their helpful comments on the manuscript. This work was supported by NIH Grant R01GM57773.

REFERENCES

1. Fu, D., Libson, A., Miercke, L. J., Weitzman, C., Nollert, P., Krucinski, J., and Stroud, R. M. (2000). Structure of a glycerol-conducting channel and the basis for its selectivity. *Science* **290**, 481–486.
2. Murata, K., Mitsuoka, K., Hirai, T., Walz, T., Agre, P., Heymann, J. B., Engel, A., and Fujiyoshi, Y. (2000). Structural determinants of water permeation through aquaporin-1. *Nature* **407**, 599–605.
3. Blattner, F. R., Plunkett, G., III, Bloch, C. A., Perna, N. T., Burland, V., Riley, M., Collado-Vides, J., Glasner, J. D., Rode, C. K., Mayhew, G. F., Gregor, J., Davis, N. W., Kirkpatrick, H. A., Goeden, M. A., Rose, D. J., Mau, B., and Shao, Y. (1997). The complete genome sequence of *Escherichia coli* K-12. *Science* **277**, 1453–1474.
4. Bateman, A., Birney, E., Durbin, R., Eddy, S. R., Howe, K. L., and Sonnhammer, E. L. (2000). The Pfam protein families database. *Nucleic Acids Res.* **28**, 263–266.
5. Krogh, A., Larsson, B., von Heijne, G., and Sonnhammer, E. L. (2001). Predicting transmembrane protein topology with a hidden Markov model: Application to complete genomes. *J. Mol. Biol.* **305**, 567–580.
6. Kunst, F., Ogasawara, N., Moszer, I., Albertini, A. M., Alloni, G., Azevedo, V., Bertero, M. G., Bessieres, P., Bolotin, A., Borchert, S., Borriss, R., Boursier, L., Brans, A., Braun, M., Brignell, S. C., Bron, S., Brouillet, S., Bruschi, C. V., Caldwell, B., Capuano, V., Carter, N. M., Choi, S. K., Codani, J. J., Connerton, I. F., and Danchin, A., *et al.* (1997). The complete genome sequence of the gram-positive bacterium *Bacillus subtilis*. *Nature* **390**, 249–256.
7. McFarland, N., McCarter, L., Artz, S., and Kustu, S. (1981). Nitrogen regulatory locus glnR of enteric bacteria is composed of cistrons ntrB and ntrC: Identification of their protein products. *Proc. Natl. Acad. Sci. USA* **78**, 2135–2139.
8. Miranda-Rios, J., Sanchez-Pescador, R., Urdea, M., and Covarrubias, A. A. (1987). The complete nucleotide sequence of the glnALG operon of *Escherichia coli* K12. *Nucleic Acids Res.* **15**, 2757–2770.
9. Heidelberg, J. F., Eisen, J. A., Nelson, W. C., Clayton, R. A., Gwinn, M. L., Dodson, R. J., Haft, D. H., Hickey, E. K., Peterson, J. D., Umayam, L., Gill, S. R., Nelson, K. E., Read, T. D., Tettelin, H., Richardson, D., Ermolaeva, M. D., Vamathevan, J., Bass, S., Qin, H., Dragoi, I., Sellers, P., McDonald, L., Utterback, T., Fleishmann, R. D., Nierman, W. C., and White, O. (2000). DNA sequence of both chromosomes of the cholera pathogen *Vibrio cholerae*. *Nature* **406**, 477–483.
10. Rudolph, J., and Oesterhelt, D. (1995). Chemotaxis and phototaxis require a CheA histidine kinase in the archaeon *Halobacterium salinarium*. *EMBO J.* **14**, 667–673.
11. Hoch, J. A., and Silhavy, T. J. (1995) Two-Component Signal Transduction, p. 488. ASM Press, Washington, DC.
12. Grebe, T. W., Stock, J. B. (1999). The histidine protein kinase superfamily. *Adv. Microbial. Physiol.* **41**, 139–227.
13. Koretke, K. K., Lupas, A. N., Warren, P. V., Rosenberg, M., and Brown, J. R. (2000). Evolution of two-component signal transduction. *Mol. Biol. Evol.* **17**, 1956–1970.
14. Artymiuk, P. J., Rice, D. W., Mitchell, E. M., and Willett, P. (1990). Structural resemblance between the families of bacterial signal- transduction proteins and of G proteins revealed by graph theoretical techniques. *Protein Eng.* **4**, 39–43.
15. Lukat, G. S., Lee, B. H., Mottonen, J. M., Stock, A. M., and Stock, J. B. (1991). Roles of the highly conserved aspartate and lysine residues in the response regulator of bacterial chemotaxis. *J. Biol. Chem.* **266**, 8348–8354.
16. Heldin, C. H. (1995). Dimerization of cell surface receptors in signal transduction. *Cell* **80**, 213–223.

17. Schlessinger, J. (2000). Cell signaling by receptor tyrosine kinases. *Cell* **103**, 211–225.
18. Kossiakoff, A. A. (1995). Structure of the growth hormone-receptor complex and mechanism of receptor signaling. *J. Nuclear Med.* **36**, 14S–16S.
19. Ottensmeyer, F. P., Beniac, D. R., Luo, R. Z., and Yip, C. C. (2000). Mechanism of transmembrane signaling: insulin binding and the insulin receptor. *Biochemistry* **39**, 12103–12112.
20. Surette, M. G., Levit, M., Liu, Y., Lukat, G., Ninfa, E. G., Ninfa, A., and Stock, J. B. (1996). Dimerization is required for the activity of the protein histidine kinase CheA that mediates signal transduction in bacterial chemotaxis. *J. Biol. Chem.* **271**, 939–945.
21. Bilwes, A. M., Alex, L. A., Crane, B. R., and Simon, M. I. (1999). Structure of CheA, a signal-transducing histidine kinase. *Cell* **96**, 131–141.
22. Tomomori, C., Tanaka, T., Dutta, R., Park, H., Saha, S. K., Zhu, Y., Ishima, R., Liu, D., Tong, K. I., Kurokawa, H., Qian, H., Inouye, M., and Ikura, M. (1999). Solution structure of the homodimeric core domain of *Escherichia coli* histidine kinase EnvZ. *Nature Struct. Biol.* **6**, 729–734.
23. Dutta, R., Qin, L., and Inouye, M. (1999). Histidine kinases: Diversity of domain organization. *Mol. Microbiol.* **34**, 633–640.
24. Yang, Y., and Inouye, M. (1991). Intermolecular complementation between two defective mutant signal- transducing receptors of *Escherichia coli*. *Proc. Natl. Acad. Sci. USA* **88**, 11057–11061.
25. Ninfa, E. G., Atkinson, M. R., Kamberov, E. S., and Ninfa, A. J. (1993). Mechanism of autophosphorylation of *Escherichia coli* nitrogen regulator II (NRII or NtrB): Trans-phosphorylation between subunits. *J. Bacteriol.* **175**, 7024–7032.
26. Taniguchi, T. (1995). Cytokine signaling through nonreceptor protein tyrosine kinases. *Science* **268**, 251–255.
27. Kolibaba, K. S., and Druker, B. J. (1997). Protein tyrosine kinases and cancer. *Biochim. Biophys. Acta* **1333**, F217–F248.
28. Milburn, M. V., Prive, G. G., Milligan, D. L., Scott, W. G., Yeh, J., Jancarik, J., Koshland, D. E., Jr., and Kim, S. H. (1991). Three-dimensional structures of the ligand-binding domain of the bacterial aspartate receptor with and without a ligand. *Science* **254**, 1342–1347.
29. Jeffery, C. J., and Koshland, D. E., Jr. (1993). Three-dimensional structural model of the serine receptor ligand- binding domain. *Protein Sci.* **2**, 559–566.
30. Yamamoto, K., and Imae, Y. (1993). Cloning and characterization of the *Salmonella typhimurium*-specific chemoreceptor Tcp for taxis to citrate and from phenol. *Proc. Natl. Acad. Sci. USA* **90**, 217–221.
31. Gardina, P. J., Bormans, A. F., Hawkins, M. A., Meeker, J. W., and Manson, M. D. (1997). Maltose-binding protein interacts simultaneously and asymmetrically with both subunits of the Tar chemoreceptor. *Mol. Microbiol.* **23**, 1181–1191.
32. Wang, J., Balazs, Y. S., and Thompson, L. K. (1997). Solid-state REDOR NMR distance measurements at the ligand site of a bacterial chemotaxis membrane receptor. *Biochemistry* **36**, 1699–1703.
33. Zhang, Y., Gardina, P. J., Kuebler, A. S., Kang, H. S., Christopher, J. A., and Manson, M. D. (1999). Model of maltose-binding protein/chemoreceptor complex supports intrasubunit signaling mechanism. *Proc. Natl. Acad. Sci. USA* **96**, 939–944.
34. Falke, J. J., and Koshland, D. E., Jr. (1987). Global flexibility in a sensory receptor: A site-directed cross-linking approach. *Science* **237**, 1596–1600.
35. Milligan, D. L., and Koshland, D. E., Jr. (1988). Site-directed cross-linking: Establishing the dimeric structure of the aspartate receptor of bacterial chemotaxis. *J. Biol. Chem.* **263**, 6268–6275.
36. Scott, W. G., Milligan, D. L., Milburn, M. V., Prive, G. G., Yeh, J., Koshland, D. E., Jr., and

Kim, S. H. (1993). Refined structures of the ligand-binding domain of the aspartate receptor from *Salmonella typhimurium*. *J. Mol. Biol.* **232**, 555–573.

37. Yeh, J. I., Biemann, H. P., Pandit, J., Koshland, D. E., and Kim, S. H. (1993). The three-dimensional structure of the ligand-binding domain of a wild-type bacterial chemotaxis receptor. Structural comparison to the cross-linked mutant forms and conformational changes upon ligand binding. *J. Biol. Chem.* **268**, 9787–992.
38. Yeh, J. I., Biemann, H. P., Prive, G. G., Pandit, J., Koshland, D. E., Jr., and Kim, S. H. (1996). High-resolution structures of the ligand binding domain of the wild-type bacterial aspartate receptor. *J. Mol. Biol.* **262**, 186–201.
39. Mizuno, T., Mutoh, N., Panasenko, S. M., and Imae, Y. (1986). Acquisition of maltose chemotaxis in *Salmonella typhimurium* by the introduction of the *Escherichia coli* chemosensory transducer gene. *J. Bacteriol.* **165**, 890–895.
40. Gardina, P. J., Bormans, A. F., and Manson, M. D. (1998). A mechanism for simultaneous sensing of aspartate and maltose by the Tar chemoreceptor of *Escherichia coli*. *Mol. Microbiol.* **29**, 1147–1154.
41. Mowbray, S. L., and Koshland, D. E., Jr. (1987). Additive and independent responses in a single receptor: Aspartate and maltose stimuli on the tar protein. *Cell* **50**, 171–180.
42. Gardina, P., Conway, C., Kossman, M., and Manson, M. (1992). Aspartate and maltose-binding protein interact with adjacent sites in the Tar chemotactic signal transducer of *Escherichia coli*. *J. Bacteriol.* **174**, 1528–1536.
43. Wanner, B. L. (1995). Signal transduction and cross regulation in the *Escherichia coli* phosphate regulon by PhoR, CreC, and acetyl phosphate. *In* "Two-Component Signal Transduction" (J. A. Hoch, and T. J. Silhavy, eds.), pp 203–221. ASM Press, Washington, DC.
44. Wanner, B. L. (1996). Phosphorus assimilation and control of the phosphate regulon. *In* "*Escherichia coli* and *Salmonella typhimurium*: Cellular and Molecular Biology" (F. C. Neidhardt, eds.) Vol. 1, pp 1357–1381. ASM, Washington, DC.
45. Scholten, M., and Tommassen, J. (1993). Topology of the PhoR protein of *Escherichia coli* and functional analysis of internal deletion mutants. *Mol. Microbiol.* **8**, 269–275.
46. Haldimann, A., Daniels, L. L., and Wanner, B. L. (1998). Use of new methods for construction of tightly regulated arabinose and rhamnose promoter fusions in studies of the *Escherichia coli* phosphate regulon. *J. Bacteriol.* **180**, 1277–1286.
47. Wigley, D. B., Davies, G. J., Dodson, E. J., Maxwell, A., and Dodson, G. (1991). Crystal structure of an N-terminal fragment of the DNA gyrase B protein. *Nature* **351**, 624–629.
48. Prodromou, C., Roe, S. M., O'Brien, R., Ladbury, J. E., Piper, P. W., and Pearl, L. H. (1997). Identification and structural characterization of the ATP/ADP-binding site in the Hsp90 molecular chaperone. *Cell* **90**, 65–75.
49. Tanaka, T., Saha, S. K., Tomomori, C., Ishima, R., Liu, D., Tong, K. I., Park, H., Dutta, R., Qin, L., Swindells, M. B., Yamazaki, T., Ono, A. M., Kainosho, M., Inouye, M., and Ikura, M. (1998). NMR structure of the histidine kinase domain of the *E. coli* osmosensor EnvZ. *Nature* **396**, 88–92.
50. Prodromou, C., Roe, S. M., Piper, P. W., and Pearl, L. H. (1997). A molecular clamp in the crystal structure of the N-terminal domain of the yeast Hsp90 chaperone. *Nature Struct. Biol.* **4**, 477–482.
51. Ban, C., Junop, M., and Yang, W. (1999). Transformation of MutL by ATP binding and hydrolysis: A switch in DNA mismatch repair. *Cell* **97**, 85–97.
52. Brino, L., Urzhumtsev, A., Mousli, M., Bronner, C., Mitschler, A., Oudet, P., and Moras, D. (2000). Dimerization of *Escherichia coli* DNA-gyrase B provides a structural mechanism for activating the ATPase catalytic center. *J. Biol. Chem.* **275**, 9468–9475.
53. Bilwes, A. M., Quezada, C. M., Croal, L. R., Crane, B. R., and Simon, M. I. (2001). Nucleotide binding by the histidine kinase CheA. *Nature Struct. Biol.* **8**, 353–360.

54. Berger, J. M., Gamblin, S. J., Harrison, S. C., and Wang, J. C. (1996). Structure and mechanism of DNA topoisomerase II. *Nature* **379**, 225–232.
55. Ban, C., Yang, W. (1998). Crystal structure and ATPase activity of MutL: Implications for DNA repair and mutagenesis. *Cell* **95**, 541–552.
56. Pratt, W. B. (1998). The hsp90-based chaperone system: Involvement in signal transduction from a variety of hormone and growth factor receptors. *Proc. Soc. Exp. Biol. Med.* **217**, 420–434.
57. Levit, M., Liu, Y., Surette, M., and Stock, J. (1996). Active site interference and asymmetric activation in the chemotaxis protein histidine kinase CheA. *J. Biol. Chem.* **271**, 32057–32063.
58. Qin, L., Dutta, R., Kurokawa, H., Ikura, M., and Inouye, M. (2000). A monomeric histidine kinase derived from EnvZ, an *Escherichia coli* osmosensor. *Mol. Microbiol.* **36**, 24–32.
59. Zhu, Y., Qin, L., Yoshida, T., and Inouye, M. (2000). Phosphatase activity of histidine kinase EnvZ without kinase catalytic domain. *Proc. Natl. Acad. Sci. USA* **97**, 7808–7813.
60. Williams, S. B., and Stewart, V. (1997). Nitrate- and nitrite-sensing protein NarX of *Escherichia coli* K-12: Mutational analysis of the amino-terminal tail and first transmembrane segment. *J. Bacteriol.* **179**, 721–729.
61. Ninfa, A. J., Ninfa, E. G., Lupas, A. N., Stock, A., Magasanik, B., and Stock, J. (1988). Crosstalk between bacterial chemotaxis signal transduction proteins and regulators of transcription of the Ntr regulon: Evidence that nitrogen assimilation and chemotaxis are controlled by a common phosphotransfer mechanism. *Proc. Natl. Acad. Sci. USA* **85**, 5492–5496.
62. McCleary, W. R., Stock, J. B., and Ninfa, A. J. (1993). Is acetyl phosphate a global signal in *Escherichia coli*? *J. Bacteriol.* **175**, 2793–2798.
63. Hsing, W., Russo, F. D., Bernd, K. K., and Silhavy, T. J. (1998). Mutations that alter the kinase and phosphatase activities of the two- component sensor EnvZ. *J. Bacteriol.* **180**, 4538–4546.
64. Pao, G. M., and Saier, M. H., Jr. (1997). Nonplastid eukaryotic response regulators have a monophyletic origin and evolved from their bacterial precursors in parallel with their cognate sensor kinases. *J. Mol. Evol.* **44**, 605–613.
65. Hall, M. N., and Silhavy, T. J. (1981). Genetic analysis of the ompB locus in *Escherichia coli* K-12. *J. Mol. Biol.* **151**, 1–15.
66. Forst, S., and Inouye, M. (1988). Environmentally regulated gene expression for membrane proteins in *Escherichia coli*. *Annu. Rev. Cell Biol.* **4**, 21–42.
67. Mizuno, T., Wurtzel, E. T., and Inouye, M. (1982). Osmoregulation of gene expression. II. DNA sequence of the envZ gene of the ompB operon of *Escherichia coli* and characterization of its gene product. *J. Biol. Chem.* **257**, 13692–13698.
68. Forst, S., Comeau, D., Norioka, S., and Inouye, M. (1987). Localization and membrane topology of EnvZ, a protein involved in osmoregulation of OmpF and OmpC in *Escherichia coli*. *J. Biol. Chem.* **262**, 16433–16438.
69. Roberts, D. L., Bennett, D. W., and Forst, S. A. (1994). Identification of the site of phosphorylation on the osmosensor, EnvZ, of *Escherichia coli*. *J. Biol. Chem.* **269**, 8728–8733.
70. Forst, S., Delgado, J., and Inouye, M. (1989). Phosphorylation of OmpR by the osmosensor EnvZ modulates expression of the ompF and ompC genes in *Escherichia coli*. *Proc. Natl. Acad. Sci. USA* **86**, 6052–6056.
71. Igo, M. M., Ninfa, A. J., and Silhavy, T. J. (1989). A bacterial environmental sensor that functions as a protein kinase and stimulates transcriptional activation. *Genes Dev.* **3**, 598–605.
72. Aiba, H., Nakasai, F., Mizushima, S., and Mizuno, T. (1989). Evidence for the physiological importance of the phosphotransfer between the two regulatory components, EnvZ and

OmpR, in osmoregulation in *Escherichia coli*. *J. Biol. Chem.* **264**, 14090–14094.
73. Kanamaru, K., Aiba, H., Mizushima, S., and Mizuno, T. (1989). Signal transduction and osmoregulation in *Escherichia coli*: A single amino acid change in the protein kinase, EnvZ, results in loss of its phosphorylation and dephosphorylation abilities with respect to the activator protein, OmpR. *J. Biol. Chem.* **264**, 21633–21637.
74. Tokishita, S., Yamada, H., Aiba, H., and Mizuno, T. (1990). Transmembrane signal transduction and osmoregulation in *Escherichia coli* II. The osmotic sensor, EnvZ, located in the isolated cytoplasmic membrane displays its phosphorylation and dephosphorylation abilities as to the activator protein, OmpR. *J. Biochem. (Tokyo)* **108**, 488–893.
75. Dutta, R., and Inouye, M. (1996). Reverse phosphotransfer from OmpR to EnvZ in a kinase–/phosphatase+ mutant of EnvZ (EnvZ.N347D), a bifunctional signal transducer of *Escherichia coli*. *J. Biol. Chem.* **271**, 1424–1429.
76. Hsing, W., and Silhavy, T. J. (1997). Function of conserved histidine-243 in phosphatase activity of EnvZ, the sensor for porin osmoregulation in *Escherichia coli*. *J. Bacteriol.* **179**, 3729–3735.
77. Igo, M. M., Ninfa, A. J., Stock, J. B., and Silhavy, T. J. (1989). Phosphorylation and dephosphorylation of a bacterial transcriptional activator by a transmembrane receptor. *Genes Dev.* **3**, 1725–1734.
78. Mizuno, T., and Mizushima, S. (1990). Signal transduction and gene regulation through the phosphorylation of two regulatory components: The molecular basis for the osmotic regulation of the porin genes. *Mol. Microbiol.* **4**, 1077–1082.
79. Tokishita, S., Kojima, A., Aiba, H., and Mizuno, T. (1991). Transmembrane signal transduction and osmoregulation in *Escherichia coli*: Functional importance of the periplasmic domain of the membrane-located protein kinase, EnvZ. *J. Biol. Chem.* **266**, 6780–6785.
80. Leonardo, M. R., and Forst, S. (1996). Re-examination of the role of the periplasmic domain of EnvZ in sensing of osmolarity signals in *Escherichia coli*. *Mol. Microbiol.* **22**, 405–413.
81. Forst, S., Delgado, J., Ramakrishnan, G., and Inouye, M. (1988). Regulation of ompC and ompF expression in *Escherichia coli* in the absence of envZ. *J. Bacteriol.* **170**, 5080–5085.
82. Matsubara, M., Kitaoka, S. I., Takeda, S. I., and Mizuno, T. (2000). Tuning of the porin expression under anaerobic growth conditions by his- to-Asp cross-phosphorelay through both the EnvZ-osmosensor and ArcB- anaerosensor in *Escherichia coli*. *Genes Cells* **5**, 555–569.
83. Bhriain, N. N., Dorman, C. J., and Higgins, C. F. (1989). An overlap between osmotic and anaerobic stress responses: A potential role for DNA supercoiling in the coordinate regulation of gene expression. *Mol. Microbiol.* **3**, 933–942.
84. Higgins, C. F., Dorman, C. J., Stirling, D. A., Waddell, L., Booth, I. R., May, G., and Bremer, E. (1988). A physiological role for DNA supercoiling in the osmotic regulation of gene expression in *S. typhimurium* and *E. coli*. *Cell* **52**, 569–584.
85. Dutta, R., Yoshida, T., and Inouye, M. (2000). The critical role of the conserved thr247 residue in the functioning of the osmosensor EnvZ, a histidine kinase/phosphatase, in *Escherichia coli*. *J. Biol. Chem.* **275**, 38645–38653.
86. Russo, F. D., and Silhavy, T. J. (1991). EnvZ controls the concentration of phosphorylated OmpR to mediate osmoregulation of the porin genes. *J. Mol. Biol.* **222**, 567–580.
87. Ishige, K., Nagasawa, S., Tokishita, S., and Mizuno, T. (1994). A novel device of bacterial signal transducers. *EMBO J.* **13**, 5195–5202.
88. Iuchi, S., and Lin, E. C. (1992). Purification and phosphorylation of the Arc regulatory components of *Escherichia coli*. *J. Bacteriol.* **174**, 5617–5623.
89. Iuchi, S., and Lin, E. C. (1993). Adaptation of *Escherichia coli* to redox environments by gene expression. *Mol. Microbiol.* **9**, 9–15.
90. Hall, M. N., and Silhavy, T. J. (1981). The ompB locus and the regulation of the major outer

membrane porin proteins of *Escherichia coli* K12. *J. Mol. Biol.* **146**, 23–43.

91. Inokuchi, K., Furukawa, H., Nakamura, K., and Mizushima, S. (1984). Characterization by deletion mutagenesis in vitro of the promoter region of ompF, a positively regulated gene of *Escherichia coli*. *J. Mol. Biol.* **178**, 653–668.
92. Pirrung, M. C. (1999). Histidine kinases and two-component signal transduction systems. *Chem. Biol.* **6**, R167–R175.
93. Thomason, P., and Kay, R. (2000). Eukaryotic signal transduction via histidine-aspartate phosphorelay. *J. Cell Sci.* **113**, 3141–3150.
94. Stock, A. M., Robinson, V. L., and Goudreau, P. N. (2000). Two-component signal transduction. *Annu. Rev. Biochem.* **69**, 183–215.
95. Ott, A., Oehme, F., Keller, H., and Schuster, S. C. (2000). Osmotic stress response in Dictyostelium is mediated by cAMP. *EMBO J.* **19**, 5782–5792.
96. Buck, V., Quinn, J., Soto Pino, T., Martin, H., Saldanha, J., Makino, K., Morgan, B. A., and Millar, J. B. (2001). Peroxide sensors for the fission yeast stress-activated mitogen-activated protein kinase pathway. *Mol. Biol. Cell* **12**, 407–419.
97. Freeman, J. A., Lilley, B. N., and Bassler, B. L. (2000). A genetic analysis of the functions of LuxN: A two-component hybrid sensor kinase that regulates quorum sensing in *Vibrio harveyi*. *Mol. Microbiol.* **35**, 139–149.
98. Island, M. D., Wei, B. Y., and Kadner, R. J. (1992). Structure and function of the uhp genes for the sugar phosphate transport system in *Escherichia coli* and *Salmonella typhimurium*. *J. Bacteriol.* **174**, 2754–2762.
99. Lina, G., Jarraud, S., Ji, G., Greenland, T., Pedraza, A., Etienne, J., Novick, R. P., and Vandenesch, F. (1998). Transmembrane topology and histidine protein kinase activity of AgrC, the agr signal receptor in *Staphylococcus aureus*. *Mol. Microbiol.* **28**, 655–662.
100. Piazza, F., Tortosa, P., and Dubnau, D. (1999). Mutational analysis and membrane topology of ComP, a quorum-sensing histidine kinase of *Bacillus subtilis* controlling competence development. *J. Bacteriol.* **181**, 4540–4548.
101. Ouchane, S., and Kaplan, S. (1999). Topological analysis of the membrane-localized redox-responsive sensor kinase PrrB from *Rhodobacter sphaeroides* 2.4.1. *J. Biol. Chem.* **274**, 17290–17296.
102. Ethier, J., and Boyd, J. M. (2000). Topological analysis and role of the transmembrane domain in polar targeting of PilS, a *Pseudomonas aeruginosa* sensor kinase. *Mol. Microbiol.* **38**, 891–903.
103. Lyon, G. J., Mayville, P., Muir, T. W., and Novick, R. P. (2000). Rational design of a global inhibitor of the virulence response in *Staphylococcus aureus*, based in part on localization of the site of inhibition to the receptor-histidine kinase, AgrC. *Proc. Natl. Acad. Sci. USA* **97**, 13330–13335.
104. Ji, G., Beavis, R. C., and Novick, R. P. (1995). Cell density control of staphylococcal virulence mediated by an octapeptide pheromone. *Proc. Natl. Acad. Sci. USA* **92**, 12055–12059.
105. Havarstein, L. S., Coomaraswamy, G., and Morrison, D. A. (1995). An unmodified heptadecapeptide pheromone induces competence for genetic transformation in *Streptococcus pneumoniae*. *Proc. Natl. Acad. Sci. USA* **92**, 11140–11144.
106. Havarstein, L. S., Gaustad, P., Nes, I. F., and Morrison, D. A. (1996). Identification of the streptococcal competence-pheromone receptor. *Mol. Microbiol.* **21**, 863–869.
107. Cheng, Q., Campbell, E. A., Naughton, A. M., Johnson, S., and Masure, H. R. (1997). The com locus controls genetic transformation in *Streptococcus pneumoniae*. *Mol. Microbiol.* **23**, 683–692.
108. Whatmore, A. M., Barcus, V. A., and Dowson, C. G. (1999). Genetic diversity of the streptococcal competence (com) gene locus. *J. Bacteriol.* **181**, 3144–3154.

109. Lefkowitz, R. J. (2000). The superfamily of heptahelical receptors Nature Cell Biol. 2, E133–E136.
110. Dixon R. A., Kobilka, B. K., Strader, D. J., Benovic, J. L., Dohlman, H. G., Frielle, T., Bolanowski, M. A., Bennett, C. D., Rands, E., and Diehl, R. E., *et al.* (1986). Cloning of the gene and cDNA for mammalian beta-adrenergic receptor and homology with rhodopsin. *Nature* **321**, 75–79.
111. Ji, T. H., and Grossmann, M., Ji, I. (1998). G protein-coupled receptors. I. Diversity of receptor-ligand interactions. *J. Biol. Chem.* **273**, 17299–17302.
112. Nagatomo, T., and Koike, K. (2000). Recent advances in structure, binding sites with ligands and pharmacological function of beta-adrenoceptors obtained by molecular biology and molecular modeling. *Life Sci.* **66**, 2419–2426.
113. Palczewski, K., Kumasaka, T., Hori, T., Behnke, C. A., Motoshima, H., Fox, B. A., Le Trong, I., Teller, D. C., Okada, T., Stenkamp, R. E., Yamamoto, M., and Miyano, M. (2000). Crystal structure of rhodopsin: A G protein-coupled receptor. *Science* **289**, 739–745.
114. Gether, U., and Kobilka, B. K. (1998). G protein-coupled receptors. II. Mechanism of agonist activation. *J. Biol. Chem.* **273**, 17979–17982.
115. Kadner, R. J. (1995). Expression of the Uhp Sugar-Phospphate Transport System of *Escherichia coli. In* "Two-Component Signal Transduction" (J. A. Hoch, T. J. Silhavy eds.), pp. 263–274. ASM Press, Washington, DC.
116. von Heijne, G. (1992). Membrane protein structure prediction: Hydrophobicity analysis and the positive-inside rule. *J. Mol. Biol.* **225**, 487–494.
117. Sonnhammer, E. L., von Heijne, G., and Krogh A. (1998). A hidden Markov model for predicting transmembrane helices in protein sequences. *Proc. Int. Conf. Intell. Syst. Mol. Biol.* **6**, 175–182.
118. Lloyd A. D., and Kadner, R. J. (1990). Topology of the *Escherichia coli* uhpT sugar-phosphate transporter analyzed by using TnphoA fusions. *J. Bacteriol.* **172** 1688–1693.
119. Island, M. D., and Kadner, R. J. (1993). Interplay between the membrane-associated UhpB and UhpC regulatory proteins. *J. Bacteriol.* **175** 5028–5034.
120. Mowbray, S. L., and Sandgren, M. O. J. (1998). Chemotaxis Receptors: A progress report on structure and function. *J. Struct. Biol.* **124**, 257–275.
121. Stock J., and Da Re, S. (1999). A receptor scaffold mediates stimulus-response coupling in bacterial chemotaxis. *Cell Calcium*, **26**, 157–164.
122. Bibikov, S. I., Biran, R., Rudd, K. E., Parkinson, J. S. (1997). A signal transducer for aerotaxis in *Escherichia coli. J Bacteriol.* **179**, 4075–4079.
123. Rebbapragada, A., Johnson, M. S., Harding, G. P., Zuccarelli, A. J., Fletcher, H. M., Zhulin, I. B., and Taylor, B. L. (1997). The Aer protein and the serine chemoreceptor Tsr independently sense intracellular energy levels and transduce oxygen, redox, and energy signals for *Escherichia coli* behavior. *Proc. Natl. Acad. Sci. USA* **94**, 10541–10546.
124. Gegner, J. A., Graham, D. R., Roth, A. F., and Dahlquist, F. W. (1992). Assembly of an MCP receptor, CheW, and kinase CheA complex in the bacterial chemotaxis signal transduction pathway. *Cell* **70**, 975–982.
125. Surette, M. G., Stock, J. B. (1996). Role of alpha-helical coiled-coil interactions in receptor dimerization, signaling, and adaptation during bacterial chemotaxis. *J. Biol. Chem.* **271**, 17966–17973.
126. Liu, Y., Levit, M., Lurz, R., Surette, M., and Stock, J. (1997). Receptor-mediated protein kinase activation and the mechanism of transmembrane signaling in bacterial chemotaxis. *EMBO J.* **16**, 7231–7240.
127. Kim, K. K., Yokota, H., and Kim, S. H. (1999). Four-helical-bundle structure of the cytoplasmic domain of a serine chemotaxis receptor. *Nature* **400**, 787–792.
128. Maddock, J. R., and Shapiro, L. (1993). Polar location of the chemoreceptor complex in the

Escherichia coli cell. *Science* **259**, 1717–1723.
129. Lybarger, S. R., and Maddock, J. R. (1999). Clustering of the chemoreceptor complex in *Escherichia coli* is independent of the methyltransferase CheR and the methylesterase CheB. *J. Bacteriol.* **181**, 5527–5529.
130. Skidmore, J. M., Ellefson, D. D., McNamara, B. P., Couto, M. M., Wolfe, A. J., and Maddock, J. R. (2000). Polar clustering of the chemoreceptor complex in *Escherichia coli* occurs in the absence of complete CheA function. *J. Bacteriol.* **182**, 967–973.
131. Sourjik, V., and Berg, H. C. (2000). Localization of components of the chemotaxis machinery of *Escherichia coli* using fluorescent protein fusions. *Mol. Microbiol.* **37**, 740–751.
132. Borkovich, K. A., Kaplan, N., Hess, J. F., and Simon, M. I. (1989). Transmembrane signal transduction in bacterial chemotaxis involves ligand-dependent activation of phosphate group transfer. *Proc. Natl. Acad. Sci. USA* **86**, 1208–1212.
133. Ninfa, E. G., Stock, A., Mowbray, S., and Stock, J. (1991). Reconstitution of the bacterial chemotaxis signal transduction system from purified components. *J. Biol. Chem.* **266**, 9764–9770.
134. Ames, P., and Parkinson, J. S. (1994). Constitutively signaling fragments of Tsr, the *Escherichia coli* serine chemoreceptor. *J. Bacteriol.* **176**, 6340–6348.
135. Hess, J. F., Bourret, R. B., and Simon, M. I. (1988). Histidine phosphorylation and phosphoryl group transfer in bacterial chemotaxis. *Nature* **336**, 139–143.
136. Mourey, L., Da Re, S., Pédelacq, J.-D., Tolstykh, T., Faurie, C., Guillet, V., Stock, J. B., and Samama, J.-P. (2001). Crystal structure of the CheA histidine phosphotransfer domain that mediates response regulator phosphorylation in bacterial chemotaxis. *J. Biol. Chem.* **276**, 81074–81082.
137. Hess, J. F., Oosawa, K., Kaplan, N., and Simon, M. I. (1988). Phosphorylation of three proteins in the signaling pathway of bacterial chemotaxis. *Cell* **53**, 79–87.
138. Wylie, D., Stock, A., Wong, C. Y., and Stock, J. (1988). Sensory transduction in bacterial chemotaxis involves phosphotransfer between Che proteins. *Biochem. Biophys. Res. Commun.* **151**, 891–896.
139. Lupas, A., and Stock, J. (1989). Phosphorylation of an N-terminal regulatory domain activates the CheB methylesterase in bacterial chemotaxis. *J. Biol. Chem.* **264**, 17337–17342.
140. Li, J., Swanson, R. V., Simon, M. I., and Weis, R. M. (1995). The response regulators CheB and CheY exhibit competitive binding to the kinase CheA. *Biochemistry* **34**, 14626–14636.
141. Welch, M., Oosawa, K., Aizawa, S. I., and Eisenbach, M. (1994). Effects of phosphorylation, Mg^{2+}, and conformation of the chemotaxis protein CheY on its binding to the flagellar switch protein FliM. *Biochemistry* **33**, 10470–10476.
142. Stock, A. M., Wylie, D. C., Mottonen, J. M., Lupas, A. N., Ninfa, E. G., Ninfa, A. J., Schutt, C. E., and Stock, J. B. (1988). Phosphoproteins involved in bacterial signal transduction. *Cold Spring Harb. Symp. Quant. Biol.* **53 Pt 1**, 49–57.
143. Hess, J. F., Bourret, R. B., Oosawa, K., Matsumura, P., and Simon, M. I. (1988). Protein phosphorylation and bacterial chemotaxis. *Cold Spring Harb. Symp. Quant. Biol.* **53 Pt 1**, 41–48.
144. Silversmith, R. E., Appleby, J. L., and Bourret, R. B. (1997). Catalytic mechanism of phosphorylation and dephosphorylation of CheY: Kinetic characterization of imidazole phosphates as phosphodonors and the role of acid catalysis. *Biochemistry* **36**, 14965–14974.
145. Falke, J. J., and Kim, S. H. (2000). Structure of a conserved receptor domain that regulates kinase activity: the cytoplasmic domain of bacterial taxis receptors. *Curr. Opin. Struct. Biol.* **10**, 462–469.
146. Armitage, J. P. (1999). Bacterial tactic responses. *Adv. Microbial. Physiol.* **41**, 229–289.
147. Stock, J., and Levit, M. (2000). Signal transduction: Hair brains in bacterial chemotaxis. *Curr. Biol.* **10**, R11–R14.
148. Berg, H. C. (2000). Motil Behavior of Bacteria. *Phys. Today* **53**, 24–29.

149. Clarke, S., and Koshland, D. E., Jr. (1979). Membrane receptors for aspartate and serine in bacterial chemotaxis. *J. Biol. Chem.* **254**, 9695–9702.
150. Hazelbauer, G. L., and Engstrom, P. (1981). Multiple forms of methyl-accepting chemotaxis proteins distinguished by a factor in addition to multiple methylation. *J. Bacteriol.* **145**, 35–42.
151. Manson, M. D., Blank, V., Brade, G., and Higgins, C. F. (1986). Peptide chemotaxis in E. coli involves the Tap signal transducer and the dipeptide permease. *Nature* **321**, 253–256.
152. Beel, B. D., and Hazelbauer, G. L. (2001). Substitutions in the periplasmic domain of low-abundance chemoreceptor trg that induce or reduce transmembrane signaling: Kinase activation and context effects. *J. Bacteriol.* **183**, 671–679.
153. Hazelbauer, G. L., Park, C., and Nowlin, D. M. (1989). Adaptational crosstalk and the crucial role of methylation in chemotactic migration by *Escherichia coli*. *Proc. Natl. Acad. Sci. USA* **86**, 1448–1452.
154. Weerasuriya, S., Schneider, B. M., and Manson, M. D. (1998). Chimeric chemoreceptors in *Escherichia coli*: signaling properties of Tar-Tap and Tap-Tar hybrids. *J. Bacteriol.* **180**, 914–920.
155. Feng, X., Lilly, A. A., and Hazelbauer, G. L. (1999). Enhanced function conferred on low-abundance chemoreceptor Trg by a methyltransferase-docking site. *J. Bacteriol.* **181**, 3164–3171.
156. Lybarger, S. R., and Maddock, J. R. (2000). Differences in the polar clustering of the high- and low-abundance chemoreceptors of *Escherichia coli*. *Proc. Natl. Acad. Sci. USA* **97**, 8057–8062.
157. Barnakov, A. N., Barnakova, L. A., and Hazelbauer, G. L. (1999). Efficient adaptational demethylation of chemoreceptors requires the same enzyme-docking site as efficient methylation. *Proc. Natl. Acad. Sci. USA* **96**, 10667–10672.
158. Wu, J., Li, J., Li, G., Long, D. G., and Weis, R. M. (1996). The receptor binding site for the methyltransferase of bacterial chemotaxis is distinct from the sites of methylation. *Biochemistry* **35**, 4984–4993.
159. Djordjevic, S., and Stock, A. M. (1998). Chemotaxis receptor recognition by protein methyltransferase CheR. *Nature Struct. Biol.* **5**, 446–450.
160. Russo, A. F., and Koshland, D. E., Jr. (1983). Separation of signal transduction and adaptation functions of the aspartate receptor in bacterial sensing. *Science* **220**, 1016–1020.
161. Krikos, A., Mutoh, N., Boyd, A., and Simon, M. I. (1983). Sensory transducers of *E. coli* are composed of discrete structural and functional domains. *Cell* **33**, 615–622.
162. Bollinger, J., Park, C., Harayama, S., and Hazelbauer, G. L. (1984). Structure of the Trg protein: Homologies with and differences from other sensory transducers of *Escherichia coli*. *Proc. Natl. Acad. Sci. USA* **81**, 3287–3291.
163. Barnakov, A. N., Barnakova, L. A., and Hazelbauer, G. L. (1998). Comparison in vitro of a high- and a low-abundance chemoreceptor of *Escherichia coli*: similar kinase activation but different methyl- accepting activities. *J. Bacteriol.* **180**, 6713–6718.
164. Springer, M. S., Goy, M. F., and Adler, J. (1979). Protein methylation in behavioural control mechanisms and in signal transduction. *Nature* **280**, 279–284.
165. Borkovich, K. A., Alex, L. A., and Simon, M. I. (1992). Attenuation of sensory receptor signaling by covalent modification. *Proc. Natl. Acad. Sci. USA* **89**, 6756–6760.
166. Lupas, A., Van Dyke, M., and Stock, J. (1991). Predicting coiled coils from protein sequences. *Science* **252**, 1162–1164.
167. Le Moual, H., and Koshland, D. E., Jr. (1996). Molecular evolution of the C-terminal cytoplasmic domain of a superfamily of bacterial receptors involved in taxis. *J. Mol. Biol.* **261**, 568–585.
168. Bass, R. B., and Falke, J. J. (1999). The aspartate receptor cytoplasmic domain: In situ

chemical analysis of structure, mechanism and dynamics. *Struct. Fold. Des.* 7, 829–840.
169. Lee, G. F., Burrows, G. G., Lebert, M. R., Dutton, D. P., and Hazelbauer, G. L. (1994). Deducing the organization of a transmembrane domain by disulfide cross-linking: The bacterial chemoreceptor Trg. *J. Biol. Chem.* **269**, 29920–29927.
170. Zhulin, I. B., Taylor, B. L., and Dixon, R. (1997). PAS domain S-boxes in Archaea, Bacteria and sensors for oxygen and redox. *Trends Biochem. Sci.* **22**, 331–333.
171. Anantharaman, V., and Aravind, L. (2000). Cache: A signaling domain common to animal Ca(2+)-channel subunits and a class of prokaryotic chemotaxis receptors. *Trends Biochem. Sci.* **25**, 535–537.
172. Wolff, C., and Parkinson, J. S. (1988). Aspartate taxis mutants of the *Escherichia coli* Tar chemoreceptor. *J. Bacteriol.* **170**, 4509–4515.
173. Björkman, A. M., Dunten, P., Sandgren, M. O., Dwarakanath, V. N., and Mowbray, S. L. (2001). Mutations that affect ligand binding to the *E. coli* aspartate receptor: Implications for transmembrane signaling. *J. Biol. Chem.* **276**, 2808–2815.
174. Lee, L., Mizuno, T., and Imae, Y. (1988). Thermosensing properties of *Escherichia coli* tsr mutants defective in serine chemoreception. *J. Bacteriol.* **170**, 4769–4774.
175. Pakula, A. A., and Simon, M. I. (1992). Determination of transmembrane protein structure by disulfide cross- linking: the *Escherichia coli* Tar receptor. *Proc. Natl. Acad. Sci. USA* **89**, 4144–4148.
176. Lee, G. F., Dutton, D. P., and Hazelbauer, G. L. (1995). Identification of functionally important helical faces in transmembrane segments by scanning mutagenesis. *Proc. Natl. Acad. Sci. USA* **92**, 5416–5420.
177. Chervitz, S. A., Lin, C. M., and Falke, J. J. (1995). Transmembrane signaling by the aspartate receptor: Engineered disulfides reveal static regions of the subunit interface. *Biochemistry* **34**, 9722–9733.
178. Eisenberg, D., Weiss, R. M., and Terwilliger, T. C. (1984). The hydrophobic moment detects periodicity in protein hydrophobicity. *Proc. Natl. Acad. Sci. U.S.A.* **81**, 140–144.
179. Jeffery, C. J., and Koshland, D. E., Jr. (1994). A single hydrophobic to hydrophobic substitution in the transmembrane domain impairs aspartate receptor function. *Biochemistry*,**33**, 3457–3463.
180. Chervitz, S. A., and Falke, J. J. (1995). Lock on/off disulfides identify the transmembrane signaling helix of the aspartate receptor. *J. Biol. Chem.* **270**, 24043–24053.
181. Jeffery, C. J., and Koshland, D. E., Jr. (1999). The *Escherichia coli* aspartate receptor: Sequence specificity of a transmembrane helix studied by hydrophobic-biased random mutagenesis. *Protein Eng.* **12**, 863–872.
182. Williams, S. B., and Stewart, V. (1999). Functional similarities among two-component sensors and methyl- accepting chemotaxis proteins suggest a role for linker region amphipathic helices in transmembrane signal transduction. *Mol. Microbiol.* **33**, 1093–1102.
183. Aravind, L., and Ponting, C. P. (1999). The cytoplasmic helical linker domain of receptor histidine kinase and methyl-accepting proteins is common to many prokaryotic signalling proteins. *FEMS Microbiol. Lett.* **176**, 111–116.
184. Butler, S. L., and Falke, J. J. (1998). Cysteine and disulfide scanning reveals two amphiphilic helices in the linker region of the aspartate chemoreceptor. *Biochemistry* **37**, 10746–10756.
185. Thompson, J. D., Higgins, D. G., and Gibson, T. J. (1994). CLUSTAL W: Improving the sensitivity of progressive multiple sequence alignment through sequence weighting, position-specific gap penalties and weight matrix choice. *Nucleic Acids Res.* **22**, 4673–4680.
186. Cochran, A. G., and Kim, P. S. (1996). Imitation of *Escherichia coli* aspartate receptor signaling in engineered dimers of the cytoplasmic domain. *Science* **271**, 1113–1116.
187. Murre, C., and Baltimore, D. (1992). The helix loop helix motif: Structure and function.

In "Transcriptional Regulation" (S. L. McKnight, and K. R. Yamamoto, eds.), Vol. 1, pp. 961–979. Cold Spring Harbor Laboratory Press, Plainview, NY.
188. Ferre-D'Amare, A. R., Prendergast, G. C., Ziff, E. B., and Burley, S. K. (1993). Recognition by Max of its cognate DNA through a dimeric b/HLH/Z domain. *Nature* **363**, 38–45.
189. Long, D. G., and Weis, R. M. (1992). Oligomerization of the cytoplasmic fragment from the aspartate receptor of *Escherichia coli*. *Biochemistry* **31**, 9904–9911.
190. Mowbray, S. L., Foster, D. L., and Koshland, D. E., Jr. (1985). Proteolytic fragments identified with domains of the aspartate chemoreceptor. *J. Biol. Chem.* **260**, 11711–11718.
191. Seeley S. K., Weis, R. M., and Thompson, L. K. (1996). The cytoplasmic fragment of the aspartate receptor displays globally dynamic behavior. *Biochemistry* **35**, 5199–5206.
192. Rudolph, J., Nordmann, B., Storch, K. F., Gruenberg, H., Rodewald, K., and Oesterhelt, D. (1996). A family of halobacterial transducer proteins. *FEMS Microbiol. Lett.* **139**, 161–168.
193. Kehry, M. R., and Dahlquist, F. W. (1982). The methyl-accepting chemotaxis proteins of *Escherichia coli*. Identification of the multiple methylation sites on methyl-accepting chemotaxis protein I. *J. Biol. Chem.* **257**, 10378–10386.
194. Terwilliger, T. C., and Koshland, D. E., Jr. (1984). Sites of methyl esterification and deamination on the aspartate receptor involved in chemotaxis. *J. Biol. Chem.* **259**, 7719–7725.
195. Nowlin, D. M., Bollinger, J., and Hazelbauer, G. L. (1987). Sites of covalent modification in Trg, a sensory transducer of *Escherichia coli*. *J. Biol. Chem.* **262**, 6039–6045.
196. Rice, M. S., and Dahlquist, F. W. (1991) Sites of deamidation and methylation in Tsr, a bacterial chemotaxis sensory transducer. *J. Biol. Chem.* **266**, 9746–9753.
197. Kehry, M. R., and Dahlquist, F. W. (1982). Adaptation in bacterial chemotaxis: CheB-dependent modification permits additional methylations of sensory transducer proteins. *Cell* **29**, 761–772.
198. Kehry, M. R., Bond, M. W., Hunkapiller, M. W., and Dahlquist, F. W. (1983). Enzymatic deamidation of methyl-accepting chemotaxis proteins in *Escherichia coli* catalyzed by the cheB gene product. *Proc. Natl. Acad. Sci. USA* **80**, 3599–3603.
199. Terwilliger, T. C., Bogonez, E., Wang, E. A., and Koshland, D. E., Jr. (1983). Sites of methyl esterification on the aspartate receptor involved in bacterial chemotaxis. *J. Biol. Chem.* **258**, 9608–9611.
200. Kehry, M. R., Doak, T. G., and Dahlquist, F. W. (1985). Sensory adaptation in bacterial chemotaxis: regulation of demethylation. *J. Bacteriol.* **163**, 983–990.
201. Bornhorst, J. A., and Falke, J. J. (2000). Attractant regulation of the aspartate receptor-kinase complex: Limited cooperative interactions between receptors and effects of the receptor modification state. *Biochemistry* **39**, 9486–9493.
202. Li, G., and Weis, R. M. (2000). Covalent modification regulates ligand binding to receptor complexes in the chemosensory system of *Escherichia coli*. *Cell* **100**, 357–365.
203. Liu, J. D., and Parkinson, J. S. (1991). Genetic evidence for interaction between the CheW and Tsr proteins during chemoreceptor signaling by *Escherichia coli*. *J. Bacteriol.* **173**, 4941–4951.
204. Ames, P., Yu, Y. A., and Parkinson, J. S. (1996). Methylation segments are not required for chemotactic signalling by cytoplasmic fragments of Tsr, the methyl-accepting serine chemoreceptor of *Escherichia coli*. *Mol. Microbiol.* **19**, 737–746.
205. Ames, P., and Parkinson, J. S. (1988). Transmembrane signaling by bacterial chemoreceptors: *E. coli* transducers with locked signal output. *Cell* **55**, 817–826.
206. Bass, R. B., Coleman, M. D., and Falke, J. J. (1999). Signaling domain of the aspartate receptor is a helical hairpin with a localized kinase docking surface: Cysteine and disulfide scanning studies. *Biochemistry* **38**, 9317–9327.
207. Morrison, T. B., and Parkinson, J. S. (1994). Liberation of an interaction domain from the phosphotransfer region of CheA, a signaling kinase of *Escherichia coli*. *Proc. Natl. Acad. Sci.*

USA **91**, 5485–5489.
208. Zhou, H., McEvoy, M. M., Lowry, D. F., Swanson, R. V., Simon, M. I., and Dahlquist, F. W. (1996). Phosphotransfer and CheY-binding domains of the histidine autokinase CheA are joined by a flexible linker. *Biochemistry* **35**, 433–443.
209. Welch, M., Chinardet, N., Mourey, L., Birck, C., and Samama, J. P. (1998). Structure of the CheY-binding domain of histidine kinase CheA in complex with CheY. *Nature Struct. Biol.* **5**, 25–29.
210. McEvoy, M. M., Hausrath, A. C., Randolph, G. B., Remington, S. J., and Dahlquist, F. W. (1998). Two binding modes reveal flexibility in kinase/response regulator interactions in the bacterial chemotaxis pathway. *Proc. Natl. Acad. Sci. USA* **95**, 7333–7338.
211. Swanson, R. V., Schuster, S. C., and Simon, M. I. (1993). Expression of CheA fragments which define domains encoding kinase, phosphotransfer, and CheY binding activities. *Biochemistry* **32**, 7623–7629.
212. Stewart, R. C., Jahreis, K., and Parkinson, J. S. (2000). Rapid phosphotransfer to CheY from a CheA protein lacking the CheY- binding domain. *Biochemistry* **39**, 13157–13165.
213. Bourret, R. B., Davagnino, J., and Simon, M. I. (1993). The carboxy-terminal portion of the CheA kinase mediates regulation of autophosphorylation by transducer and CheW. *J. Bacteriol.* **175**, 2097–2101.
214. Griswold, I. J., Zhou, H., Matison, M., Swanson, R. V., McIntosh, L. P., Simon, M. I., and Dahlquist, F. W. (2002). The solution structure and interactions of CheW from *Thermotoga maritima*. *Nat. Struct. Biol.* **9**, 121–125.
215. Varughese, K. I., Madhusudan, Zhou, X. Z., Whiteley, J. M., and Hoch, J. A. (1998). Formation of a novel four-helix bundle and molecular recognition sites by dimerization of a response regulator phosphotransferase. *Mol. Cell* **2**, 485–493.
216. Zapf, J., Sen, U., Madhusudan, Hoch, J. A., and Varughese, K. I. (2000). A transient interaction between two phosphorelay proteins trapped in a crystal lattice reveals the mechanism of molecular recognition and phosphotransfer in signal transduction. *Struct. Fold. Des.* **8**, 851–862.
217. Levit, M. N., Grebe, T. W., and Stock, J. B. (2001). Organization of the receptor-kinase signaling array that regulates *E. coli* chemotaxis. Submitted.
218. Gegner, J. A., and Dahlquist, F. W. (1991). Signal transduction in bacteria: CheW forms a reversible complex with the protein kinase CheA. *Proc. Natl. Acad. Sci. USA* **88**, 750–754.
219. Levit, M. N., Liu, Y., and Stock, J. B. (1999). Mechanism of CheA protein kinase activation in receptor signaling complexes. *Biochemistry* **38**, 6651–6658.
220. Tatsuno, I., Homma, M., Oosawa, K., and Kawagishi, I. (1996). Signaling by the *Escherichia coli* aspartate chemoreceptor Tar with a single cytoplasmic domain per dimer. *Science* **274**, 423–425.
221. Gardina, P. J., and Manson, M. D. (1996). Attractant signaling by an aspartate chemoreceptor dimer with a single cytoplasmic domain. *Science* **274**, 425–426.
222. Borkovich, K. A., and Simon, M. I. (1990). The dynamics of protein phosphorylation in bacterial chemotaxis. *Cell* **63**, 1339–1348.
223. Morrison, T. B., and Parkinson, J. S. (1997). A fragment liberated from the *Escherichia coli* CheA kinase that blocks stimulatory, but not inhibitory, chemoreceptor signaling. *J. Bacteriol.* **179**, 5543–5550.
224. Oosawa, K., and Simon, M. (1986). Analysis of mutations in the transmembrane region of the aspartate chemoreceptor in *Escherichia coli*. *Proc. Natl. Acad. Sci. USA* **83**, 6930–6934.
225. Brooun, A., Zhang, W., and Alam, M. (1997). Primary structure and functional analysis of the soluble transducer protein HtrXI in the archaeon *Halobacterium salinarium*. *J. Bacteriol.* **179**, 2963–2968.
226. Harrison, D. M., Skidmore, J., Armitage, J. P., and Maddock, J. R. (1999). Localization and

environmental regulation of MCP-like proteins in *Rhodobacter sphaeroides*. *Mol. Microbiol.* **31**, 885–892.

227. Chervitz, S. A., and Falke, J. J. (1996). Molecular mechanism of transmembrane signaling by the aspartate receptor: A model. *Proc. Natl. Acad. Sci. USA* **93**, 2545–2550.
228. Bibikov, S. I., Barnes, L. A., Gitin, Y., and Parkinson, J. S. (2000). Domain organization and flavin adenine dinucleotide-binding determinants in the aerotaxis signal transducer Aer of *Escherichia coli*. *Proc. Natl. Acad. Sci. USA* **97**, 5830–5835.
229. Repik, A., Rebbapragada, A., Johnson, M. S., Haznedar, J. O., Zhulin, I. B., and Taylor, B. L. (2000). PAS domain residues involved in signal transduction by the Aer redox sensor of *Escherichia coli*. *Mol. Microbiol.* **36**, 806–816.
230. Utsumi, R., Brissette, R. E., Rampersaud, A., Forst, S. A., Oosawa, K., and Inouye, M. (1989). Activation of bacterial porin gene expression by a chimeric signal transducer in response to aspartate. *Science* **245**, 1246–1249.
231. Jin, T., and Inouye, M. (1993). Ligand binding to the receptor domain regulates the ratio of kinase to phosphatase activities of the signaling domain of the hybrid *Escherichia coli* transmembrane receptor, Taz1. *J. Mol. Biol.* **232**, 484–492.
232. Baumgartner, J. W., Kim, C., Brissette, R. E., Inouye, M., Park, C., and Hazelbauer, G. L. (1994). Transmembrane signalling by a hybrid protein: Communication from the domain of chemoreceptor Trg that recognizes sugar-binding proteins to the kinase/phosphatase domain of osmosensor EnvZ. *J. Bacteriol.* **176**, 1157–1163.
233. Jin, T., and Inouye, M. (1994). Transmembrane signaling. Mutational analysis of the cytoplasmic linker region of Taz1-1, a Tar-EnvZ chimeric receptor in *Escherichia coli*. *J. Mol. Biol.* **244**, 477–481.
234. R. Dutta, and M. Inouye, personal communication (2001).
235. Koshland, D. E., Jr., Goldbeter, A., and Stock, J. B. (1982). Amplification and adaptation in regulatory and sensory systems. *Science* **217**, 220–225.
236. Stock, J. (1999). Sensitivity, cooperativity and gain in chemotaxis signal transduction. *Trends Microbiol.* **7**, 1–4.
237. Segall, J. E., Block, S. M., and Berg, H. C. (1986). Temporal comparisons in bacterial chemotaxis. *Proc. Natl. Acad. Sci. USA* **83**, 8987–8991.
238. Bray, D., Levin, M. D., and Morton-Firth, C. J. (1998). Receptor clustering as a cellular mechanism to control sensitivity. *Nature* **393**, 85–88.
239. Duke, T. A., Bray, D. (1999). Heightened sensitivity of a lattice of membrane receptors. *Proc. Natl. Acad. Sci. USA* **96**, 10104–10108.
240. Alon, U., Camarena, L., Surette, M. G., Aguera y Arcas, B., Liu, Y., Leibler, S., and Stock, J. B. (1998). Response regulator output in bacterial chemotaxis. *EMBO J.* **17**, 4238–4248.
241. Cluzel, P., Surette, M., and Leibler, S. (2000). An ultrasensitive bacterial motor revealed by monitoring signaling proteins in single cells. *Science* **287**, 1652–1655.
242. Kim, C., Jackson, M., Lux, R., and Khan, S. (2001). Determinants of chemotactic signal amplification in *Escherichia coli*. *J. Mol. Biol.* **307**, 119–135.
243. Sourjik, V., and Schmitt, R. (1998). Phosphotransfer between CheA, CheY1, and CheY2 in the chemotaxis signal transduction chain of *Rhizobium meliloti*. *Biochemistry* **37**, 2327–2335.
244. Falke, J. J., Bass, R. B., Butler, S. L., Chervitz, S. A., and Danielson, M. A. (1997). The two-component signaling pathway of bacterial chemotaxis: A molecular view of signal transduction by receptors, kinases, and adaptation enzymes. *Annu. Rev. Cell. Dev. Biol.* **13**, 457–512.
245. Jacobs, C., Domian, I. J., Maddock, J. R., and Shapiro, L. (1999). Cell cycle-dependent polar localization of an essential bacterial histidine kinase that controls DNA replication and cell division. *Cell* **97**, 111–120.
246. Stock, J. B., Surette, M. G., Levit, M., and Park, P. (1995). Two-component signal transduc-

tion systems: Structure-function relationships and mechanisms of catalysis. *In*: Two-Component Signal Transduction (J. A. Hoch, and T. J. Silhavy, eds.), pp. 25–51. ASM Press, Washington, D.C.

247. Dunten, P., and Koshland, D. E., Jr. (1991). Tuning the responsiveness of a sensory receptor via covalent modification. *J. Biol. Chem.* **266**, 1491–1496.
248. Lin, L. N., Li, J., Brandts, J. F., and Weis, R. M. (1994). The serine receptor of bacterial chemotaxis exhibits half-side saturation for serine binding. *Biochemistry* **33**, 6564–6570.
249. Levit, M. N., and Stock, J. B. (2002). Dynamic sensitivity in a receptor-kinase signaling array. Submitted.

CHAPTER 6

Structure–Function Relationships: Chemotaxis and Ethylene Receptors

H. JOCHEN MÜLLER-DIECKMANN AND SUNG-HOU KIM

Department of Chemistry, University of California, Berkeley, Berkeley, California 94720

The survival and well-being of living organisms critically depend on their ability to adapt to changes in their surroundings. An elaborate network of environmental sensors and response regulators enable the cell to probe their milieu. In bacteria, fungi, yeasts, and plants, two distinct but related sensor proteins modulate specific phosphorelay circuits referred to as the "two-component system" (TCS): methyl-accepting chemotaxis proteins (MCPs) and the membrane bound histidine protein kinases (HPKs) [1, 2]. The basic biochemical events of two component signal transduction comprise autophosphorylation of a His protein kinase (HPK) and the subsequent transfer of the phosphoryl moiety to a response regulators (RR). For MCPs, kinase activity resides in a separate but closely related HPK. Response regulators induce changes by altering the pattern of gene expression or, in the case of chemotaxis, the swimming behavior. Over 30 distinct HPK-RR circuits have been found in *Escherichia coli* alone, regulating metabolic uptake, virulence, and osmolarity among other vital aspects. Surprisingly, this ubiquitous prokaryotic system was eventually also found in a limited number of eukaryotic organisms, such as fungi, slime molds, and plants, but not in humans [3, 4]. The first example of a eukaryotic two-component system was the ethylene receptor of *Arabidopsis thaliana* [5].

INTRODUCTION

Most sensor proteins of the bacterial two component system (TCS) are located in the cytoplasmic membrane. Their periplasmic domains are structurally unrelated, reflecting the diverse receptor function. The periplasmic and cytoplasmic domains are connected by transmembrane (TM) helices. Ligand binding of the sensor protein is expected to induce conformational changes that are transduced to the cytosol by TM helices and bias HPK activity. Methyl-accepting chemotaxis proteins (MCPs) and membrane-bound histidine protein kinases (HPKs) from prokaryotes are both functionally related. Hybrid receptors, created by fusing periplasmic and transmembrane domains of a MCP with the cytoplasmic domain of an environmental sensor, are functional [6, 7]. This strongly suggests that the same movement of membrane-spanning helices modulates HPK activity.

The eukaryotic membrane bound HKP of the ethylene receptor has no significant periplasmic domain [5]. Rather, the gaseous ligand ethylene is bound within the membrane by the TM helices of the receptor [8], a reaction that is enhanced in the presence of Cu^+ ions [9]. Ethylene binding conceivably induces conformational changes that alter the autophosphorylation activity of the cytosolic HPK [10].

In contrast to many mammalian receptors, which signal by oligomerization on ligand binding [11], chemoreceptors and eukaryotic ethylene receptors are dimeric even in the absence of their ligands, and their signaling does not depend on a monomer–dimer equilibrium [8, 12]. A dimerization of the HPK is required to allow *trans* autophosphorylation of the conserved His [13, 14].

CHEMOTAXIS AND CHEMORECEPTORS

Methyl-accepting chemotaxis proteins, or chemoreceptors, allow bacteria to detect concentration gradients of attractants, e.g., nutrients and repellents, e.g., toxins. In response to such a gradient, bacteria change their swimming behavior accordingly. In the absence of stimulants, a bacterium moves in a random walk by alternating between a swimming and a tumbling motion. The swimming motion corresponds to the counterclockwise (CCW) rotation of its flagella and the tumbling motion to a clockwise (CW) rotation (Fig. 1A). Tumbling, i.e., CW mode is triggered by phosphorylated CheY, which receives a phosphoryl group from the soluble HPK CheA. The autophosphorylation activity of CheA is controlled by the cytoplasmic domain of chemoreceptors. Increasing concentration of an attractant or

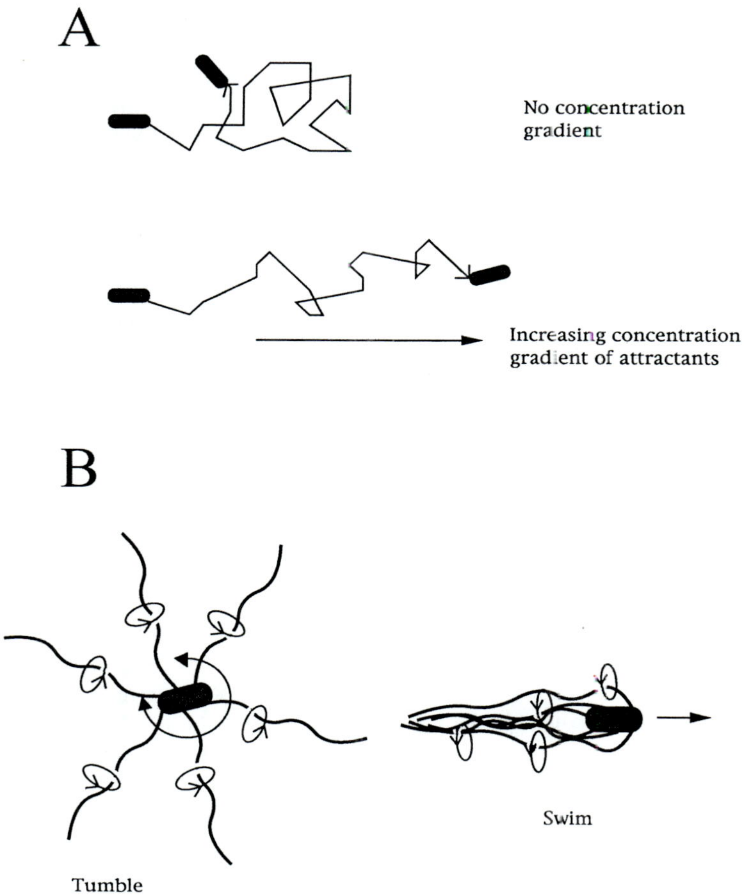

FIGURE 1 (A) Clockwise rotation of the flagella causes a bacterium to tumble and counterclockwise rotation to swim. (B) Random walk of a bacterium in the absence of stimulants and biased random walk along a gradient.

decreasing concentration of a repellant inhibits autophosphorylation activity of CheA and therefore tumbling. This enables the bacterium to swim toward nutrition and away from toxic substances (Fig. 1B).

Chemoreceptors have a simple topology, with an N-terminal sensory domain linked to a cytoplasmic-signaling domain by a membrane-spanning helix. Each subunit of a homodimeric receptor has a molecular mass of ~60 kDa and is organized in three different regions of the cell: periplasm, membrane, and cytoplasm. The periplasmic domain constitutes the ligand-binding sensory domain, which has been crystallized and forms a 36-kDa dimer of symmetric four-helix bundles [15]. The long N- and C-terminal helices of this bundle, termed α1 or TM1 and α4 or TM2, respectively, extend into the membrane, where they form a 12-kDa transmembrane domain. Disulfide-mapping studies [16, 17] and the crystal structure of the ligand-binding domain suggest that TM1 and TM2 from each monomer form a quasi four-helix bundle in the membrane. The N-terminal helices (TM1 and TM1', the prime distinguishes different subunits) lie closer to each other, stabilizing the subunit interfaces of the periplasmic and TM domains through extensive coiled coil interactions [18, 19]. The cytoplasmic domain is a four-helix bundle formed by the helix hairpins of α4, one from each subunit. TM2 and TM2', respectively, interconnect the periplasmic domain with the cytoplasmic domain, which has a molecular mass of 72 kDa. This domain can be subdivided into a linker, a methylation and the signaling domain. The major part of the cytoplasmic domain of the Ser chemotaxis receptor of *Escherichia coli* has been solved by X-ray analysis and revealed a ~200-Å long four-helix bundle [20].

The best characterized bacterial chemoreceptor is Tar, the protein that conveys chemotaxis in response to apartate in *E. coli* (Tar_E) and *Salmonella typhimurium* (Tar_S) [2]. Ligand-free Tar enhances autophosphorylation of the HPK CheA dimer and the aspartate-bound receptor inhibits CheA activity [21]. However, ligand binding also induces a feedback loop in which the methylation of specific glutamate residues resets the activity of CheA. All known bacterial chemotaxis receptors share a highly conserved cytoplasmic domain, which unites signals from different ligand domains into the same signaling pathway that alters the swimming behavior. The cytoplasmic domain of the dimeric receptor provides the structural framework for a multifunctional receptor kinase complex [2]. The wide variety of signals recognized by the periplasmic ligand-binding domain, which bind to proteins or small ligand molecules, includes the apartate receptor and aspartate, maltose-binding protein; the serine receptor and serine; the ribose and galactose receptor and ribose and galactose/glucose-binding protein, respectively; the dipeptide receptor and dipeptide-binding protein; and the citrate receptor and citrate or citrate-binding protein.

LIGAND-BINDING DOMAIN

Usually, the ligand-binding domain of the receptor is located in the periplasm, where it interacts with a small molecule or the binding protein of a small molecule. An exception to this rule was described for aerotaxis. In this receptor the FAD-containing sensory domain is placed at the N terminus, which is located in the cytoplasm [22]. Communication between the periplasm and the cytoplasm is initiated by a ligand-induced conformational change in the ligand binding domain. The crystal structures of the disulfide-linked ligand-binding domains of the aspartate receptor from *S. typhimurium* [pTar$_S$) with and without ligand gave the first insight into chemoreceptors on a molecular level [15]. This construct was engineered to lie just outside the membrane spanning parts of the receptor. The disulfide-linked dimer was chosen for this initial crystallographic study because wild-type protein yielded only poorly diffracting crystals. Disulfide-linked dimers bind aspartate with near-native affinity [23]. Evidence that the engineered disulfide is only minimally perturbing transmembrane signaling also came from methylation rate assays [24] and assays of transmembrane regulation of kinase activity [12]. Later, the apo and aspartate complexed forms of the wild-type ligand-binding domains were solved by X-ray analysis at medium resolution [25]. A comparison of the different structures confirmed that the effects of the disulfide bond are confined to the vicinity of the disulfide bond in the N-terminal helix. Peculiarly, the disulfide-linked protein might be a valid model for the membrane bound form of the ligand-binding domain as it tethers together N-terminal helices $\alpha 1$ and $\alpha 1'$ (i.e., TM1 and TM1′), which have been shown to be in close contact in the intact receptor [18, 19]. In the wild-type form of the protein, the N-terminal ends of TM1 are slightly further apart. The anticipated structural differences between apo and ligand-bound pTar$_S$ were expected to reveal the conformational changes responsible for signal transduction through the membrane. However, the observed conformational changes are rather small and are a cause for ongoing discussions on their true effects on TM-helix movement.

Each subunit of the cross-linked ligand-binding domain forms an antiparallel four-helix bundle with dimensions of ~70 Å in height and ~20 Å in diameter (Fig.2). At the dimer interface, the long N- and C-terminal helices of each monomer form a "quasi" four-helix bundle, which is less compact than the regular four-helix bundle of each individual subunit. In this "quasi" four-helix bundle, which is composed from helices of two different subunits, TM1 and TM1' are in closer contact than TM2 and TM2'. The monomers of the unligated form are related by crystallographic diads, a symmetry relation that is abolished in the ligand-bound form. The crystallized constructs are framed by the two membrane-spanning segments of TM1 (residues 7 to 30 in Tar$_S$)

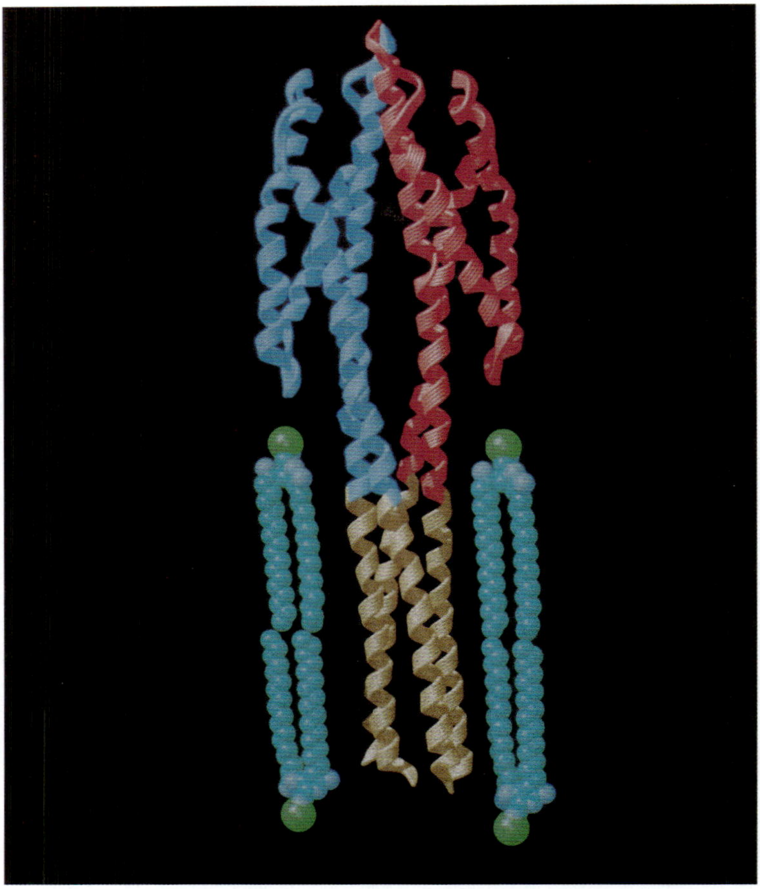

FIGURE 2 Ribbon drawing of the dimeric ligand binding domain of Tar_S. The transmembrane parts of TM1 and TM2 are modeled and depicted as thin ribbons. Lipid molecules are placed according to the beginning and ending of the putative hydrophobic region of the receptor. The model of the dimer is >120 Å long and ~40 Å in diameter.

and TM2 (residues 189 to 212 in Tar_S), respectively [26, 27]. TM1 and TM2 were predicted to form α helices throughout and when modeled and energy minimized as such, continue to form the aforementioned quasi four-helix bundle (Fig. 2), consistent with various disulfide cross-linking studies. The domain organization of $pTar_S$ is identical with the ligand-binding domain of the aspartate receptor from *E .coli* ($pTar_E$) [28].

Only one aspartate molecule was found per ligand-binding dimer, despite attempts to obtain full occupancy by soaking crystalls with a 42-fold excess of aspartate. The single aspartate is buried deep at the dimer interface, with 80% of its accessible surface buried. Monomer A contributes about 60%, monomer B about 40% to this area, and both monomers form strong salt brides with the ligand (Fig. 3). The aspartate-binding site is situated at the periplasmic end of the sensory domain and about 60 Å away from the membrane. The related second binding pocket showed some residual electron density that was interpreted as a sulfate ion because the crystallization buffer contained ammonium-sulfate as a precipitant. However, it was not possible to exclude the possibility that residual density could be the result of a disordered or partially occupied aspartate. Both crystal structures (wild type and disulfide linked) of ligand-bound $pTar_S$ showed that aspartate-binding sites become asymmetric once one of them is occupied. This was indicated by the loss of the 2-fold crystallographic symmetries, as well as by the orientation of the ligand-bound dimers in the asymmetric unit, presenting one fully occupied site and one site with little electron density. Subsequently, negative cooperativity between the two binding sites of the aspartate receptor was demonstrated [29]. Similar signal binding cooperativities have been observed for Tar_S, Tar_E and Tsr_E [30],

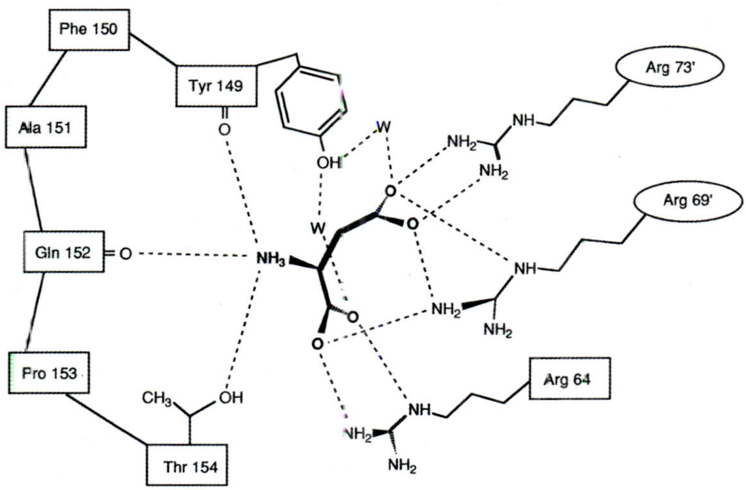

FIGURE 3 Architecture of the hydrogen-bonding network at the aspartate-binding site of Tar_S. The boxed amino acid residues are from subunit B. The two water molecules are labeled "W". Presumed hydrogen bonds are depicted as dashed lines along with the distance.

making negative cooperativity a general feature of attractant binding in chemoreceptors. Genetic studies indicate that the maltose-binding protein (MBP), the other specific signal of Tar, also docks at the extreme periplasmic end of the receptor [31]. MBP interacts with both subunits in an asymmetric fashion and close proximity to the aspartate binding site. This is consistent with a model of MBP docked computationally to the ligand binding domain [32]. In the modeled complex, one aspartate-binding site is blocked while the other one is intact. This complies with the observation that MBP binding and aspartate binding are independent and additive in effect [33].

Considering that TM2 is the only connection between the ligand binding domain and the cytoplasmic-signaling domain of the receptor, the possibilities to communicate across the bilayer are limited. The two mechanisms in discussion comprise an intrasubunit "piston-like" movement of TM2 relative to TM1 and an intersubunit rotation resulting in "scissor" or "unwinder-like" rearrangement. Biochemical data from cross-linking studies [16, 34, 35], either fusing different helix interfaces and measuring the reduced helical sliding on receptor activity or measuring the sensitivity of cystein pairs to oxidation *in vivo* and ^{19}F NMR studies [23] led to the suggestion of the first model. Another strong argument for an intrasubunit mechanism was the finding that hybrid dimers where a full-length subunit is dimerized with subunits that lack various parts or the entire cytoplasmic domain still mediate aspartate dependent receptor methylation [36, 37]. Independent of the exact nature of signal transduction, these findings suggested that signal transduction takes place within one subunit while the other subunit remains static. In other words, signal transduction was an intrasubunit mechanism.

The comparisons of several ligand-free proteins with their ligand-bound forms revealed a rotation of subunit A relative to B, i.e., an intersubunit conformational change of 4° to 8°. This was true for the disulfide-linked and wild-type receptor. It is also true for the sulfate-free crystal form of $pTar_E$. One critique of crystallographic structures has been that the apo structures contained sulfate ions, which were found to bind to the unoccupied binding pockets, acting as pseudo ligands. This was found for the receptor structures of $pTar_S$ and $pTar_E$ [28]. This pseudo ligand could conceivably interfere with or obscure the true nature of conformational changes on ligand binding. Superposition of sulfate-free and of $pTar_S$ complexed with aspartate revealed an intersubunit rotational angle of 8.3°, the largest rotation obserevd [38].

By means of a different distance analysis, the disulfide-linked apo- and aspartate-occupied ligand-binding domains can be superimposed in a way that confines the entire conformational rearrangement to the C-terminal helix (α4/TM2) of the occupied subunit. Here, helix α4 undergoes a piston-like movement, together with a 5° tilt, relative to the rigid intersubunit interface [12]. Conversely, a comparable rearrangement could not be found in the native, i.e., disulfide link-free structures. A translation of 1.6 Å toward the membrane on ligand binding is small enough in size to maintain the inter-helix registration and therefore small enough energetically to be caused by the binding of a small ligand-like aspartate (39]. It is also consistent with many of the aforementioned biochemical data. However, even though a small translation like that can be compensated by interhelix side chain flexibility, it is still expected to be large enough more than 200 Å away. A 4° to 8° rotation between subunits around a pivot axis parallel to the membrane and perpendicular to the twofold relating the subunits could easily translate into a much larger cytoplasmic movement. The structure-based observation, therefore, gave rise to the idea of a scissor-like movement of the receptor or an unwinding of the negatively wound cytosolic coiled coil. Both of the latter two mechanisms constitute intersubunit rearrangements.

Cytoplasmic Domain

The cytoplasmic domain of chemotactic receptors provides a platform that confers activity regulation of its cognate histidine kinase as well as adaptive responses to prolonged exposure to stimulants. While the activity increasing interaction between receptor and CheA requires the presence of the regulator protein CheW, the inhibiting interaction does not, suggesting a direct interaction between receptor and histidine kinase [40, 41]. One class of receptors also provides C-terminal interaction sites for CheR and CheB, enzymes of adaptational modification [2].

As mentioned earlier, the activity of the receptor itself is also controlled by a cytoplasmic feedback loop that covalently modifies four highly conserved glutamate residues. The carboxylate side chains of these glutamates are methylated by the methyltransferase CheR, which binds tightly to the C terminus of the receptor [42]. Methylated receptors display increased kinase activation as compared to its native form. Demethylation is controlled by CheB, another RR the activity of which is also regulated by CheA. As a result, the receptor controls its own methylation state and therefore tunes its own activity. This feedback adaption enables the cell to respond to gradients on top of a large, constant level of a stimulus. For example, MBP and aspartate bind to the same receptor and their effects are additive (see earlier discussion). Methylation rates also serve as a memory, where a high methylation rate indicates that attractant concentration has been high in the recent past and low methylation rates indicate the opposite. Thus, current swimming behavior can be compared with past experience and corrected accordingly.

The crystal structure of a cytoplasmic domain of the serine receptor of *E. coli* ($cTsr_E$) has been solved in which all four methyl-sensitive glutamates have been mutated to glutamine (cTsr$_E$Q) (Fig. 2) [20]. This form of the receptor corresponds to the methylated and therefore activated from of the receptor. Due to the high degree of conservation among bacterial cytoplasmic domains, this structure is expected to be essentially the same for all bacterial MCPs. The structure of cTsr$_E$Q is a dimer with a partial noncrystallographic dyad (Fig. 4). Each monomer is a 200-Å-long coiled-coil of two antiparallel helices connected by a "U turn." Two monomers form a supercoiled four-helix bundle. Most of the N-terminal helix of a monomer is in contact with two C-terminal helices: one from the same monomer and the second from the other monomer. Similarly, each C-terminal helix is in contact with two N-terminal helices. As predicted, the third and seventh residue of a heptad

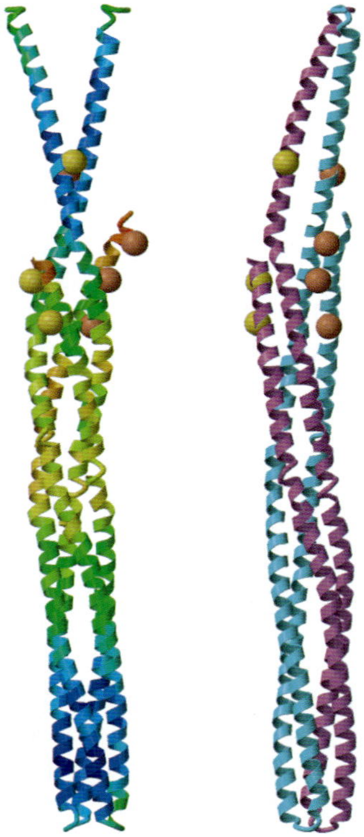

FIGURE 4 Two views of the cTsr$_E$Q dimer structure related by a 90° rotation around the noncrystallographic C_2 axis along the length of the molecules. Methylation sites are shown as yellow balls in one monomer and as orange balls in the other monomer. (Right) One monomer is shown in purple and the other is light blue. (Left) Residues with high temperature factors are shown in red and those with low temperature factors in blue.

repeat starting from residue 302 to the end of the N-terminal helix and from residue 397 to the end of the C-terminal helix are buried at the helix interface. A second, formerly undiscovered, repeat starts at residue 305 of the N-terminal helix and from residue 396 of the C-terminal helix. Such a pattern should be a common motif in other multihelix bundles because each individual helix forms interfaces with two other helices. Most residues of the interface are hydrophobic with the exception of some hydrophilic residues that form hydrogen bonds with other hydrophilic residues across the interface. Such a large hydrophobic interface is unlikely to be solvent exposed, and the observation of active hybrid MCPs with one partially truncated subunit might need further examination.

The methylation region of Tsr_E comprises glutamates 297, 304, and 311 in the N-terminal helix and glutamate 493 of the C-terminal helix, which were all mutated to glutamines in the current structure. In the crystal structure, the N-terminal part of this region is very flexible and none of the corresponding glutamine side chains are visible beyond the β carbon. Side chain conformations, based on the β carbon positions, indicate that these glutamine side chains can form hydrogen bonds with glutamine and glutamate side chains of the C-terminal helix in the other dimer. Conceivably, the existence of a clustered hydrogen-bonding network at the methylation domain affects interhelix plasticity, thereby modulating signal transduction along the helix. The difference in activity between the fully methylated and demethylated forms of the receptor is up to 50 fold [42a].

A Model of the Chemoreceptor

A model of the entire Tsr receptor was built, based on the crystallographic models of the periplasmic domain of the aspartate receptor of *E. coli* and of the cytoplasmic domain of the serine receptor of *E. coli*. Together, these parts amount to about 72% of the total number of amino acids (Fig. 5). Missing parts of the structure, like the four-helix bundle region of the linker, were modeled using the four-helix bundle architecture of $cTar_E$ as a template. This was based on the observation of a heptad repeat of hydrophobic residues similar to the coiled-coil regions of the structure. The entire model measures 380 Å in length, with an 80-Å-long periplasmic domain, a 40-Å-long transmembrane region, and a 260-Å-long cytoplasmic domain. The model consists entirely of a four-helix bundle besides a short stretch in the linker region (residues 222–241] that forms a two-helix coiled coil. The high conservation among bacterial cytoplasmic domains suggests this model to reflect the general feature of MCPs with the exception of the variable periplasmic ligand-binding domain.

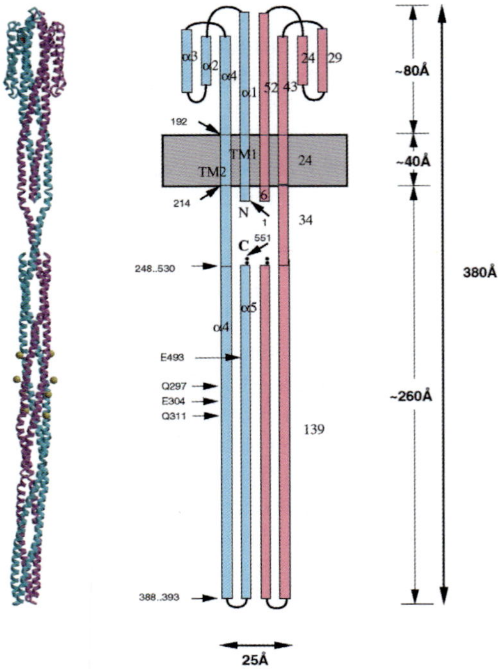

FIGURE 5 Model of an intact receptor dimer of Tsr_E. (Right) Diagram of the entire Tsr receptor dimer, with one molecule in blue and the other in pink. The presumed membrane bilayer is represented as a gray bar. Landmark residues are labeled in smaller font, and the number of residues in different helical sections are shown in larger font. The length of each domain is indicated. (Left) Ribbon representation of the intact Tsr dimer model viewed perpendicular to the crystallographic C_2 axis. The dimensions are chosen to match the scale at right. One monomer is in purple and the other in cyan. Methylation sites are marked by yellow and orange balls, respectively, for each monomer, and the ligand serine is shown as a red ball partially hidden at the upper left corner. The computer-modeled parts of the receptor are less reliable and include residue 1 to the end of TM1, TM2, and up to residue 293 and residues 521 to 551, especially the CheR-binding region at the C terminus.

The "U-turn" region of $cTsrQ_E$ is part of the signaling region, which lies at the cytosolic tip of the MCP and contains residues that are completely conserved among all chemoreceptors [43]. In the crystal, three dimers related by a crystallographic threefold axis form an interface burying 970 $Å^2$ of the accessible surface area. Because this interface involves residues that are strictly conserved, dimer trimerization of chemoreceptors may be an intrinsic property. Such clustering may reflect the observed patchy localization of receptor clusters *in vivo* [44] and oligomerization of signaling domains *in vitro* [45]. Such a higher order interaction can also explain the ability of receptors with a C-terminal-binding site for adaptational proteins such as CheR or CheB to facilitate the adaptation of receptors that lack such interaction sites [46] (Fig. 6).

THE ETHYLENE RECEPTOR

Climacteric plants use ethylene as a hormone to regulate a variety of developmental and physiological processes. Gaseous ethylene is primarily known for its role in fruit ripening, but it also controls seed germination, flower development, senescence, and adaptive responses to stress, such as heat, flooding, or

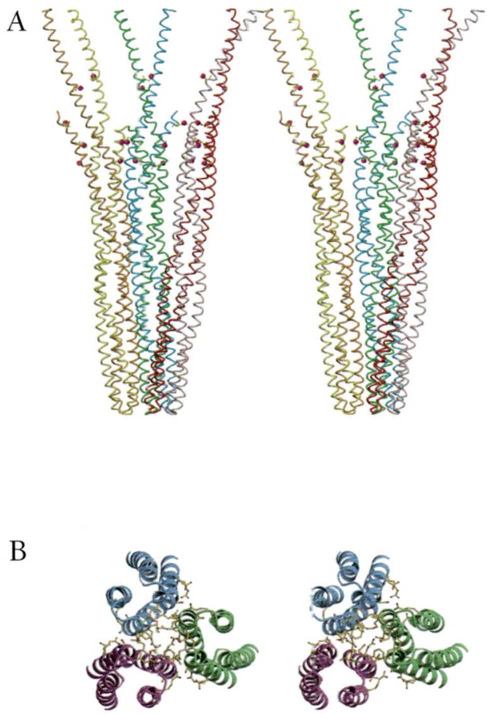

FIGURE 6 Trimer of $cTsr_E$ dimers. (A) Stereogram of a trimer of the $cTsr_EQ$ dimer in the crystal. Each monomer is colored differently. Methylation sites are shown as small balls. (B) Stereo view of the trimer interface in detail. Each dimer is shown in a different color.

pathogen attacks [47–49]. In the higher plant *Arabidopsis thaliana*, ethylene is perceived by a family of five receptors (ETR1, ETR2, ERS1, ERS2, and EIN4) [5, 50–52]. ETR1 was the first member of this receptor family identified. While the N terminus of ETR1 showed no detectable sequence similarity, the cytosolic C-terminal domain revealed similarity with other prokaryotic members of the two-component system. ETR1 was the first example of a two-component system in plants. The membrane-bound HPK domain of ETR1 contains all the sequence elements necessary for histidine kinase activity and is capable of autophosphorylation *in vivo* [53]. The C-terminal domain of ETR1 also revealed sequence similarity to classical bacterial RR. RR that do not contain an output domain whose activity is manipulated by the phosphorylation state of RR are also termed receiver domains (RD). Such a RD is fused to the C-terminal end of ETR1. Similar hybrid two-component systems are found in a number of bacterial systems [4].

All members of the family of ethylene receptors share the highest similarity among each other in their N-terminal sensor domains, reflecting their common task of binding ethylene. This domain consists of three TM helices, contains a Cu^{+}-dependent ethylene-binding site, and forms a disulfide linked dimer [8, 54]. This five-membered family can be divided into two subfamilies. ETR1, ERS1, and ETR2 contain a C-terminal RD, whereas ERS2 and EIN4 do not. Interestingly, even though ETR2 and ERS2 share many sequential features of HPKs, they lack the essential phosphoryl-accepting His at the usual His box. The structures of a prokaryotic histidine kinase and the RD of a eukaryotic TCS are both available. The histidine kinase consists of two domains, one containing the conserved histidine, the site of autophosphorylation as well as of transphosphorylation to the conserved aspartate of the RR, and a second domain containing several highly conserved regions. The second domain from the osmosensor EnvZ from *E.* coli was solved by nuclear magnetic resonance and revealed novel kinase fold [55] as well as the structure of CheA [56]. Several structures of bacterial RRs are known, and the first three-dimensional structure of the eukaryotic RD of ETR1 ($ETR1_{RD}$) displayed the expected fold homology with prokaryotic RDs [57]. $ETR1_{RD}$ forms a dimer in solution and in the crystal, and the corresponding dimer interface was predicted to be dependent on the phosphorylation state of the protein.

CHEMORECEPTORS AND MEMBRANE-BOUND HISTIDINE PROTEINS KINASES

Prokaryotic MCPs and eukaryotic ethylene receptors are both members of the two-component system. Whereas MCPs modulate the activity of a separate soluble HPK, the eukaryotic ethylene receptor is a membrane-bound HKP, as realized in numerous examples in bacteria. This section highlights some of the remarkable differences and similarities they share.

A high degree of sequence similarity and the conservation of several key motifs, as well as functional (HPK) and structural (RD) conservation, have clearly demonstrated that prokaryotic and eukaryotic TCSs are related. Additionally, MCPs and ETR1 form dimers and signal transduction is not dependent on a monomer–dimer equilibrium. Rather, binding of a ligand to a dimeric sensory domain controls the activity of the HPK.

In contrast to many other receptors, ethylene receptors have a very small periplasmic domain. Ethylene binding, as demonstrated for ETR1 *in vivo*, occurs in the membrane by TM helices in a Cu^{+}-dependent way. This makes sense, considering that ethylene is 14 times more soluble in lipids than it is in water [58]. MCPs, however, display a variety of structurally unrelated large periplasmic domains, reflecting the diversity of bound ligands. Despite the fact that more than 70% of an entire MCP is known structurally together with the conformational changes that occur on ligand binding in the periplasmic domain and mass of biochemical data, the true nature of HPK activation by its cognate MCP is still the subject of controversial discussions (see earlier discussion). This is due in part to the rather small conformational change, as well of course to the lack of three-dimensional data on an entire MCP. Activation of the histidine kinase domain of ETR1 is direct by the anticipated conformational change of the TM helices on ethylene binding. The details of such a conformational change await structural examination.

What about the downstream targets of these receptors? In prokaryotic TCSs, the HPK, membrane bound or soluble, directs the activity of a RR via phosphoryl transfer. MCPs are a special case in the sense that the output activity of its HPK are RRs, whose activity either changes the swimming behavior of the bacteria or the activity level of the receptor. Most membrane-bound bacterial HPK modulate RR that display transcriptional activity. Hence, the usual output of prokaryotic TCSs controls gene expression. This is achieved directly or via a phospho-relay cascade involving several histidine kinases and RRs. Such a cacscade increases the points of regulation. The only known downstream target of ETR1 is CTR1, which has been suggested to be a MAPK pathway regulating Ser/Thr kinase because of its sequence homology to the Raf kinase [59]. CTR1 constitutively activates the ethylene pathway in loss-of-function mutations. None of the known eukaryotic RRs resembles a transcription factor; it appears that they feed into the distinctly eukaryotic MAPK pathway [60]. However, although transcription factors (e.g., ERF1 [61]) controlled by the ethylene pathway were found, nothing is known on the intermediate steps leading to their activation. The trend seems to be that the direct output activities of eukaryotic HPKs lie further upstream from the eventual regulators of gene expression. In ETR1, a RD is fused to the C terminus of the HPK, whereas other ethylene receptors do not have a fused RD. Some bacterial systems use this modular arrangement as a competing substrate of phosphotransfer or as a relay station in a series of phosphotransfer steps. Another possibility was mentioned earlier where the possibly phophorylation dependent dimerization of the RD might control the activity of ETR1 itself or its interaction with CTR1. This scenario functionally resembles the prokaryotic MCP, where phosphorylated CheY triggers CW rotation of the flagella and phosphorylated CheB alters its sensitivity. Besides this superficial resemblance, the mechanistic details of these events are certainly quite different.

REFERENCES

1. Parkinson, J. S., and Kofoid, E. C. (1992). Communication modules in bacterial signaling proteins. *Annu. Rev. Genet.* **26**, 71–112.
2. Falke, J. J., Bass, R. B., Butler, S. L., Chervitz, S. A., and Danielson, M. A. (1997). The two-component signaling pathway of bacterial chemotaxis: A molecular view of signal transduction by receptors, kinases and adaptation enzymes. *Annu. Rev. Cell Dev. Biol.* **13**, 457–512.
3. Loomis, W. F., Shaulsky, G., and Wang, N. (1997). Histidine kinases in signal transduction pathways of eukaryotes. *J. Cell Sci.* **110**, 1141–1145.
4. Wurgler-Murphy, S. M., and Saito, H. (1997). Two-component signal transducers and MAPK cascades. *Trends Biochem. Sci.* **22**, 172–176.
5. Chang, C., Kwok, S. F., Bleeker, A. B., and Meyerowitz, E. M. (1993). *Arabidopsis thaliana* ethylene-response gene *ETR1*: Similarity of product to two-component regulators. *Science* **262**, 539–544.
6. Utsumi, R., Brissette, R. E., Rampersaud, A., Forst, S. A., Oosawa, K., and Inouye, M. (1989). Activation of bacterial porin gene expression by a chimeric signal transducer in response to aspartate. *Science* **245**, 1246–1249.
7. Baumgartner, J. W., Kim, C., Brissette, R. E., Inouye, M., Park, C., and Hazelbauer, G.L. (1994). Transmembrane signaling by a hybrid protein: Communication from the domain of chemoreceptor Trg that recognizes sugar-binding proteins to the kinase/phosphatase domain of osmosensor EnvZ. *J. Bacteriol.* **176**, 1157–1163.
8. Schaller, G. E., and Bleeker, A. B. (1995). Ethylene-binding sites created in yeast expressing *Arabidopsis* ETR1 gene. *Science* **270**, 1809–1811.
9. Rodriguez, F. I., Esch, J. J., Hall, A. E., Binder, B. M., Schaller, E., and Bleeker, A. B. (1999). *Science* **283**, 996–998.
10. Bleeker, A. B., Esch, J. J., Hall, A. E., Rodriguez, F. I., and Binder, B. M. (1998). *Phil. Trans. R. Soc. Lond.* **B 353**, 1405–1412.
11. Ullrich, A., and Schlessinger, J. (1990). Signal transduction by receptors with tyrosine kinase activity. *Cell* **61**, 203–212.
12. Cherwitz, S. A., and Falke, J. J. (1996). Molecular mechanism of transmembrane signaling by the aspartate receptor: A model. *Proc. Natl. Acad. Sci. USA* **93**, 2545–2550.
13. Yang, Y., and Inouye, M. (1991). Intermolecular complementation between two defective mutant signal-transducing receptors of *E. coli*. *Proc. Natl. Acad. Sci. USA* **88**, 11057–11061.
14. Swanson, R. V., Bourret, R. B., and Simon, M. I. (1993). Intermolecular complementation of

the kinase activity of CheA. *Mol. Microbiol.* 8, 435–441.
15. Milburn, M. V., Gilbert, G. G., Milligan, D. L., Scott, W. G., Yeh, J., Jancarik, J., Koshland D. E. Jr., and Kim, S. H. (1991). Three dimensional structure of the ligand binding domain of the bacterial aspartate receptor with and without a ligand. *Science* **254**, 1342–1347.
16. Lynch, B. A., and Koshland D. E. Jr. (1991). Disulfide cross-linking studies of the transmembrane regions of the aspartate sensory receptor of *Escherichia coli*. *Proc. Natl. Acad. Sci. USA* **88**, 10402–10406.
17. Scott, W. G., and Stoddard, B. L. (1994). Transmembrane signalling and the aspartate receptor. *Structure* **2**, 877–887.
18. Pakula, A. A., and Simon, M. I. (1992). Determination of transmembrane structure by disulfide cross-linking: The *Escherichia coli* Tar receptor. *Proc. Natl. Acad. Sci USA* **89**, 4144–4148.
19. Lee, G. F., Burrows, G. G., Lebert, M. R., Dutton, D. P., and Hazelbauer, G. L. (1994). Deducing the organization of a transmembrane domain by disulfide cross-linking: The bacterial chemoreceptor Trg. *J. Biol. Chem.* **269**, 29920–29927.
20. Kim, K. K., and Kim, S. H. (1999). Four helix-bundle structure of the cytoplasmic domain of a serine chemotaxis receptor. *Nature* **400**, 787–792.
21. Stock, J. B., and Surette, M. (1996). *In* "*Escherichia coli* and *Salmonella typhimurium*: Cellular and Molecular Biology" (F. C. Neidhardt, *et al*, eds.) pp. 1103–1129. ASM Press, Washington, DC.
22. Bibikov, S. I, Biran, R. Rudd, K. E., and Parkinson, J. S. (1997). A signal transducer for aerotaxis in *E. coli*. *J. Bacteriol.* **179**, 4075–4079.
23. Danielson, M. A., Biemann, H. P., Koshland, D. E., Jr., and Falke, J. J. (1994). Attractant- and disulfide-induced conformational changes in the ligand-binding domain of the chemotaxis aspartate receptor: A F-[19]NMR study. *Biochemistry* **33**, 6100–6109.
24. Milligan, D. L., and Koshland, D. E., Jr. (1991). Intrasubunit signal transduction by the aspartate chemoreceptor. *Science* **254**, 1651–1654.
25. Yeh, J. I, Biemann, H. P., Pandit, J., Koshland, D. E., Jr. and Kim, S. H. (1993). The three-dimensional structure of the ligand-binding domain of a wild-type bacterial chemotaxis receptor. *J. Biol. Chem.* **268**, 9787–9792.
26. Russo, A. F., and Koshland, D. E., Jr. (1983). Separation of signal transduction and adaptation functions of the aspartate receptor in bacterial sensing. *Science* **220**, 1016–1019.
27. Krikos, A., Mutoh, N., Boyd, A., and Simon, M. I. (1983). Sensory transducers of *E.coli* are composed of discrete structural and functional domains. *Cell* **33**, 615–622.
28. Bowie, J. U., Pakula, A. A., and Simon, W. I. (1995). The 3-dimensional structure of the aspartate receptor from *Escherichia coli*. *Acta Crystallogr.* **D51**, 306–312.
29. Bieman, H.-D., and Koshland, D. E., Jr. (1994). Aspartate receptor of *Escherichia coli* and *Salmonella typhimurium* bind ligand with negative cooperativity and half-of-the-sites cooperativity. *Biochemistry* **33**, 629–634.
30. Lin, L. N., Li, J. Y., Brandts, J. F., and Weis, R. M. (1994). The serine receptor of bacterial chemotaxis exhibits half-site saturation for serine binding. *Biochemistry* **33**, 6564–6570.
31. Gardina, P. J., Bormans, A. F., Hawkins, M. A., Meeker, J. W., and Manson, M. D. (1997). Maltose-binding protein interacts simultaneously and asymmetrically with both subunits of the Tar chemoreceptor. *Mol. Microbiol.* **23**, 1181–1191.
32. Stoddard, B. L., and Koshland, D. E., Jr. (1992). Prediction of the structure of a receptor protein complex using a binary docking method. *Nature* **358**, 774–776.
33. Mowbray, S. L., and Koshland, D. E., Jr. (1987). Additive and independent responses in a single receptor: Aspartate and maltose stimuli on the Tar protein. *Cell* **50**, 171–180.
34. Hughson, A. G., and Hazelbauer, G. L. (1996). Detecting the conformational changes of transmembrane signaling in a bacterial chemoreceptor by measuring effects on disulfide crosslinking in vivo. *Proc. Natl. Acad. Sci. USA* **93**, 11546–11551.

35. Beel, D. B., and Hazelbauer, G. L. (2001). Signalling substitutions in the periplasmic domain of chemoreceptor Trg induce or reduce helical sliding in the transmembrane domain. *Mol. Microbiol.* **40**, 824–834.
36. Tatsuno, I., Homma, M., Oosawa, K., and Kawagishi, I. (1996). Signaling by the *Escherichia coli* aspartate chemoreceptor Tar with a single cytoplasmic domain per dimmer. *Science* **274**, 423–425.
37. Gardina, P. J., and Manson, M. D. (1996). Attractant signaling by an aspartate chemoreceptor dimer with a single cytoplasmic domain. *Science* **274**, 425–426.
38. Chi, Y. I, Yokota, H., and Kim, S. H. (1997). Apo structure of the ligand binding domain of aspartate receptor from *Escherichia coli* and its comparison with ligand-bound or pseudoligand-bound structures. *FEBS Lett.* **414**, 327–332.
39. Falke, J. J., and Hazelbauer, G. L. (2001). Transmembrane signaling in bacterial chemoreceptors. *Trends Biochem. Science* **26**, 257–265.
40. Ames, P., and Parkinson, J. S. (1994). Constitutively signaling fragments of Tsr, the *Escherichia coli* serine chemoreceptor. *J. Bacteriol.* **176**, 6340–6348.
41. Morrison, T. B., and Parkinson, J. S. (1997). A fragment liberated from *Escherichia coli* CheA kinase that blocks stimulatory, but not inhibitory, chemoreceptor signaling. *J. Bacteriol.* **179**, 5543–5550.
42. Wu, J. R., Li, J. Y., Li, G. Y., Long, D. G., and Weis, R. M. (1996). The receptor binding site for the methyltransferase of bacterial chemotaxis is distinct from the sites of methylation. *Biochemistry* **35**, 3056–3065.
42a. Borkovich, K. A., Alex, L. A., and Simon, M. I. (1992). Attenuation of sensory signaling by covalent modification. *Proc. Natl. Acad. Sci. USA* **89**, 6756–6760.
43. LeMoual, H., and Koshland, D. E. (1996). Molecular evolution of the C-terminal cytoplasmic domain of a superfamily of bacterial receptors involved in taxis. *J. Mol. Biol.* **261**, 568–585.
44. Maddock, J. R., and Shapiro, L. (1993). Polar location of the chemoreceptor complex in the *Escherichia coli* cell. *Science* **259**, 1717–1723.
45. Liu, Y., Levit, M., Lurz, R., Surett, M. G., and Stock, J. B. (1997). Receptor mediated protein kinase activation and the mechanism of transmembrane signaling in bacterial chemotaxis. *EMBO J.* **16**, 7231–7240.
46. LeMoual, H., Quang, T., and Koshland, D. E. (1997). Methylation of the *E. coli* chemotaxis receptors: intra- and interdimer mechanisms. *Biochemistry* **36**, 13441–13448.
47. Ecker, J. R., and Theologis, A. (1994). *In* "Ethylene: A Unique Signaling Molecule" (C., Sommerville, and E., Meyerowitz, eds.). Cold Spring Harbor Laboratory Press, Cold Spring Harbor, NY.
48. Kieber, J. J. (1997). The ethylene pathway in *Arabidopsis*. *Annu. Rev. Plant Physiol. Plant Mol. Biol.* **48**, 277–296.
49. Bleeker, A. B., and Kende, H. (2000). Ethylene: A gaseous signal molecule in plants. *Annu. Rev. Cell Dev. Biol.* **16**, 1–18.
50. Hua, J., Chang, C., Sun, Q., and Meyerowitz, E. M. (1995). Ethylene insensitivity conferred by Arabidopsis *ERS* gene. *Science* **269**, 1712–1714.
51. Hua, J., Sakai, H., Nourizadeh, S., Chen, Q. G., and Bleeker, A. B. (1998). *EIN4* and *ERS2* are members of the putative ethylene receptor family in *Arabidopsis*. *Plant Cell* **10**, 1321–1332.
52. Sakai, H., Hua, J., Chen,. Q. G., Chang, C., Medrano, L. J. et al. (1998). *ETR2* is an *ETR1*-like gene involved in ethylene signaling in *Arabidopsis*. *Proc. Natl. Acad. Sci USA* **95**, 5812–5817.
53. Gamble, R. L., Coonfield, M. L., and Schaller, G. E. (1998). Histidine kinase activity of the ETR1 ethylene receptor from *Arabidopsis*. *Proc. Natl. Acad. Sci. USA* **95**, 7825–7829.
54. Hirayama, T., and Alonso, J. M. (2000). Ethylene captures a metal! Metal ions are involved in ethylene perception and signal transduction. *Plant Cell Physiol.* **41**, 548–555.

55. Tanaka, T., Saha, K. S., Tomomori, C., Ishima, R., Liu, D., Tong, *et al.* (1998). NMR structure of the histidine kinase domain of the *E. coli* osmosensor EnvZ. *Nature* **396**, 88–92.
56. Bilwes, , A. M., Alex, L. A., Crane, B. R., and Simon, M. I. (1999). Structure of CheA, a signal transducing histidine kinase. *Cell* **96**, 131–141.
57. Müller-Dieckmann, H. J., Grantz, A. A., and Kim, S. H. (1999). The structure of the signal receiver domain of the *Arabidopsis thaliana* ethylene receptor ETR1. *Struct. Fold. Des.* **7**, 1547–1565.
58. Abeles, F. B., Morgan, P. W., and Saltveit, M. E. (1995). *In* "Ethylene in Plant Biology" (2nd Ed.) Academic Press, New York.
59. Clark, K. L, Larsen, P. B., Wang, X., and Chang, C. (1998). Association of the *Arabidopsis* CTR1 Raf-like kinase with the ETR1 and ERS ethylene receptors. *Proc. Natl. Acad. Sci. USA* **95**, 5401–5406.
60. Chang, C., and Stewart, R. S. (1998). The two-component system. *Plant Physiol.* **117**, 723–731.
61. Solano, R., Stepanova, A., Chao, Q. M., and Ecker, J. R. (1998). Nuclear events in ethylene signaling: A transduction cascade mediated by ETHYLENE-INSENSITIVE3 and ETHYLENE-RESPONSIVE-FACTOR1. *Genes Dev.* **12**, 3703–3714.

CHAPTER 7

New Insights into the Mechanism of the Kinase and Phosphatase Activities of *Escherichia coli* NRII (NtrB) and Their Regulation by the PII Protein

PENG JIANG, AUGEN PIOSZAK, MARIETTE R. ATKINSON, JAMES A. PELISKA, AND ALEXANDER J. NINFA

Department of Biological Chemistry, University of Michigan Medical School, Ann Arbor, Michigan 48109

The dimeric two-component system transmitter protein NRII (NtrB) contributes to the nitrogen regulation of gene expression by catalyzing phosphorylation and dephosphorylation of the NRI (NtrC) receiver protein. NRII

Histidine Kinases in Signal Transduction

dimers consist of three types of protein domains: N-terminal domains involved in intramolecular signal transduction; a central domain mediating dimerization involved in kinase, phosphotransfer, and phosphatase catalytic activities; and C-terminal ATP-binding domains. Data indicate that (1) the kinase and phosphatase activities of the central domain of NRII are regulated by the binding of the PII protein to the C-terminal ATP-binding domains of NRII and (2) the N-terminal domains of NRII are involved in stabilizing the "phosphatase" conformation of the NRII central domain. The two subunits of the NRII dimer act in a highly concerted manner during the autophosphorylation reaction. An "alternating sites" hypothesis is used to explain the autophosphorylation mechanism of NRII and the regulation of NRII activities by PII.

INTRODUCTION

The NRI/NRII two-component system controls the expression of nitrogen-regulated (Ntr) genes in response to signals of carbon and nitrogen status. The "response regulator" or "receiver" protein, NRI (NtrC), is an enhancer-binding transcription factor that activates transcription from sigma 54-dependent promoters when it is in its active, phosphorylated form. The "modulator" or "transmitter" protein, NRII (NtrB), brings about the phosphorylation and dephosphorylation of NRI in response to cellular signals of nitrogen status. These intracellular signals control the activity of the related PII and GlnK signal transduction proteins, which, upon binding to NRII, inhibit its NRI kinase activity and stimulate its NRI~P phosphatase activity. The structure/function relationships of the "receiver" NRI (NtrC) and of the "transmitter" NRII (NtrB) and other related proteins have been reviewed extensively [1–6]. This chapter summarizes results concerning the function of the NRI/NRII two-component system. Most of these results concern regulation of the activities of the transmitter protein, NRII. We first describe in a general way the conclusions of the recent work and the hypotheses that have been developed from this work. We then review the experiments that have led to these conclusions and hypotheses.

Studies with purified components have shown that NRII exerts rheostat-like control of the extent of phosphorylation of NRI in response to signals of carbon and nitrogen status (specifically, 2-ketoglutarate and glutamine) [7–11]. These signals are sensed by accessory proteins (PII and UTase/UR) and are transmitted to NRII by PII. Specifically, the signals have the overall effect of controlling the availability of the active form of PII, i.e., the conformation of PII that is able to bind to NRII [12]. When PII binds to NRII, it inhibits the kinase activity of NRII and activates the phosphatase activity of

NRII [13]. PII appears to play a strictly regulatory role in these activities; it does not appear to play a catalytic role in kinase and phosphatase activities. For example, mutant forms of NRII have been identified that have phosphatase activity in the absence of PII [14, 15], and the isolated central domain of NRII has phosphatase activity in the absence of PII [5]. In addition, cells contain another PII-like protein, GlnK, that interacts with NRII under certain conditions [16, 17]. It is possible that additional signals regulating NRII remain to be discovered.

The NRII protein is a dimer consisting of three types of domains (shown schematically in Fig. 1). The N-terminal domain of NRII subunits is unrelated to the N-terminal domains of other two-component system transmitter

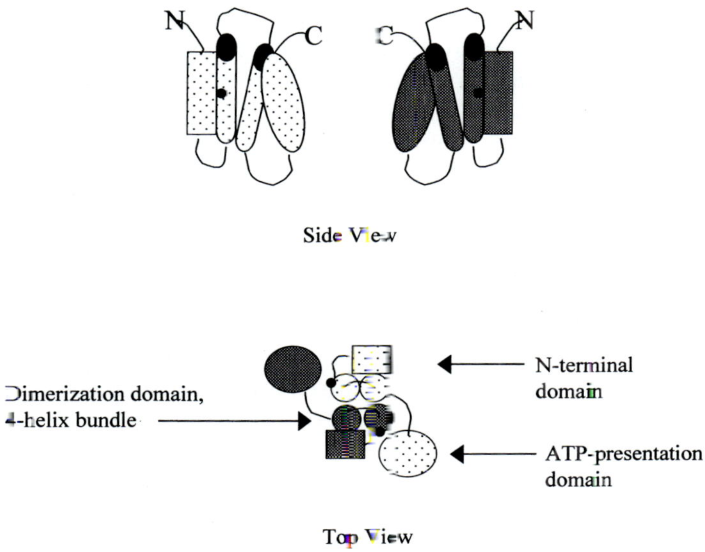

FIGURE 1 Diagram of the hypothesized structure of NRII. (Top) A side view in which the two subunits of NRII have been separated and are placed side by side. (Bottom) A top view of the intact dimer. The N-terminal domains of NRII are depicted as a rectangle, helices forming the central domain are depicted as tubes, and C-terminal ATP-presenting domains are depicted as ovals. The active site histidine for autophosphorylation (His-139) is depicted as a black circle. Linkers between domains are depicted as thin lines. The helix containing this site is encoded by the "H box" of the transmitter module, whereas the adjacent helix is encoded by the "X box" of the transmitter module [5, 21]. The arrangement of the domains in the dimer is surmised based on structural data from related proteins [22–24] and functional data discussed in this review. The asymmetry of the NRII dimer is not depicted. In the top view of the intact dimer, N-terminal domains are connected to the H-box helix below the plane of the page, and the linker connecting the X-box helix to the C-terminal ATP-presenting domains exits the X-box helix below the plane of the page.

proteins, but is shared by a variety of bacterial NRII proteins. This domain contains a PAS motif, which has been shown previously to be involved in binding sensory ligands in some proteins and to mediate the interactions between domains in other proteins [18–20]. Results suggest that this domain in NRII is not involved directly in binding PII, but is involved in intramolecular signal transduction events that are required for PII to inhibit the kinase activity of NRII and activate the phosphatase activity of NRII [21]. That is, the N-terminal domain is involved in interactions with other domains that serve to transmit the changes occurring upon binding of the PII protein. The isolated N-terminal domain of NRII is monomeric, and enzymological data suggest that the two N-terminal domains in the dimer interact with the other domains of NRII [21].

The central domain of NRII consists of a four-helix bundle that mediates dimerization of the protein. This conclusion is based on structural data for other transmitter proteins and related proteins [22–24] and by the properties of the purified central domain of NRII [21]. Two of the helices are provided by each subunit of the NRII dimer (Fig. 1). Although the four-helix bundle provides the primary dimerization determinant of NRII, the stability of the dimer is affected by the presence of the other domains. In particular, the presence of N-terminal and C-terminal domains seems to destabilize the dimer [21]. The central domain appears to contain all of the known catalytic activities of NRII, including autophosphorylation, phosphotransfer to NRI, and NRI~P phosphatase activities. Autophosphorylation of the central domain requires the presence of a suitably aligned ATP molecule, which provided by the C-terminal ATP-binding domains. New evidence also suggests that a particular conformation of the central domain is required for the phosphatase activity and that the N-terminal domain plays a key role in favoring this conformation [21]. For example, one may think of the N-terminal domain as the anvil against which the central domain is pressed in order to force it into the phosphatase conformation.

The C-terminal domain of NRII is the ATP-binding domain [25]. Although this domain has been called the "kinase" domain in the past, a more appropriate name may simply be the "ATP-binding" domain. Our reason for drawing this distinction is presented later. This domain is monomeric when separated from the rest of the protein [21]. The conformation of the ATP-binding domain influences the conformation of the central domain of NRII, allowing the binding of ATP or nonhydrolyzable ATP analogs to regulate the phosphatase activity of NRII [21]. The PII protein activates the phosphatase activity of the central domain by binding to the ATP-binding domain (Fig. 2) [26]. Thus, the two ATP-binding domains of the NRII dimer are also the sensory domains (with regard to PII and GlnK). All contacts of PII with NRII appear to be localized within this domain.

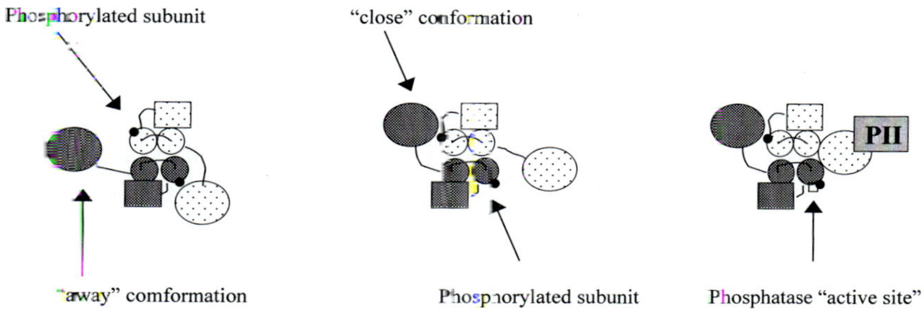

FIGURE 2 Alternating sites hypothesis for the autophosphorylation of NRII and hypothetical mechanism for the regulation of NRII by PII. The first two diagrams depict the proposed conformational changes occurring during one complete cycle of NRII autophosphorylation. Symbols are as in the top view shown in Fig 1. The third diagram depicts the mechanism of PII activation of NRII phosphatase activity. PII, depicted as a rectangle labeled PII, binds to the C-terminal ATP-presenting domain of one subunit. This forces the adjacent H-box helix into a conformation with phosphatase activity. The proposed "active site" for phosphatase activity on the H-box-encoded helix is depicted by a small white square.

The domains of NRII are connected by short linkers. The two N-terminal domains, the central domain, and the two ATP-binding domains of the NRII dimer all interact such that the conformations of all domains are changed in a highly concerted manner in response to signals and phosphorylation state (Fig. 2) [21, 27].

A speculative hypothesis that is nevertheless consistent with all existing data is as follows: The phosphorylation of NRI by NRII dimers may use an "alternating sites" mechanism in which first one active site histidine of the central domain is phosphorylated and put into position for phosphotransfer to NRI (Fig. 2). The equilibrium constant for phosphorylation of the histidine by ATP is far in favor of the phosphoryl group being on ATP [27]. However, an ~70 to 80-fold enhancement of the equilibrium constant for histidine phosphorylation is obtained by conformational changes that result in movement of the phosphorylated histidine residue away from the active site and into position for phosphotransfer to NRI (Fig. 2). In the hemiphosphorylated molecule, the other active site histidine residue of the central domain and its ATP-binding domain (which is from the opposing subunit) are in close association and the phosphoryl group is transferred back and forth rapidly between ATP and the histidine residue, with the equilibrium distribution greatly favoring ATP (Fig. 2). In the hemiphosphorylated NRII dimer, the "second" histidine residue is not able to undergo the conformational changes that culminate in movement away from its ATP-binding domain, and thus it is mainly unphosphorylated.

When NRI dephosphorylates the phosphorylated histidine in hemiphosphorylated NRII, forming NRI~P, the "second" histidine residue becomes phosphorylated rapidly and the two subunits of NRII trade conformations, placing the "first" active site histidine residue in close association with its ATP-binding domain (from the opposing subunit) and the "second" phosphorylated subunit into position for phosphotransfer to NRI (Fig. 2). This "alternating sites" hypothesis thus proposes that NRII works somewhat like a two-cylinder engine.

Regulation by PII involves forcing the central domain of NRII into the conformation with potent phosphatase activity. Conceivably, binding of PII to one of the ATP-binding domains of NRII forces the dimer into a particular conformation, where an unphosphorylated active site histidine region is available for NRI~P dephosphorylation (Fig. 2). Figure 2 depicts the "phosphatase active site" as mapping adjacent to the site of histidine phosphorylation. Phosphatase activity does not involve reverse transfer of phosphoryl groups from NRI~P to the NRII active site histidine but seems to require nearby residues [13, 14, 29]. The phosphatase activity of NRII may represent the activation of autophosphatase activity of NRI~P by this portion of NRII. Even though PII binds to ATP-binding domains, the N-terminal domain is necessary for the NRII dimer to obtain the "phosphatase conformation" [21, 26].

In the context of the hypothesis, PII could prevent the exchange of positions by NRII subunits that typically occurs after dephosphorylation of hemiphosphorylated NRII by NRI, while simultaneously perturbing somewhat the conformation of the exposed, unphosphorylated, active site histidine region. One could imagine that this regulation evolved in steps, with PII originally acting simply as an inhibitor of the conformational inversion between the two domains of NRII on dephosphorylation of the hemiphosphorylated form, and later both molecules were selected to have an altered conformation of the exposed histidine region with high phosphatase activity.

MECHANISM OF NRII AUTOPHOSPHORYLATION AND REGULATION OF THIS ACTIVITY BY PII

The highly concerted nature of the conformational changes occurring in NRII on autophosphorylation or on the binding of PII is illustrated by the strong asymmetry of the NRII autophosphorylation reaction [27]. Earlier results have indicated that NRII autophosphorylation proceeds exclusively by a *trans-intramolecular* mechanism, in which the ATP bound to an ATP-binding domain is used to phosphorylate the active site histidine from the opposing subunit of the dimer (within the central domain) [25]. NRII autophosphorylation is highly asymmetric, which is due to an ~ 70 to 80-fold difference in

[NRII] = [MBP-NRII], increasing the concentration of both species led to longer $t_{1/2}$. Thus, both simple associative and simple dissociative models appear to be inconsistent with what was seen in the experiments (unpublished data). A dissociative model in which tetramers are a dead-end intermediate was consistent with the experimental results. Further studies will be required to conclusively state the mechanism. Current data fail to exclude the possibility that heterodimers are formed from monomers.

The presence of N-terminal domains and C-terminal ATP-binding domains affects the subunit exchange activity of NRII greatly [21]. In particular, species lacking the N-terminal domain (CT111, CT126, Fig. 3) do not appear to undergo subunit exchange. Similarly, fusions of MBP to the central domain of NRII result in a dimeric fusion protein that does not appear to undergo subunit exchange with intact NRII. Finally, the purified dimeric central domain of NRII (HX119-189; Fig. 3) does not appear to undergo subunit exchange with intact NRII. These observations suggest that these polypeptides may form a more stable dimer than intact NRII. The NT189 polypeptide, consisting of just the N-terminal domain of NRII and the central domain (Fig. 3), is able to undergo subunit exchange with NRII and MBP-NRII, but the accumulation of the heterodimer is very slow and never reaches the point where an equal proportion of each subunit is present in homo- and heterodimers. This observation may signify that the NT189 homodimer is more stable than the NRII homodimer or the NRII::NT189 heterodimer, but only slightly so.

Although the NRII subunit exchange reaction provides a unique opportunity to investigate the role of the various domains of NRII, PII, ATP, and so on on dimer stability, the physiological significance of the subunit exchange reaction is not clear at this time. Experiments measuring this activity are typically performed in our laboratory with micromolar concentrations of proteins, while the intracellular concentration of NRII is in the nanomolar range [14]. Because the NRII autophosphorylation reaction proceeds by an obligate trans-intramolecular mechanism and the potent phosphatase activity appears to require a particular conformation of the dimer, factors that affect the dimerization of NRII may have a profound effect on the level of the NRII activities *in vivo*.

MAPPING THE INTERACTION OF PII WITH NRII

Early work established that PII interacts with NRII to activate the phosphatase activity of NRII [28, 29]. Structure/function analysis of PII indicated that an exposed loop of PII known as the T loop was responsible for the interactions of PII with NRII and its other receptors [30, 31]. For example, deletion of the apex of the T loop results in a stable trimeric PII protein that binds its regulatory ligands normally, but fails to interact with any of its known receptors [31]. Also, the point mutation A49P in the T loop of PII specifically eliminates the interaction of PII with NRII, while having only a modest effect on the interaction of PII with its other receptors [31]. Interestingly, a heterotrimeric PII containing a single functional T loop was able to interact with NRII to activate NRII phosphatase activity [32]. Based on the conditions at which the experiment was performed, it appears that the interaction of a single T loop of PII with an NRII dimer is sufficient to cause NRII to adopt the phosphatase conformation. This result is consistent with titration experiments where phosphatase activity was measured as a function of the ratio of NRII and PII (unpublished data). Such experiments suggested that a 1:1 complex of PII trimers to NRII dimers was responsible for the phosphatase activity.

The ability of PII to interact with NRII is regulated by the modification state of PII and by the regulatory ligands that bind to PII and regulate its activity allosterically [12]. Specifically, ATP and 2-ketoglutarate are regulatory ligands that control PII activity, apparently by controlling conformation of the T loop. These regulatory ligands bind to PII synergistically. Also, the binding of 2-ketoglutarate to PII exhibits negative cooperativity, such that the binding of one molecule of 2-ketoglutarate to the trimer disfavors the binding of additional molecules of this effector. The form of PII with optimal ability to bind to NRII appears to be the form with three molecules of ATP and one molecule of 2-ketoglutarate bound per trimer. This distinctive pattern of allosteric regulation has served as a convenient control for the physiological relevance of binding events observed in experiments with purified components.

As noted previously, enzymological studies of the activities of truncated forms of NRII and their regulation by PII suggested that PII interacted with the transmitter module of NRII and not with the N-terminal domain of NRII, as had been thought previously. For example, the stoichiometry of CT111 and CT126 autophosphorylation is regulated dramatically by PII even though these polypeptides completely lack the N-terminal domain of NRII. Similarly, PII weakly inhibited the transphosphorylation of the purified central domain by CT190 in reaction mixtures that completely lack the NRII N-terminal domain. These results make it obvious that PII must interact with a site within the transmitter module of NRII.

In order to further study the interaction of PII with NRII, a cross-linking approach was used [26] that is similar to the approach used to study the interaction of cAMP-CAP with RNA polymerase [33]. The PII protein contains a single cysteine residue at position 73, and this cysteine residue was mutated to serine, with no significant effect on the ability of PII to interact with NRII. Unique cysteine residues were then placed at three positions on the T loop of PII so as to take advantage of the unique chemistry of cysteine.

Heterobifunctional cross-linkers were anchored at the unique cysteine residues. These cross-linkers contain photoactivatable groups at their free end that cross-link nonspecifically with numerous sites on proteins. The photoactivatable cross-linking of the modified PII proteins to NRII was then examined. Cross-linking of PII to NRII was observed to depend on photoactivation in the presence of the PII ligands at the appropriate concentrations [26]. The ligand dependency of the reaction suggests that the observed cross-linking was physiologically relevant. Another suggestion that the cross-linking reaction was physiologically relevant came from the observation that the presence of a vast excess of Bovine Serum Albumin (BSA) in the cross-linking reaction mixtures did not affect ligand-dependent cross-linking of PII to NRII, and furthermore, no apparent cross-linking of PII to BSA occured [26].

Further support for the physiological relevance of cross-linking experiments comes from purification of the cross-linked complex and examination of its activities [26]. The PII::NRII cross-linked complex is larger than either PII or NRII and is purified from the starting materials by gel-filtration chromatography. On gel filtration, the complex runs slightly faster than unmodified NRII, which is to say it elutes at a volume that is larger than what is expected based its molecular mass [26]. Thus, the complex appears to be "compact." Examination of the constitution of the complex was performed using material that was purified directly from nondenaturing gels as well as by examination of the complex purified by gel-filtration chromatography. In both cases, the complex appeared to consist of PII trimers linked to NRII dimers by a single covalent attachment [26]. For example, upon denaturation with SDS, approximately equal numbers of cross-linked and uncross-linked NRII subunits were evident. Larger complexes consisting of two molecules of PII cross-linked to the NRII dimer were not detected even in reactions that had been subjected to exhaustive photoactivation.

The purified cross-linked PII::NRII complex had potent NRI~P phosphatase activity, further supporting the idea that the observed cross-linking represented a physiologically relevant interaction between the two proteins [26]. Interestingly, the phosphatase activity of the complex was stimulated by PII regulatory ligands, but did not absolutely require these ligands as is the case with uncross-linked proteins [26]. This observation is significant for two reasons. First, ligand-independent phosphatase activity is a novel activity found only with cross-linked species, eliminating the possibility that somehow uncross-linked normal NRII dimers and PII trimers were formed in our reaction mixtures by subunit exchange reactions. (Earlier experiments had shown that the PII trimer is far more stable than the NRII dimer [32].) Second, the observation suggests that once PII is tethered to NRII, the necessity for the regulatory ligands is eliminated, i.e., the ligands primarily affect the ability of PII to bind to NRII or the stability of the complex.

To map the site of interaction, the ligand-dependent cross-linking of PII to the isolated domains of NRII and various truncated species was examined [26]. To summarize the results, PII became cross-linked to any polypeptide that contained the ATP-binding domain of NRII, including a polypeptide that essentially consists of just this domain (CT190). Thus, PII binds to the ATP-binding domain of NRII. By qualitative estimation of the efficiency of cross-linking, based on visual examination of gels, it seems that CT190 cross-linked to PII just as efficiently as intact NRII. This suggests that all the binding determinants are found within the CT190 polypeptide.

The finding that the ATP-presenting domain of NRII constitutes its sensory domain for binding PII represents a significant change in the way we think about sensation by transmitter proteins. In most such proteins, an N-terminal transmembrane domain is present that is thought to be involved in sensation. In some cases, there is compelling evidence that signaling involves transmembrane signaling [34, 35]. By analogy to those proteins, we expected that PII would interact with the N-terminal domains of the NRII dimer. However, a sensory activity of the C-terminal portions of transmitter proteins is not without precedent. The FixT protein of *R. meliloti* is thought to regulate the FixL transmitter by binding to the transmitter module of FixL [36]. Also, the activity of CheA is regulated by receptors that, along with an adaptor protein, interact with a site located C-terminal to the ATP-binding domain of CheA [37]. Thus, it appears that signals controlling transmitter activities may be transduced to the transmitter module in different ways. Perhaps many transmitter proteins are able to sense multiple stimuli by different mechanisms and function as processors to integrate the different signals. For example, an unknown signal may regulate NRII by interacting with its N-terminal domain.

MAPPING THE ACTIVITIES OF NRII

Previous studies had shown clearly that the site of NRII autophosphorylation is His-139 and that mutations affecting the highly conserved "G box" motif affected the ability to bind ATP [25, 38]. We now know from the structures of CheA and EnvZ that the latter assignment was correct.

Mapping Phosphatase Activity

Work by Kramer and Weiss [5] showed that the phosphatase activity of NRII was obtained with a small peptide consisting of residues 122–221 of NRII. This peptide contains the central domain of NRII. We have shown that a slightly smaller peptide (HX119-189; Fig. 3) also contains phosphatase activity [21].

Thus it seems that the phosphatase activity "active site" is found within the central domain of NRII. However, the activity of the isolated central domain is very weak compared to that of intact NRII in the presence of PII [21]. The phosphatase activity of the isolated central domain is not stimulated by PII.

A slightly larger polypeptide, consisting of the N-terminal domains of NRII along with the central domain (NT189; Fig. 3), displays fairly strong phosphatase activity that is not regulated by PII [21]. We hypothesize that interactions between N-terminal domains and the central domain are involved in stabilizing the phosphatase conformation of the central domain.

As already noted, the CT111 and CT126 polypeptides (Fig. 3) have very low phosphatase activity, even in the presence of PII. The phosphatase activity of these polypeptides is significantly lower than that of the HX119-189 polypeptide. Because the CT111 and CT126 polypeptides have an intact central domain, exist as dimers, have autophosphorylation and NRI kinase activity, and bind PII, it seems that they lack phosphatase activity because they are unable to obtain the conformation of the central domain with potent phosphatase activity. The N-terminal domain apparently is necessary for the central domain to assume the active conformation.

Mapping ATP-Cleaving Activity

Because the ATP-binding domain brings about transphosphorylation of the central domain, it seems that this domain behaves as a kinase. However, preliminary results from our laboratory suggest that this domain functions only to present ATP and that catalysis of the autophosphorylation reaction may be accomplished by the central domain of NRII. In particular, we have observed that the construct MBP-HX103-189, but not MBP-HX119-189, is able to become autophosphorylated with very low but clearly discernible stoichiometry upon incubation with ATP (unpublished data). Further studies are required to know for certain how the MBP-HX103-189 protein becomes autophosphorylated. An admittedly speculative hypothesis to explain the curious observations is that the construction (or a subsequent mutation resulting in protein microheterogeneity) somehow created a very low-affinity ATP-binding site on the surface of MBP or comprised by the interface of MBP and the central domain. This low-affinity ATP-binding site is sufficient to permit autophosphorylation of the central domain of NRII found in the fusion protein. If this hypothesis is correct, it would signify that the central domain itself has autophosphorylation activity and that the sole role of the C-terminal domain of NRII is to present ATP in a suitable conformation. Thus, we have chosen to refer to the C-terminal domain as an ATP-binding domain (as opposed to a kinase domain) in this review.

If it is true that the central domain of NRII has a major role in catalyzing the autophosphorylation reaction, then one can imagine that the two-component system histidine kinases evolved by an ancient gene fusion bringing together the four-helix bundle and a suitable ATP-binding domain. The possible "evolution" of an autophosphorylatable form of MBP-HX103-189 during fermentations to purify the protein may be thought of as a reiteration of this process, with a greatly shortened time scale. The absence of clear examples of transmitter domains in higher eukaryotes may reflect the possibility that the four-helix bundle, not recognized easily in homology searches due to its small size, in these organisms may have become fused to a different type of ATP-binding domain.

EXPLAINING THE ACTIVITIES OF MUTANT FORMS OF NRII

The ultimate proof of the hypotheses presented so far in this review will require elucidation of the structure of NRII and of the NRII::PII complex, as well as further studies of the NRII activities, such as by rapid quench flow methods or rapid kinetics spectroscopic methods. However, at this point in time it will be useful to review the known properties of mutant forms of NRII and examine whether the properties of these mutant forms and the locations of the mutations can be rationalized in the context of the working hypothesis.

Numerous mutant forms of NRII are available that have the effect of reducing the phosphatase activity of NRII *in vivo* under conditions where PII activity is high. For example, starting with cells lacking the UTase/UR and thus unable to bring about the uridylylation of PII under nitrogen-limiting conditions, mutations altering NRII were isolated that permitted the expression of Ntr genes requiring a high intracellular concentration of NRI~P [39]. These mutations are mainly clustered within the NRII central domain and the linker connecting the central domain to the N-terminal domain. Most of the central domain mutations map near the active site histidine, with only one mutation mapping within the second (X-box-encoded) helix. A few mutations were mapped within the N terminal domain, including mutations at the extreme N-terminus of NRII.

Two questions should be considered: how could mutations in these positions reduce the phosphatase activity of NRII and why were no mutations found within the C-terminal ATP-presenting domain, where the PII-binding site is contained? In addressing the first of these questions, we suggest that mutations in the central domain that reduce phosphatase activity may do so by directly preventing the phosphatase conformation or by altering the contacts with NRI~P that are necessary to bring about its dephosphorylation. Given the highly concerted nature of the conformations of the domains of NRII, it would not be surprising if mutations that block adoption of the phosphatase conformation by the central domain also reduce PII binding by the ATP-binding domain. Because the central domain provides the main dimerization determinant of NRII, mutations in this part of NRII may also affect phosphatase activity by altering the orientation of the subunits in the dimer. For example, if a contact between the central domain and N-terminal domain is required for the central domain to assume the conformation with potent phosphatase activity, altering the orientation between domains slightly may affect phosphatase activity.

Mutations in the N-terminal domain of NRII and the linker connecting the N-terminal domain to the central domain may affect the interaction between these two domains. We have suggested that the N-terminal domain may serve as the anvil against which the central domain must rest in order for the latter to be forced into phosphatase conformation by PII binding to the ATP-presenting domain. In that context, we can explain how mutations in this part of the protein may reduce phosphatase activity by hypothesizing that these proteins are altered in the interaction between the N-terminal domain and the central domain. If the conformations of the domains of NRII change in a completely concerted fashion, these mutant proteins may prove to be defective in binding PII.

A more difficult question to address is why the initial selections for relief from PII regulation did not pick up mutations in the C-terminal domain where the direct binding studies indicate that PII binds. However, a reasonable explaination for this comes from the study of cells lacking PII and GlnK [17]. These cells have a severe growth defect on minimal medium, and this growth defect is due to the unregulated activity of NRII, which leads to unregulated expression of the Ntr regulon. Unregulated expression of one or more Ntr genes seems to cause the severe growth defect. Thus, in the experiments reported previously, the complete absence of NRII phosphatase activity was selected against, as those experiments involved growth of the cells on minimal medium [39]. We might imagine that a mutation altering the PII/GlnK site within the C-terminal domain would cause unrestrained Ntr expression. To address this issue, we have repeated the selection for relief from PII regulation and isolated a set of mutations under condtions where the unregulated activity of NRII is not lethal (i.e., growing cells on rich medium and using the expression of gene fusions to identify mutations). These experiments may reveal mutations in the PII-binding site as well as the classes of mutations obtained previously.

REFERENCES

1. Ninfa, A. J., Jiang, P., Atkinson, M. R., and Peliska, J. A. (2000). Integration of antagonistic signals in the regulation of nitrogen assimilation in *Escherichia coli*. *Curr. Top. Cell. Regul.* **36**, 31–75.
2. Volz, K. (1995). Structural and functional conservation in response regulators. *In* "Two-Component Signal Transduction" (J. A. Hoch and T. J. Silhavy, eds.), Chapter 4, pp. 53–64. ASM, Washington, DC.
3. Stock, J. B., Surette, M. G., Levit, M., and Park, P. (1995). Two-component signal transduction systems: Structure-function relationships and mechanisms of catalysis. *In* "Two-Component Signal Transduction" (J. A. Hoch and T. J. Silhavy, eds.), Chapter 3, pp. 25–52. ASM, Washington, DC.
4. Ninfa, A. J., Atkinson, M. R., Kamberov, E. S., Feng, J., and Ninfa, E. G. (1995). Control of nitrogen assimilation by the NRI-NRII two-component system of enteric bacteria. *In* "Two-component Signal Transduction" (J. A. Hoch and T. J. Silhavy, eds.), Chapter 5, pp. 67–88. ASM, Washington, DC.
5. Kramer, G., and Weiss, V. (1999). Functional dissection of the transmitter module of the histidine kinase NtrB in *Escherichia coli*. *Proc. Natl. Acad. Sci. USA* **96**, 604–609.
6. Rombel, I., North, A., Hwang, I., Wyman, C., and Kustu, S. (1998). The bacterial enhancer-binding protein NtrC as a molecular machine. *Cold Spring Harb. Symp. Quant. Biol.* **63**, 157–166.
7. Jiang, P., Peliska, J. A., and Ninfa, A. J. (1998). Enzymological characterization of the signal-transducing uridylyltransferase/uridylyl-removing enzyme (E.C. 2.7.7.59) of *Escherichia coli* and its interaction with the PII protein. *Biochemistry* **37**, 12782–12794.
8. Jiang, P., Peliska, J. A., and Ninfa, A. J. (1998). Reconstitution of the signal-transduction bicyclic cascade responsible for regulation of Ntr gene expression in *Escherichia coli*. *Biochemistry* **37**, 12795–12801.
9. Atkinson, M. R., Kamberov, E. S., Weiss, R. L., and Ninfa, A. J. (1994). Reversible uridylylation of the *Escherichia coli* PII signal; transduction protein regulates its ability to stimulate the dephosphorylation of the transcription factor nitrogen regulator I (NRI or NtrC). *J. Biol. Chem.* **269**, 28288–28293.
10. Kamberov, E. S., Atkinson, M. R., and Ninfa, A. J. (1995). The *Escherichia coli* PII signal transduction protein is activated upon binding 2-ketoglutarate and ATP. *J. Biol. Chem.* **270**, 17797–17807.
11. Liu, J., and Magasanik, B. (1995). Activation of the dephosphorylation of nitrogen regulator I-phosphate of *Escherichia coli*. *J. Bacteriol.* **177**, 926–931.
12. Ninfa, A. J., and Atkinson, M. R. (2000). Bacterial PII proteins. *Trends. Microbiol.* **8**, 172–179.
13. Jiang, P., and Ninfa, A. J. (1999). Regulation of the autophosphorylation of *Escherichia coli* nitrogen regulator II by the PII signal transduction protein. *J. Bacteriol.* **181**, 1906–1911.
14. Atkinson, M. R., and Ninfa, A. J. (1993). Mutational analysis of the bacterial protein kinase/phosphatase NRII. *J. Bacteriol.* **175**, 7016–7023.
15. Kamberov, E. S., Atkinson, M. R., Chandran, P., and Ninfa, A. J. (1994). Effect of mutations in *Escherichia coli glnL* (*ntrB*), encoding nitrogen regulator II (NRII or NtrB), on the phosphatase activity involved in bacterial nitrogen regulation. *J. Biol. Chem.* **269**, 28294–28299.
16. van Heeswijk, W., Hoving, S., Molenaar, D., Stegman, B., Kahn, D., and Westerhoff, H. V. (1996). An alternative PII protein in the regulation of glutamine synthetase in *Escherichia coli*. *Mol. Microbiol.* **21**, 133–146.
17. Atkinson, M. R., and Ninfa, A. J. (1998). Role of the GlnK signal transduction protein in the regulation of nitrogen assimilation in *Escherichia coli*. *Mol. Microbiol.* **29**, 431–447.
18. Pointing, C. P., and Aravind, L. (1997). PAS: A multifunctional domain family comes to light.

Curr. Biol. 7, R674–677.

19. Gu, Y. Z., Hogenesch, J. B., and Bradfield, C. A. (2000). The PAS superfamily: Sensors of environmental and developmental signals. *Annu. Rev. Pharmacol. Toxicol.* 40, 519–561.
20. Gong, W., Hao, B., Mansey, S. S., Gonzalez, G., Gilles-Gonzalez, M. A., and Chan, M. K. (1998). Structure of a biological oxygen sensor: A new mechanism for heme-driven signal transduction. *Proc. Natl. Acad. Sci. USA* 95, 15177–15182.
21. Jiang, P., Srisawat, C., Sun, Q., and Ninfa, A. J. (2000). Functional dissection of the dimerization and enzymatic activities of *Escherichia coli* nitrogen regulator II and their regulation by the PII protein. *Biochemistry* 39, 13433–13449.
22. Tomomori, C., Tanaka, T., Dutta, R., Park, H., Saha, S. K., Zhu, Y., Ishima, R., Liu, D., Tong, K. I., Kurokawa, H., Qian, H., Inouye, M., and Ikura, M. (1999) Solution structure of the homodimeric core domain of *Escherichia coli* histidine kinase EnvZ. *Nature Struct. Biol.* 6, 729–734.
23. Bilwes, A. M., Alex, L. A., Crane, B. R., and Simon, M. I. (1999). Structure of CheA, a signal-transducing histidine kinase. *Cell* 96, 131–141.
24. Varughese, K. I., Madhusudan, Zhou, X. Z., Whiteley, J. M., and Hoch, J. A (1998). Formation of a novel four-helix bundle and molecular recognition sites by dimerization of a response regulator phosphotransferase. *Mol. Cell* 2, 485–493.
25. Ninfa, E. G., Atkinson, M. R., Kamberov, E. S., and Ninfa, A. J. (1993). Mechanism of autophosphorylation of *Escherichia coli* NRII: Trans-phosphorylation between subunits. *J. Bacteriol.* 175, 7024–7032.
26. Pioszak, A. A., Jiang, P., and Ninfa, A. J. (2000). The *Escherichia coli* PII signal transduction protein regulates the activities of the two-component system transmitter protein NRII (NtrB) by direct interaction with the kinase domain of the transmitter module. *Biochemistry* 39, 13450–13461.
27. Jiang, P., Peliska, J. A., and Ninfa, A. J. (2000). Asymmetry in the autophosphorylation of the two-component system transmitter protein NRII (NtrB) of *Escherichia coli*. *Biochemistry* 39, 5057–5065.
28. Ninfa, A. J., and Magasanik, B. (1986). Covalent modification of the *glnG* product, NRI, by the *glnL* product, NRII, regulates the transcription of the *glnALG* operon in *Escherichia coli*. *Proc. Natl. Acad. Sci. USA* 83, 5909–5913.
29. Kamberov, E. S., Atkinson, M. R., Feng, J., Chandran, P., and Ninfa, A. J (1994). Signal transduction components controlling bacterial nitrogen assimilation. *Cell. Mol. Biol. Res.* 40, 175–191.
30. Carr, P. D., Cheah, E., Suffolk, P. M., Vasudevan, S. G., Dixon, N. E., and Ollis, D. L. (1996). X-ray structure of the *Escherichia coli* signal transduction protein PII. *Structure* 2, 931–990.
31. Jiang, P., Zucker, P., Atkinson, M. R., Kamberov, E. S., Tirasophon, W. Chandran, P., Schefke, B. R., and Ninfa, A. J. (1997). Structure/function analysis of the PII signal transduction protein of *Escherichia coli*: Genetic separation of interactions with receptors. *J. Bacteriol.* 179, 4342–4353.
32. Jiang, P., Zucker, P., and Ninfa, A. J. (1997). Probing interactions of the homotrimeric PII signal transduction protein with its receptors by use of PII heterotrimers formed *in vitro* from wild-type and mutant subunits. *J. Bacteriol.* 179, 4354–4561.
33. Chen, Y., Ebright, Y. W., and Ebright, R. H. (1994). Identification of the target of a transcription activator protein by protein-protein photocrosslinking. *Science* 265, 90–92.
34. Williams, S. B., and Stewart, V. (1997). Discrimination between structurally related ligands nitrate and nitrite controls autokinase activity of the NarX transmembrane signal transducer of *Escherichia coli* K-12. *Mol. Microbiol.* 26, 911–925.
35. Cavicchioli, R., Chiang, R. C., Kalman, L. V., and Gunsalus, R. P. (1996). Role of the periplasmic domain of the *Escherichia coli* NarX sensor-transmitter protein in nitrate-

dependent signal transduction and gene regulation. *Mol. Microbiol.* **21**, 901–911.

36. Garnerone, A. M., Cabanes, D., Foussard, M., Boistard, P., and Batut, J. (1999). Inhibition of the FixL sensor kinase by the FixT protein in *Sinorhizobium meliloti*. *J. Biol. Chem.* **274**, 32500–32506.
37. Bourret, R. B., Davagnino, J., and Simon, M. I. (1993). The carboxy-terminal portion of the CheA kinase mediates regulation of autophosphorylation by transducer and CheW. *J. Bacteriol.* **175**, 2097–2101.
38. Ninfa, A. J., and Bennett, R. L. (1991). Identification of the site of autophosphorylation of the bacterial protein kinase/phosphatase NRII. *J. Biol. Chem.* **266**, 6888–6893.
39. Atkinson, M. R., and Ninfa, A. J. (1992). Characterization of *Escherichia coli glnL* mutations affecting nitrogen regulation. *J. Bacteriol.* **174**, 4538–4548.

CHAPTER 8

Role of the Histidine-Containing Phosphotransfer Domain (HPt) in the Multistep Phosphorelay through the Anaerobic Hybrid Sensor, ArcB

TAKESHI MIZUNO AND MASAHIRO MATSUBARA
Laboratory of Molecular Microbiology, School of Agriculture, Nagoya University, Chikusa-ku, Nagoya 464-8601, Japan

Histidine Kinases in Signal Transduction

A His-kinase is a central player of a His→Asp phosphorelay signal transduction system. In some cases, however, another common histidine-containing phosphotransfer domain (or factor) plays a crucial role in a sequential His→Asp→His→Asp signaling event that is generally referred to as a "multistep His→Asp phosphorelay." This chapter discusses characteristic features of the HPt domain with special reference to the *Escherichia coli* ArcB hybrid His-kinase that contains the first discovered HPt domain. In *E. coli* physiology, this particular His-kinase is involved in the complex transcriptional regulatory network that allows *E. coli* cells to respond to various aerobic and anaerobic growth conditions. General views as to the widespread occurrence of HPt domains are also discussed.

INTRODUCTION

Histidine-to-aspartate (His→Asp) phosphorelay (or two-component) systems are very common signal transduction mechanisms that are implicated in a wide variety of cellular responses to environmental stimuli [1–6]. To date, numerous instances of such His→Asp phosphorelay signaling systems have been uncovered not only in many prokaryotic species [7–9], but also certain eukaryotic species [10–14]. A classical His→Asp phosphorelay system consists of two types of common signal transducers, a sensor containing a transmitter domain that exhibits a histidine (His)-kinase activity and a response

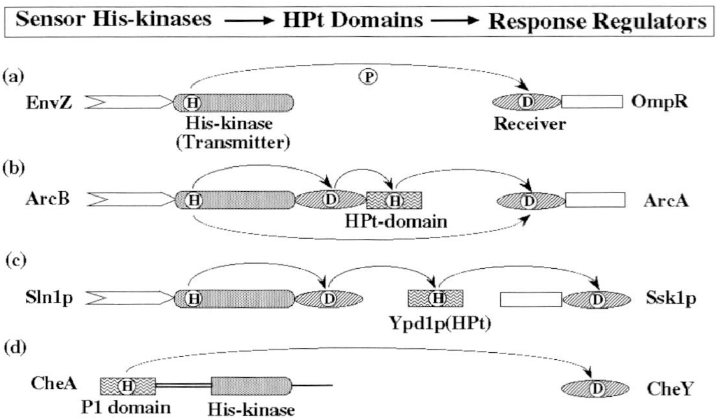

FIGURE 1 Schematic representation of typical His→Asp phosphorelay signaling between the sensor His-kinase and the response regulator. The EnvZ→OmpR, ArcB→ArcA, CheA→CheY systems operate in *E. coli*, and Sln1p→Ypd1p→Ssk1p operates in *S. cerevisiae*. Other details are given in the text.

regulator containing a phospho-accepting aspartate (Asp) in its receiver domain (Fig. 1a, see EnvZ sensor→OmpR regulator) [4]. A crucial event underlying this signal transduction mechanism is a His→Asp phosphorelay from a His-kinase to its cognate response regulator. In a more sophisticated case, however, a histidine-containing phosphotransfer (HPt) domain plays an essential role as a mediator (or alternative transmitter) of phosphorelay (Fig. 1b, see ArcB sensor→ArcA regulator). In this case, a phosphoryl group moves from a His-kinase to a receiver, then to an HPt domain, and finally to another receiver in a given phosphorelay signaling pathway [15–17]. This sequential His→Asp→His→Asp signaling event is referred to as a "multistep His→Asp phosphorelay," in which the HPt domain plays a crucial role [16]. This chapter discusses characteristic features of the HPt domain with special reference to the *Escherichia coli* ArcB hybrid sensor that contains the first discovered HPt domain. In *E. coli* physiology, this particular His-kinase is involved in the complex transcriptional regulatory network that allows *E. coli* cells to respond to various aerobic and anaerobic growth conditions.

HPt DOMAIN

Many instances of HPt domains has been identified and each is assumed to play an important role in some (but not all) His→Asp phosphorelay systems [15–17]. A typical HPt domain was first discovered in the *E. coli* ArcB sensor His-kinase [19, 20]. For a long time, ArcB was considered to contain a His-kinase domain, followed by a receiver domain in its primary amino acid sequence [21]. However, it was later found that this hybrid sensor possesses another phosphorylated histidine site in its very C-terminal region that has never been noticed previously (Fig. 1b) [19]. It was then demonstrated *in vitro* that this C-terminal region containing a crucial histidine site can acquire a phosphoryl group, and thus is capable of serving as an alternative phosphotransmitter domain [20]. This domain was generally termed the "HPt domain." A plausible scheme can be proposed for a complex circuitry of the ArcB→ArcA phosphorelay signaling [20, 22]. First of all, like in other authentic His-kinases, His-292 in the ArcB His-kinase domain acquires a γ-phosphoryl group from ATP through its own catalytic function (*i.e.*, autophosphorylation). Then, the phosphoryl group on His-292 moves onto the phospho-accepting aspartate (Asp-576) site in the intrinsic ArcB receiver domain. Subsequently, His-717 in the ArcB HPt domain is also modified by phosphorylation, in which both His-292 and Asp-576 play crucial roles. The final destination of the phosphoryl group on His-717 is Asp-54 in the ArcA receiver domain. This stream is a typical example of multistep phosphorelays. Interestingly, however, ArcA can also acquire a phosphoryl group directly from His-292 of ArcB, at least under certain *in vitro* conditions (Fig. 1b) [20].

The discovery of the HPt domain in ArcB revealed soon that such an HPt domain is not unique for ArcB. Inspection of the entire nucleotide sequence of the *E. coli* genome revealed that this bacterium has four more hybrid sensors, each containing an HPt domain (BarA, EvgS, TorS, and YojN) (Fig. 2) [8]. It is now known that many other bacteria also have a number of hybrid sensors that contain a common HPt domain. For example, the *Bordetella pertussis* BvgS hybrid sensor has a structural design very similar to ArcB and contains a typical HPt domain, of which the functional importance was demonstrated experimentally [23, 24]. Another striking example of HPt domains was found in the eukaryotic microorganism *Saccharomyces cerevisiae*. In the well-documented osmoregulatory response of this yeast [25], the Sln1p→Ypd1p→Ssk1p three components represent another example of multistep phosphorelay strategies in which Ypd1p comprising only an HPt domain plays a crucial role as a mediator of phosphorelay (Fig. 1c) [26]. More recently, many examples of Ypd1-like HPt factors were found even in the higher plant, *Arabidopsis* [27]. Taking all these examples together, it is clear that the HPt domain in a number of signal transducers serves as a common device, which most likely plays an important role as an intermediate for a given multistep His→Asp phosphorelay signal.

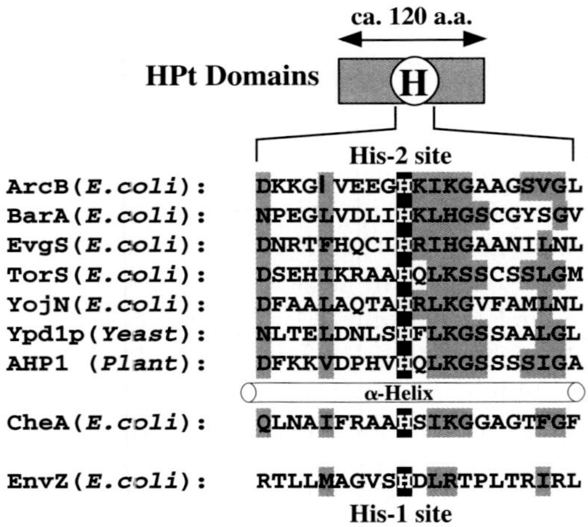

FIGURE 2 Arrangement of conserved amino acid sequences of HPt domains of the hybrid sensor His-kinase family members. The phosphorylated histidine sites (His-2) are highlighted. The authophosphorylated histidine site (His-1) of the authentic His-kinase, EnvZ, is also shown as a reference.

STRUCTURE AND FUNCTION OF COMMON HPt DOMAINS

From the structural viewpoint, even if one has an amino acid sequence of a presumed His→Asp phosphorelay signal transducer, the occurrence of the HPt domain is hard to recognize in the sequence, as characteristic stretches of amino acid sequences of HPt domains are relatively short and quite variable among them (Fig. 2). They do not resemble those of the authentic histidine sites in His-kinase domains (e.g., EnvZ) [19]. Moreover, the conventional BLAST and FASTA search programs are not always helpful in finding such an HPt domain. However, an invariant phosphorylated histidine site can be found in a certain context of amino acid sequence (often referred to as the "His-2 site" in comparison with the "His-1 site" in an authentic His-kinase) (Fig. 2). This crucial His-2 site is surrounded by a short characteristic stretch of conserved amino acids. The X-ray crystal structure of the ArcB-HPt domain, consisting of about 120 amino acids, revealed that it contains six α helices, including a long four-helix bundle with a kidney-like shape (Fig. 3) [28]. Essentially, the similar structure was determined for the eukaryotic HPt domain, Ypd1p of *S. cerevisiae* [29, 30]. It is worth mentioning that these determined HPt structures are considerably similar to that of the P1 domain of CheA (see Fig. 1d), which is the first discovered autophosphorylated His-kinase involved in the chemotactic CheA→CheY phosphorelay system [31]. Indeed, the amino acid sequence surrounding the phosphorylated His site of

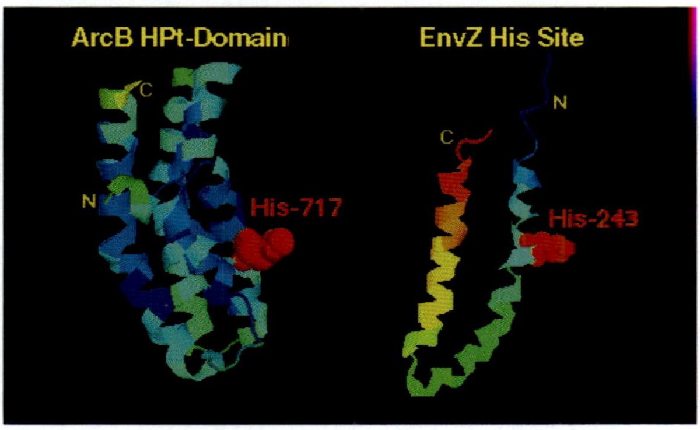

FIGURE 3 Representation of the three-dimensional structure. The crystal structure of the HPt domain of ArcB containing the phosphorylated histidine site (His-2 site = His-717) is compared with the nuclear magnetic resonance structure of the autophosphorylated histidine site (His-1 site = His-243) of EnvZ (for references, see Kato *et al.* [28] and Tomomori *et al.* [32], respectively). Note that the sizes of these two are not proportional. These pictures were made by RasMac v2.6.

CheA significantly resembles those of HPt domains (Fig. 2). In this sense, CheA can be categorized as an unorthodox sensor containing an HPt domain. In any case, these determined structures supported further the reliability of the short consensus sequence that is proposed as the signature of the HPt family of domains [19, 28]. Indeed, based on this consensus sequence, it has allowed us to identify many HPt domains even in the genomic database of the higher plant, *Arabidopsis* [27].

It should be briefly noted that the nuclear magnetic resonance (NMR) structure has been determined for an EnvZ domain encompassing the His-1 site, which consists of a two-helix bundle (Fig. 3) [32, 33]. At first glance, this structure may resemble that of the HPt domain of ArcB (note that this EnvZ domain most likely forms a homodimer). In any case, a detailed comparison of these two structures, each containing the active His-1 and His-2 sites, respectively, must await further structural analyses. The results should provide us with clues for understanding the mechanistic basis of the His→Asp phosphorelay and answering the question of how the specificity of a given His→Asp phosphorelay is determined.

From a functional viewpoint, unlike authentic His-kinases, HPt domains may not exhibit any catalytic function. HPt domains appear to serve solely as a passive intermediate molecule (or substrate) in a given His→Asp phosphorelay pathway by acquiring/transferring a phosphoryl group from/to another signaling domain (*e.g.*, cognate receivers). Rather, such a cognate receiver itself appears to function as an enzyme capable of transferring/acquiring a phosphoryl group to/from a HPt domain. The phosphoryl group incorporated into the isolated HPt domain of ArcB is relatively stable in solution, at least *in vitro*. Collectively, results from intensive studies support the general view that the HPt domain is a widespread structural and functional motif, involved in many (but not all) His→Asp phosphorelay systems in both prokaryotes and eukaryotes.

Here, a critical question arises. When one considers the classical case (*e.g.*, the single-step EnvZ→OmpR phosphorelay), it is very curious why a phosphoryl group should travel along a long railroad with extra stations. To address this intriguing issue, the multistep ArcB→ArcA phosphorelay is one of the best-characterized paradigms. To this end, the physiological (or *in vivo*) relevance of a complicated multistep phosphorelay has been elucidated extensively, as discussed later.

MULTISTEP ArcB→ArcA PHOSPHORELAY SYSTEM IN *Escherichia coli* ANAEROBIOSIS

E. coli is a facultative anaerobe, which can adopt different metabolic pathways, fermentation, anaerobic respiration, and aerobic respiration for the processes of energy transduction, depending on the availability of external oxygen and anaerobic electron acceptors such as nitrate, trimethylamine-*N*-oxide (TMAO), dimethyl sulfoxide (DMSO), and fumarate (with regard to the following views, see the relevant chapters in Neidhardt [34] and references therein). *E. coli* must adopt sophisticated regulatory mechanisms that enable this bacterium to exploit adroitly energy sources to their greatest possible advantage, depending on the external growth conditions, including the availability of external oxygen. In such adaptive regulatory responses, two global transcriptional regulators, Fnr and ArcA, are known to play central roles in concert with other specific regulatory proteins. A large number of *E. coli* genes involved in the different energy metabolism pathways are under the coordinate control of either Fnr or Arc regulons (often both). With regard to these issues, a number of comprehensive reviews have appeared previously [34]. This chapter does not address such physiological issues in detail. Rather, it focuses on the multistep ArcB→ArcA phosphorelay mechanism with special emphasis on the function of the HPt domain.

The tricarboxylic acid (TCA) cycle is highly operative only in aerobically grown cells, with the key regulatory control responsible for determining the levels of TCA cycle enzymes. As a typical example, expression of succinate dehydrogenase (SDH; an enzyme complex of the TCA cycle), encoded by the *sdhCDAB* operon, is elevated markedly in the presence of external oxygen and is suppressed severely during growth under anaerobic conditions [34, 36]. This particular event can be followed conventionally by monitoring the expression of an *sdh*::*lacZ* fusion gene on the *E. coli* chromosome under both aerobic and anaerobic growth conditions (Fig. 4, bottom). The expression of *sdh*::*lacZ* (or β-galactosidase activity) is markedly high in cells grown aerobically and is severely repressed under anaerobic (or microaerobic) growth conditions. In an *arcB* null mutant (Δ*arcB*) background, such an anaerobic repression of *sdh*::*lacZ* is completely abolished, suggesting the crucial role of ArcB in this regulatory event. It is thus clear that the ArcB→ArcA two components are the central control elements of this regulation at the level of transcription. In this regard, a well-defined scenario as to the molecular mechanism underlying the ArcB→ArcA signaling system has previously been proposed inductively from a series of intensive studies of Lin and colleagues [37–49]. ArcB functions as an anaerobic sensor and is activated under certain anaerobic growth conditions. ArcB exhibits its His-kinase activity specifically toward the ArcA response regulator. The resulting phospho-ArcA functions as a DNA-binding transcriptional repressor for the *sdhCDAB* operon. This whole scenario is seemingly a simple and classical example of the common two-component regulatory systems. However, results from other studies, including the discovery of the HPt domain of ArcB, revealed that the reality is more complex. The ArcB→ArcA system

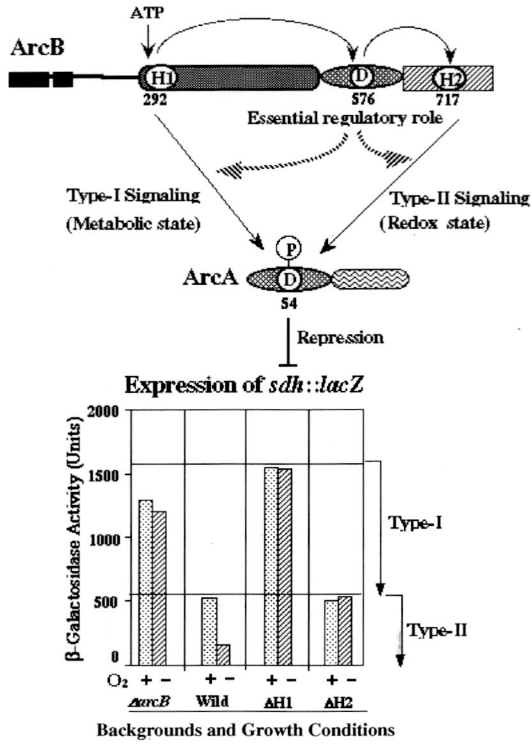

FIGURE 4 Phosphorelay circuitry of the multisignal sensor ArcB. A proposed model used to explain the dual-signaling mechanism underlying signal transduction in the ArcB→ArcA multistep phosphorelay is shown. This model is based on *in vivo* and *in vitro* findings, the crucial example of which is shown in the lower part. Details are given in the text.

can operate in a more sophisticated manner than thought previously, as discussed later.

ADVANTAGE OF MULTISTEP PHOSPHORELAY

ArcB is a hybrid sensor having multiple (at least three) phosphorelay domains, including the newly uncovered HPt domain. As emphasized earlier, this raised the general question of what is the advantage of such a multistep phosphorelay mechanism through the additional HPt domain? Because such a phosphorelay is a simple (mechanistically reversible) flow of a phosphoryl group among certain amino acid residues on polypeptides, it does not necessarily serve to amplify signals, in contrast to the common eukaryotic signal transduction cascades, involving a number of catalytic Ser/Thr-kinases, Tyr-kinases, and phosphoprotein phosphatases. Why should a phosphoryl group travel along a long railroad with extra stations? Is there any mechanistic advantage? Such an extra His→Asp phosphorelay component (or step) may serve as a regulatory checkpoint in a given signaling pathway. It may also provide the potential for an integration of multiple signals at the intermediate step. Alternatively, the HPt intermediate makes it possible to link together two (or more) distinct phosphorelay pathways through a cross-regulation mechanism. These issues have long been the subjects of debate. Results from extensive studies on the ArcB→ArcA phosphorelay have begun to shed light on these general issues as to the function of hybrid His-kinases, as can be seen in the following sections.

MULTISIGNALING CIRCUITRY OF THE ArcB→ArcA PHOSPHORELAY

To gain an insight into the physiological relevance of the *in vitro* observed multistep phosphotransfer circuitry of ArcB→ArcA, a set of plasmids were constructed and characterized, each of which carries a critical mutant of the *arcB* gene (Fig. 4) [36, 50, 51]. ArcB consists of 778 amino acids, among which His-292, Asp-576, and His-717 are crucial for phosphotransfer circuitry. Each of these amino acids was replaced by an altered one to create a set of mutant ArcB proteins: ArcB-ΔH1 (His-292 to Leu), ArcB-ΔD (Asp-576 to Gln), and ArcB-ΔH2 (His-717 to Leu). By monitoring the anaerobic regulation of the *sdhCDAB* operon (*shd::lacZ*), this set of ArcB mutants allowed us to intensively examine *in vivo* the ArcB→ArcA signaling system, with special reference to the multistep phosphotransfer circuitry. Results showed that both the phosphorylated His-292 and His-717 sites are essential for the anaerobic repression of the *sdhCDAB* operon (Fig. 4, bottom). Interestingly, however, in contrast to the His-292 mutant, the ArcB mutant lacking the crucial His-717 does not necessarily exhibit a null phenotype. Rather, this HPt mutant still maintains a certain ability to signal ArcA, particularly under aerobic growth conditions. Namely, even under fully aerobic growth conditions, expression of the *sdhCDAB* operon is repressed to a considerable extent in the HPt mutant background, as compared with the case of an *arcB* null (Δ*arcB*) mutant. These and other *in vivo* results led us to propose a mechanism by which ArcB functions as a multisignaling sensor that is capable of propagating two types of stimuli through two distinct phosphotransfer pathways, as shown schematically in Fig. 5. ArcA is phosphorylated through the two distinct phospho-

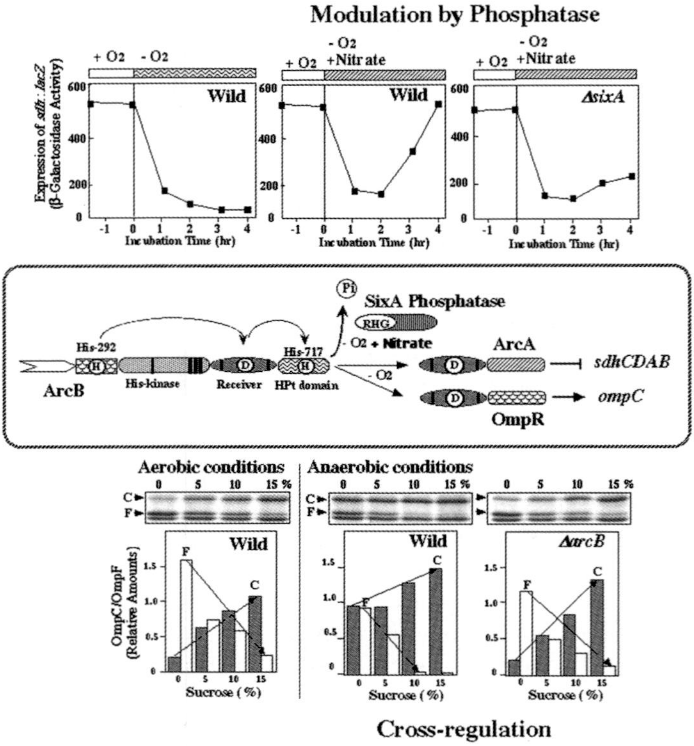

FIGURE 5 Phosphorelay circuitry through ArcB→ArcA implying modulation by the SixA phosphohistidine phosphatase, and cross-phosphorelay between ArcB→OmpR. A proposed model used to explain the sophisticated task, exerted by the hybrid sensor, ArcB is shown. This model is based on *in vivo* and *in vitro* findings, the crucial examples of which are shown in both the upper and the lower parts. Details are given in the text.

transfer pathways, one directly from His-292 of ArcB, and the other through the multistep His→Asp phosphotransfer mediated by the HPt domain (His-717). In any case, the resulting phospho-ArcA functions as the transcriptional repressor for the *sdhCDAB* operon. *In vivo* results were best interpreted by assuming that the HPt-dependent (type-II signaling) pathway is responsible for the response to anoxic conditions, whereas the short-cut His-292 to ArcA (type I signaling) pathway appears to operate in a manner that modulates the *shdCDAB* expression even under the fully aerobic conditions.

To evaluate this model from the physiological viewpoint, one should ask the question of what is the primary (or physiological) signal(s) that is perceived by ArcB? This is a long-standing puzzle, and the answer is not yet clear

[48]. Results from physiological and genetic studies excluded oxygen itself as the signal. It has been proposed that a redox state (*e.g.*, the ratio of intracellular NADH/NAD$^+$) may be a primary signal for the His-kinase activity of ArcB. An element of the electron transport chain (or perhaps proton motive force) may also be a signal. It has also been hypothesized that a certain set of cytosolic metabolites, such as D-lactate, acetate, and pyruvate, may directly affect the kinase activity of ArcB. Our model will shed light on such a long-standing puzzle by assuming that ArcB can respond to two (or more) distinct stimuli by functioning as a multisignaling sensor that is capable of propagating these presumed stimuli through two distinct phosphotransfer pathways, as mentioned earlier (Fig. 4). This view was supported by extensive physiological studies on anaerobic regulation of the TCA cycle [52]. In any event, a following hypothetical view can be envisaged. Upon activated by an anoxic stimulus, ArcB signals ArcA through the type II pathway involving the HPt domain. Alternatively, the type I (or shortcut) pathway may be responsible for the response to the presumed intracellular metabolic state, operating even under fully aerobic conditions. In this model, at present, the function of the internal ArcB receiver domain containing the phosphorylated Asp-576 site is not clear. However, Asp-576 plays an essential role in both type I and type II signaling, as demonstrated that the ArcB-ΔD mutation resulted in a complete loss of both aerobic and aerobic regulations [36]. This ArcB receiver domain may function as an essential self-controlling molecular switch in such a manner that it makes interplay between the two signaling pathways possible.

In short, the proposed model for the multistep ArcB→ArcA phosphorelay implies the general view that a hybrid sensor with an HPt domain may exert a sophisticated task by which it makes possible to propagate multiple signals, depending on different external/internal stimuli.

PHOSPHO-HPt PHOSPHATASE IS INVOLVED IN THE ArcB→ArcA SIGNALING CIRCUITRY

In general, it is tempting to assume that a His→Asp phosphorelay component may serve as a target of a certain phosphatase that functions as a regulator of phosphorelay. Indeed, certain phosphatases have been implicated in some two-component systems. The KinA→Spo0F→Spo0B→Spo0A four-component system is the first known multistep His→Asp phosphorelay that is involved in regulation of *Batilius subtilis* sporulation (note that no HPt domain is implicated in this particular phosphorelay) [53, 54]. Two phosphoprotein phosphatases (RpaA and RpaB) were identified as the ones specific toward the Spo0F response regulator [55]. In *E. coli*, dephosphorylation of phosphoaspartate of the well-known chemotactic CheY response regulator is modulated by CheZ [56]. It was also reported that the stress-activated *E. coli* CpxR→CpxA pathway is modulated by apparent phosphatase activities of PrpA and PrpB [57]. It should also be emphasized that it is generally known that certain His-kinases have both activities of "kinase and phosphatase" toward their cognate receivers [15]. In any case, each target of these known phosphatases is a specific response regulator. Similarly, one can envisage *a priori* that a phosphohistidine in an HPt domain would also be an alternative target of such a regulatory phosphatase. Such a presumed phosphohistidine phosphatase, if present, should also serve as a modulator for a given His→Asp phosphorelay. Based on this rationale, we searched for a phosphohistidine phosphatase that affects the function of the HPt domain of ArcB [58]. The *E. coli sixA* gene product was identified as a candidate. SixA was first identified as a factor that seemed to have an *in vivo* ability to facilitate dephosphorelation from the phospho-HPt domain of ArcB. Indeed, the purified SixA protein exhibits an *in vitro* ability to release the phosphoryl group from the phospho-HPt domain (His-717) of ArcB, but neither from His-kinase (His-292) nor receiver (Asp576) domains (Fig. 5). As far as we know, SixA is the first phosphohistidine phosphatase that is implicated in a certain His→Asp phosphorelay signaling.

The SixA phosphotase consists of 161 amino acids, in which a noticeable sequence motif, an arginine–histidine–glycine (RHG) signature, is located at its N-terminal end (Fig. 5) [58]. Such an RHG signature sequence that is presumably important for a nucleophilic phosphoacceptor is commonly found in a set of divergent enzymes, including eukaryotic fructose-2-,6-bisphosphatase, *E. coli* periplasmic phosphatase, and ubiqutous phosphoglycerate mutase [58]. The three-dimensional structure of SixA has been determined by X-ray crystal analysis [T., Hakoshima, unpublished data]. Results revealed a fine structure analogous to those of fructose-2,6-bisphosphatase and phosphoglycerate mutase. This structural analysis further supported that SixA belongs to a family of phosphatases. It is also interesting to note that proteins homologous to SixA are predicted to be in certain other bacteria, including *Haemophilus influenzae*, *Vibrio cholerae*, *Pseudomona aeruginosa*, and *Synechocystis* sp. [58–60]. This is compatible with the idea that a SixA-like protein may be function as a modulator of a His→Asp phosphorelay in other bacterial species.

PHYSIOLOGICAL ROLE OF SixA-PHOSPHATASE IN RESPONSE TO ANAEROBIC RESPIRATORY CONDITIONS

As mentioned earlier, the SixA phosphatase was identified through an artificial *in vivo* screening strategy [58]. No direct evidence has been provided for that SixA is indeed involved in a signal transduction circuitry of the ArcB→ArcA phosphorelay system *per se* [58]. To address this, a *sixA* null (Δ*sixA*) mutant was isolated. When the Δ*sixA* mutant was characterized in terms of the anaerobic repression of the *shdCDAB* operon in comparison with the wild type, no difference was observed between them (Fig. 5) [51]. This indicated that SixA has apparently nothing to do with ArcB→ArcA signaling. This observation was somewhat disappointing. However, results of further studies showed that this phosphatase is an essential regulatory factor that modulates ArcB→ArcA signaling, particularly under certain anaerobic respiratory growth conditions [61].

It is known that the ArcB signaling pathway is implicated in a more complex regulatory network that allows *E. coli* cells to respond not only to external oxygen, but also certain anaerobic respiratory conditions [35, 42]. As mentioned previously, expression of the *sdhCDAB* operon is typically relevant to the aerobic (TCA cycle) metabolism. Nevertheless, it is also recruited for anaerobic respiration in the absence of oxygen, which instead is mediated by anaerobic electron acceptors, such as nitrate, TMAO, DMSO, and fumarate. This physiologically meaningful event is also regulated through ArcB→ArcA signaling, at least partly [35, 38]. Such a regulatory event can be observed by a simple experiment (Fig. 5, top). First, *E. coli* cells were grown exponentially under aerobic conditions and then the cells were grown under anaerobic conditions in a fresh medium supplemented with and without nitrate. Upon the onset of anoxic conditions, expression of the *sdh* operon is repressed rapidly and severely both in the presence and in the absence of nitrate. Strikingly, however, if nitrate is present, the once repressed expression of *sdh::lacZ* is derepressed markedly after a while. This regulation does make sense from the physiological viewpoint that whenever an exogenous electron acceptor is available in medium, *E. coli* cells tend to curtail its fermentation process in favor of anaerobic respiration, even under anoxic conditions. The Δ*sixA* mutant is defective in this particular induction of the *shdCDAB* operon under such anaerobic respiratory conditions (Fig. 5).

These results are best interpreted by assuming that SixA plays an important role in down regulation of the ArcB→ArcA phosphorelay under certain anaerobic respiratory conditions by exhibiting its phosphatase activity toward the HPt domain. Even under anaerobic conditions, SixA can drain a phosphoryl group from the HPt domain, thereby resulting in down regulation of the ArcB→ArcA phosphorelay at the intermediate step. This down regulation results in derepression of the *sdhCDAB* operon even under anaerobic grown conditions, provided that an anaerobic electron acceptor is available. This regulatory mechanism does make sense and is genius. As a whole, the ArcB hybrid His-kinase, in concert with SixA phosphatase, can propagate certain anaerobic respiratory signals in a sophisticated manner. In this mechanism, one can see an advantage of the multistep His→Asp phosphorelay. Namely, the HPt domain can provide a means, by which a given His→Asp phosphorelay is modulated at an intermediate step by a specific phosphatase.

CROSS-PHOSPHORELAY OCCURS ON OmpR THROUGH EnvZ-OSMOSENSOR AND ArcB ANAEROSENSOR

Inspection of the entire genomic sequence of *E. coli* revealed the occurrence of 28 His-kinases and 32 response regulators in this single species [8]. This means that as many as 30 distinct His→Asp phosphorelay signaling pathways operate in response to a wide variety of environmental stimuli. Does each signaling pathway operate specifically and independently or do some of them together make a network of signaling pathways by unknown mechanisms, such as "cross-regulation" [8, 62, 63]? In fact, a number of *in vivo* and *in vitro* observations show that a certain response regulator can acquire a phosphoryl group from not only its cognate His-kinase, but also heterologous ones (even from low molecular weight substrates in some cases) [64–69]. However, most of these observations were made with artificial reactant stoichiometories. Thus, any physiologically meaningful cross-regulation may prove difficult, and this is a long-standing subject of debate [8, 62, 63]. We addressed this issue with special reference to the osmoresponsive EnvZ→OmpR and anaeroresponsive ArcB→ArcA phosphorelay systems.

As well documented previously, expression of the major outer membrane OmpC and OmpF proteins (or porins) is regulated coordinately at the transcriptional level in response to the medium osmolarity (Fig. 5, bottom) [70, 71]. Both EnvZ (osmosensor) and OmpR (transcriptional regulator) are crucially involved in this particular osmoregulation. EnvZ exhibits a typical His-kinase activity specific toward OmpR (Fig. 1a) [72, 73]. The resulting phospho-OmpR is an active form of the DNA-binding transcriptional activator for both *ompC* and *ompF* promoters. EnvZ exhibits a higher kinase (or a lower phosphatase) activity in response to a higher osmotic stimulus [75–78]. Consequently, the relative amount of phospho-OmpR in cells varies in response to the medium osmolarity, which in turn results in the differential activation of *ompC* and *ompF*, depending on the level of phospho-OmpR. A higher level of OmpC is expressed at a higher osmolarity, whereas a higher level of OmpF is expressed at a lower osmolarity (Fig. 5). This EnvZ→OmpR system is one of the best-characterized examples of single-step phosphorelays [76–78]. During the course of studies on this classical EnvZ→OmpR phosphorelay, it was found that the HPt domain of ArcB is capable of functioning as an alternative phosphodonor for OmpR *in vivo* [19]. This may be indicative that ArcB has an ability to signal not only the cognate ArcA regulator, but also the noncognate OmpR regulator under certain physiological growth conditions, although a multicopy *arcB* gene was employed in the experiment. However, it was of interest to examine the expression profile of OmpC and OmpF under anaerobic growth conditions [79]. The results was intriguing in that, under anaerobic growth conditions, *E. coli* cells exhibit a markedly altered expression profile of OmpC and OmpF, as compared with in the case of standard aerobic conditions (Fig. 5, bottom). Under anaerobic conditions, a significantly larger amount of OmpC is produced even under low osmolarity conditions. Results of extensive genetic studies showed that, under such anaerobic growth conditions, the *arcB* gene serves as an auxiliary genetic determinant that regulates the expression profile of porins (Fig. 5, note that the altered osmoregulatory profile of OmpC and OmpF under anaerobic growth conditions was reverted in a Δ*ArcB* mutation to the same as that observed under aerobic conditions). Results of further *in vivo* and *in vitro* studies supported the following conclusion. Under certain anaerobic growth conditions, porin expression is tuned not only by the authentic osmoresponsive EnvZ sensor, but also by the anaeroresponsive ArcB sensor in an OmpR-dependent manner, thus suggesting that the presumed ArcB→OmpR cross-phosphorelay plays a physiological role by integrating anoxic signals into the osmoregulation of porins [79].

According to Wanner's definition of "cross-regulation," the term refers to control of a response regulator of one phosphorelay system by another [62, 63]. In general, it is attractive to assume that such cross-regulation is an important and common tactic of global control that can link a given phosphorelay system with another to constitute a signaling network. In this regard, the revealed ArcB→OmpR cross-phosphorelay is a clear example of such an interplay of two distinct His→Asp phosphorelay signaling pathways, which results in a multisignal integration into a single response regulator (i.e., EnvZ→OmpR, ArcB→OmpR). It should be remembered that *E. coli* alone has four more hybrid sensors that contain an HPt domain (BarA, EvgS, TorS, and YojN). Each of these other hybrid sensors may also be implicated in each unknown cross-regulatory network among 30 phosphorelay systems in *E. coli*. YojN is discussed further later in this respect.

ATYPICAL HPt FACTOR IS INVOLVED IN THE MULTISTEP RcsC→YojN→RcsB PHOSPHORELAY

Among the *E. coli* His→Asp phosphorelay systems, particularly puzzling is the RcsC (sensor His-kinase)→RcsB (response regulator) system, which is involved in the regulation of polysaccharide synthesis. *E. coli* and other enteric microorganisms are capable of synthesizing an extracellular polysaccharide capsule, called colanic acid, under certain environmental conditions, which include exposures to low temperature, high osmotic strength, and desiccation [80, 81]. Detailed genetic studies on this adaptive response have revealed a very complex regulatory circuitry that modulates expression of the capsular polysaccharide synthesis (*cps*) operon [82–84]. This regulatory network is built up by a certain number of positive and negative regulators, including RcsA, RcsF, and Lon protease, together with the two components (RcsC→RcsB). By analogy with other His→Asp phosphorelay systems, a simple model has been formulated previously [82–84]. The previous model postulated that RcsC senses environmental stimuli and then this His-kinase (together with RcsF) acts to facilitate the phosphorylation of RcsB. Consequently, phospho-RcsB (together with RcsA) stimulates expression of the *cps* operon. However, this model is somewhat puzzling. Like ArcB, RcsC is a hybrid sensor. Unlike ArcB, however, this sensor is unique in that it consists of a His-kinase domain and a receiver domain, but without an HPt domain (Fig. 5). RcsB may acquire a phosphoryl group directly from the His-kinase domain of RcsC or an unusual Asp→Asp phosphorelay may occur between both the receivers of RcsC and RcsB. In any case, it is of interest to address this issue for two reasons. First, (or generally), many hybrid sensors (particularly eukaryotic ones) have a structural design very similar to RcsC in that they consist of a His-kinase domain and a receiver domain, but lack an HPt domain (see Fig. 6). Second (or specifically), a thorough understanding of the *E. coli* Rcs-signaling mechanism underlying activation of the capsular synthesis pathway would resolve the question of why many other virulent and/or pathogenic bacteria have the homologous Rcs-signaling system [85–88]. Here we propose an alternative model of the Rcs-signaling system, in which a novel and unique His-containing phosphotransmitter (named YojN) is implicated [84a] (Fig. 6).

Both the *rcsC* and *rcsB* genes are located next to each other in a divergent orientation at the *E. coli* genome coordinates of approximately 2500 kb [85]. Upstream of the *rcsB* gene, there is another gene, named *yojN*, the deduced amino acid sequence of which (890 amino acids) shows a considerable similarity to that of RcsC, particularly in the His-kinase domain. Nevertheless, YojN may not be a sensor because the crucial autophosphorylation (His-1) site is missing in the corresponding YojN sequence. Furthermore, YojN contains no receiver domain at its C-terminal portion; rather it does contain an about 100 amino acid sequence, in which a putative HPt motif is found. A short stretch of amino acid sequence in the C-terminal region of YojN is highly similar to that of the ArcB HPt motif (see Fig. 3). Thus, YojN appears to have a unique structural design in that it consists of a pseudo-His-kinase domain, followed by a HPt domain. These facts together led us to hypothesize

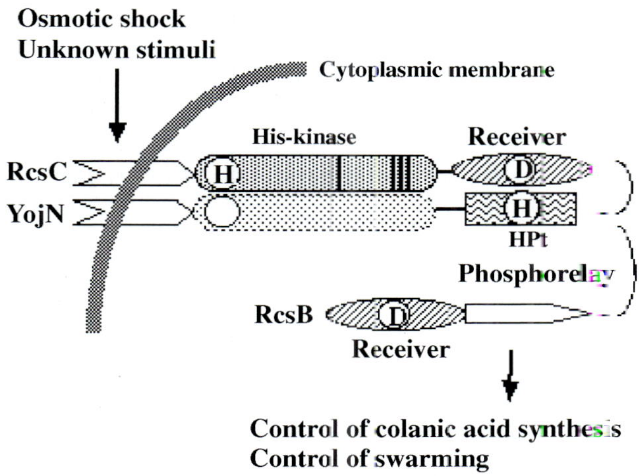

FIGURE 6 A revised model for the mechanism of multistep RcsC→YojN→RcsB phosphorelay, which is responsible for regulating the swarming behavior, as well as the colanic acid synthesis in *E. coli*. Note that this phosphorelay system is somewhat unique, as compared with the classical ones shown in Fig. 1. Other details are given in the text.

that YojN might be involved in the RcsC→RcsB phosphorelay as a histidine-containing phosphotransmitter. Results of extensive studies showed that this is indeed the case.

Here a revised model can be proposed in which both *yojN* and *rcsC* genes are essentially involved in adaptive induction of the *cps* operon through the multistep RcsC (His→Asp)→YojN (His)→RcsB (Asp) phosphorelay signaling (Fig. 6). In this unique mechanism by which the *E. coli* capsular synthesis is modulated, a new member YojN plays a crucial role in the presumed multistep His→Asp→His→Asp phosphorelay. YojN containing an HPt domain at its C terminus serves as a phosphotransfer intermediate to link between RcsC and RcsB. Because YojN has a hydrophobic domain(s) at its N-terminal region, like RcsC, these two proteins may be located together in the cytoplasmic membrane by forming a heterodimer. When stimulated, such a RcsC/YojN heterodimer as a whole may function as a sensor for an as yet unknown external stimuli. An intermolecular His→Asp→His phosphorelay may be allowed to occur, in which RcsC serves as a primary His-kinase. Consequently, RcsB acquires a phosphoryl group from the HPt domain of YojN through the multistep phosphorelay.

Among His→Asp phosphorelay systems in *E. coli*, the proposed RcsC→YojN→RcsB framework is unique in that there is no such precedent. This supports the current view that the common His→Asp phosphorelay strategy is highly plastic in mechanistic designs. Interestingly, YojN is common in the sense that there are several YojN homologues in the databases of other microorganisms, which include *Salmonella typhi*, *Vibrio cholerae*, *Pseudomonas aeroginosa*, *Klebsiella pneumoniae*, *Erwinia amylovora*, and *Proteus mirabilis* [86–88]. For example, in the recently released genome sequences of both *V. cholerae* and *P. aeruginosa*, not only the RcsC and RcsB homologues, but also the YojN homologues are found, although their regulatory functions are entirely elusive [59, 60]. In *P. mirabilis*, a YojN homologue (named RsbA) appears to be involved in a coordinate regulation of colony-swarming migration [88]. In the report, RsbA was characterized as an "atypical His-kinase." However, our inspection of the RsbA sequence revealed that it has indeed an HPt domain at its C-terminal end, as the *E. coli* YojN counterpart does. These instances are indicative of that such an atypical multistep RcsC→YojN→RcsB phosphorelay system is commonly conserved in a wide variety of bacterial species in which this sophisticated adaptive response system often associates with virulence and/or pathogenesis. The framework proposed for *E. coli* RcsB→YojN→RcsB phosphorelay will provide a general basis for understanding the analogous adaptive responses in many other bacteria.

HPt DOMAIN IN HIGHER PLANTS

In the higher plant *Arabidopsis thaliana*, results of intensive studies suggest that His→Asp phosphorelay mechanisms are involved presumably in the propagation of environmental stimuli, such as phytohormones (*e.g.*, ethylene and cytokinin), as has been demonstrated through molecular genetic approaches (for a review, see D'Agostino and Kieber [89], and references therein). These facts suggest that the bacterial type of signal transduction mechanism is common in higher plants and plays fundamental roles in adaptive responses to environments (Fig. 7). Indeed, a further inspection of *Arabidopsis* databases revealed that this model plant has at least 11 sensor His-kinases. Five (ETR1, ETR2, ERS1, ERS2, and EIN4) have been demonstrated to be ethylene receptors [90–92], two (CKI1 and CKI2) were assumed to be involved in a cytokinin response [93], and one (ATHK1) was proposed to be a putative osmosensor [94]. The real receptor His-kinase for cytokinin has been uncovered as AHK4 [95, 96]. These *Arabidopsis* His-kinases have structural designs very similar to those of RcsC and Sln1p in that they consist of a His-kinase domain, followed by a receiver domain, lacking any HPt domain. Furthermore, it has been demonstrated that *Arabidopsis* has a number of response regulators (named ARR series, *Arabidopsis* response regulators), each containing a typical phospho-accepting receiver domain [14, 97–100].

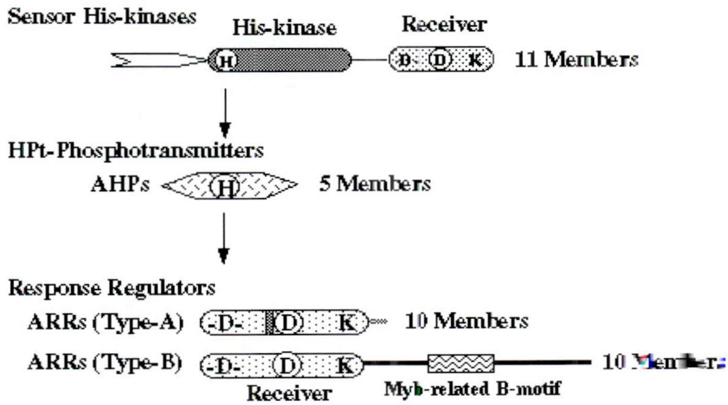

FIGURE 7 Schematic representations of structural designs of the signal transducers involved in the presumed *Arabidopsis* His-to-Asp phosphorelay network. They include His-kinases, HPt phosphotransmitters, and response regulators, each of which contains either histidine (H) or aspartate (D), both of which are crucial for the presumed phosphorelay interaction between these signal transducers, as shown schematically.

This plant has at least 20 members of the family of response regulators that can be classified into two distinct subtypes (type A and type B), as judged from their structural designs and expression profile. The type-A family of response regulators (10 members) resembles CheY (see Fig. 1d) in that each of them comprises only a receiver domain without any output domain. Type-B family members (10 members) are presumably transcriptional factors, each of which has a Myb-related DNA binding domain as well as a nuclear localization signal (NLS). Interestingly, type-A family members are induced by cytokinin treatment of plants at the level of transcription, but type-B family members are not [101]. These facts suggest that the bacterial type of signal transduction mechanism is very common in this higher plant. Then, one can easily envisage that *Arabidopsis* must have genes each encoding a HPt domain.

Indeed, it has been reported that *Arabidopsis* has at least five genes each encoding a typical HPt phosphotransmitter [27, 102]. Like Ypd1p of the budding yeast, each of these HPt factors (named AHP series, *Arabidopsis HPt* factors) contains only a HPt domain consisting of about 150 amino acids, and their amino acid sequences are significantly similar to that of Ypd1p (see Fig. 2]. Several lines of evidence support that these AHPs have an ability to interact with ARRs through a phosphorelay reaction. Provided that these HPt factors play a fundamental biological role in a manner that is common (or general) among higher plants, one can expect that there must be homologous (or orthologous) proteins in many other plants, if not all. A further inspection to this end revealed that this is indeed the case. When a search was done for cDNA sequences each encoding a protein similar to AHPs in the currently available plant EST databases, a wide variety of plant species, including both dicots (*e.g.*, cotton and tomato) and monocots (*e.g.*, maize and rice), appear to have genes, each of which specifies a protein strikingly similar to AHPs in their amino acid sequences. Collectively, such widespread occurrences of His→Asp phosphorelay components in higher plants are best interpreted by assuming that multiple His→Asp phosphorelay pathways are involved in a variety of fundamental biology of higher plants. Nevertheless, it should be emphasized that elucidation of their biology and physiology is at a very early stage.

CONCLUDING REMARKS

As overviewed briefly here, a multistep His→Asp phosphorelay mechanism exerts a more sophisticated task than thought previously, in which a common HPt domains acts in concert with the classical two components: His-kinases and response regulators. Numerous instances of HPt domains can be predicted to occur in the current databases for both prokaryotes and eukaryotes, and their numbers are growing very rapidly. Nevertheless, their biological (or physiological) roles are virtually unknown, except for the cases mentioned here. Because His→Asp phosphorelay signaling systems are so common and global in both prokaryotes and eukaryotes, they are the best paradigms of choice to explore through taking the newly developing postsequencing approaches, such as DNA microarray and proteome analyses.

ACKNOWLEDGMENT

This study was supported by a grant-in-aid for scientific research on a priority area [Tokutei (B) to 12142201 to TM] from the Ministry of Education, Science, Sports and Culture of Japan.

REFERENCES

1. Stock, J. B., Ninfa, A. D., and Stock, A. M. (1989). Protein phosphorylation and regulation of adaptive response in bacteria. *Microbiol. Rev.* 53, 450–490.
2. Stock, J. B., Stock, A. M., and Mottonenn, J. M. (1990). Signal transduction in bacteria. *Nature* 344, 395–400.
3. Bourret, R. B., Borkovich, K. A., and Simon, M. I. (1991). Signal transduction pathways involving protein phosphorylation in prokaryotes. *Annu. Rev. Biochem.* 60, 401–441.

4. Parkinson, J. S., and Kofoid, E. C. (1992). Communication modules in bacterial signalling proteins. *Annu. Rev. Genet.* **26**, 71–112.
5. Parkinson, J. S. (1993). Signal transduction schemes of bacteria. *Cell.* 73, 857–871.
6. Hoch, J. A., and Silhavy, T. J. (1995). "Two-Component Signal Transduction." ASM Press, Washington, DC.
7. Mizuno, T., Kaneko, T. and Tabata, S. (1996). Compilation of all genes encoding bacterial two-component signal transducers in the genome of the cyanobacterium, *Synechocystis* sp. strain PCC 6803. *DNA Res.* **3**, 407–414.
8. Mizuno, T. (1997). Compilation of all genes encoding two-component phosphotransfer signal transducers in the genome of *Escherichia coli*. *DNA Res.* **4**, 161–168.
9. Farbret, C., Feher, V. A., and Hoch, J. A. (1999). Two-component signal transduction in *Bacillus subtilis*: How one organism sees its world. *J. Bacteriol.* **181**, 1975–1983.
10. Chang, C., Kwok, S. F., Bleeker, A. B., and Meyerowitz, E. M. (1993). *Arabidopsis* ethylene-response gene ETR1: Similarity of product to two-component regulators. *Science* **262**, 539–544.
11. Alex, L. A., Borkovich, K. A., and Simon, M. I. (1996). Hyphal development in *Neurospora crassa*: Involvement of a two-component histidine kinase. *Proc. Natl. Acad. Sci. USA* **93**, 3416–3421.
12. Schuster, S. C., Noegel, A. A., Oehmen, F., Gerisher, G., and Simon, M.I. (1996). The hybrid histidine kinase DokA is part of the osmotic response system of *Dictyostelium*. *EMBO J.* **15**, 3880–3889.
13. Wurgler-Murphy, S. M., and Saito, H. (1997). Two-component signal transducers and MAPK cascades. *Trends Biochem. Sci.* **22**, 172–176.
14. Imamura, A., Hanaki, N., Nakamura, A., Suzuki, T., Taniguchi, M., Kiba, T., Ueguchi, V., Sugiyama, T., and Mizuno, T. (1999). Compilation and characterization of *Arabidopsis* response regulators implicated in His-Asp phosphorelay signal transduction. *Plants Cell Physiol.* **40**, 733–742.
15. Inouye, M. (1996). His-Asp phosphorelay; two components or more? *Cell* **85**, 13–14.
16. Appleby, J. L., Parkinson, J. S., and Bourret, R. B. (1996). Signal transduction via the multi-step phosphorelay: Not necessarily a road less traveled. *Cell* **86**, 845–848.
17. Mizuno, T. (1998). His-Asp phosphotranfer signal transduction. *J. Biochem.* (Tokyo) **123**, 555–563.
18. Perraud, A. L., Weiss, V., and Gross, R. (1999). Signalling pathways in two-component phosphorelay systems. *Trends Microbiol.* **7**, 115–120.
19. Ishige, K., Nagosawa, S., Tokishita, S. and Mizuno, T. (1994). A novel device of bacterial signal transducers. *EMBO J.* **13**, 5195–5202.
20. Tsuzuki, M., Ishige, K., and Mizuno, T. (1995). Phosphotransfer circuitry of the putative multi-signal transducer, ArcB, of *Escherichia coli*: *In vitro* studies with mutants. *Mol. Microbiol.* **18**, 953–962.
21. Iuchi, S. (1993). Phosphorylation/dephosphorylation of the receiver module at the conserved aspartate residue controls transphosphorylation activity of histidine kinase in sensor protein ArcB of *Escherichia coli*. *J. Biol. Chem.* **268**, 23972–23980
22. Georgellis, D., Lynch, S., and Lin, E. C. C. (1997). *In vitro* phosphorylation study of the Arc two-component signal transduction system of *Escherichia coli*. *J. Bacteriol.* **179**, 5429–5435.
23. Arico, B., Miller, J. F., Roy, C., Stibitz, S., Monack, D., Falkow, S., Gross, R., and Rappuoli, R. (1989). Sequences required for expression *Bordetella pertussis* virulence factors share homology with prokaryotic signal transduction proteins. *Proc Nat. Acad. Sci. USA* **86**, 6671–6675.
24. Uhl, M. A., and Miller, J. F. (1996). Integration of multiple domains in a two component sensor protein, the *Bordetella pertussis* BvgAS. *EMBO J.* **15**, 1028–1036.

25. Maeda, T., Wurgler-Murphy, S. M., and Saito, H. (1994). A two-component system that regulates an osmosensing MAP kinase cascade in yeast. *Nature* **369**, 242–245.
26. Posas, F., Wurgler-Murphy, S. M., Maeda, T., Witten, E .A., Thai, T. C., and Saito, H. (1996). Yeast HOG1 MAP kinase cascade is regulated by a multistep phosphorelay mechanism in the SLN1-YPD1-SSK1 two-component osmosensor. *Cell* **86**, 865–875.
27. Suzuki, T., Imamura, A., Ueguchi, C., and Mizuno, T. (1998). Histidine-containing phosphotransfer (HPt) signal transducers implicated in His-to-Asp phosphorelay in *Arabidopsis*. *Plant Cell Physiol.* **39**, 1258–1268.
28. Kato, M., Mizuno, T., Shimizu, T. and Hakoshima, T. (1997). Insights into multistep phosphorelay from the crystal structure of the C-terminal HPt domain of ArcB. *Cell* **88**, 717–723.
29. Song, H. K., Lee, J. Y., Lee, M. G., Moon, J., Min, K., Yang, J. K., and Suh, S. W. (1999). Insights into eukaryotic multistep phosphorelay signal transduction revealed by the crystal structure of Ypd1 from *Saccharomyces cerevisiae*. *J. Mol. Biol.* **293**, 753–761.
30. Xu, Q., and West, A. H. (1999). Conservation of structure and function among histidine-containing phosphotransfer (HPt) domains as revealed by the crystal structure of YPD1. *J. Mol. Biol.* **292**, 1039–1050.
31. Zhou, H., Lowry, D.F., Swanson, R.V., Simon, M.I., and Dahlquist, F.W. (1995). NMR studies of the phosphotransfer domain of the histidine kinase CheA from *Eschierchia coli*: Assignments, secondary structure, general fold, and backbone dynamics. *Biochemistry* **34**, 1358–1370.
32. Tomomori, C., Tanaka, T., Dutta, R., Park, H., Saha, S. K., Zhu, Y., Ishima, R., Liu, D., Tong, K. I., Kurokawa, H., Qian, H., Inouye, M., and Ikura, M. (1999). Solution structure of the homodimeric core domain of *Escherichia coli* histidine kinase EnvZ. *Nature Struct. Biol.* **6**, 729–34.
33. Tanaka, T., Saha, S. K., Tomomori, C., Ishima, R., Liu, D., Tong, K. I., Park, H., Dutta, R., Qin, L., Swindells. M. B., Yamazaki, T., Ono, A. M., Kainosho, M., Inouye, M., and Ikura, M. (1998). NMR structure of the histidine kinase domain of the *E. coli* osmosensor EnvZ. *Nature* **396**, 88–92.
34. Neidhardt, F. C. (editor in Chief) (1996). *Escherichia coli* and *Sakmonella*. ASM Press, Washington, DC.
35. Park, S.-J., Tseng, C.-P., and Gunsalus, R. P. (1995). Regulation of succinate dehydrogenase (sdhCDAB) operon expression in *Escherichia coli* response to carbon supply and anaerobiosis: Role of ArcA and Fnr. *Mol. Microbiol.* **15**, 473–482.
36. Matsushika, A., and Mizuno, T. (1998). A dual-signaling mechanism mediated by the ArcB hybrid sensor kinase containing the histidine-containing phosphotransfer domain in *Escherichia coli*. *J. Bacteriol.* **180**, 3973–3977.
37. Iuchi, S., and Lin, E. C. C. (1988). *arcA* (*dye*), a global regulatory gene in *Escherichia coli* mediating repression of enzymes in aerobic pathways. *Proc. Natl. Acad. Sci. USA* **85**, 1888–1892.
38. Iuchi, S., Matsuda, Z., Fujiwara, T., and Lin, E. C. C. (1990). The *arcB* gene of *Escherichia coli* encodes a sensor-regulator protein for anaerobic repression of the *arc* modulon. *Mol. Microbiol.* **4**, 715–727.
39. Iuchi, S., Chepuri, V., Fy, H.-A., Gennis, R. B., and Lin, E. C. C. (1990). Requirement for the terminal cytochromes in generation of the aerobic signal for the arc regulatory system in *Escherichia coli*: Study utilizing deletions and *lac* fusions of *cyo* and *cyd*. *J. Bacteriol.* **172**, 6020–6025.
40. Iuchi, S., and Lin, E. C. C. (1992). Mutational analysis of signal transduction by ArcB, a membrane sensor protein responsible for anaerobic repression of operons involved in the central aerobic pathways in *Escherichia coli*. *J. Bacteriol.* **174**, 3972–3980.
41. Iuchi, S., and Lin, E. C. C. (1992). Purification and phosphorylation of the Arc regulatory

components of *Escherichia coli*. *J. Bacteriol.* **174**, 5617–5623.
42. Iuchi, S., and Lin, E. C. C. (1993). Adaptation of *Escherichia coli* to redox environments by gene expression. *Mol. Microbiol.* **9**, 715–727.
43. Iuchi, S., and Lin, E. C. C. (1995). Signal transdution in the Arc system for control of operons encoding aerobic respiratory enzymes. *In* "Two-Component Signal Transduction" J. (A., Hoch, and T. J. Silhavy, eds.). ASM Press, Washington, DC.
44. Lynch, A. S., and Lin, E. C. C. (1996). Responses to molecular oxygen. *In* "*Escherichia coli* and *Salmonella*: Cellular and Molecular Biology" (F. C., Neidhardt, editor in chief). ASM Press, Washington, DC.
45. Lynch, A. S., and Lin, E. C. C. (1996). Transcriptional control mediated by the arcA two-component response regulator protein of *Escherichia coli*: Characterization of DNA binding at target promoters. *J. Bacteriol.* **178**, 6238–6249.
46. Georgellis, D., Kwon, O., and Lin, E. C. C. (1999). Amplification of signaling activity of the arc two-component system of *Escherichia coli* by anaerobic metabolites: An *in vitro* study with different protein modules. *J. Biol. Chem.* **274**, 35950–35954.
47. Kwon, O., Georgellis, D., and Lin, E. C. C. (2000). Phosphorelay as the sole physiological route of signal transmission by the *arc* two-component system of *Escherichia coli*. *J. Bacteriol.* **182**, 3858–3862.
48. Kwon, O., Georgellis, D., Lynch, A. S., Boyd, D., and Lin, E. C. C (2000). The ArcB sensor kinase of *Escherichia coli*: Genetic exploration of the transmembrane region. *J. Bacteriol.* **182**, 2960–2966.
49. Georgellis, D., Kwon, O., and Lin, E. C. C. (1999). Amplification of signaling activity of the arc two-component system of *Escherichia coli* by anaerobic metabolites: An *in vitro* study with different protein modules. *J. Biol. Chem.* **274**, 35950–35954.
50. Matsushika, A. and Mizuno, T. (1998). The structure and function of the histidine-containing phosphotransfer (HPt) signaling domain of the *Escherichia coli* ArcB sensor. *J. Biochem.* (*Tokyo*) **124**, 440–445.
51. Matsushika, A., and Mizuno, T. (1998). Mutational analysis of the histidine-containing phosphotransfer (HPt) signaling domain of the ArcB sensor in *Escherichia coli*. *Biosci. Biotechnol. Biochem.* **62**, 2236–2238.
52. Alexeeva, S., de Kort, B., Sawers, G., Hellingwerf, K. J., and de Mattos, M. J. (2000). Effects of limited aeration and of the ArcAB system on intermediary pyruvate catabolism in *Escherichia coli*. *J. Bacteriol.* **182**, 4934–4940.
53. Burbulys, D., Trach, K. A., and Hock, J. A. (1991). Initiation of sporulation in *B. subtilis* is controlled by a multicomponent phosphorelay. *Cell* **65**, 545–552.
54. Perego, M., C. Hanstein, K. M. Welsh, T. Djavakhishvili, P. Glaser, and Hoch, J. A. (1994). Multiple protein-aspartate phosphatases provide a mechanism for the integration of diverse signals in the control of development in *B. subtilis*. *Cell* **79**, 1047–1055.
55. Blat, Y., Gillespie, B., Bren, A., Dahlquist, F. W., and Eisenbach, M. (1998). Regulation of phosphatase activity in bacterial chemotaxis. *J. Mol. Biol.* **284**, 1191–1199.
56. Missiakas, D., and Raina, S. (1997). Signal transduction pathways in response to protein misfolding in the extracytoplasmic compartments of *E. coli*: Role of two new phosphoprotein phosphatases PrpA and PrpB. *EMBO J.* **16**, 1670–1685.
57. Tzeng, Y. L., Feher, V. A., Cavanagh, J., Perego, M., and Hoch, J. A. (1998). Characterization of interactions between a two-component response regulator, Spo0F, and its phosphatase, RapB. *Biochemistry* **37**, 16538–16545.
58. Ogino, T., Matsubara, M., Kato, N., Nakamura, Y., and Mizuno, T. (1998). An *Escherichia coli* protein that exhibits phosphohistidine phosphatase activity towards the HPt domain of the ArcB sensor involved in the multistep His-Asp phosphorelay. *Mol. Microbiol.* **27**, 573–585.
59. Heidelberg, J. F., Eisen, J. A., Nelson, W. C., Clayton, R. A., Gwinn, M. L., et al. (2000).

DNA sequence of both chromosomes of the cholera pathogen *Vibrio cholerae*. *Nature* **406**, 477–483.

60. Stover, C. K., Pham, X. Q., Erwin, A. L., Mizoguchi, S. D., Warrener, P., *et al.* (2000). Complete genome sequence of *Pseudomonas aeruginosa* PA01, an opportunistic pathogen. *Nature* **406**, 959–964.
61. Matsubara, M., and Mizuno, T. (2000). The SixA phospho-histidine phosphatase modulates the ArcB phosphorelay signal transduction in *Escherichia coli*. *FEBS Lett.* **47**, 118–124.
62. Wanner, B. L., and Wilmes-Riesenberg, M. R. (1992). Involvment of phosphotransacetylase, acetate kinase, and acetyl phosphate synthesis in control of the phosphate regulon in *Escherichia coli*. *J. Bacteriol.* **174**, 2124–2130.
63. Wanner, B. L. (1992). Is cross regulation by phosphorylation of two-component response regulator proteins important in bacteria? *J. Bacteriol.* **174**, 2053–2058.
64. Ninfa, A. J., Ninfa, E. G., Lupas, A. N., Stock, A., and Magasanik, B. (1988). Crosstalk between bacterial chemotaxis signal transductionproteins and regulators of transcription of the ntr regulon: Evidence that nitrogen assimilation and chemotaxis are controlled by a common phosphotransfer mechanism. *Proc. Natl. Acad. Sci. USA* **85**, 5492–5500.
65. Amemura, M., Makino, K., Shinagawa, H., and Nakata, A. (1990). Cross talk to the phosphate regulon of *Escherichia coli* by PhoM protein: PhoM is a histidine protein kinase and cataltzes phosphorylation of PhoB and PhoM-open reading frames. *J. Bacteriol.* **172**, 6300–6307.
66. Nagasawa, S., Tokishita, S., Aiba, H., and Mizuno, T. (1992). A novel sensor-regulator protein that belongs to the homologous family of signal transduction proeins involved in adaptive responses in *Escherichia coli*. *Mol. Microbiol.* **6**, 799–807.
67. Feng, J., Atkinson, M. R., McCleary, W., Stock, J. B., Wanner, B. L., and Ninfa, A. J. (1992). Role of phosphorylated metabolic intermediates in the regulation of glutamine synthetase synthesis in *Escherichia coli*. *J. Bacteriol.* **174**, 6061–6070.
68. Nagasawa, S., Ishige, K., and Mizuno, T. (1993). Novel members of the two-components signal transduction genes in *Escherichia coli*. *J. Biochem.* (*Tokyo*) **114**, 350–357.
69. McCleary, W. R., and Stock, J. B. (1994). Acetyl phosphate and the activation of two-component response regulators. *J. Biol. Chem.* **269**, 31567–31572.
70. Mizuno, T., and Mizushima, S. (1990). Signal trnsduction and gene regulation through the phosphorylation of two regulatory components: The molecular basis for the osmotic regulation of porin genes. *Mol. Microbiol.* **4**, 1077–1082.
71. Mizuno, T., Chou, M.-Y., and Inouye, M. (1983). A comparative study on the genes for three porins of the *Escherichia coli* outer membrane: DNA sequence of the osmoregulated ompC gene. *J. Biol. Chem.* **258**, 6932–6940.
72. Tokishita, S., Yamada, H., Aiba, H., and Mizuno, T. (1990). Transmembrane signal transduction and osmoregulation in *Escherichia coli*. II. The osmotic sensor, EnvZ, located in the isolated cytoplasmic membrane displays its phosphorylation and dephosphorylation abilities as to the activator protein, OmpR. *J. Biochem.* (*Tokyo*) **108**, 488–493.
73. Mizuno, T., and Tanaka, I. (1997). Structure of the DNA-binding domain of the OmpR family of response regulators. *Mol. MIcrobiol.* **24**, 665–667.
74. Aiba, H., Mizuno, T., and Mizushima, S. (1989). Transfer of phosphoryl group between two regulatory proteins involved in osmoregualtory expression of the ompF and ompC genes in *Escherichia coli*. *J. Biol. Chem.* **264**, 8563–8567.
75. Aiba, H., Nakasai, F., Mizushima, S., and Mizuno, T. (1989). Evidence for the physiological importance of the phosphotransfer between the two regulatory components, EnvZ and OmpR, in osmoregulation in *Escherichia coli*. *J. Biol.Chem.* **264**, 14090–14094.
76. Aiba, H., and Mizuno, T. (1990). Phosphorylation of a bacterial activator protein, OmpR, by a protein kinase, EnvZ, stimutates the transcription of the ompF and ompC gene in

Escherichia coli. *FEBS Lett*. **261**, 19–22.

77. Forst, S., Delgado, J., and Inouye, M. (1989). Phosphorylation of OmpR by the osmosensor EnvZ modulates the expression of the *ompF* and *ompC* genes in *Escherichia coli*. *Proc. Natl. Acad. Sci. USA* **86**, 6052–6056.
78. Igo, M. M., Ninfa, A. J., Stock, J.B., and Silhavy, T. J. (1989). Phosphorylation and dephosphorylation of a bacterial transcriptional activator by a transmembrane receptor. *Genes Dev.* **3**, 1725–1734.
79. Matsubara, M., Kitaoka, S., Takeda, S., and Mizuno, T. (2000). Tuning of the porin expression under anaerobic growth conditions by His-to-Asp cross-phosphorelay through both the EnvZ-osmosensor and ArcB-anaerosensor in *Escherichia coli*. *Genes Cells* **5**, 555–569.
80. Gottesman, S., Trisler, P., and Torres-Cabassa, A. (1985). Regulation of capsular polysaccharide synthesis in *Escherichia coli* K-12: Characterization of three regulatory genes. *J. Bacteriol*. **162**, 1111–1119.
81. Gottesman, S., and Stout, V. (1991). Regulation of capsular polysaccharide synthesis in *Escherichia coli* K-12. *Mol. Microbiol*. **5**, 1599–1606.
82. Stout, V., and Gottesman, S. (1990). RcsB and RcsC: A two-component regulator of capsule synthesis in *Escherichia coli*. *J. Bacteriol*. **172**, 659–669.
83. Stout, V. (1994). Regulation of capsule synthesis includes interactions of the RcsC/RcsB regulatory pair. *Res. Microbiol*. **145**, 389–392.
84. Stout, V. (1996). Identification of the promoter region for the colanic acid polysaccharide biosynthetic genes in *Escherichia coli* K-12. *J. Bacteriol*. **178**, 4273–4280

84a Takeda, S., Fujisawa, Y., Matsubara, M., and Mizuno, T. (2001). A novel feature of the multistep phosphorelay in *Escherichia coli*: A revised model of the RcsC→YojN→RcsB signaling pathway implicated in capsular synthesis and swarming behaviour. *Mol. Microbiol*. **40**, 440–450.

85. Brill, J.A., Quinlan-Walshe, C., and Gottesman, S. (1988). Fine-structure mapping and identification of two regulators of capsule synthesis in *Escherichia coli* K-12. *J. Bacteriol*. **170**, 2599–2611.
86. Bereswill, S., and Geider, K. (1997). Characterization of the rcsB gene from *Erwinia amylovora* and its influence on exoploysaccharide synthesis and virulence of the fire blight pathogen. *J. Bacteriol*. **179**, 1354–61.
87. Arricau, N., Hermant, D., Waxin, H., Ecobichon, C., Duffey, P. S., and Popoff, M. Y. (1998). The RcsB-RcsC regulatory system of *Salmonella typhi* differentially modulates the expression of invasion proteins, flagellin and Vi antigen in response to osmolarity. *Mol. Microbiol*. **29**, 835–850.
88. Belas, R., Schneider, R., and Melch, M. (1998). Characterization of *Proteus mirabilis* precocious swarming mutants: Identification of *rsbA*, encoding a regulator of swarming behavior. *J. Bacteriol*. **180**, 6126–6139.
89. D'Agostino, I. B., and Kieber, J. J. (1999). Phosphorelay signal transduction: The emerging family of plant response regulators. *Trends Biol. Sci*. **24**, 452–456.
90. Hua, J., Chang, C., Sun, Q., and Meyerowitz, E. M. (1995). Ethylene insensitivity conferred by *Arabidopsis* ERS gene. *Science*, **269**, 1712–1714.
91. Hua, J., and Meyerowitz, E. M. (1998). Ethylene responses are negatively regulated by a receptor gene family in *Arabidopsis thaliana*. *Cell*. **94**, 261–271.
92. Sakai, H., Hua, J., Chen, Q. G., Chang, C., Medrano, L. J., Bleecker, A. B., and Meyerowitz, E. M. (1998). ETR2 is an ETR1-like gene involved in ethylene signaling in *Arabidopsis*. *Proc. Natl. Acad. Sci. USA* **95**, 5812–5817.
93. Kakimoto, T. (1996) CKI1, a histidine kinase homolog implicated in cytokinin signal transduction. *Science* **274**, 982–985.
94. Urao, T., Yakubov, B., Satoh, R., Yamaguchi-Shinozaki, K., Seki, M., Hirayama, T., and

Shinozaki, K. (1999). A transmembrane hybrid-type histidine kinase in *Arabidopsis* functions as an osmosensor. *Plant Cell* **11**, 1743–1754.

95. Suzuki, T., Miwa, K., Ishikawa, K., Yamada, H., Aiba, H., and Mizuno, T. (2001). The *Arabidopsis* sensor His-kinase, AHK4, can respond to cytokinins. *Plant Cell Physiol.* **42**, 107–113.
96. Inoue, T., Higuchi, M., Hashimoto, Y., Seki, M., Kobayashi, M., Kato, T., Tabata, S., Shinozaki, K., and Kakimoto, T (2001). Dentiofication of CRE1 as a cytokinin receptor from *Arabidopsis. Nature* **409**, 1060–1063.
97. Brandstatter, I., and Kieber, J. J. (1998). Two genes with similarity to bacterial response regulators are rapidly and specifically induced by cytokinin in *Arabidopsis. Plant Cell,* **10**, 1009–1019.
98. Imamura, A., Hanaki, N., Umeda, H., Nakamura, A., Suzuki, T., Ueguchi, C., and Mizuno, T. (1998). Response regulators implicated in His-to-Asp phosphotransfer signaling in *Arabidopsis. Proc. Natl. Acad. Sci. USA* **95**, 2691–2696.
99. Sakai, H., Aoyama, T., Bono, H., and Oka, A. (1998). Two-component response regulators from *Arabidopsis thaliana* contain a putative DNA-binding motif. *Plant Cell Physiol.* **39**, 1232–1239.
100. Urao. T., Yakubov, B., Yamaguchi-Shinozaki, K., and Schinozaki, K. (1998). Stress-responsive expression of genes for two-component response regulator-like proteins in *Arabidopsis thaliana. FEBS Lett.* **427**, 175–178.
101. Taniguchi, M., Kiba, T., Sakakibara, H., Ueguchi, C., Mizuno, T., and Sugiyama, T. (1998). Expression of *Arabidopsis* response regulator homologs is induced by cytokinins and nitrate. *FEBS Lett.* **429**, 259–262.
102. Miyata, S.-I., Urao, T., Yamaguchi-Shinozaki, K., and Shinozaki, K. (1998). Characterization of genes for two-component phosphorelay mediators with a single HPt domain in *Arabidopsis thaliana. FEBS Lett.* **437**, 11–14.

CHAPTER 9

Genome-Wide Analysis of *Escherichia coli* Histidine Kinases

TAKESHI MIZUNO,* HIROFUMI AIBA,* TAKU OSHIMA,[†] HIROTADA MORI[†]
AND BARRY L. WANNER[‡]

**Laboratory of Molecular Microbiology, School of Agriculture, Nagoya University, Chikusa-ku, Nagoya 464-8601, Japan and [†]CREST, Research and Education Center for Genetic Information, Nara Institute of Science and Technology, Ikoma 630-0101, Japan and [‡]Department of Biological Sciences, Purdue Univeristy, West Lafayette, Indiana 47907*

With special reference to the His→Asp phosphorelay system, now is the time to open up new fields for better understanding of *Escherichia coli* biology by means of systematic genomics, proteomics, and metabolomics. To this end, we first need the compiled map of His→Asp phosphorelay signal transducers of *E. coli*, including all histidine kinases (HKs). This chapter briefly provides a genome-wide view of *E. coli* HKs. Genome-wide analysis was carried out in *E. coli* to identify all genes encoding histidine kinases (HKs) and response regulators (RRs). We demonstrated that *E. coli* contains a total of 29 HKs, 32 RRs, and 1 HPt (histidine-containing phosphotransfer factor). Except for 2

RRs (FimZ and RssB), all the other 30 RRs are assigned to cognate HKs, as "two components." Including 5 hybrid HKs, we propose that *E. coli* is equipped with a total of 31 His→Asp phosphorelay signal transduction systems involved in various stress responses and adaptations. We have constructed deletion mutations for every HP-RR operon, as well as several RR genes and the HPt gene. A preliminary result for DNA microarray analysis of these 36 deletion strains is presented, suggesting that there is a complex network among individual His→Asp phosphorelay signaling pathways to globally regulate *E. coli* physiology.

INTRODUCTION

Analysis of the current genome databases of many organisms reveals that there is a large family of histidine kinases (HKs). Such organisms, possessing HKs, include most of bacteria, archaea, certain lower eukaryotes such as yeasts, and even higher eukaryotes such as plants. In 1982, when the nucleotide sequences of the *Escherichia coli ompR-envZ* operon were determined [1], the deduced amino acid sequences of these gene products showed no homology to any other proteins. As more *E. coli* genes have been sequenced, it did not take more than 5 years to learn that some amino acid sequences of OmpR and EnvZ are well conserved in a group of *E. coli* regulatory proteins that respond to environmental stimuli [2]. These *E. coli* proteins were classified into two groups: EnvZ, CheA, and NtrB belong to one family, and OmpR, CheY, and NtrC belong to the other. Members of the former group share a common sequence of about 240 amino acid residues, whereas those of the latter share another common sequence of about 120 amino acid residues. The question then arose as to what are the common biochemical functions of these two conserved domains. Soon afterward, it was revealed that both CheA and NtrB have a unique *in vitro* ability to be autophosphorylated at a certain histidine residue [3, 4]. It was also found that CheY and NtrC acquire a phosphoryl group from CheA and NtrB, respectively, at a certain aspartate residue. The same biochemical events were subsequently demonstrated for the EnvZ→OmpR pair [5–7]. Autophosphorylated domains containing the crucial histidine resuides are generally referred to "transmitters," whereas domains containing the phospho-accepting aspartate resuides are "receivers" [8]. To appreciate their biological (rather than biochemical) roles, proteins with a transmitter are referred to as "sensor histidine kinases (HK)," whereas those with a receiver are "response regulators" (RR) [9, 10]. They have simply been termed as "two-component regulatory systems," each consisting of a pair of sensor→regulator two components that respond to a certain environmental stimulus [11].

In 1994, however, the systems were found to be more complex than "two components," as another common phosphotransfer domain containing a phosphorylated histidine residue was uncovered and demonstrated to play a role in certain sensor→regulator phosphorelay systems [12]. This is called the HPt domain consisting of about 150 amino acid residues, which can function as a phosphotransfer intermediate by acquiring/transferring a phosphoryl group from/to aspartate residues in receivers. Such HPt domains are common particularly in the eukaryotic His→Asp phosphorelay systems. Thus, the concept of such two-component systems was extended to as "multistep phosphorelays" [13]. As the multistep His→Asp phosphorelay systems became recognized to play a major role in signal transduction required for response and adaptation to environmental changes in bacteria, one started to wonder how many HKs *E. coli* has. In 1997, determination of the entire genomic sequence of *E. coli* allowed us to conclude that this gram-negative bacterium has 29 HKs [15]. Subsequently, total numbers of HKs have been determined for *Bacillus subtilis* (gram-positive bacterium) (33 HKs) [16] and *Synechocystis* (photosynthetic cyanobacterium) (42 HKs) [17]. Notably, *Pseudomonas aeruginosa* has as many as 63 HKs [18]. Taking advantage of *E. coli* as a model unicellular microorganism, now is the time to open up new fields for the better understanding of *E. coli* biology, with special reference to the His→Asp phosphorelay two-component systems by means of systematic genomics, proteomics, and metabolomics. To move on, we need the complete guide map of phosphorelay signal transducers of *E. coli*. This chapter analyzes all of the genes for *E. coli* histidine kinases and their cognate genes for response regulators and discusses their roles in *E. coli* physiology by constructing deletion mutant analysis for these HKs and RRs.

HISTIDINE KINASE GENES IN THE *E. Coli* GENOME

The entire genomic sequence of *E. coli* allows us to compile a complete list of genes encoding *E. coli* HKs, as well as other members of His→Asp phosphorelay two-component signal transducers, such as RRs and HPt factors. Total 29 ORFs are identified to encode putative HKs, as shown in Fig. 1 (for an alternative quick overview, see http://www.genome.ad.jp/dbget-bin/get_htext?E.coli.kegg+B). Among them, 23 HKs have a structural feature common for orthodox HKs. In other words, they each contain an HK domain that is preceded by an N-terminal signal-input domain. However, each signal-input domain of these HKs is unique, as compared with each other in their amino acid sequences and lengths, suggesting that they serve individually as specific signal transducers. It may also be noteworthy that the amino acid sequences

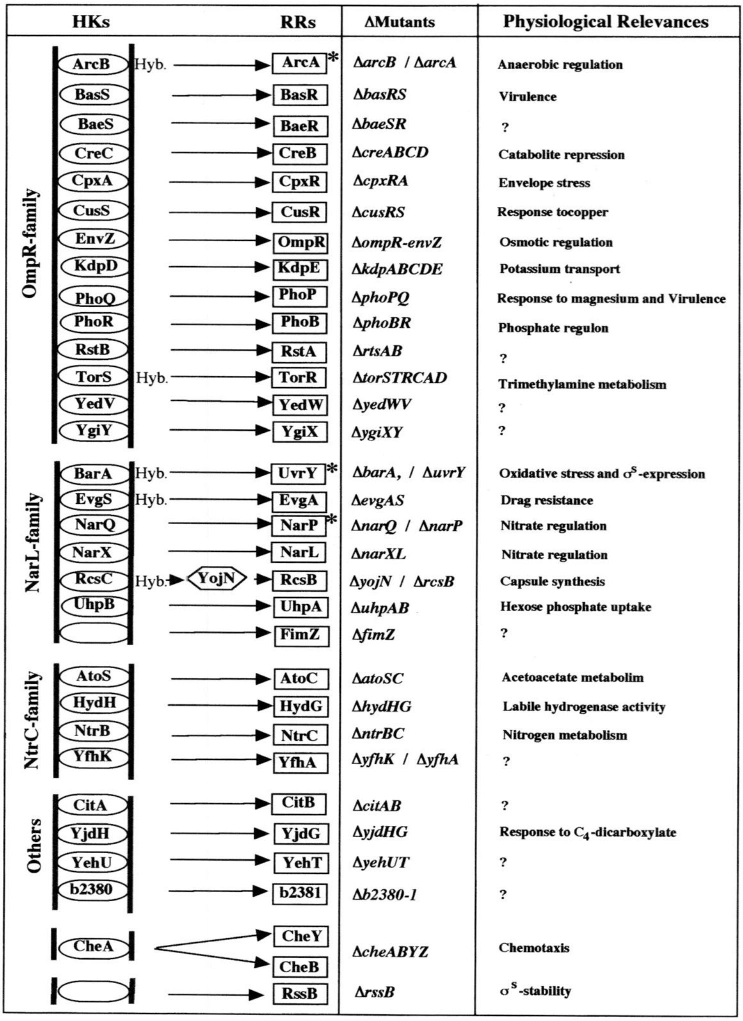

FIGURE 1 Compiled list of genes involved in *E. coli* His→Asp phosphorelay two-component systems. Cognate pairs of histidine kinase (HK) / response regulator (RR) are depicted by assuming that each HK is located on the cell surface (or in the cytoplasmic membrane) and that each HK specifically phosphorylates its cognate RR, as indicated by an arrow. These His→Asp phosphorelay two-component systems are classified into several groups, according to the sub-families of RRs. Hybrid HKs are denoted by "Hyb." RRs, denoted by asterisks, reside apart from HK partners on the chromosome. Appropriate deletion mutant strains were constructed for individual systems as indicated. For each, the possible physiological role(s) is remarked. For details of the chromosomal locations of these genes, see Mizuno [15].

of the HK domains in YehU and b2380 are considerably divergent from others. However, a close inspection reveals that they contain a set of consensus motifs common to authentic HKs, although highly divergent, including the autophosphorylating histidine residue. Furthermore, on the *E. coli* genome, they reside next to YehT and b2381, respectively, each of which encodes a typical RR. It should also be noted that the well-characterized chemotaxis CheA HK has a unique structural design in which the autophosphorylated histidine resiude is located closely to the N terminus, followed by a presumed signal-input domain. In fact, the CheA phosphorylation domain is more like an HPt domain.

In addition to the 24 HKs mentioned earlier, *E. coli* has five hybrid HKs (Hyb.), including ArcB, BarA, EvgS, RcsC, and TorS (Fig. 1). Among these, RcsC contains an HK domain, followed by a receiver domain, whereas all the others have an additional HPt domain at the C-terminal end. Thus these four hybrid HKs contain all three types of common phosphorelay domains in a single polypeptide (i.e., HK, RR, and HPt). One can thus envisage that they must be involved in multistep phosphorelays, as indeed well documented for ArcB [19]. It should be further mentioned about a unique ORF, named YojN. In the C-terminal region of the YojN sequence, a typical HPt domain is found. Interestingly, a region of about 200 residues, upstream of the HPt domain, is somewhat similar to that of an HK (particularly RcsC). However, no autophosphorylated histidine resiude could be assigned in it. Indeed, this unique factor has been demonstrated to function as an HPt phosphotransfer intermediate downstream of RcsC and upstream of RcsB, thereby constructing the RcsC→YojN→RcsB multistep phosphorelay pathway [20]. In any case, it is tempting to speculate that these sophisticated hybrid HKs might act in concert with another His→Asp phosphorelay two-component system(s) to create a higher order of signaling network (or cross-regulation), as has been proposed for the ArcB system that cross-regulates the EnvZ→OmpR system [21]. In this context, it may also be noted that such hybrid HKs are very common in eukaryotes. For example, the higher plant *Arabidopsis thaliana* has nine hybrid HKs, three of which (named AHK2–AHK4) have been shown to act as plant hormone (cytokinin) receptors [22]. Interestingly, the *Arabidopsis* gene for AHK4 can complement the mutational lesion of the *E. coli* rcsC HK gene in such a manner that the complemented *E. coli* cells can propagate the AHK4→YojN→RcsB multistep phosphorelay pathway in response to the external plant hormone, cytokinin [23], clearly indicating the universality of HKs.

The chromosomal positions of *E. coli* HK-coding genes are scattered randomly throughout the genome (Fig. 2). In many instances (25 out of 29 HKs), a pair of HK and RR genes are located next (or closely) to each other on the chromosome. One can reasonably assume that a given physical pair on

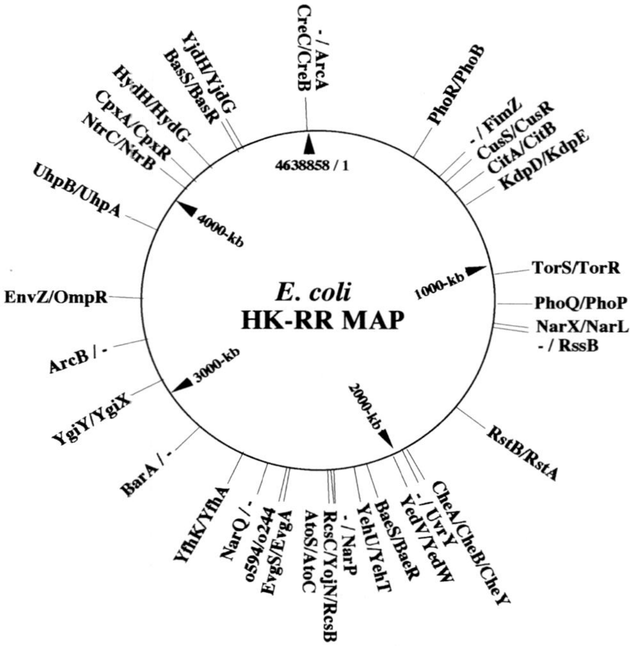

FIGURE 2 Mapping of the chromosomal positions of *E. coli* His→Asp phosphorelay two-component genes. This is based on the physical map of strain MG1655. HK/RR (a pair of two components), HK/-(solo His-kinase), -/RR (solo response regulator). Revised from Mizuno [15].

the chromosome act together as a functional pair in a signaling pathway. The well-characterized pairs (i.e., ArcB→ArcA, NarQ→NarP) are exceptional in that each HK gene resides apart from its cognate RR at a different location on the chromosome, although each pair is known to act together in the same signaling pathway. Although BarA (hybrid HK) has been known as an orphan without its functional RR partner, it has been suggested that its missing partner appears to be the UvrY response regulator, the gene for which is also located apart from *barA* [24].

E. coli has 32 genes encoding RRs (Fig. 1). With regard to their presumed functions, most of them are characterized as DNA-binding transcription factors (29 members). Exceptions are CheB and CheY, involved in the chemotaxis regulation, and RssB that is involved in the regulation of σ^S-stability. These presumed transcription factors are classified into several distinct subgroups, as judged by their amino acid sequence similarities of the C-terminal halves that contain DNA-binding domains. On the basis of such homologies, they are classified into four subgroups: the OmpR family (14 members), NarL family (7 members), NtrC family (4 members), and another group for the remaining four. In any case, each of these RRs has its own single HK partner, as mentioned earlier (Fig. 1). Note that CheA is the only known HK that directs two RRs: CheY and CheB. However, there seems to be other such instances, as two RRs (FimZ and RssB) still remain as orphans (see Fig. 1). As the function of the latter is known to be a regulator of σ^S-stability, it may not take so long to find its cognate HK.

VERSATILITY OF *E. Coli* HISTIDINE KINASES

Needless to say, *E. coli* is the organism of choice for comprehensive understanding the physiological roles and the molecular mechanisms underlying all the His→Asp phosphorelay two-component systems in which each HK is considered to play a crucial role as a specific environmental (or signal) sensor. Among 29 *E. coli* two-component systems, their physiological roles of 22 systems are assigned to those involved in certain adaptive systems, as documented experimentally (Fig. 1). From physiological viewpoints, some HKs are associated with transport systems (cations, anions, and others), others regulate intracellular metabolisms (e.g., nitrogen) and macromolecule syntheses (e.g., capsule), whereas some of them appear to be somehow responsible for global stress responses and/or virulence of *E. coli*. At present, physiological functions are not known for the remaining nine instances. In any case, based on the fact that most RRs appear to be DNA-binding transcription factors, with a few exceptions (CheB, CheY, and RssB), one can easily imagine that most of the two-component systems are somehow directly implicated in gene regulation. The well-known exception is the CheA→CheY (and CheB) system, which regulates bacterial motility (chemotaxis). Collectively, it seems to be evident that the His→Asp phosphorelay two-component systems have been evolved as very successful and powerful means of signal transduction in response to a wide variety of environmental stimuli (or stresses) for *E. coli* (or organisms in general).

DELETION ANALYSIS OF EVERY HISTIDINE KINASE GENE IN THE *E. Coli* GENOME

On the basis of the entire genomic sequence of *E. coli*, and also with the aid of the newly developed elegant methods of genome engineering [25, 26], null mutants have been isolated not only for all of the 29 HK genes but also for 7 RR genes, as also listed in Fig. 1 (H. Aiba and B. L. Wanner, unpublished data). These 36 distinct mutants were constructed systematically by the unified method developed by Datsenko and Wanner [26] by means of one-step inactivation with appropriate polymerase chain reaction-amplified DNA segments. As shown in Fig. 1, for most of them (24 mutants), both HK and RR genes were disrupted at the same time, as they appear to constitute an operon. Otherwise, each separated HK and/or RR gene was knocked out one by one. Interestingly, none of the 36 mutations was lethal, and the mutants grew well in the standard Luria's broth liquid medium and on agar plates. They were viable on M9 minimal synthetic medium at temperature ranging from 16 to 42°C. These results indicated that none of the *E. coli* HKs is essential under the conditions used for growth. In addition to the Δ*cheA* mutant, some other mutants (Δ*arcA* and Δ*atoSC*) also lost (or reduced) the ability to swim on a soft agar plate. In any event, this complete set of deletion mutant strains of *E. coli* should be highly useful, particularly for systematic analyses (genomics and/or proteomics), with special reference to the molecular physiology of the *E. coli* His→Asp phosphorelay two-component systems.

DNA MICROARRAY ANALYSIS OF HISTIDINE KINASES FOR GENE REGULATION

Now, several *E. coli* whole genome, high-density microarrays are available, which have already been used for addressing a number of issues crucial for understanding global *E. coli* physiology (for examples, see Refs. [27–30]). We have also been employing a high-density microarray (*E. coli* GeneChip, Takara, Kyoto, Japan), which contains 4000 independent and duplicated *E. coli* ORFs, in the hope of gaining new insights into the roles of every HK and RR in *E. coli* physiology. Through this approach, we may be able to identify downstream target genes for each His→Asp phosphorelay system. In particular, one may be able to deduce the physiological roles for those yet uncharacterized His→Asp phosphorelay systems (see Fig.1). This approach may also provide new insights into the question of how all the His→Asp phosphorelay signaling pathways are connected with each other to regulate global *E. coli* physiology. Taking advantage of the complete set of HK mutants, to this end, we have just begun microarray analysis on all these deletion mutants.

In a preliminary experiment, each of the 36 deletion mutants was grown under aerobic conditions in Luria's broth medium, and then RNA samples were prepared from cells harvested at the logarithmic growth phase. Each Cy-5-labeled RNA was hybridized with the Takara's microarray with reference to Cy-3-labeled RNA from wild-type cells (BW25113). All these data were analyzed statistically with appropriate softwares developed for chip bioinfomatics. An example of these analyses is presented in Fig. 3. Although no special stress condition was used in the present experiment, a number of

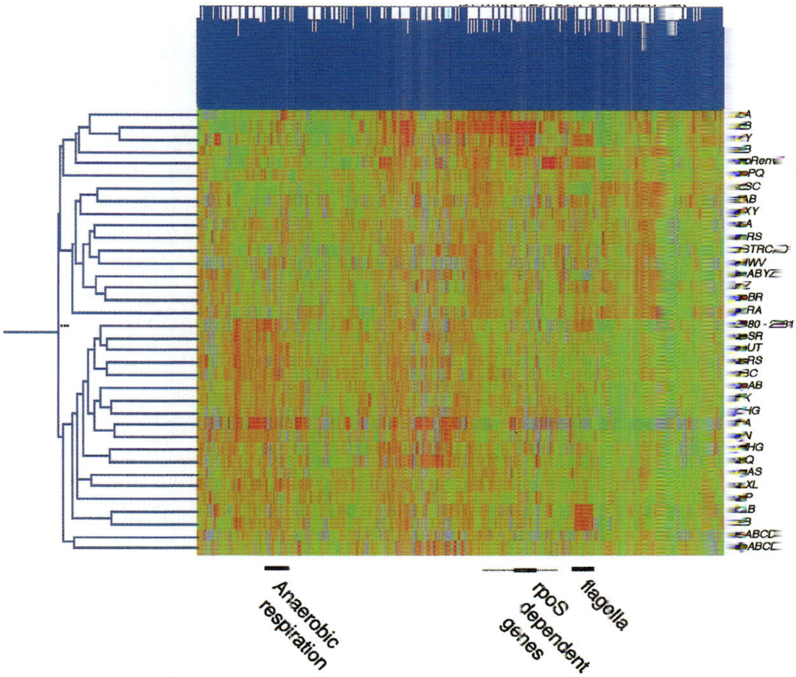

FIGURE 3 Pairwise correlation analysis with expression profiles of 36 *E. coli* two-component mutants. A pairwise correlation analysis was performed using microarray signals that were detected for about 930 genes. A thin vertical line represents each *E. coli* gene and is color coordinated so as to mean that red ones are upregulated, whereas green ones are downregulated. The 36 mutated genes were also clustered more closely to each other, if a given mutant showed an overall profile more similar to another. The resultant profile gives us several implications. For example, the profile shows that certain red genes, which are relevant to σ^S-dependent regulation, are gathered in a specific area that is composed by the horizontal columns of Δ*arcB*, Δ*uvrY*, and Δ*rssB*, suggesting that a set of σ^S-dependent genes are upregulated in a similar manner in these His-kinase mutants (see text). Similarly, genes relevant to flagella formation are also gathered in a specific area, and also for genes involving anaerobic respiration. For another example, *hydHG*, *narQ*, *engAS*, *narXL*, and *narP* were clustered closely (see the tree on the left-hand side), suggesting that the overall (i.e., up and down) profiles of the about 930 genes in these mutants were apparently similar to each other. This may suggest that these His→Asp phosphorelay two-component systems might play a related physiological role(s). In short, such two-dimensional information provides us with several insights into the global networks of the His→Asp phosphorelay two-component systems in *E. coli*.

interesting observations were made, as follows. In some deletion mutants, considerable changes were observed in a large number of genes. Over 100 of the genes were either upregulated or downregulated. Such pleiotropic mutants include Δ*arcA*, Δ*arcB*, Δ*ompR-envZ*, Δ*uvrY*, and Δ*yfhA*. Interestingly, when these genes thus influenced were categorized from physiological viewpoints, there is an intriguing tendency. Namely, many energy synthesis genes were affected in Δ*arcA*, Δ*arcB*, Δ*uvrY*, and Δ*yfhA*, similarly, certain amino acid biosynthesis genes in Δ*ompR-envZ*, Fe^{3+} transport genes in Δ*arcB* and Δ*ompR-envZ*, and also many motility genes in Δ*arcA* and Δ*ompR-envZ*. This fact may suggest that these His→Asp phosphorelay systems might play much more complex roles than what has been thought, even under standard growth conditions. It would also be worth mentioning that many σ^S-regulated genes were upregulated in Δ*arcB*, Δ*uvrY*, and Δ*rssB*. The result for Δ*rssB* is consistent with its known role because RssB is directly responsible for σ^S stability. In the other mutants, the effect may be explained by the fact that σ^S gene expression itself is derepressed in these mutants. Similarly, a certain set of chemotaxis genes (mostly flagella genes) were also widely affected in some deletion mutants (downregulated in ΔarcA or ΔatoSC, whereas were upregulated in Δ*ompR-envZ*, Δ*rcsB*, Δ*uvrY*, Δ*citAB*, or Δ*b2380-1*). Consequently, the former type of mutants lost motility. These results may be indicative of an occurrence of interplays of multiple two-component systems for σ^S-dependent gene regulation, composing a very fundamental transcription circuitry. Similarly, signals through multiple His→Asp phosphorelay systems might also be integrated in order to properly control *E. coli* motility, which is a very energy-consuming process.

Although the present results are preliminary, one can already see a significant impact of the microarray analysis of all the deletion mutants on our understanding of the global network of the His→Asp phosphorelay signal transduction systems in *E. coli*. In any event, the era of genomics has come. In this connection, such results of further microarray analyses should shed further light on the issues addressed earlier. One may access our data (or information) as to the microarray analysis of the *E. coli* His→Asp phosphorelay two-component systems at *http://ecoli.aist-nara.ac.jp/genobase/2_component/xp_analysis/all.html*.

ACKNOWLEDGMENTS

We apologize that a number (or most) of relevant and original works could not be cited because of the limited space. Thanks are also due to Masayori Inouye for his critical reading of our manuscript. This study was supported by a grant-in-aid for scientific research on a priority area [12142201 to TM] from the Ministry of Education, Science, Sports and Culture of Japan.

REFERENCES

1. Mizuno, T., Chou, M.-Y., and Inouye, M. (1982). Osmoregulation of gene expression. II. DNA sequence of the envZ gene of the ompB operon of *Escherichia coli* and characterization

of its gene product. *J. Biol. Chem.* **257**, 13692–13698.

2. Ronson, C. W., Nixon, B. T., and Ausubel, F. M. (1987). Conserved domains in bacterial regulatory proteins that respond to environmental stimuli. *Cell* **49**, 579–581.
3. Ninfa, A. J., and Magasanik, B. (1986). Covalent modification of the *glnG* product, NRI, by the glnL product, NRII, regulates the transcription of the *glnALG* operon in *Escherichia coli*. *Proc. Natl. Acad. Sci. USA* **83**, 5909–5913.
4. Hess, J. F, Oosawa, K., Matsumura, P., and Simon, M. I. (1987). Protein phosphorylation is involved in bacterial chemotaxis. *Proc. Natl. Acad. Sci. USA* **84**, 7609–7613.
5. Aiba, H., Mizuno, T., and Mizushima, S. (1989). Transfer of phosphoryl group between two regulatory proteins involved in osmoregualtory expression of the ompF and ompC genes in *Escherichia coli*. *J. Biol. Chem.* **264**, 8563–8567
6. Forst, S., Delgado, J., and Inouye, M. (1989). Phosphorylation of OmpR by the osmosensor EnvZ modulates the expression of the *ompF* and *ompC* genes in *Escherichia coli*. *Proc. Natl. Acad. Sci. USA* **86**, 6052–6056.
7. Igo, M. M., Ninfa, A. J., Stock, J.B., and Silhavy, T. J. (1989). Phosphorylation and dephosphorylation of a bacterial transcriptional activator by a transmembrane receptor. *Genes Dev.* **3**, 1725–1734.
8. Parkinson, J. S., and Kofoid, E. C. (1992). Communication modules in bacterial signalling proteins. *Annu. Rev. Genet.* **26**, 71–112.
9. Stock, J. B., Ninfa, A. D., and Stock, A. M. (1989). Protein phosphorylation and regulation of adaptive response in bacteria. *Microbiol. Rev.* **53**, 450–490.
10. Bourret, R. B., Borkovich, K. A., and Simon, M. I. (1991). Signal transduction pathways involving protein phosphorylation in prokaryotes. *Annu. Rev. Biochem.* **60**, 401–441.
11. Hoch, J. A., and Silhavy, T. J. (1995). "Two-Component Signal Transduction." ASM Press, Washington, DC.
12. Mizuno, T. (1998). His-Asp phosphotranfer signal transduction. *J Biochem. (Tokyo)* **123**, 555–563.
13. Appleby, J. L., Parkinson, J. S., and Bourret, R. B. (1996). Signal transduction via the multi-step phosphorelay: Not necessarily a road less traveled. *Cell* **86**, 845–848.
14. Blattner, F. R., Plunkett, G., III, Bloch, C. A., Perna, N. T., Burland, V., Riley, M., Collado-Vides, J., Glasner, J. D., Rode, C. K., Mayhew, G. F., Gregor, J., Davis, N. W., Kirkpatrick, H. A., Goeden, M. A., Rose, D. J., Mau, B., and Shao, Y. (1997). The complete genome sequence of *Escherichia coli* K-12. *Science* **277**, 1453–1474.
15. Mizuno, T. (1997). Compilation of all genes encoding two-component phosphotransfer signal transducers in the genome of *Escherichia coli*. *DNA Res.* **4**, 161–168.
16. Farbret, C., Feher, V. A., and Hoch, J. A. (1999). Two-component signal transduction in *Bacillus subtilis*: How one organism sees its world. *J. Bacteriol.* **181**, 1975–1983.
17. Mizuno, T., Kaneko, T., and Tabata, S. (1995). Compilation of all genes encoding bacterial two-component signal transducers in the genome of the cyanobacterium. *Synechocystis* sp. strain PCC 6803. *DNA Res.* **3**, 407–414.
18. Galperin, M. Y., Nikolskaya, A. N., and Koonin, N. E. (2001). Novel domains of the prokaryotic two-component signal transduction systems. *FEMS Microbiol. Lett.* **203**, 11–21.
19. Matsushika, A., and Mizuno, T. A dual-signaling mechanism mediated by the ArcB hybrid sensor kinase containing the histidine-containing phosphotransfer domain in *Escherichia coli*. *J. Bacteriol.* **180**, 3973–3977.
20. Takeda, S., Fujisawa, Y., Matsubara, M., and Mizuno, T. (2001). A novel feature for the multistep phosphorelay in *Escherichia coli*: A revised model of the RcsC→YojN→RcsB signaling pathway implicated in capsular synthesis and swarming behavior. *Mol. Microbio.* **40**, 440–450.
21. Matsubara, M., Kitaoka, S., Takeda, S., and Mizuno, T. (2000). Tuning of the porin expres-

sion under anaerobic growth conditions by His-to-Asp cross-phosphorelay through both the EnvZ-osmosensor and ArcB-anaerosensor in *Escherichia coli*. *Genes Cells* **5**, 555–569.

22. Pernestig, A. K., Melefors, O., and Georgellis, D. (2001). Identification of UvrY as the cognate response regulator for the BarA sensor kinase in *Escherichia coli*. *J. Biol. Chem.* **276**, 225–231.
23. Suzuki, T., Miwa, K., Ishikawa, K., Yamada, H., Aiba, H., and Mizuno, T. (2001). The *Arabidopsis* sensor His-kinase, AHK4, can respond to cytokinins. *Plant Cell Physiol.* **42**, 107–113.
24. Yamada, H., Suzuki, T., Terada, K., Takei, K., Ishikawa, K., Miwa, K., Yamashino, T., and Mizuno, T. (2001). The *Arabidopsis* AHK4 histidine kinase is a cytokinin-binding receptor that transduces cytokinin signals across the membrane. *Plant Cell Physiol.* **42**, 1017–1023.
25. Yu, D., Ellis, H. M., Lee, E.-C., Jenkins, N. A., Copeland, N. G., and Court, D. L. (2000). An efficient recombination system for chromosome engineering in *Escherichia coli*. *Proc. Natl. Acad. Sci. USA* **97**, 5978–5983.
26. Datsenko, K. A., and Wanner, B. L. (2000). One-step inactivation of chromosomal genes *in Escherichia coli* K-12 using PCR products. *Proc. Natl. Acad. Sci. USA* **97**, 6640–6645.
27. Richmond, C. S., Glasner, J. D., Mau, R., Jin, H., and Blattner, F. R. (1999). Genome-wide expression profiling in *Escherichia coli* K-12. *Nucleic Acids Res.* **27**, 3821–3835.
28. Zimmer, D. P., Soupene, E., Lee, H. L., Wendisch, V. F., Khodursky, A. B., Peter, B. J., Bender, R. A., and Kustu, S. (200). Nitrogen regulatory protein C-controlled genes of *Escherichia coli*: Scavenging as a defense against nitrogen limitation. *Proc. Natl. Acad. Sci. USA* **97**, 14674–14679.
29. Khodursky, A. B., Peter, B. J., Cozzarelli, N. R., Botstein, D., Brown, P. O., and Yanofsky C. (2000). DNA microarray analysis of gene expression in response to physiological and genetic changes that affect tryptophan metabolism in *Escherichia coli*. *Proc. Natl. Acad. Sci. USA* **97**, 12170–12175.
30. Courcelle. J., Khodursky, A., Peter, B., Brown, P. O., and Hanawalt. P. C. (2001). Comparative gene expression profiles following UV exposure in wild-type and SOS-deficient *Escherichia coli*. *Genetics* **158**, 41–64.

CHAPTER 10

Signal Transmission and Specificity in the Sporulation Phosphorelay of *Bacillus subtilis*

KOTTAYIL I. VARUGHESE

Department of Molecular and Experimental Medicine, The Scripps Research Institute, La Jolla, California 92037

Bacteria, many lower eukaryotes and some plants utilize the two-component/phosphorelay systems to monitor environmental signals and respond to it. In this process, the transmission of information is accomplished through the exchange of a phosphoryl group from one protein. In order for the signal to stay on track, the proteins must specifically recognize their partners and transfer the phosphoryl group. The problem of specificity is particularly severe in bacteria, which possess 30–40 two-component pairs with considerable similarities. Structural characterization of the components of the sporulation phosphorelay in *Bacillus subtilis* has shown how a response regulator interacts with a phosphotransferase domain. The interaction has provided clues on how discrimination and specificity are achieved in response regulator: histidine kinase interactions. The interacting surface of a response regulator consists of conserved core residues surrounded by variable residues. Conserved residues appear to initiate the binding, whereas variable residues give rise to specificity.

INTRODUCTION

The survival of a bacterium depends on its ability to adapt to a changing environment. *Bacillus subtilis* sporulates in response to poor growth conditions and then, under more favorable conditions, the genetic matter in the spore is used to produce a new bacterium. It uses a very complex molecular machinery to decide whether to sporulate or to divide. The backbone of this complex machinery is a phosphorelay [1], which is an expanded version of a two-component system commonly used by bacteria for monitoring and responding to the environment (Fig. 1a). A typical two-component system consists of a histidine kinase and a response regulator, which is activated by phosphorylation to carry out a specific mission, usually transcription. The histidine kinase, in the majority of cases, acts as a signal sensor that responds to the initiating signal by autophosphorylating a histidine residue by transferring a γ-phosphoryl group from bound ATP. Sensor kinases are generally divisible into two domains: an N-terminal signal detection domain connected to a kinase domain. The kinase domain is made up of a phosphotransferase subdomain containing the active histidine and an ATP-binding subdomain. Response regulator transcription factors also consist of two domains: the N-terminal receiver domain that accepts the phosphoryl group and a C-terminal DNA-binding domain. The histidine kinase dephosphorylates by transferring the phosphoryl group to an aspartate on the N-terminal domain of the response regulator. Phosphorylation of the receiver domain generally enhances the DNA-binding affinity of the second domain. In the two-component system, phosphorylation of the substrate is thus a two-step process, whereas in Ser/Thr/Tyr kinases, the phosphoryl group is transferred directly from ATP to the substrate proteins. In the case of phosphorelay, phosphotransfer becomes an even more elaborate process involving four steps.

The phosphorelay that controls the initiation of sporulation of *B. subtilis* is depicted in Fig. 1b. In this phosphorelay, histidine kinases respond to the

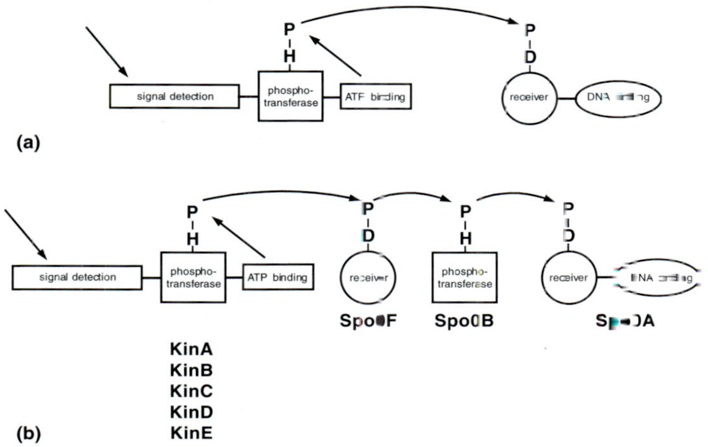

FIGURE 1 Domain organization of a typical two-component system and the sporulation phosphorelay signal transduction system. (a) Signal recognition by the kinase induces transfer of the γ-phosphate of ATP and phosphorylation of the phosphotransferase domain. In a two-component system, the phosphoryl group is donated to an aspartate on a response regulator/transcription factor by the kinase. (b) In this phosphorelay, histidine kinases pass the phosphoryl group to an intermediate response regulator, Spo0F, and subsequently to the response regulator/transcription factor, Spo0A, via a phosphotransferase, Spo0B. The sporulation pathway makes use of five different histidine kinases: KinA, KinB, KinC, KinD, and KinE.

incoming signal by autophosphorylating a His residue that is then dephosphorylated by a common response regulator, Spo0F. Phosphorylated Spo0F in turn becomes the substrate for phosphotransferase Spo0B, which serves to mediate phosphoryl transfer from Spo0F to Spo0A, the ultimate transcription factor [2]. Spo0B phosphorylation occurs on a histidine residue while it transfers a phosphoryl group. In this multistep phosphoryl transfer reaction, the sequence of phosphorylated amino acids is His-Asp-His-Asp (Fig. 1b). Phosphorelay signal transduction systems are widespread and have also been found to regulate important pathways such as pathogenesis in *Bordetella pertussis* [3], anaerobic gene expression in *Escherichia coli* [4] and osmosensing in *Saccharomyces cerevisiae* [5].

Five histidine kinases are known to participate in the sporulation phosphorelay (Fig. 1b) of *B. subtilis*; however, most of the signal input is mediated through KinA. Structural studies on histidine kinases are still in the preliminary stages, but the structures of the other three components — Spo0F [6, 7] and Spo0B [8] from *B. subtilis* and Spo0A from a closely related species, *B. stearothermophilus* [9] — have been reported. In addition, structural analysis of the complex between Spo0F and Spo0B [10] and the transcription factor

Spo0A in complex with DNA [11] has been carried out. This chapter deals with the structural aspects of these molecules in relevance to molecular recognition and phosphotransfer.

STRUCTURAL CHARACTERIZATION OF THE PHOSPHORELAY COMPONENTS

Response regulators are generally two domain proteins, but Spo0F and CheY are simpler in construction and contain only a receiver domain. Structural studies on two-component systems started with the analysis of CheY [12, 13]. This was followed by the structure solution of seven other response regulators by crystallography and nuclear magnetic resonance (NMR); Spo0F [6, 7, 14], Spo0A [15], NtrC [16], NarL [17], CheB [18], PhoB [19], and FixJ [20]. In general, these receiver domains appear very similar in overall structure, despite differences in primary amino acid sequences. In addition, active site catalytic residues are invariant or highly conserved in all response regulators and the geometries of the active sites are identical. Crystal structures of the phosphorylated N-terminal domains of FixJ [20] and Spo0A [21] have been solved. Structures of the DNA-binding domains of NarL [17], Spo0A [9], OmpR [22], and PhoB [23] have been reported, and these structures show that effector domains are structurally diverse in contrast to receiver domains, which are similar.

Spo0F Structure

Spo0F is a single domain protein and has an α/β fold (Fig. 2a) [7]. The structure is made up of a central β sheet consisting of five parallel β strands and five α helices that are situated two on one side and three on the other side of the β sheet. The active site pocket is situated at the C-terminal end of the β sheet, and this small pocket is lined by five residues highly conserved among response regulators. They are Asp10, Asp11, Asp54, Thr82, and Lys104. Asp54, the site of phosphorylation, is located at the bottom of the shallow pocket and the reactive carboxylate is accessible to solvent. Active site aspartates are flanked by five loops that connect β strands to the α helices. These loops are labeled L1–L5 corresponding to the β strand and α helix they connect.

Metal Binding of Spo0F

Phosphotransfer reactions require the participation of divalent cations, and response regulators bind cations with varying affinities. The crystal structures

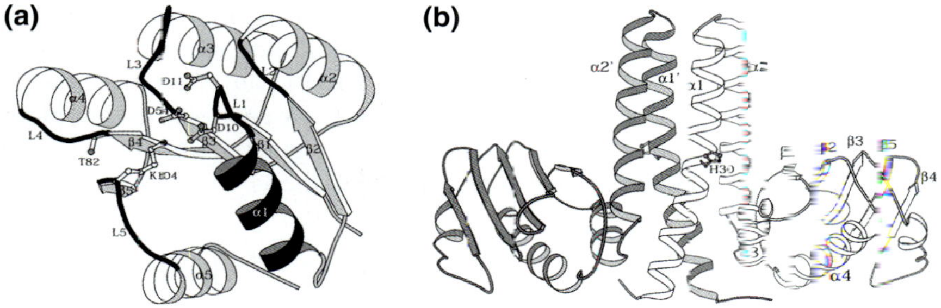

FIGURE 2 Structure of Spo0F and Spo0B. (a) Ribbon representation of the structure of Spo0F. The central β sheet consists of five parallel β strands. There are five α helices α1 to α5. The five catalytic residues in and around the active site, Asp10, Asp11, Asp54, Thr82, and Lys104, are shown. The β–α loops that surround the active site are labeled L1 to L5. These loops and helix α1, which form the interacting interface with Spo0B, are shown in black. (b) Ribbon representation of the Spo0B dimer. One of the protomers is shaded. A protomer comprises two domains; the N-terminal α-helical hairpin made up of helices α1 and α2 and the C-terminal α/β domain. His30 is the site of phosphorylation.

of Spo0F in calcium-bound and metal-free forms have been elucidated. The crystal structure of the calcium complex of Spo0F revealed that the metal is coordinated by the carboxylates of Asp11 and Asp54 and the carbonyl of Lys56 [7]. The mode of coordination is the same as in the magnesium complex of CheY [24]. In the metal-free form of Spo0F, Asp11 points away from the active site, but metal binding reorients this side chain toward the active site. On the contrary, in CheY, the corresponding aspartate is already positioned in the proper geometry for metal coordination in the metal-free form. In fact, CheY binds magnesium at least an order of magnitude tighter than Spo0F. For Spo0F, the affinity for magnesium is rather low with a K_d of 20 mM while it binds the bigger metal calcium ion more strongly with a K_d of 6 mM.

Phosphorylation Induces Changes in Response Regulators

Crystal structures of the phosphorylated N-terminal domains of Spo0A [21] and FixJ [20] have been solved and both structures exhibit similar conformational changes on phosphorylation. The phosphorylated response regulators retain their overall structure; however, there are some significant changes at the active site after phosphorylation. The most significant change is the rearrangement of loop L4. The side chain of conserved Thr82 (84 in Spo0A), which points away from the active site in the unphosphorylated state, now turns toward the site of phosphorylation to form a hydrogen bond with a phosphoryl oxygen. On the whole there are significant rearrangements in the active site to provide stability to the phosphoryl group. Despite knowing the nature of phosphorylation-induced conformation changes, the question of how phosphorylation activates the response regulator remains a more complex question and remains in the realm of hypothesis.

Autophosphatase Activity of Response Regulators Vary to Suit Their Specific Roles

Response regulators are only active when phosphorylated and they have an autophosphatase activity that ensures that they do not stay permanently activated. Despite the structural and sequence similarities in the superfamily of response regulators [25], the hydrolysis rates of phosphorylated response regulators differ by a thousandfold, with Spo0F~P being one of the most "stable" phosphorylated response regulators. The magnitude of autophosphatase activity seems to be appropriately geared for its biological roles of the regulators. For example, for the chemotaxis response regulator, the half-life of CheY~P is of order of seconds [26], about the same amount of time bacteria take to change the direction of swimming. In contrast, sporulation in *B. subtilis* is initiated over at least an hour, and Spo0F autodephosphorylates over the course of several hours. It is, however, not easy to define how the composition of amino acids around the active site has evolved to suit the required autophosphatase activity. Examination of the active site of Spo0F shows that Lys56 is located on the edge of the active site covering Asp54, providing a partial shield from external water molecules (Fig. 3). This could therefore play a role in enhancing the stability. The substitution of Lys56 by Met does not give rise to any significant increase in autophosphatase activity, whereas substitution by Ala increases the autophosphatase activity 3-fold. CheY has an Asn residue at this position, and substitution of Lys56 by Asn results in a 23-fold increase in autophosphatase activity. When the phosphoryl group is shielded from solvent and when it is involved in strong interactions at the active site, the phosphoryl group must be stable. However, when the active site can position a water molecule suitable for nucleophilic attack at the phosphate atom, autophosphatase activity must be high. The presence of divalent cations increases the autophosphatase activity by a factor of 10 [27]. It appears that the Asn at position 56 can act in parallel with the divalent cations to promote hydrolysis of the acyl phosphate by the positioning of the water molecule for nucleophilic attack on the phosphate atom. In addition to autophosphatase activity, response regulators are dephosphorylated by specific phosphatases as part of the regulatory mechanism [28].

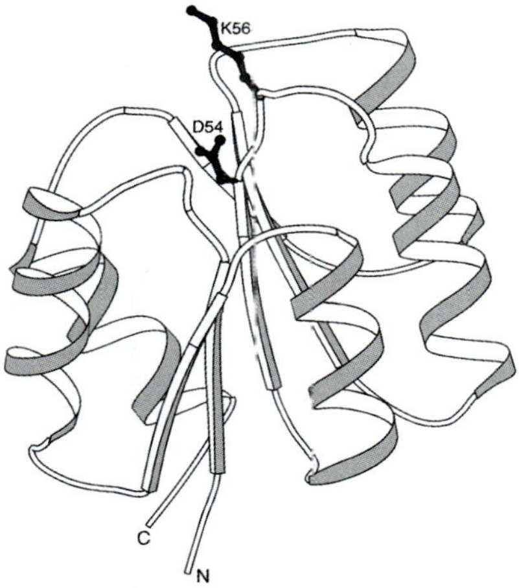

FIGURE 3 The positioning of Lys56 near the active site partially shields the site of phosphorylation from external water molecules.

The crystal structure of the DNA-binding domain of Spo0A has been solved [9] and it is an α helical domain consisting of six α helices linked by short loops. The structure contains a helix-turn-helix motif known to bind the DNA. The overall fold of this domain is very different from the other effector domains whose structures have been determined — MarI, PhoB, and OmpR. Despite the difference in structure, the nature of the helix-turn-helix is very similar in all these structures.

Spo0B Phosphotransferase

Spo0B catalyzes specific phosphoryl transfer between Spo0F and Spo0A at high rates and the reaction is freely reversible. Spo0B exists as a dimer in solution as well as in the crystal structure (Fig. 2b). The monomer is made up of two domains: an N-terminal α-helical hairpin and a C-terminal domain with a α/β fold. The protein dimerizes by association of the helical hairpin domains from two protomers to form a four-helix bundle. The dimer assumes the shape of an "anchor" with the four-helix bundle resembling the stem and the C-terminal domains appearing as hooks [8]. The site of phosphorylation is His30, and its side chain protrudes from the four-helix bundle toward the solvent. There are two active sites per dimer, and each active site is formed by residues from both protomers.

NMR studies of this domain in the *E. coli* histidine kinase EnvZ reveal a structure very similar to the four-helix bundle of Spo0B serving as a dimerization surface and as a histidine phosphorylation site [29]. The EnvZ four-helix bundle also superimposes well onto the Spo0B four-helix bundle. Because the majority of histidine kinases in bacteria have the general structure of EnvZ, it is likely that a Spo0B-like domain provides a major dimerization surface in each and is the phosphotransferase domain of these histidine kinases. This class of kinases differs in organization from CheA, the histidine kinase involved in the chemotaxis signaling system. The structure of the phosphotransferase domain of the CheA has been determined [30]. In addition, the Hpt domains of ArcB [31] and Ypd1 [32] have also been determined. Although Spo0B is functionally similar to these phosphotransferase domains, it differs in construction. The four-helix bundle of the P1 domain of CheA and the Hpt domain are formed from the folding of a single monomer polypeptide chain, whereas the four-helix bundle of Spo0B results from protomer:protomer interactions of the dimer.

INTERACTIONS OF THE RESPONSE REGULATOR WITH THE PHOSPHOTRANSFERASE DOMAIN

The crystal structure of the complex between response regulator Spo0F and phosphotransferase Spo0B [10] reveals how phosphotransfer domains and regulatory domains associate together to transfer the phosphoryl group. Because the four-helix bundle of the Spo0B dimer has two active sites, the cocrystal contains two Spo0F molecules per Spo0B dimer (Fig. 4). Each Spo0F molecule is arranged such that its α1 helix is aligned nearly parallel with the four-helix bundle (Fig. 5). In addition, the α1 helix and loop L5 make close interactions with the α1 helix of Spo0B. These interactions align the histidine of Spo0B with the aspartate of Spo0F, providing precise geometry for phosphotransfer. Spo0F also contacts the four-helix bundle of Spo0B via the residues in loop 4 that interact with the α2 helix of the second protomer (Fig. 6]. In view of the close similarity of the four-helix bundles of Spo0B and sensor kinases, the Spo0F:Spo0B structure could be a paradigm for response regulator–sensor kinase interaction, with the exclusion of the chemotaxis system.

MOLECULAR RECOGNITION AND SPECIFICITY

Bacteria, finding the two-component system to be a useful tool for adaptation and precise regulation, expanded it to perform various specialized functions

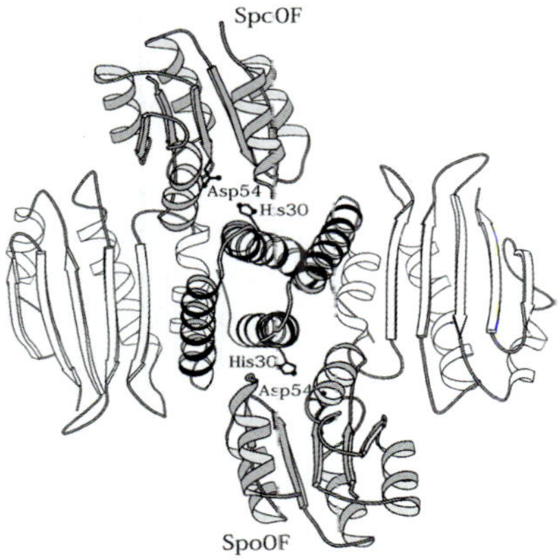

FIGURE 4 A view of the Spo0B:Spo0F complex down the axis of the four-helix bundle (shown in black). The sites of phosphorylation, His30 of Spo0B and Asp54 of Spo0F, are in close proximity for phosphoryl transfer.

by gene duplication and mutation. Bacteria such as *E. coli* or *B. subtilis* possess 30 to 40 different pairs of two-component systems, each dedicated to unique signals and unique responses [33, 34]. Careful sequence comparisons of the sensor kinases and response regulators showed the presence of two major and several minor families of two-component systems within each bacterium [33, 34]. Within a family, a high degree of amino acid identity and

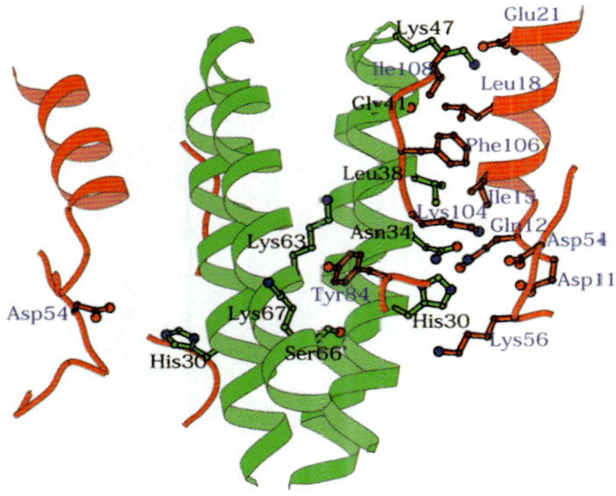

FIGURE 5 A view of the association of helix α1 of Spo0F (red) with the four-helix bundle (green). Residues Gln12, Ile15, and Leu18 from α1 of Spo0F interact with α1 of Spo0B. Loop 5 contains Lys104, Phe106, and Ile108, and these residues also interact with helix α1 of Spo0B. The sites of phosphorylation His30 and Asp54 are shown. The labels for Spo0F residues are in blue and for Spo0B in black.

similarity exists, yet these highly similar systems must process different signals, interact only with their partner, and activate unique genes. The question arises of how is fidelity achieved in such signal transduction systems when there is such close sequence similarity?

The genome sequence of *B. subtilis* revealed that it has 34 histidine kinase:response regulator pairs. The receiver domains of the response regulators are structurally very similar and have nearly identical active sites. The four-helix bundle, which mediates most of the interactions, is also expected to be very similar. Hence, it is intriguing how a particular kinase specifically recognizes its partner and activates it to produce the correct response when there are a large number of pairs.

The interaction of Spo0F with the four-helix bundle of Spo0B provides insight into the mechanism of recognition and association. The helix $\alpha 1$ and the five loops L1 to L5 form the interaction surface of Spo0F. Five hydrophobic residues — two from helix $\alpha 1$ (Leu15 and Ile18) and three from the loop L5 (Pro105, Phe106, and Ile108) — form a hydrophobic patch on the interaction surface of Spo0F (Fig. 6). This patch interacts with the histidine containing helix $\alpha 1$ of Spo0B. These five residues are fairly well conserved in the 34 response regulators. With the exclusion of the NarL family, they are conserved from 88 to 100%. This hydrophobic patch appears to form the core of the interacting surface and can be thought of as the "initiator of binding."

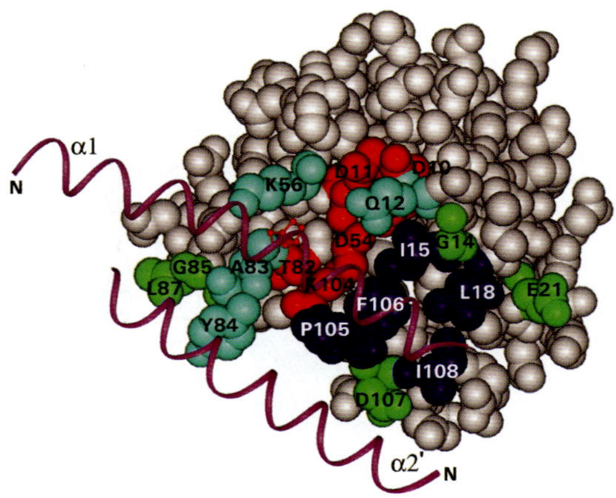

FIGURE 6 Interaction surface of Spo0F. The five residues that form the hydrophobic patch are shown in dark green. Four additional residues interacting with the four-helix bundle that are conserved within the OmpR family are colored cyan. Residue K104, which interacts with the Spo0B helix bundle, is invariant in all response regulators. Five residues that interact with the four-helix bundle are highly variable within the OmpR family and are colored green.

20. Birck, C., Mourey, L., Gouet, P., Fabry, B., Schumacher, J., Rousseau, P., Kahn, D., and Samama, J.-P. (1999). Conformational changes induced by phosphorylation of the FixJ receiver domain. *Structure* 7, 1505–1515.
21. Lewis, R. J., Brannigan, J. A., Muchova, K., Barak, I., and Wilkinson, A. J. (1999). Phosphorylated aspartate in the structure of a response regulator protein. *J. Mol. Biol.* 294, 9–15.
22. Martinez-Hackert, E., and Stock, A. M. (1997). The DNA-binding domain of OmpR: Crystal structure of a winged helix transcription factor. *Structure* 5, 109–124.
23. Okamura, H., Hanaoka, S., Nagadoi, A., Makino, K., and Nishimura, Y. (2000). Structural comparison of the PhoB and OmpR DNA-binding/transactivation domains and the arrangement of PhoB molecules on the phosphate box. *J. Mol. Biol.* 295, 1225–1236.
24. Stock, A. M., Martinez-Hackert, E., Rasmussen, B. F., West, A. H., Stock, J. B., Ringe, D., and Petsko, G. A. (1993). Structure of the $Mg^{(2+)}$-bound form of CheY and mechanism of phosphoryl transfer in bacterial chemotaxis. *Biochemistry* 32, 13375–13380.
25. Volz, K. (1993). Structural conservation in the CheY superfamily. *Biochemistry* 32, 11741–11753.
26. Lukat, G. S., Lee, B. H., Mottonen, J. M., Stock, A. M., and Stock, J. B. (1991). Roles of the highly conserved aspartate and lysine residues in the response regulator of bacterial chemotaxis. *J. Biol. Chem.* 266, 8348–8354.
27. Zapf, J., Madhusudan, Grimshaw, C. E., Hoch, J. A., Varughese, K. I., and Whiteley, J. M. (1998). A source of response regulator autophosphatase activity: The critical role of a residue adjacent to the Spo0F autophosphorylation active site. *Biochemistry* 37, 7725–7732.
28. Perego, M. (1999). Self-signaling by Phr peptides modulates *Bacillus subtilis* development. *In* "Cell-Cell Signaling in Bacteria" (G. M. Dunny *et al.* eds.), pp. 243–258. American Society for Microbiology, Washington, DC.
29. Tomomori, C., Tanaka, T., Dutta, R., Park, H., Saha, S. K., Zhu, Y., Ishima, R., Liu, D., Tong, K. I., Kurokawa, H., Qian, H., Inouye, M., and Ikura, M. (1999). Solution structure of the homodimeric core domain of *Escherichia coli* histidine kinase EnvZ. *Nature Struct. Biol.* 6, 729–734.
30. Zhou, H. and Dahlquist, F. W. (1997). Phosphotransfer site of the chemotaxis-specific protein kinase CheA as revealed by NMR. *Biochemistry* 36, 699–710.
31. Kato, M., Mizuno, T., Shimizu, T., and Hakoshima, T. (1997). Insights into multistep phosphorelay from the crystal structure of the C-terminal HPt domain of ArcB. *Cell* 88, 717–723.
32. Xu, Q., and West, A. H. (1999). Conservation of structure and function among histidine-containing phosphotransfer (HPt) domains as revelated by the crystal structure of YPD1. *J. Mol. Biol.* 292, 1039–1050.
33. Fabret, C., Feher, V. A., and Hoch, J. A. (1999). Two-component signal transduction in *Bacillus subtilis*: How one organism sees its world. *J. Bacteriol.* 181, 1975–1983.
34. Mizuno, T. (1997). Compilation of all genes encoding two-component phosphotransfer signal transducers in the genome of *Escherichia coli*. *DNA Res.* 4, 161–168.
35. Hoch, J. A., and Varughese, K. I. (2001). Keeping signals straight in phosphorelay signal transduction. *J. Bacteriol.* 183(17), 4941–4949.
36. Tzeng, Y.-L., and Hoch, J. A. (1997). Molecular recognition in signal transduction: The interaction surfaces of the Spo0F response regulator with its cognate phosphorelay proteins revealed by alanine scanning mutagenesis. *J. Mol. Biol.* 272, 200–212.
37. Shan, S. O., and Herschlag, D. (1996). The change in hydrogen bond strength accompanying charge rearrangement: Implications for enzymatic catalysis. *Proc. Natl. Acad. Sci. USA* 93, 14474–14479.
38. Zapf, J. W., Hoch, J. A., and Whiteley, J. M. (1996). A phosphotransferase activity of the *Bacillus subtilis* sporulation protein Spo0F that employs phosphoramidate substrates. *Biochemistry* 35, 2926–2933.

39. Knowles, J. R. (1980). Enzyme-catalyzed phosphoryl transfer reactions. *Annu. Rev. Biochem.* **49**, 877–919.
40. Lukat, G. S., Stock, A. M., and Stock, J. B. (1990). Divalent metal ion binding to the CheY protein and its significance to phosphotransfer in bacterial chemotaxis. *Biochemistry* **29**, 5436–5442.

CHAPTER 11

Histidine Kinases: Extended Relationship with GHL ATPases

Wei Yang

Laboratory of Molecular Biology, National Institute of Diabetes and Digestive and Kidney Diseases, National Institutes of Health, Bethesda, Maryland 20892

Identification of sequence similarity is the most effective way to date to predict structural and mechanistic relationships between proteins with diverse biological functions. Three-dimensional structures provide unequivocal verification of such predictions, but mutagenesis and functional studies are essential for interpretation of the structural results and for application of structural information to biology. Identification of three small conserved sequence motifs led to the discovery of a conserved ATP-binding site among DNA gyrase, which is a member of the topoisomerase II family, Hsp90, MutL, and histidine kinases. The relationship among these four protein families has propelled the revelation of a weak yet intrinsic ATPase activity of both MutL, and Hsp90 protein and demonstrated a previously unappreciated similarity between these ATPases and histidine kinases. Studies of these ATPases and kinases provide a compelling example of the interdependence of structural and functional studies and illustrate an ever-growing protein superfamily that utilizes ATP to carry out diverse cellular functions.

INTRODUCTION

In 1997, Eugene Koonin and colleagues discovered that histidine kinases (HK) share conserved amino acid sequence motifs with members of the DNA topoisomerase II family, the hsp90 family, and the MutL DNA mismatch repair protein family [1]. Three peptide sequence motifs, N, G1, and G2, were initially found to be conserved in these four seemingly unrelated protein families (Fig. 1). In the crystal structure of the ATPase fragment of DNA gyrase subunit B (NgyrB) [2], a member of the topoisomerase II family, these three sequence motifs form an ATP-binding site (Fig. 1) [2]. Although MutL was not known to bind or hydrolyze ATP and the ATPase activity of Hsp90 was controversial at that time, Koonin and colleagues [1] proposed that these sequence motifs may play a similar role in histidine kinases, hsp90, and MutL proteins.

The three-dimensional structures of a fragment of Hsp90, MutL, and EnvZ and CheA histidine kinase, which contain the identified conserved sequence motifs, were determined by either X-ray crystallography or nuclear magnetic resonance (NMR) shortly afterward [3–7]. The location of these conserved sequence motifs is topologically similar among these three structures and DNA gyrase (Fig. 2). Upon biochemical, mutagenesis, and structural studies, these sequence motifs have been confirmed to form a nucleotide-binding pocket [7–9]. MutL and Hsp90 proteins are indeed proven ATPases [8, 10, 11]. Based on the structural similarity, a forth sequence motif, motif IV, in the ATP-binding pocket was also identified (Fig. 1) [8]. Because DNA gyrase, hsp90, and MutL all hydrolyze ATP, they are collectively called the GHL ATPase family [8]. Although histidine kinases do not hydrolyze ATP, they contain an ATP-binding site similar to those of the GHL ATPases and use ATP to phosphorylate a histidine side chain [12, 13]. Most ATPases known to date contain a conserved sequence motif, GXXXXGKS/T, known as the Walker A motif. Such a sequence motif is absent in GHL ATPases. Similarly, histidine

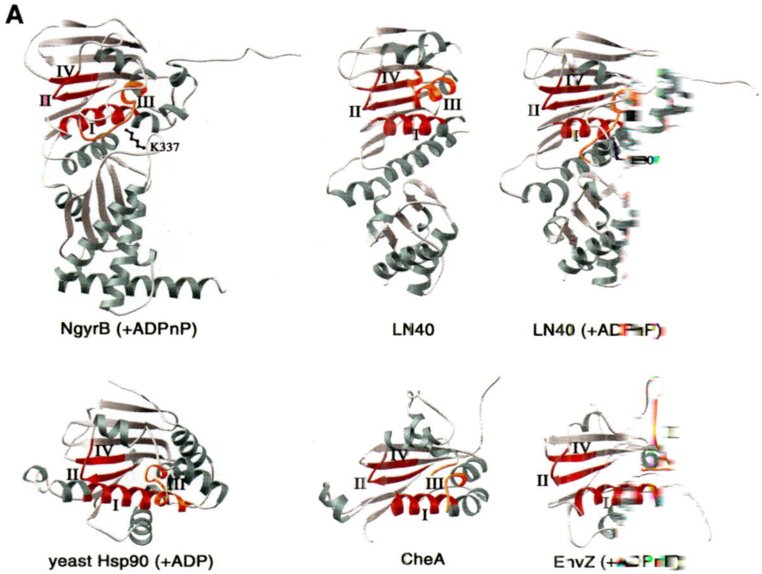

B

Motif	I (N)	II (G1)	III (G2)	IV
MutL	KElveNsiDA	DNGxGi	slGFRGEAL	GT[illegible]v
Gyrase	xEvvdNaiDe	DdGxGi	sgGxxGvGx	G[illegible]f
Hsp90	RELISNsSDA	DtGiGM	mIGQFGVGF	G[illegible]L
CheA(HK)	xhllxNaidh	DxGxGi	xxG-xGvGm	G[illegible]i
EnvZ(HK)	xnmvxNaary	DxGxGi	xxG-xGlGl	G[illegible]a
Function:	sugar-phos	base	ATP-lid	base

FIGURE 1 Structural comparison of GHL ATPases and histidine kinases. (A) Ribbon diagrams of NgyrB (ADPNP complex), LN40 either in apo form or in complex with ADPNP, yeast Hsp90 (ADP complex), CheA, and EnvZ (ADPNP complex). α helices are shown in pale green, β strands in gray, the conserved motifs I, II, and IV in red, and motif III (the ATP lid) in yellow. These structures were superimposed prior to ribbon drawing. The ATP lid adopts similar structures in NgyrB and LN40 when ADPNP is bound, which perhaps represents the "closed" form of the entire ATPase and kinase superfamily. In the structures of LN40 and CheA apoprotein and Hps90 in complex with ADP, the ATP lid seems to share similar structural features, which may represent the "open" conformation. (B) The alignment of four sequence motifs conserved among GHL ATPases and histidine kinases. Shown in uppercase are residues invariable in the topoisomerase II, Hsp90, MutL, EnvZ, or CheA family; shown in lowercase are residues conserved in each protein family; nonconserved residues are shown as an x. Sequences of DNA gyrase, MutL and EnvZ from *E. coli*, Hsp90 from human, and CheA from *T. maritima* are used as examples. Compared with previously published sequence alignments [1, 4], alterations have been made based on structure superposition. The conserved Glu serving as a general base in ATP hydrolysis by gyrase, Hsp90, and MutL is replaced by His or Asn in CheA and EnvZ, all of which are highlighted. The function of each conserved motif is summarized below the sequence alignment.

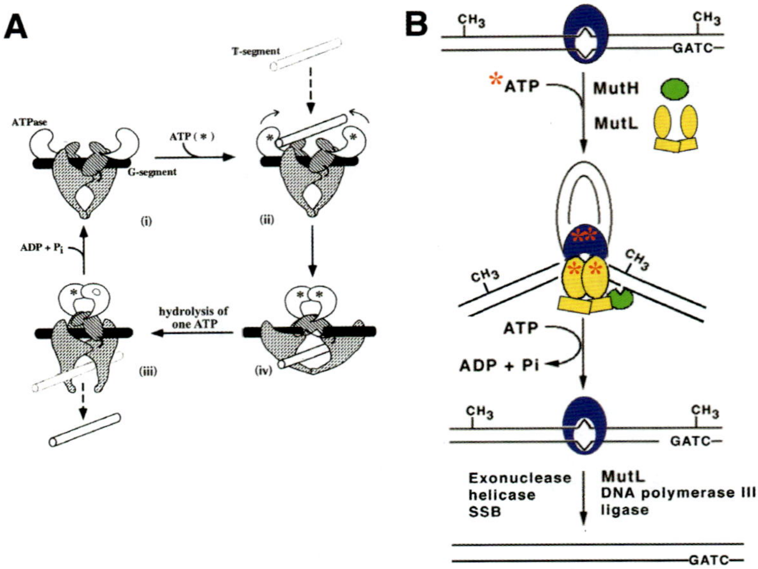

FIGURE 2 Diagram of the biological functions of DNA topoisomerases and *E. coli* MutL. (A) DNA topoisomerases are dimeric. The double-stranded DNA G segment is shown as a black rod and the T segment as a white rod. Transient cleavage of the G segment occurs independently of ATP. The ATPase region undergoes large conformational changes upon binding of ATP and traps the T segment. ATP hydrolysis opens the protein and DNA gate and drives the T segment to pass through. (B) DNA mismatch repair process in *E. coli*. MutS, represented by an elongated disc, recognizes a mismatch site and recruits MutL. In the presence of ATP, MutL activates nuclease MutH to cleave the daughter strand 5' to the unmethylated GATC sequence. After MutH nicking, MutS and MutL actively recruit DNA helicase, exonuclease, DNA polymerase III, and so on to remove the mismatch and resynthesize the daughter strand.

kinases lack typical sequence motifs found in the majority of Ser/Thr and Tyr kinases. The relationship between these unusual ATPases and protein kinases is the subject of this chapter.

DIVERSE FUNCTIONS SUPPORTED BY A CONSERVED ATP-BINDING SITE

DNA Topoisomerases II

DNA gyrase was isolated from *Escherichia coli* when an enzyme that could convert a relaxed circular DNA to a negatively supercoiled one was sought by Drs. M. Gellert, K. Mizuuchi, and H. Nash [14]. This "magic" factor was purified from *E. coli* cell lysates and was shown to require ATP and Mg^{2+} to introduce negative superhelical turns into a circular relaxed DNA, which was required for integration of phage λ into the host DNA [14]. Isolation of DNA gyrase was followed rapidly by the discovery of DNA topoisomerases in yeast and higher eukaryotes [15]. DNA topoisomerases such as DNA gyrase can change the superhelicity and topology of DNA. Changes of DNA superhelicity are essential for all cellular processes that involve DNA, such as DNA replication, recombination, repair, and transcription. Isolation of DNA gyrase also led to the discovery that it is the target of a number of very effective antibiotic drugs [16, 17]. Currently, several effective anticancer drugs are designed to target human DNA topoisomerases [18].

Type II DNA topoisomerases catalyze the reaction that moves one double-stranded DNA segment through another (see review by Wang [19]). These enzymes contain two catalytic active sites: one for ATPase activity and another for cleavage and religation of double stranded DNA (Fig. 2A) [20]. The two activities reside in either two separate structural domains of the same polypeptide chain (A subunit) or two different protein subunits as in the case of DNA gyrase (A and B subunits) (Fig. 2A) [19]. Functionally, they form A_2 dimers or A_2B_2 tetramers, respectively. The reaction catalyzed by members of the DNA topoisomerase II family includes four steps. In the first step, an enzyme binds to and transiently cleaves a dsDNA segment, which is called the G segment for gate. The cleaved DNA phosphodiester bonds are replaced by phosphotyrosyl bonds formed between tyrosyl residues of the protein dimer and the two cleaved 5' ends of DNA. In the second step, the enzyme entraps a second DNA segment on binding of ATP. In the third step, the enzyme actively transports the second DNA, which is called T for transport segment, through the cleaved G segment. This step is accelerated by hydrolyzing bound ATP molecules [21]. In the last step, the enzyme rejoins the 3' hydroxyl and the 5' phosphate groups of DNA and frees the tyrosyl residues (Fig. 2A).

Type II topoisomerases hydrolyze 5–10 ATP molecules per second on average and under an optimal condition hydrolyze 2 ATPs for changing the DNA linking number by one [22]. The ATPase activity of topoisomerases is stimulated by DNA and is inhibited by various antibiotic and anticancer drugs. ATP can be viewed in this case as an energy source to allow proteins to do "work" to change the DNA topology by either introducing supercoils into relaxed circular DNA or taking them out.

Hsp90

Hsp90 is a ubiquitous protein found in the cytoplasm of all eukaryotic cells. Homologues have also been found in prokaryotes and endoplasmic reticulum (see reviews [23, 24]). The functions of Hsp90 are diverse and not fully characterized. Hsp90 is clearly essential for cell viability, and yeast cells without Hsp90 die. The general understanding is that Hsp90 is a molecular chaperone. When cells are under stress, Hsp90 works with Hsp70 and other chaperone proteins to help to fold proteins correctly [25]. There are multiple isoforms of Hsp90 in a single cell. In addition to the Hsp90 that is induced upon stress, some are expressed constitutively and some fluctuate with the cell cycle [23]. Under normal cell growth conditions, Hsp90 plays essential roles in cell signaling. For instance, Hsp90 is necessary for stabilizing steroid hormone receptors in the cytosol, for enhancing binding of steroids, and for transporting them to the nucleus afterward [26]. Hsp90 also assists the proper folding of certain receptor-coupled kinases that function in signal transduction cascades [26].

Whether Hsp90 has ATPase activity was a topic of much debate. Yeast, rat, and human Hsp90 was reported to have low but measurable ATPase activity [27]. ATP-induced conformational changes and ATP-dependent protein–protein interactions were also reported [28, 29]. However, the ATPase activity of Hsp90 was not detected when the ATP-binding assay was carried out side by side with both positive and negative controls [30]. In addition, an *in vitro* assay showed that Hsp90 could bind to a denatured protein and prevent it from aggregation in the absence of ATP, which led to the conclusion that Hsp90 could perform the chaperone function without consuming ATP [31]. When the crystal structure of the fragment of human hsp90 was determined in complex with the antitumor drug geldanamycin, the geldanamycin molecule occupied the site composed of many conserved residues in the Hsp90 family [6]. The native ligand of this binding site was not immediately appreciated. Only when the sequence and structural similarity between Hsp90 and the ATPase fragment of the DNA gyrase (NgyrB) became apparent shortly afterward [1, 9, 32] did the investigation of the ATPase activity of Hsp90 take center stage.

The ATPase activity assay combined with mutagenesis studies and *in vivo* and *in vitro* functional analyses finally confirmed a low but intrinsic and essential ATPase activity of Hsp90 [10, 11, 33]. Yeast cells devoid of native Hsp90 and supplemented with an Hsp90 protein carrying a point mutation that renders it defective in ATP binding are nonviable [10]. It is suggested that ATP binding is essential for Hsp90 to release unfolded proteins and pass them on to Hsp70 for refolding, even though binding of unfolded proteins to Hsp90 is independent of ATP [10, 33, 34]. The ATPase activity of Hsp90 probably plays an even more important role in steroid hormone regulation [33]. ATP binding seems to be essential for the association of Hsp90 with the steroid hormone receptors, and ATP hydrolysis is essential for the dissociation, thereby facilitating the cycling of these receptors from cytoplasm to nucleus.

MutL

The discovery of the ATPase activity of MutL, a DNA mismatch repair protein, also followed a circuitous path and serves as another example of structure-based functional discovery. MutL, together with MutS and MutH, is essential for initiating DNA mismatch repair in *E. coli* (Fig. 2B) [35]. Mismatches in DNA duplexes most often occur due to misincorporation by DNA polymerases. MutS detects mismatches, including mispaired and unpaired bases in a DNA duplex. In *E. coli*, MutS, with the assistance of MutL, activates the endonuclease MutH to cleave the newly synthesized daughter strand, which is unmethylated but paired with a methylated template strand [35]. After activating MutH to cleave the daughter strand, MutL, together with MutS, also recruits DNA helicase (UvrD), exonuclease, and DNA polymerase III to remove the daughter strand from the nick to beyond the mismatch site and resynthesize the daughter strand (Fig. 2B). Although the MutL protein was purified to homogeneity [36] and an *in vitro* mismatch repair assay was reported in 1989 [37], MutL was regarded as having no ATPase activity, no DNA nuclease, or any enzymatic activities and MutL was thought to be merely an adapter to mediate the interaction between MutS and MutH. The requirement of hydrolyzable ATP in the reconstituted DNA mismatch repair initiation assay was explained by the ATPase activity of MutS protein and the DNA helicase, both of which contain a bona fide Walker A motif and well-established ATPase activity [38, 39]. The other reason for the failure to appreciate the ATPase activity of MutL is that the ATP turnover rate by MutL at ~0.4/min is very low [3], as compared with the rate of 50–100/s of DNA topoisomerases or even the ~2/min of the MutS ATPase.

The ATPase activity of MutL was not studied systematically until the ATP-dependent activation of the endonuclease MutH by MutL was detected and MutL was found to exhibit striking primary, secondary, and tertiary structural homology to DNA gyrase [3]. The ATPase activity of MutL was finally confirmed by site-directed mutagenesis, which showed the correspondence between mutating a suspected catalytic residue and loss of ATPase activity [8]. The crystal structures of MutL in complex with both a nonhydrolyzable ATP analog, ADPNP, and ATP, which had become hydrolyzed to ADP, showed the nucleotides bound in the predicted active site [8]. It has been shown that binding of ATP enables MutL to activate MutH endonuclease activity and that ATP hydrolysis is essential for MutL to mediate the mismatch detection by MutS and a full activation of MutH [40].

There are three homologues of MutL, MLH1, PMS1, and PMS2, in eukaryotes. Prokaryotic MutL proteins are functionally homodimers and eukaryotic MutL proteins form MLH1–PMS1 and MLH1–PMS2 heterodimers [41]. Inactivation of MutL homologues in humans, hMLH1 in particular, by either mutation or epigenetic silencing, results in the susceptibility of humans to hereditary nonpolyposis colorectal cancer (HNPCC) [42]. Several of the missense mutations found in MutL homologues of HNPCC kindreds are located in or near the ATP-binding pocket, demonstrating the importance of ATPase activity [3].

Histidine Kinases

Hisditine kinases were identified to be the first component of a so-called two-component phosphorelay signal transduction pathway, which enables bacteria, fungi, and plants to sense and respond to their environment [13, 43]. Histidine kinases regulate bacterial chemotaxis, osmoregulation, photosesitivity, sporulation, and plant responses to ethylene and microbiol pathogenesis. For example, during bacterial chemotaxis, histidine kinase CheA phosphorylates a histidine residue in response to attractants or repellants in the environment detected by the sensory unit. The phosphoryl moiety is later transferred to an Asp residue of the second component, response regulator, which upon phosphorylation activates motor proteins and results in cell movement. Thereby, the histidine kinase relays messages from the chemical sensory unit to cell movement regulated by the response unit. Two-component systems can be viewed as a primitive version of a G-protein coupled eukaryotic signal transduction cascade. Knowledge of histidine kinases is reviewed thoroughly in this book in every chapter other than this one.

Histidine kinases were known to phosphorylate the histidine side chain utilizing the γ-phosphate of ATP [44]. However, the structure of the ATP-binding site of histidine kinases remained unpredictable for a number of years due to the lack of kinase signature motifs. Detection of the sequence similarity between bacterial histidine kinases and GHL ATPases predicted that the kinase active site would be similar to that of GHL ATPases [1]. The three-dimensional structures of EnvZ and CheA [4, 7], which represent class I and II histidine kinases, respectively, confirmed that the ATP-binding sites are conserved among the family members and bear a close relationship to that of GHL ATPases. The structural similarity suggests a possible mechanistic resemblance between ATPases and histidine kinases.

FEATURES OF THE ATP-BINDING SITE

The Mobile ATP Lid

The four sequence motifs conserved among GHL ATPases and histidine kinases are located in a single structural domain. Motif I, which is also known as the N box, is located on an α helix. Immediately adjacent to the helix, a pair of anti parallel β strands contains motifs II (the G1 box) and IV (Figs. 1 and 3). The remaining motif III (the G2 box) forms a mobile ATP lid, whose conformation depends on the presence of a ligand. Without a bound ATP the lid is protracted and partly disordered as observed in LN40, Hsp90, and CheA (Fig. 1A) [3, 4, 6]. In the structures of LN40 and NgyrB complexed with ADPNP, a nonhydrolyzable ATP analog, this mobile region forms a stable and extended structure that interacts with the phosphate moiety of the ATP molecule, thus covering the otherwise exposed binding site (Fig. 1) [2, 3].

The N-terminal ATPase fragments of MutL (LN40) and DNA gyrase (NgyrB) consist of over 350 residues and form a two-domain structural unit [2, 8]. The first domain contains the four conserved sequence motifs, and its structural features are maintained among GHL ATPases and histidine kinases. The second domain plays an essential role in ATPase activity as well. A Lys residue of the second domain (K337 of NgyrB and K307 of LN40) coordinates the γ-phosphate of ATP and helps enclose the nucleotide-binding site (Figs. 1 and 3) [2, 3]. The N-terminal fragment of Hsp90, which contains only the first structural domain and lacks the second (Fig. 1) [24], is inca-

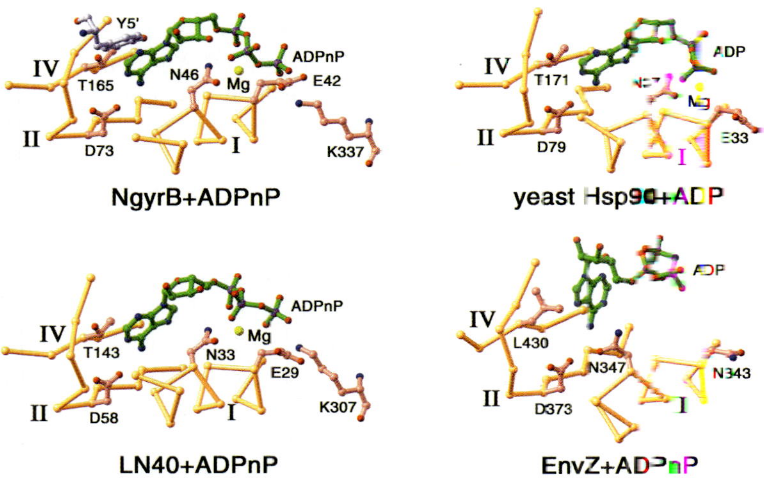

FIGURE 3 Comparison of the ATP-binding site of GHL ATPases and EnvZ. Motifs I, II, and IV are shown as a yellow stick model following Cα traces; the bound nucleotide is shown in green, and side chains of the most conserved residues are shown as a brown stick model. Tyr5 (Y5') supplied by the adjacent second subunit of NgyrB in coordinating ADPNP binding is shown in purple. ADPNP and ADP molecules in the complex with NgyrB, LN40, and Hsp90 are oriented very similarly to the conserved motifs and are very different from that in the complex with EnvZ. Motif III (the ATP lid) is omitted for clarity.

pable of hydrolyzing ATP. Even though the four conserved sequence motifs are present and form the ATP-binding site, the γ-phosphate of the bound ATP is too mobile to be visible to X-rays [9, 10]. The structure of this fragment of the Hsp90–ATP complex is identical to that of the Hsp90–ADP complex. Even when the Hsp90 fragment is associated with ATP or ADP, the ATP lid of Hsp90 is semiopen and similar to that of the LN40 apo-protein structure rather than that of the LN40–ADPNP and NgyrB–ADPNP complexes [Fig. 1]. In agreement with what was observed with Hsp90, the ATP lid and the K307-containing loop of the second domain are less ordered in the LN40–ADP complex than those in the LN40–ADPNP complex and are disordered in the LN40 apo-protein structure. Proper closing of the ATP-lid, therefore, seems to depend on stable binding of the entire ATP, including the γ-phosphate moiety.

MutL as a Paradigm of the GHL ATPase Cycle

Structural studies of the ATPase fragment of MutL were particularly fruitful. The ATPase fragment of MutL (LN40), which resembles that of DNA gyrase (NgyrB), was crystallized in three different forms: apo-protein, protein–ATP analog (ADPNP) complex, and protein–ADP complex [3, 8]. The three-dimensional structure of NgyrB is available only in the form of the protein–ADPNP complex. Structural transformation induced by binding of ATP or ADPNP was suggested by solution studies of both DNA topoisomerases and MutL [19, 41]. However, the individual residues that undergo conformational changes and the magnitude of the movement became known in detail only after the crystal structures of MutL were determined. Nearly 70 residues, which were disordered in the apo-MutL structure, became ordered in the MutL–ATP complex (Fig. 4) [8].

A common feature of members of the GHL superfamily is that they are dimeric ATPases. The C-terminal region of these proteins is responsible for dimeric interactions, and the N-terminal region contains ATPase motifs (Fig. 5A) [19, 24, 41]. The two copies of the ATPase fragment within a dimeric MutL or DNA topoisomerase are dissociated in the absence of ATP, but become associated upon binding of ATP (Figs. 4 and 5). Unlike LN40 and NgyrB, the smaller N-terminal fragment of Hsp90 does not dimerize upon binding of ATP, but it has been shown to become associated in the context of the full-length intact protein [45]. Therefore, the C-terminal dimer interface is constitutive, and the interaction at the N terminus is induced by ATP binding (Fig. 5).

ATP binding induces formation of the new structural elements and association of LN40 and NgyrB. Among the 70 residues in MutL that undergo dramatic structural transformation, only some are involved in ATP binding,

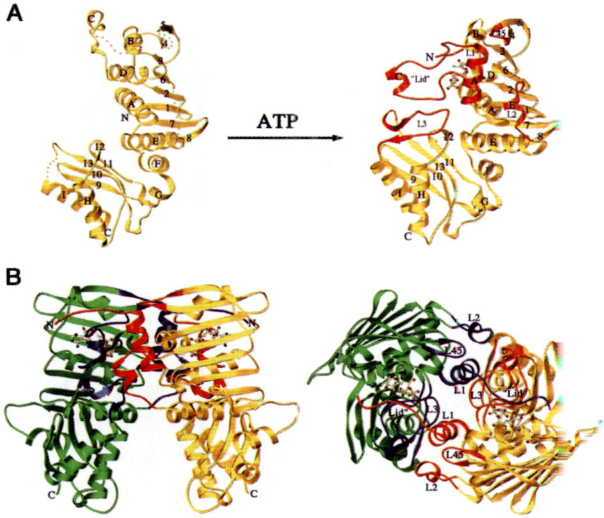

FIGURE 4 Structural transformation of LN40 on ATP binding. (A) Structures of LN40 in the absence and presence of ADPNP are shown as ribbon diagrams. Structural elements that are formed only when ADPNP is bound are shown in orange, which account for ~70 residues. (B) Two orthogonal views of the LN40 dimer when it is bound to ADPNP. The dimer interface is formed exclusively between the newly formed structural elements shown in orange and blue.

such as in the ATP lid. The majority of them, however, contribute to the newly formed dimer interface (Fig. 4). In DNA gyrase, the N-terminal residues of one subunit, Tyr-5 (Y5) in particular, coordinate ATP binding to the adjacent second subunit (Fig. 3). Therefore, dimerization is both structurally and functionally required for the ATPase activity of DNA gyrase. Even though the N-terminal inducible dimer interface and the ATPase active site are physically separate in MutL, mutations that weaken the association of LN40 also decrease the ATPase activity of MutL [8]. Apparently, dimerization is still essential for MutL ATPase activity, probably because it reinforces and stabilizes the protein structure in an active form. A similar correspondence between ATPase activity and association of the N-terminal domains of Hsp90 has also been observed [45].

The cycle of GHL ATPases may be simplified in two major structural states: ATP bound and ATP free. The structural transformation induced by ATP binding greatly alters the molecular shape and surface of MutL, DNA topoisomerases, and Hsp90. These changes probably facilitate MutL and Hsp90 to switch protein partners to coordinate cellular processes and enable DNA topoisomerases to capture and transport the T segment DNA through the transiently cleaved G segment (Fig. 2) [20, 23, 41].

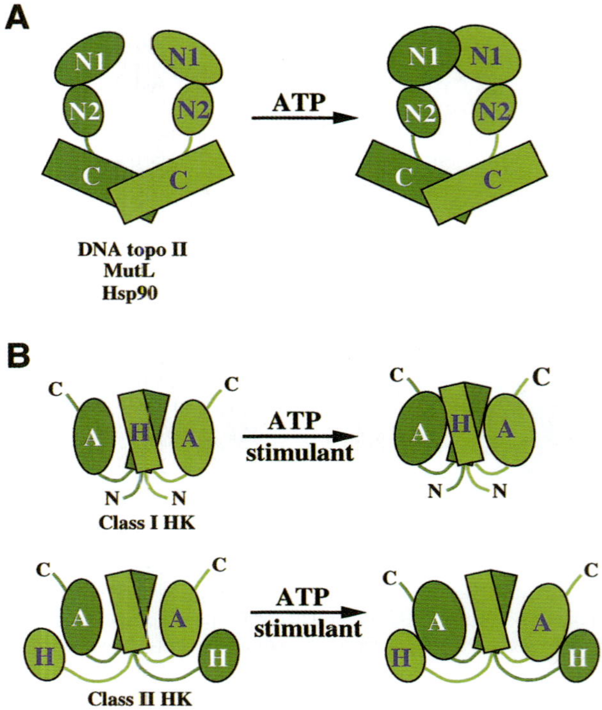

FIGURE 5 A common theme in the GHL ATPase superfamily and proposed models for histidine kinases. (A) All proteins in the GHL ATPase superfamily contain two homologous subunits, shown in light and dark green. The ATPase fragment is composed of two structural domains, N1 and N2, which are immediately next to one another in the primary sequence in the cases of DNA topoisomerases and MutL, but are likely separated by the intervening sequence in Hsp90. ATP binding induces association of the ATPase fragment of DNA topoisomerases, MutL and Hsp90. (B) All histidine kinases are dimeric as well. The ATP-binding domain (A), which is equivalent to the N1 domain, is often located at the C terminus to the histidine-containing domain (H), which is designated to be phosphorylated and functionally may be equivalent to the N2 domain. In the class I histidine kinase, the dimerization domain is also the histidine-containing domain. In class II histidine kinases, the dimerization domain is located between the histidine-containing and the ATP-binding domain. Upon activation, the ATP-binding domain and histidine-containing domain become associated across the dimer interface and the two protein subunits phosphorylate each other.

Comparison of Histidine Kinases with GHL ATPases

The structure of the ATP-binding domain of histidine kinase EnvZ was determined in the presence of a nonhydrolyzable ATP analog, ADPNP, using NMR

[7]. On the whole, ADPNP is bound in a similar site to those observed in the structures of DNA gyrase, MutL, and Hsp90. However, orientation of the ATP is quite different from that bound to NgyrB, LN40, or Hsp90 (Fig. 3). The adenine base is flipped in EnvZ, and the phosphate moiety is not in its usual binding position. In addition, the ATP lid of EnvZ is unlike that observed in GHL ATPase structures either in the presence or in the absence of nucleotide and antibiotics. On the contrary, the crystal structure of CheA containing the ATP-binding domain, but without a bound ligand [4], is quite similar to that of the LN40 apo-protein structure (Fig. 1).

Motif I contains a nonvariable Asn in all four protein families, which is the reason that it is also known as the N box (Fig. 1B). This conserved Asn is essential for chelating a Mg^{2+} ion that interacts with the β and γ phosphates of the nucleotide bound to NgyrB, LN40, and Hsp90 (Fig. 3). Motifs II and IV, which are fairly well conserved in the HK as well as in the GHL family, determine the specificity for ATP through multiple hydrogen bonds and van der Waals contacts between protein side chain and main chain atoms and the adenine base. These three motifs form rather stable and similar structures regardless of the ligand-binding status in all the GHL and HK structures determined so far. Based on the structural conservation of motifs I, II, and IV and their consistent interactions with the base and sugar-phosphate moiety observed in GHL ATPases, it is unlikely that these same motifs would be used to interact with ATP in a very different manner. Whether the EnvZ structure reported represents an authentic mode of ATP binding by histidine kinases in general, an isolated case, or a result of limited distance restraints measured by the NMR technique awaits future experiments.

MECHANISTIC IMPLICATIONS

The structural similarity in the ATP-binding site of DNA gyrase, Hsp90, MutL, and bacteria histidine kinases suggests that these four protein families may have a deeper relationship than the common dependence of ATP. GHL ATPases and histidine kinases carry out a similar phosphoryl transfer reaction, i.e., is to break the phosphodiester bond between the β and the γ phosphate of ATP. The difference is that the nucleophile is a hydroxyl group in the case of GHL ATPases, whereas it is an imidazole side chain in histidine kinases

CONVERTING AN ATPASE TO HISTIDINE KINASE

A conserved Glu residue in motif I of GHL ATPases serves as a general base to deprotonate the nucleophilic water molecule (Fig. 3). Mutation of this Glu residue to Ala eliminates ATP hydrolysis by DNA gyrase (E42A), Hsp90 (E33A), and MutL (E29A) [8, 10, 11, 33, 46]. This Glu is replaced by a polar residue in the HK family (Fig. 1B), which probably enables histidine kinases to avoid inappropriately hydrolyzing ATP. In addition, the leaving γ-phosphate of ATP is coordinated by a conserved Lys residue donated from the second structural domain in MutL and DNA gyrase, K307 and K337, respectively (Figs. 1 and 3). Mutation of this Lys residue diminishes or abolishes ATP binding and hydrolysis in both enzymes [8, 47]. Structures of both yeast and human Hsp90 fragments contain only one structural domain with the four conserved sequence motifs. This domain alone binds ATP weakly and fails to hydrolyze it. It is most likely that another domain of Hsp90 C-terminal to the ATP-binding domain plays a similar role as the second domain of LN40 and DNA gyrase and is required for optimal ATP binding and hydrolysis.

Similar to the case of Hsp90, the structure of the EnvZ and CheA fragment contains only the catalytic ATP-binding (CA) domain. The histidine residue to be phosphorylated is located on a separate structural domain from the ATP-binding domain. Structurally, this domain is dissimilar to the second domain of the NgyrB and LN40 that contains the Lys residue and facilitates ATP binding and hydrolysis in MutL and DNA gyrase [48, 49]. Functionally, however, it has to serve a similar role in complementing the ATP-binding domain in order for phosphorylation to occur. No structural information is available of the complete histidine kinase active site, in which the His residue is poised to attack the γ-phosphate. The ATP-binding domain and the histidine-containing domain are either tethered by a flexible linker or separated by other protein domains. Separation of the ATP-binding and the histidine-containing domains suggests that for the histidine kinases a complete active site to carry out phosphotransfer is not assembled until large conformational changes occur and the two structural domains are brought together [13]. The assembly of these structural domains depends not only on the binding of a proper nucleotide, but often on association of a chemical stimulus to the sensory unit [50]. Separation of the histidine-containing domain from the ATP-binding phosphate donor domain also affords the phosphoryl transfer from the histidine residue to the downstream response regulatory unit.

Conservation between GHL ATPases and Histidine Kinases

We have established that structural transformation on ligand binding, which can be ATP or a chemical stimulus, is the hallmark of both GHL ATPases and histidine kinases. In parallel to the activation of histidine kinases by the sensory unit, the ATPase activity of both DNA topoisomerases and MutL is increased in the presence of DNA. In the case of DNA topoisomerase, DNA is the substrate and serves as a positive feedback [51]. In the case of MutL, DNA seems to play a regulatory role in coordinating DNA repair. In the presence of ssDNA, both K_m and k_{cat} of the MutL ATPase activity are increased, which means less MutL is in the ATP-bound form compared with in the absence of ssDNA. ssDNA is generated after MutH cleaves the daughter strand. By then the interaction between MutL and MutH, which requires MutL to associate with ATP, is no longer needed. The shift of equilibrium between ATP-bound and apo-protein forms by ssDNA may enable MutL to change interacting partners and streamline the process of DNA mismatch repair [41]. To generalize, GHL ATPases can be regulated like histidine kinases by both ATP and additional ligands.

All histidine kinases known to date are dimeric proteins, which is remarkably similar to GHL ATPases [13]. In class I histidine kinases, the dimerization domain is also the phosphoryl histidine-containing domain [52] (Fig. 5B). Mutagenesis experiments show that two histidine kinase mutants defective in either the ATP-binding domain or the histidine-containing domain can complement each other and complete the phosphotransfer relay reaction, while each defective protein alone does not [53]. This result suggests that the two histidine kinase subunits within a dimer phosphorylate each others His residues across the dimer interface, which can be categorized as transautophosphorylation. In class II histidine kinases, the dimerization domain intervenes the histidine-containing and the ATP-binding domain, but transautophosphorylation also occurs (Fig. 5B) [54]. Similar to GHL ATPases, dimerization of histidine kinase is thus an essential part of kinase activity.

Finally, ATP hydrolysis by GHL ATPases and phosphorylation by histidine kinases occur at one active site a time even though the proteins are dimeric. Studies of an EnvZ derivative comprising a single polypeptide chain with two dimerization domains linked in tandem and followed by a single ATP binding domain indicate that phosphorylation of one His residue is sufficient for the two-component system to carry out signal transduction [55]. Similarly, presteady-state kinetic studies of yeast DNA topoisomerase II [21] and mutagenesis studies of DNA gyrase [46] conclude that hydrolysis of only one of the two bound ATP is sufficient for these enzymes to change DNA topology.

In conclusion, GHL ATPases and histidine kinases share a conserved ATP-binding site and are dimeric proteins. They probably depend on a dimer interface to effectively change conformation in response to ligand binding. An active site of these protein enzymes may be composite and comprises residues from both protein subunits as observed in histidine kinases and DNA gyrase. However, the two active sites of a dimeric protein can act independently, and under certain circumstances, enzymatic reaction in one active site is sufficient to support the biological functions.

CLOSING REMARKS

Jacob Monod once said that " what is true for *E. coli* is true for the elephant." Many genes are conserved from *E. coli* to humans and even more so with enzymatic active sites and reaction mechanisms. It is not unreasonable to expect that other protein enzymes will share some features common to GHL ATPases and histidine kinases in utilizing nucleotide triphosphate. On the one hand, the binding site for ATP is conserved among the GHL and HK, but it may be slightly altered to accommodate a different type of nucleotide, such as GTP, CTP, or UTP in other cases. On the other hand, ATP bound to the similarly constructed pocket can use different nucleophiles to break the phosphodiester bond in ATP, either a water molecule as ATPases or a protein side chain as kinases, to support different biological processes.

REFERENCES

1. Mushegian, A. R., Bassett, D. E., Boguski, M. S., Bork, P., and Koonin, E. V. (1997). Positionally cloned human disease genes: Patterns of evolutionary conservation and functional motifs. *Proc. Natl. Acad. Sci. USA* **94**, 5831–5836.
2. Wigley, D. B., Davies, G. J., Dodson, E. J., Maxwell, A., and Dodson, G. (1991). Crystal structure of an N-terminal fragment of the DNA gyrase B protein. *Nature* **351**, 624–629.
3. Ban, C., and Yang, W. (1998). Crystal structure and ATPase activity of MutL: Implications for DNA repair and mutagenesis. *Cell* **95**, 541–552.
4. Bilwes, A. M., Alex, L. A., Crane, B. R., and Simon, M. I. (1999). Structure of CheA, a signal-transducing histidine kinase. *Cell* **96**, 131–141.
5. Prodromou, C., Roe, S. M., Piper, P. W., and Pearl, L. H. (1997). A molecular clamp in the crystal structure of the N-terminal domain of the yeast Hsp90 chaperone. *Nature Struct. Biol.* **4**, 477–482.
6. Stebbins, C. E., Russo, A. A., Schneider, C., Rosen, N., Hartl, F. U., and Pavletich, N. P. (1997). Crystal structure of an Hsp90-geldanamycin complex: Targeting of a protein chaperone by an antitumor agent. *Cell* **89**, 239–250.
7. Tanaka, T., *et al.* (1998). NMR structure of the histidine kinase domain of the *E. coli* osmosensor EnvZ. *Nature* **396**, 88–92.
8. Ban, C., Junop, M., and Yang, W. (1999). Transformation of MutL by ATP binding and hydrolysis: A switch in DNA mismatch repair. *Cell* **97**, 85–97.
9. Prodromou, C., Roe, S. M., O'Brien, R., Ladbury, J. E., Piper, P. W., and Pearl, L. H. (1997). Identification and structural characterization of the ATP/ADP-binding site in the Hsp90 molecular chaperone. *Cell* **90**, 65–75.
10. Obermann, W. M., Sondermann, H., Russo, A. A., Pavletich, N. P., and Hartl, F. U. (1998). In vivo function of Hsp90 is dependent on ATP binding and ATP hydrolysis. *J. Cell Biol.* **143**, 901–910.
11. Panaretou, B., *et al.* (1998). ATP binding and hydrolysis are essential to the function of the Hsp90 molecular chaperone in vivo. *EMBO J.* **17**, 4829–4836.
12. Dutta, R., and Inouye, M. (2000). GHKL, an emergent ATPase/kinase superfamily. *Trends Biochem. Sci.* **25**, 24–28.
13. Stock, A. M., Robinson, V. L., and Goudreau, P. N. (2000). Two-component signal transduction. *Annu. Rev. Biochem.* **69**, 183–215.

14. Gellert, M., Mizuuchi, K., O'Dea, M. H., and Nash, H. A. (1976). DNA gyrase: An enzyme that introduces superhelical turns into DNA. *Proc. Natl. Acad. Sci. USA* **73**, 3872–3876.
15. Champoux, J. J. (1978). Proteins that affect DNA conformation. *Annu. Rev. Biochem.* **47**, 449–479.
16. Gellert, M., O'Dea, M. H., Itoh, T., and Tomizawa, J. (1976). Novobiocin and coumermycin inhibit DNA supercoiling catalyzed by DNA gyrase. *Proc. Natl. Acad. Sci. USA* **73**, 4474–4478.
17. Maxwell, A. (1997). DNA gyrase as a drug target. *Trends Microbiol.* **5**, 102–109.
18. Malonne, H., and Atassi, G. (1997). DNA topoisomerase targeting drugs: Mechanisms of action and perspectives. *Anticancer Drugs* **8**, 811–822.
19. Wang, J. C. (1998). Moving one DNA double helix through another by a type II DNA topoisomerase: The story of a simple molecular machine. *Q. Rev. Biophys.* **31**, 107–144.
20. Berger, J. M., and Wang, J. C. (1996). Recent developments in DNA topoisomerase II structure and mechanism. *Curr. Opin. Struct. Biol.* **6**, 84–90.
21. Baird, C. L., Harkins, T. T., Morris, S. K., and Lindsley, J. E. (1999). Topoisomerase II drives DNA transport by hydrolyzing one ATP. *Proc. Natl. Acad. Sci. USA* **96**, 13685–13690.
22. Lindsley, J. E., and Wang, J. C. (1993). On the coupling between ATP usage and DNA transport by yeast DNA topoisomerase II. *J. Biol. Chem.* **268**, 8096–8104.
23. Csermely, P., Schnaider, T., Soti, C., Prohaszka, Z., and Nardai, G. (1998). The 90-kDa molecular chaperone family: Structure, function, and clinical applications. A comprehensive review. *Pharmacol. Ther.* **79**, 129–168.
24. Pearl, L. H., and Prodromou, C. (2000). Structure and in vivo function of Hsp90. *Curr. Opin. Struct. Biol.* **10**, 46–51.
25. Hohfeld, J. (1998). Regulation of the heat shock conjugate Hsc70 in the mammalian cell: The characterization of the anti-apoptotic protein BAG–1 provides novel insights. *Biol. Chem.* **379**, 269–274.
26. Nathan, D. F., and Lindquist, S. (1995). Mutational analysis of Hsp90 function: Interactions with a steroid receptor and a protein kinase. *Mol. Cell. Biol.* **15**, 3917–3925.
27. Nadeau, K., Das, A., and Walsh, C. T. (1993). Hsp90 chaperonins possess ATPase activity and bind heat shock transcription factors and peptidyl prolyl isomerases. *J. Biol. Chem.* **268**, 1479–1487.
28. Csermely, P., *et al.* (1993). ATP induces a conformational change of the 90-kDa heat shock protein (hsp90). *J. Biol. Chem.* **268**, 1901–1907.
29. Sullivan, W., *et al.* (1997). Nucleotides and two functional states of hsp90. *J. Biol. Chem.* **272**, 8007–8012.
30. Jakob, U., Scheibel, T., Bose, S., Reinstein, J., and Buchner, J. (1996). Assessment of the ATP binding properties of Hsp90. *J. Biol. Chem.* **271**, [illegible].
31. Jakob, U., Lilie, H., Meyer, I., and Buchner, J. (1995). Transient interaction of Hsp90 with early unfolding intermediates of citrate synthase. Implications for heat shock in vivo. *J. Biol. Chem.* **270**, 7288–7291.
32. Bergerat, A., de Massy, B., Gadelle, D., Varoutas, P.-C., Nicolas, A., and Forterre, P. (1997). An atypical topoisomerase II from archaea with implications for meiotic recombination. *Nature* **386**, 414–417.
33. Grenert, J. P., Johnson, B. D., and Toft, D. O. (1999). The importance of ATP binding and hydrolysis by hsp90 in formation and function of protein heterocomplexes. *J. Biol. Chem.* **274**, 17525–17533.
34. Prodromou, C., *et al.* (1999). Regulation of Hsp90 ATPase activity by tetratricopeptide repeat (TPR)-domain co-chaperones. *EMBO J* **18**, 754–762.
35. Modrich, P., and Lahue, R. (1996). Mismatch repair in replication fidelity, genetic recombination, and cancer biology. *Annu. Rev. Biochem.* **65**, 101–133.

36. Grilley, M., Welsh, K. M., Su, S.-S., and Modrich, P. (1989). Isolation and characterization of the *Escherichia coli mutL* gene product. *J. Biol. Chem.* **264**, 1000–1004.
37. Lahue, R. S., Au, K. G., and Modrich, P. (1989). DNA mismatch correction in a defined system. *Science* **245**, 160–164.
38. Haber, L. T., and Walker, G. C. (1991). Altering the conserved nucleotide binding motif in the *Salmonella typhimurium* MutS mismatch repair protein affects both its ATPase and mismatch binding activities. *EMBO J* **10**, 2707–2715.
39. Oeda, K., Horiuchi, T., and Sekiguchi, M. (1982). The uvrD gene of *E. coli* encodes a DNA-dependent ATPase. *Nature* **298**, 98–100.
40. Junop, M. S., Obmolova, G., Rausch, K., Hsieh, P., and Yang, W. (2001). Composite active site of an ABC ATPase: MutS uses ATP to verify mismatch recognition and authorize DNA repair. *Mol. Cell* 7, 1–12.
41. Yang, W. (2000). Structure and function of mismatch repair proteins. *Mutat. Res.* **460**, 245–256.
42. Kolodner, R. D., and Marsischky, G. T. (1999). Eukaryotic DNA mismatch repair. *Curr. Opin. Genet. Dev.* **9**, 89–96.
43. Aizawa, S. I., Harwood, C. S., and Kadner, R. J. (2000). Signaling components in bacterial locomotion and sensory reception. *J. Bacteriol.* **182**, 1459–1471.
44. Cozzone, A. J. (1993). ATP-dependent protein kinases in bacteria. *J. Cell Biochem.* **51**, 7–13.
45. Prodromou, C., *et al.* (2000). The ATPase cycle of hsp90 drives a molecular 'clamp' via transient dimerization of the N-terminal domains. *EMBO J.* **19**, 4383–4392.
46. Kampranis, S. C., and Maxwell, A. (1998). Hydrolysis of ATP at only one GyrB subunit is sufficient to promote supercoiling by DNA gyrase. *J. Biol. Chem.* **273**, 26305–26309.
47. Smith, C. V., and Maxwell, A. (1998). Identification of a residue involved in transition-state stabilization in the ATPase reaction of DNA gyrase. *Biochemistry* **37**, 9658–9667.
48. Tomomori, C., *et al.* (1999). Solution structure of the homodimeric core domain of *Escherichia coli* histidine kinase EnvZ. *Nature Struct. Biol.* **6**, 729–734.
49. Zhou, H., Lowry, D. F., Swanson, R. V., Simon, M. I., and Dahlquist, F. W. (1995). NMR studies of the phosphotransfer domain of the histidine kinase CheA from *Escherichia coli*: Assignments, secondary structure, general fold, and backbone dynamics. *Biochemistry* **34**, 13858–13870.
50. Levit, M. N., Liu, Y., and Stock, J. B. (1999). Mechanism of CheA protein kinase activation in receptor signaling complexes. *Biochemistry* **38**, 6651–6658.
51. Williams, N. L., and Maxwell, A. (1999). Locking the DNA gate of DNA gyrase: Investigating the effects on DNA cleavage and ATP hydrolysis. *Biochemistry* **38**, 14157–14164.
52. Dutta, R., Qin, L., and Inouye, M. (1999). Histidine kinases: Diversity of domain organization. *Mol. Microbiol* **34**, 633–640.
53. Yang, Y., and Inouye, M. (1991). Intermolecular complementation between two defective mutant signal-transducing receptors of *Escherichia coli*. *Proc. Natl. Acad. Sci. USA* **88**, 11057–11061.
54. Swanson, R. V., Bourret, R. B., and Simon, M. I. (1993). Intermolecular complementation of the kinase activity of CheA. *Mol. Microbiol.* **8**, 435–441.
55. Qin, L., Dutta, R., Kurokawa, H., Ikura, M., and Inouye, M. (2000). A monomeric histidine kinase derived from EnvZ, an *Escherichia coli* osmosensor. *Mol. Microbiol.* **36**, 24–32.

CHAPTER 12

Response Regulator Proteins and Their Interactions with Histidine Protein Kinases

ANN M. STOCK* AND ANN H. WEST[†]

*Center for Advanced Biotechnology and Medicine, Howard Hughes Medical Institute, Robert Wood Johnson Medical School, Piscataway, New Jersey 08854 and [†]Department of Chemistry and Biochemistry, University of Oklahoma, Norman, Oklahoma 73019

Histidine Kinases in Signal Transduction

Response regulator proteins function together with histidine protein kinases in two-component signal transduction pathways. Response regulators are typically composed of a conserved regulatory domain that controls the activity of a variable effector domain that mediates the specific output response. The conserved regulatory domain functions as a phosphorylation-regulated switch. The domains exist in at least two conformations, designated "inactive" and "active," with phosphorylation shifting the equilibrium toward the active state. Structural studies have defined a conserved mechanism through which phosphorylation induces a propagated conformational change altering one face of the regulatory domain. Subsets of this altered surface are used for regulatory protein–protein interactions. This versatile strategy allows a large variety of regulatory mechanisms to be utilized by different response regulator proteins. The level of active phosphorylated response regulator protein ultimately determines the output response and, consequently, the level of phosphorylation is highly regulated within two-component systems. Aside from the intrinsic autophosphatase activity of response regulators and/or the activity of auxiliary phosphatases, the major locus for the regulation of response regulator phosphorylation is the histidine kinase. Histidine kinases provide the phosphoryl groups for phosphotransfer reactions catalyzed by response regulators. In some systems, the autophosphorylation rate of the histidine kinase determines the level of response regulator phosphorylation. In other systems the level of response regulator phosphorylation is controlled by a response regulator phosphatase activity of the histidine kinase. Aided by three-dimensional structures, the interactions between histidine kinases and response regulators are beginning to be probed.

INTRODUCTION

No book on histidine protein kinases (HKs) would be complete without a chapter on response regulator (RR) proteins, as these two components are obligatorily coupled within phosphotransfer signaling pathways. In a typical "two-component" system, the RR lies at the end of the pathway, eliciting the specific output response of the system. In this minimal scheme, the role of the HK is to control the level of activation of the RR in a stimulus-dependent manner.

Two-component systems are prevalent in prokaryotes. Throughout the many two-component systems that have been characterized to date, there is an enormous diversity in the types of stimuli and responses that are coupled through the common intracellular phosphotransfer signaling strategy. This diversity depends on the modular nature of both HKs and RRs. Just as HKs are typically composed of variable sensing domains and conserved kinase cores, RRs typically consist of a conserved N-terminal regulatory domain and a variable C-terminal effector domain. The regulatory domain catalyzes phosphotransfer from the HK and controls the activity of the effector domain in a phosphorylation-dependent manner. The effector domain produces the output response of the system, often the modulation of gene expression.

The modular architecture allows domains of HKs and RRs to be mixed and matched into a variety of system architectures. The conserved domain of RRs, sometimes referred to as a "receiver" domain, can function in capacities other than in regulating the activity of an effector domain within a conventional RR. For example, the conserved Asp-containing domain, either as an isolated protein or as a domain of a hybrid HK, can participate in multistep phosphorelay pathways in which its primary function is the shuttling of phosphoryl groups between His-containing domains.

A recent BLAST search identified over 600 RRs in the nonredundant database and the number is growing rapidly as microbial genomes are being sequenced at an increasing pace. It is not possible to provide a comprehensive review or catalog of RRs within the scope of this chapter. Fortunately, the modular architecture of RRs allows them to be fairly completely analyzed and understood in terms of the conserved and variable functions of each domain. It is this theme that provides the organizational basis for this chapter.

An attempt will be made to summarize the fundamental structural and functional aspects of regulatory and effector domains, as well as some of the variations that provide great versatility to this family of proteins. This information has been gathered through extensive studies of numerous different RRs by many investigators focusing on a variety of different two-component systems. Perhaps because of the relative ease of dealing with soluble RRs as opposed to transmembrane HKs, characterization of RRs has advanced at a faster rate. At this time, the fundamental molecular basis of RR function is well understood. However, the great versatility of two-component systems stems from diversity in specific regulatory mechanisms and activities of RRs, and important details still remain to be determined for individual proteins. RRs are robust signaling modules on which a seemingly limitless number of variations can be imposed. The scope of this chapter will be necessarily narrow, focusing on conserved features illustrated by representative RRs. There are many excellent reviews of "two-component" systems and of RRs in particular [1–15]. The reader is encouraged to consult these sources for both historical and up-to-date perspectives on the field.

REGULATORY DOMAINS

ACTIVITIES

The conserved regulatory domain found at the N terminus of typical RRs has three activities: phosphotransfer, autodephosphorylation, and regulation of effector domain function. These activities involve common elements in all RRs. However, the activities of different proteins also show substantial variations that provide optimization of individual RRs for the specific pathways in which they function.

Although the conserved N-terminal domain of RRs is commonly referred to as a "receiver" domain, it is not a passive partner in phosphotransfer. The regulatory domain has enzymatic activity and actively catalyzes the transfer of a phosphoryl group from the phosphoHis of the HK to one of its own Asp residues. Phosphotransfer does not require a HK and small molecules containing high-energy phosphoryl groups can serve as phosphodonors for RRs in *in vitro* reactions [16]. Molecules that function as small molecule phosphodonors include acetyl phosphate, carbamoyl phosphate, and phosphoramidate. Under some conditions, acetyl phosphate can regulate RR activity *in vivo* [17], but a major role for cellular metabolites in regulating two-component systems has not been established [18, 19].

RRs also regulate their own dephosphorylation. Autophosphatase activity varies greatly among different RRs. Half-lives of phosphorylated RRs range from seconds to hours, the latter approximating the lifetime of a typical acyl phosphate in aqueous solution. The level of autophosphatase activity appears to be fine-tuned to the specific system. For instance, RRs that mediate the second to second swimming responses in bacterial chemotaxis have half-lives in the range of seconds [20], whereas RRs that mediate life cycle events, such as those in the *Bacillus subtilis* sporulation system, have half-lives of hours [21].

Perhaps the most important activity of the regulatory domain is modulation of the effector domain that determines the output response of the signaling system. Although a fundamentally similar strategy is used by all RRs to couple phosphorylation to regulation, the mechanisms of regulation themselves are diverse. Both unphosphorylated and phosphorylated forms of regulatory domains can participate in protein–protein interactions. Different protein interactions dictated by the two states form the basis for many diverse regulatory schemes (see later).

Structure

The regulatory domain consists of approximately 125 amino acids with a $(\beta\alpha)_5$ fold (Fig. 1). The β strands of the central parallel sheet have a topology 2-1-3-4-5, with helices α1 and α5 on one face of the sheet and helices α2, α3, and α4 on the other. The structure of the single-domain chemotaxis RR CheY has long served as a model for regulatory domains [22, 23]. X-ray crystal and/or nuclear magnetic resonance (NMR) solution structures of 10 different

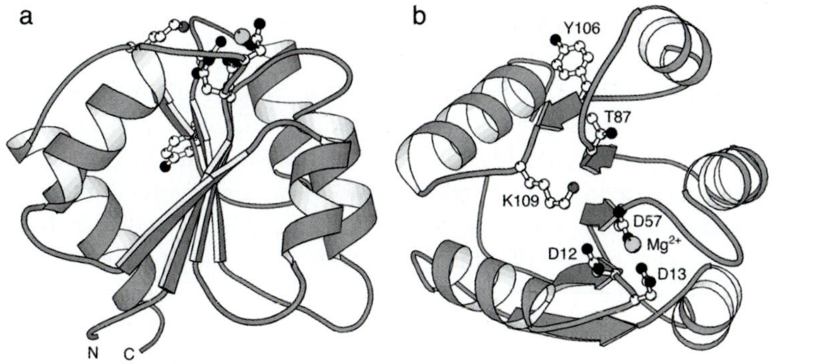

FIGURE 1 The RR regulatory domain. (a) A ribbon diagram of a representative regulatory domain (CheY bound to Mg^{2+} [39]) is shown with the side chains of conserved residues shown in a ball-and-stick representation. (b) Rotation of the molecule 90° from that shown in Fig. 1a provides a view looking directly down the β sheet from its C-terminal edge. A cluster of conserved carboxylate side chains (Asp12, Asp13, and Asp57) coordinate a Mg^{2+} ion. Lys109 is located near the active site, but does not participate in any ionic interactions in the presence of Mg^{2+}. Conserved residues Thr87 and Tyr106, which play roles in the phosphorylation-induced conformational change, are located on a path extending from the active site. Carbon, oxygen, nitrogen, and magnesium atoms are colored white, black, dark gray, and light gray, respectively.

regulatory domains show similar features (CheY [24]: Spo0F [25], CheB [26], PhoB [27], NarL [28], NtrC [29], Spo0A [30], FixJ [31], Etr1 [32], PhoP [33], and DrrD [34].

Regulatory domains display approximately 20–30% amino acid sequence identity with a small number of highly conserved residues that contribute to their common activities. One of these residues, an almost invariant Asp at the C-terminal end of β3 (Asp57 in CheY), is the site of phosphorylation [35]. This residue is clustered with two additional highly conserved carboxylate-containing side chains at the end of β1 (Asp12 and Asp13 in CheY), forming a divalent metal ion-binding site, the active site of the regulatory domain. A highly conserved Lys residue in the β5-α5 loop (Lys109 in CheY) is in close proximity and forms a salt bridge with the carboxylate of the phosphorylatable Asp in the metal-free unphosphorylated protein [24]. Two other conserved residues, which, like Lys, play roles subsequent to phosphorylation, are located on a diagonal path extending away from the active site. A hydroxyl-containing residue Ser/Thr is located on the β4-α4 loop (Thr87 in CheY) and an aromatic residue Phe/Tyr is located on β5 (Tyr106 in CheY).

Mechanism of Catalysis

Both phosphotransfer and dephosphorylation require a divalent metal ion [35–37]. Mg^{2+}, presumably the physiologically relevant metal ion, binds to CheY with a $K_D \sim 0.5$ m*M* [37, 38]. The crystal structure of CheY bound to Mg^{2+} shows octahedral coordination of the metal ion involving three protein oxygens (carboxylate oxygens of Asp13 and Asp57 and the backbone carbonyl oxygen of Asn59) and three water molecules (one of which is positioned by Asp12) [39] (Fig. 2a).

Metal ions are commonly involved in phosphoryl transfer, providing transition-state templates and charge shielding. Studies of Mg^{2+}-catalyzed phosphoryl transfer in small molecule mimics of the HK–RR pair suggest a transition state involving simultaneous coordination of the metal ion to two oxygens of a pentavalent phosphorus intermediate and an oxygen of the attacking carboxylate [40]. Such a transition state can be accommodated readily within the CheY–Mg^{2+} structure by simple rotation of the side chain of Asp57. Thus phosphotransfer is presumed to proceed by an SN2 mechanism with the carboxylate oxygen of Asp57 serving as the nucleophile for in-line attack at the axial position of a trigonal bipyramidal pentavalent phosphorus intermediate. Autodephosphorylation presumably involves a reversal of this mechanism with water serving as the nucleophile [37]. The nature of the side chain two residues beyond the phosphorylated Asp (Asn59 in CheY) has been correlated with the level of phosphatase activity [21]. It is likely that additional modulation of the exact geometry and nucleophilicity of the attacking water molecule are contributed by other active site residues as well.

Activation by Phosphorylation

Trapping the Active Conformation

The short lifetime of phosphorylated RRs has been a hindrance to their characterization. In recent years, a variety of approaches have been taken to capture the phosphorylated state for structural studies. One approach has been the use of proteins from thermophilic bacteria, which have unusually stable acyl phosphates [41]. An alternative approach has been to reduce the catalysis of dephosphorylation by the removal of metal ions [42]. Another strategy has been to determine structures by NMR in solution where proteins can be maintained in a steady-state equilibrium of phosphorylation by the presence of small molecule phosphodonors [43].

Perhaps the biggest advance in stabilizing the active state of RRs has been the use of phosphate analogs. Early attempts at creating covalent mimics of

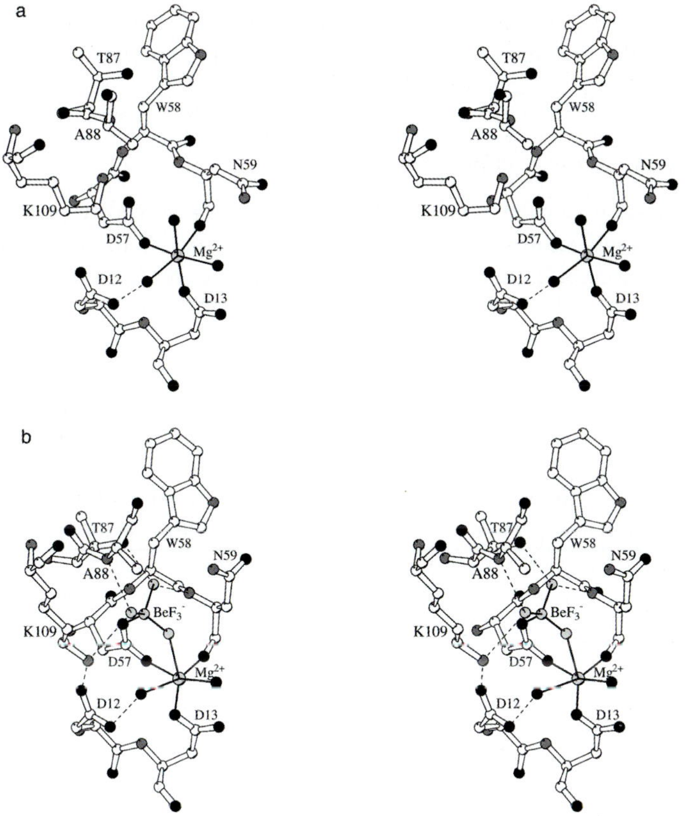

FIGURE 2 The active site of the RR regulatory domain. Stereo views of the active sites of (a) CheY bound to Mg^{2+} [39] and (b) an active CheY–BeF_3^-–Mg^{2+}–FliM complex [48] illustrate changes associated with phosphorylation. In both structures, Mg^{2+} has a similar octahedral coordination involving three protein ligands (the side chains of Asp13 and Asp57 and the backbone carbonyl oxygen of Asn59) and two water molecules (one of which is positioned by Asp12). In the CheY–BeF_3^-–Mg^{2+}–FliM complex, a fluorine atom (a mimic of a phosphate oxygen) replaces a water molecule as the sixth ligand. The two other fluorine atoms each participate in two hydrogen bonds (one with the side chain hydroxyl of Thr87 and the backbone amide of Asn59; the other with the side chain amino group of Lys109 and the backbone amide of Ala88). The octahedral coordination of Mg^{2+} and the tetrahedral coordination of BeF_3^- are indicated with solid lines. Hydrogen bonds are depicted with dashed lines. Carbon, oxygen, and nitrogen atoms are colored white, black, and dark gray, respectively. Beryllium, fluorine, and magnesium atoms are colored light gray.

the phosphoAsp involved modifications of unique active site cysteine residues with thiophosphate [44] and iodomethylphosphonate [45, 46]. While these analogs were stable, neither appears to have produced a fully active RR. More recently, the noncovalent complex BeF_3^- has been shown to be an activator of numerous RRs [47]. In several RRs, complexes with BeF_3^- have been shown to be both biochemically and structurally indistinguishable from the phosphorylated proteins [48, 49].

Propagation of the Conformational Change

A conserved molecular mechanism appears to be involved in propagation of the long-range conformational changes that accompany the phosphorylation of RRs. Phosphorylation of the conserved Asp at the C-terminal end of β3 results in minor, but significant, reorganization of the active site. The crystal structure of a CheY–BeF_3^-–Mg^{2+}–FliM peptide complex [48] and a CheY–Mg^{2+} complex [39] provides a basis for comparing the active sites of the presumed active and inactive forms of the regulatory domain (Fig. 2). In the active structure, an octahedral coordinated metal ion and the tetrahedral phosphate provide multivalent centers for ordering the side chains of active site residues. The metal ion retains many of the same ligands as in the unphosphorylated protein, specifically, the side chain carboxylate oxygens of Asp13 and Asp57, the backbone carbonyl oxygen of Asn59, and two water molecules. The additional ligand, which replaces a water molecule in the unphosphorylated structure, is a fluorine atom of BeF_3^-, a mimic of a phosphate oxygen. A second fluorine atom forms a salt bridge with Lys109 and a hydrogen bond to the backbone amide of Ala88. The third fluorine forms hydrogen bonds with the hydroxyl group of Thr87 and the backbone amides of Trp58 and Asn59. Although the metal ion is involved in numerous interactions, data suggest that it is not essential for maintaining an active conformation. Notably, phosphorylated CheY can bind to its target in the absence of metal ions [50]. Furthermore, the structure of metal-free phosphoFixJ [42] has an active site geometry similar, although not identical, to the active CheY complex described earlier.

Changes at the active site are propagated to an opposite surface of the regulatory domain. The hydrogen bond formed between the phosphate and the side chain of Thr87 requires a significant repositioning of this residue. Coincident with the altered orientation of Thr87 in the active structure, the side chain of Tyr106 is flipped from an "outward" to an "inward" orientation, burying this residue in the cavity vacated by the repositioned Thr87 side chain. The reoriented side chains of the conserved Ser/Thr residue on the β4-α4 loop and the Phe/Tyr on β5 form a contiguous path that stretches diagonally from the active site through the hydrophobic core to the opposite surface of the domain (Fig. 3). These rearrangements have been observed in all high-resolution structures of activated regulatory domains [41, 42, 46, 48] and suggest a conserved mechanism for the phosphorylation-induced conformational change.

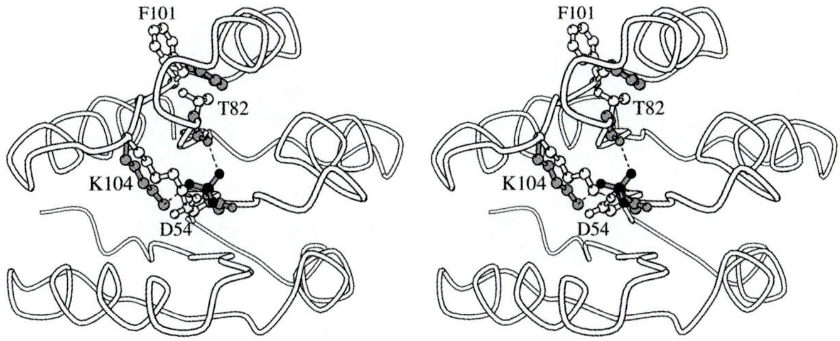

FIGURE 3 Mechanism of the phosphorylation-induced propagated conformational change. A stereo view of the regulatory domain of FixJ [42] is shown in an orientation similar to that of the domain in Fig. 1b. Ball-and-stick representations of side chains of conserved residues involved in the phosphorylation-induced conformational change (Asp54, Thr82, Phe101, and Lys104) are shown in orientations observed in structures of unphosphorylated (white) and phosphorylated (gray) FixJ. For clarity, only the backbone of unphosphorylated FixJ is shown in coil representation. Oxygen and phosphorus atoms of the phosphate are colored black. In phosphorylated FixJ, the side chain hydroxyl of Thr82 is positioned to form a hydrogen bond with a phosphate oxygen and Phe101 adopts an inward orientation, filling the space vacated by the reoriented Thr82 side chain. Lys104 forms a salt bridge with Asp54 in unphosphorylated FixJ and with the phosphate in phosphorylated FixJ. Salt bridges and hydrogen bonds are shown as dashed lines.

Phosphorylation-Induced Conformational Changes

Phosphorylation of active site Asp results in conformational changes that extend over a large surface of the molecule. In regulatory domains that have been characterized structurally [41–43, 46, 48], these changes are localized to subsets of the α3-β4-α4-β5-α5 regions of the protein (Fig. 4 and reviewed in Robinson *et al.* [130] and West and Stock [51]. The magnitude of backbone displacements ranges from <1 to 6 Å and varies significantly in different regulatory domains, with the α4 region of the NtrC regulatory domain exhibiting the greatest changes of any domain. The exact regions affected by phosphorylation also vary, with some domains exhibiting more localized changes than others. With the minor exception of helix α4 of NtrC, the changes do not affect secondary structure, but rather involve subtle repositioning of secondary structural elements. Not surprisingly, some of the largest changes occur in loops connecting β strands and α helices. Despite the small changes in backbone positions, the molecular surface of the domains is altered substantially.

The α3-β4-α4-β5-α5 surface that is altered upon phosphorylation correlates well with surfaces of different regulatory domains that have been identified by a large number of genetic and biophysical methods to be involved in protein–protein interactions that are regulated by phosphoryla-

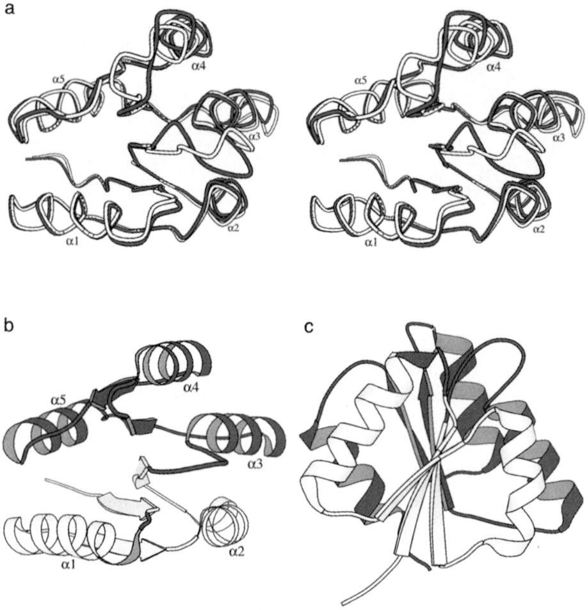

FIGURE 4 Phosphorylation-induced conformational changes. (a) Stereo representations of the backbone conformations of unphosphorylated and phosphorylated FixJ [42] are shown in white and gray, respectively. (b) A composite of all regions that have been observed to change upon phosphorylation in the regulatory domains of FixJ [42], NtrC [43], Spo0A [41] and CheY [48] are displayed as dark gray on a ribbon diagram of FixJ. (c) A view of the image in (b) is shown after a rotation of 90°.

tion (discussed later). This correlation provides the basis for the functioning of these phosphorylation-activated switch domains. Phosphorylation induces an altered conformational surface, subsets of which can be used for differential regulatory interactions in both unphosphorylated and phosphorylated states. This basic scheme provides great versatility and allows for an enormous array of different regulatory strategies.

Dynamics

RRs, like all proteins, are dynamic. A variety of biophysical data suggest that RRs exist in equilibrium between at least two conformational states, with the inactive conformation predominating in the unphosphorylated protein and an active conformation being favored in the phosphorylated protein. Multiple conformational states of CheY [39, 52–55] and FixJ [31] have been observed in crystal structures that provide static pictures of accessible states of proteins. NMR analyses have probed dynamics in Spo0F [56, 57], CheY [58], and the NtrC regulatory domain [59, 60]. In Spo0F, motions in the micro- to millisecond time scale correlate with surfaces that are involved in protein–protein interactions [61, 62] and that have been shown in other regulatory domains to be altered by phosphorylation. These motions may reflect the sampling of functionally relevant conformational states. NMR studies of NtrC mutant proteins likewise indicate the simultaneous presence of two conformations, with a correlation between the level of activation of the mutant protein and the distribution of conformational states [59]. Furthermore, NMR analyses of unphosphorylated and phosphorylated NtrC regulatory domains indicate that the region that exhibits the greatest conformational change is quite dynamic in the unphosphorylated domain and more rigid in the phosphorylated state [43]. Data suggest that active conformation is accessible to unphosphorylated regulatory domains and may explain the residual low level of activity of unphosphorylated RRs.

Additional biochemical data support the hypothesis of two conformational states, with an active state favored by phosphorylation or by binding to a target. Phosphorylation of OmpR increases the affinity of OmpR for DNA [63–65] and, conversely, binding of OmpR to DNA increases the level of OmpR phosphorylation by stimulating phosphotransfer [66] and/or by decreasing the rate of EnvZ-mediated dephosphorylation [67]. A similar reciprocal activation has been seen with phosphorylation of CheY and CheY binding to its target FliM [68].

EFFECTOR DOMAINS

ACTIVITIES

RRs typically mediate the output responses of signaling pathways. The large variation in output responses results from the diversity of effector domains. This diversity is apparent with respect to both function and structure (Fig. 5). In prokaryotic two-component systems, RRs function most commonly as transcription factors activating and/or repressing transcription of a specific set of genes. The effector domains of these RRs are capable of binding to DNA and interacting with other components of the transcriptional machinery. However, not all RR effector domains have DNA binding activity. Some effector domains have enzymatic activities, whereas others regulate downstream targets through protein–protein interactions.

A survey of the *Escherichia coli* genome reveals that only 3 of the total 32 RRs do not have recognizable or putative DNA-binding domains [6]. Two of these are components of the chemotaxis system: CheY, which consists of an

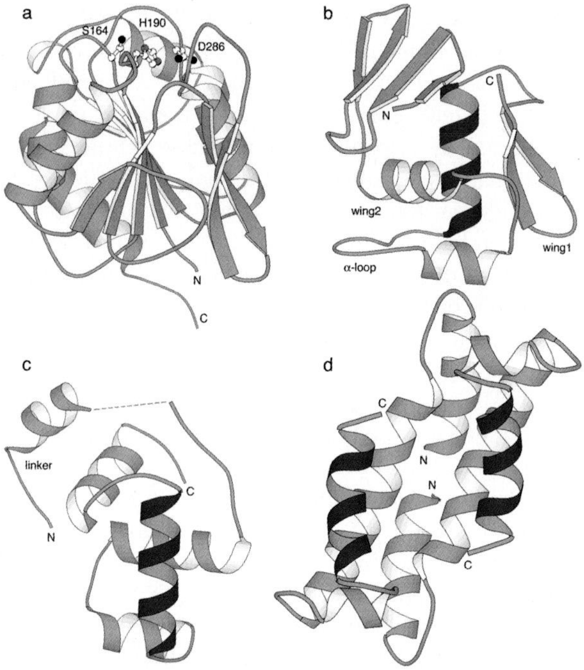

FIGURE 5 RR effector domains. Ribbon diagrams are shown for representative members of all subfamilies of RRs for which structural information is currently available. (a) The C-terminal catalytic domain of the chemotaxis methylesterase CheB [169] is shown with ball-and-stick representations of the side chains of the active site catalytic triad (Ser164, His190, and Asp286). Carbon, oxygen, and nitrogen atoms are colored white, black, and dark gray, respectively. (b) The winged-helix fold of the C-terminal DNA-binding domain of OmpR [87, 88] is shown with the recognition helix colored dark gray. Wings on either side of the recognition helix presumably participate in minor groove contacts, and the α loop is thought to be important for interaction with the C-terminal domain of the a subunit of RNA polymerase. (c) The C-terminal DNA-binding domain of NarL [97] is joined to the N-terminal regulatory domain by a short helix and flexible linker region that is disordered in the crystal structure and is represented here by a dashed line. The recognition helix, colored dark gray, is the third helix of the four-helix DNA-binding domain. (d) The central domain of NtrC [107] functions both in DNA binding and in dimerization. The first two helices of each monomer participate in dimerization, forming a four-helix bundle. The third and fourth helices form a classic helix-turn helix motif, with the fourth helix, colored dark gray, functioning as the recognition helix.

isolated regulatory domain that interacts in an intermolecular fashion with the flagellar motor, and methylesterase CheB (Fig. 5a), an enzyme that catalyzes demethylation of chemoreceptor carboxylmethylglutamate residues [69]. Another RR, RssB, regulates σ^S proteolysis through interactions with σ^S and ClpX [70, 71]. The remaining 29 RRs in *E. coli* are known, or presumed, to function as transcription factors and can be divided into subfamilies based on sequence similarity within their DNA-binding domains. Three major subfamilies emerge: the OmpR subfamily (14 members), the NarL subfamily (7 members), and the NtrC subfamily (4 members). Similar representation of these subfamilies is found in other prokaryotic genomes.

In contrast to the prevalence of DNA-binding domains within prokaryotic RRs, in eukaryotic RRs, DNA-binding effector domains are the exception rather than the norm. Only one transcription factor, *Saccharomyces cerevisiae* Skn7 [72], has been identified among eukaryotic RRs. Instead, it appears that in many eukaryotic systems, RRs interact with other proteins or have enzymatic activities that allow them to interface with other more conventional eukaryotic signaling pathways, such as cyclic nucleotide or MAP kinase cascades [7, 8, 73].

Structure/Function

OmpR/PhoB Subfamily

The OmpR/PhoB family is the largest subfamily of RRs, accounting for almost half of all two-component transcription factors. OmpR/PhoB family members function as activators or repressors of σ^{70} promoters. In cases where binding sites have been defined, the sites are located within or upstream of the promoters. Recognition sequences appear to consist of 10-bp half-sites oriented as direct repeats, rather than with dyad symmetry, which is typical of most other transcription factor-binding sites. There is great variation in the arrangement of sites, both with respect to the number of sites and the spacing between them [74–79]. It has been found for some proteins that DNA binding is insufficient for transcriptional activation, suggesting that regulation requires interactions with components of polymerase. Within the OmpR/PhoB family, the loci of these interactions vary. PhoB interacts with the σ^{70} subunit [80, 81], whereas OmpR interacts with the C-terminal domain of the α subunit of polymerase [82–86].

DNA-binding domains of OmpR/PhoB family members consist of approximately 100 residues with sequence identity between members ranging from 20 to 65%. Crystal structures of the DNA-binding domain of OmpR [87, 88] have established the fold of this family (Fig. 5b). The OmpR DNA-binding domain consists of three α helices flanked on two sides by antiparallel β sheets, an N-terminal four-stranded sheet, and a C-terminal β hairpin. Sequence analysis clearly indicates that all OmpR/PhoB family members have a similar fold, and conserved sequences within the DNA-binding domains have been correlated with functions [89, 90]. Indeed, the structure of the

DNA-binding domain of PhoB is very similar to that of OmpR [91]. The helices and C-terminal hairpin of the OmpR fold are common to a large family of transcription factors known as winged helix-turn-helix proteins. In winged helix proteins such as CAP, BirA, and HNF3-γ, helices $\alpha2$ and $\alpha3$ constitute the helix-turn-helix motif corresponding to the positioning and recognition helix, respectively. The loop connecting the β hairpin is considered a wing (W1). In some winged-helix proteins, a second wing (W2) sits on the opposite side of the recognition helix. In interactions with DNA, the recognition helix lays within the major groove while the wing(s) makes backbone contacts in the minor groove.

The OmpR domain represents a variation on the winged-helix fold and establishes a novel subfamily within winged helix-transcription factors. Two features distinguish OmpR/PhoB DNA-binding domains: the N-terminal four-stranded antiparallel β sheet and an unusually large loop connecting helices $\alpha2$ and $\alpha3$. In OmpR, the large loop has been designated the "α loop" because of its presumed role in interactions with the α subunit of RNA [86]. The N-terminal β sheet forms a platform oriented tangentially to the recognition helix and has been postulated to provide a surface for interaction with the regulatory domain. Structural analysis of an intact OmpR family member indicates that this is indeed the case [34].

NarL/FixJ Subfamily

Transcriptional regulation by the NarL/FixJ subfamily is the least understood of any of the RR subfamilies. NarL/FixJ family members function as transcription activators of σ^{70} promoters. While some members have well-conserved recognition sites for binding, there is great variation in the arrangement and positioning of binding sites, even for a single RR. For instance, NarL heptamer-binding sites occur in almost every possible configuration: in isolation, in pairs oriented as direct repeats (head to tail and tail to head), as inverted repeats (head to head), and in multiple, closely spaced direct repeats [92, 93]. These arrangements suggest that NarL may be capable of binding to DNA in many different modes, as a monomer, a dimer, or an oligomer. In several cases, promoters regulated by NarL/FixJ family members also contain additional transcription factor-binding sites such as those for IHF and FNR upstream of NarL-regulated promoters [94, 95] and CAP-binding sites upstream of UhpC-regulated promoters [96].

The structure of intact NarL [28, 97] has established both the fold of the effector domain and its relation to the regulatory domain (Fig. 5c, see also Fig. 6b). The two domains of NarL are connected by a small helix and a 13 residue flexible linker that is mostly disordered in the crystal structure. The DNA-binding domain of NarL consists of approximately 60 residues folded

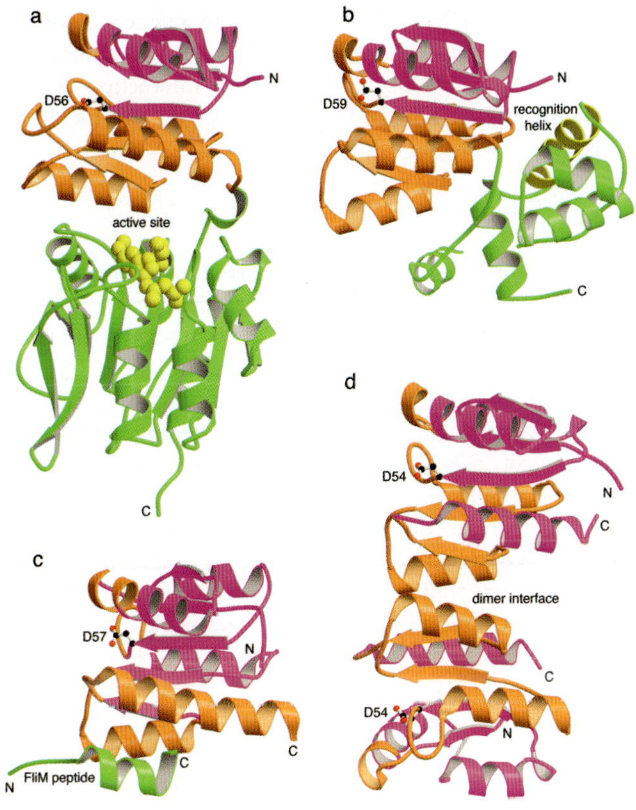

FIGURE 6 Protein–protein interactions of regulatory domains. Ribbon diagrams of intact response regulators, or regulatory domains bound to their targets are shown with the regulatory domains (magenta) in similar orientations. The site of phosphorylation is shown as a ball-and stick-representation with carbon and oxygen atoms shown in black and red, respectively. In each regulatory domain, regions that have been shown to be altered upon phosphorylation (CheY and FixJ) or the composite of regions that have been shown to be altered in other response regulators (CheB and NarL, regions defined in Fig. 4) are colored gold. (a) In methylesterase CheB [26], the regulatory domain packs against the active site of the catalytic domain (green), blocking access of the receptor substrates. Residues of the catalytic triad are shown as yellow spheres. (b) In NarL [28], the regulatory domain blocks access of DNA to the recognition helix (yellow) of the DNA-binding domain (green). Note the different relative orientations of the effector and regulatory domains in NarL and CheB. In both NarL and CheB, activation by phosphorylation is presumed to involve a repositioning of the regulatory and effector domains to relieve inhibitory interactions observed in the unphosphorylated structures. (c) The chemotaxis response regulator CheY activated by BeF_3^- [48] interacts with a peptide of the flagellar motor switch protein FliM (green) through the α4-β5-α5 face. (d) The phosphorylated regulatory domains of the transcription factor FixJ [42] dimerize through interactions of helix α4 and strand β5. The interactions depicted for both CheY and FixJ are enhanced by phosphorylation.

into a bundle of four α helices, with the central two helices forming a conventional helix-turn-helix motif. Based on the knowledge of interactions of such motifs with DNA, sequence-specific NarL–DNA interactions have been proposed. Importantly, the packing of the regulatory and effector domains in the unphosphorylated protein sterically interferes with DNA binding (see later).

NtrC/DctD Subfamily

The NtrC/DctD subfamily is the most structurally and mechanistically complex group of RR transcription factors. Members of the NtrC/DctD family activate σ^{54}-dependent promoters by binding to transcriptional enhancers and interacting with polymerase through DNA loop formation [98]. The effector regions of NtrC/DctD family members consist of two domains: an ATPase domain and a DNA-binding domain. ATPase activity is required for catalyzing the isomerization of closed transcription complexes to transcriptionally productive open complexes. In the case of NtrC, the most extensively characterized member of this family, phosphorylation of the regulatory domain controls ATPase activity. NtrC exists as a dimer that is capable of binding to DNA, recognizing a hexameric sequence arranged with dyad symmetry [99]. Upon phosphorylation, NtrC oligomerizes into octamers [100], resulting in the stimulation of ATP hydrolysis [101, 102] and allowing for open complex formation [103].

The 240 residue central ATPase domain and 90 residue C-terminal DNA-binding domain are common to all NtrC/DctD family members. The structure of the ATPase domain has not been determined experimentally, but a model has been proposed based on the structure of the GTPase Ef-Tu [104]. The C-terminal domain of NtrC functions both as a DNA-binding domain [105] and as a dimerization domain for full-length NtrC [106]. The NMR solution structure of this domain has revealed a fold similar to that of the factor for inversion stimulation (FIS) [107]. Approximately 20 residues at the N terminus of the domain were not assigned and are presumed to be flexible. Beyond this region, each monomer consists of four helices (Fig. 5d). The first two helices are involved in dimerization, pairing with those of a second monomer to form a four-helix bundle. The third and fourth helices comprise a classic helix-turn-helix motif [108] at opposite ends of the dimer. The spacing of recognition helices suggests that a dimer may be capable of bending DNA [107].

Regulation of Activity by Regulatory Domains

Given the diversity of effector domains, it is perhaps not surprising that there are numerous strategies for phosphorylation-dependent regulation of their activities. In all cases, the basis of regulation is different protein–protein interactions of the unphosphorylated and phosphorylated regulatory domains, but the nature of the interactions varies significantly. Presumably, any regulatory strategy based on protein–protein interactions can be utilized by RRs. A large number of different regulatory mechanisms have been described and, almost certainly, more remain to be discovered.

One broad classification of regulatory strategies has been based on the role of the regulatory domain as either an activator or an inhibitor of effector domain function. For some RRs, such as DctD [109], removal of the regulatory domain results in an active RR, implying that the unphosphorylated regulatory domain plays an inhibitory role, with phosphorylation presumably relieving this inhibition. For other RRs, such as NtrC [110], removal of the regulatory domain does not activate the RR. Instead, the phosphorylated regulatory domain plays a positive role in generating RR activity. These examples illustrate that even among subfamily members, basic regulatory strategies are not conserved. Several RRs, such as OmpR [63, 111] and CheB [112, 113], combine mechanisms of both negative and positive regulation. Removal of the regulatory domain gives partial activation, but full activation requires the presence of the phosphorylated regulatory domain.

The structural basis for inhibition of effector domain activity by regulatory domains has been revealed by crystal structures of NarL [28] and CheB [26]. In both proteins, the regulatory domains sterically block access to the functional regions of the effector domains, the recognition helix of NarL, and the active site catalytic triad of CheB (Figs. 6a and 6b). Steric collisions with the regulatory domain occur when modeling interactions of NarL with its target DNA or methylesterase CheB with its substrate, the chemotaxis receptor. In both proteins, it seems likely that phosphorylation results in a repositioning of the regulatory and effector domains that relieves the inhibitory interactions.

Many different strategies are used for positive regulation by phosphorylated regulatory domains. RRs that regulate transcription typically interact as dimers with DNA. In some RRs, such as PhoB [114] and FixJ [115], phosphorylation induces dimerization of the regulatory domain, which promotes DNA binding and transcriptional activation. In OmpR, phosphorylation of the regulatory domain does not result in detectable dimerization, but phosphorylation enhances the affinity of OmpR for DNA and binding occurs as a dimer [116]. In other RRs, such as NtrC [99], dimerization and binding to DNA occurs independently of phosphorylation, although phosphorylation is required for further oligomerization and transcriptional activation. In RRs that are not transcription factors, phosphorylation-regulated protein interactions of the regulatory domain involve heterologous partners, such as the interdomain interaction in methylesterase CheB [113] or the binding of CheY to its target, the flagellar motor protein FliM [117].

Different phosphorylation-regulated interactions of regulatory domains do not necessarily utilize common surfaces of the regulatory domain. While all such interactions appear to involve a subset of the surface altered by phosphorylation, there is no single locus within this region (Fig. 6). Regulatory domain interactions in structures of intact NarL, intact CheB, the phosphoFixJ receiver domain dimer, and a BeF_3^-–CheY–FliM peptide complex involve distinct surfaces of the α3-β4-α4-β5-α5 face of the regulatory domain. In many cases, a single RR regulatory domain may be involved in several different interactions. For instance, in the case of CheY, phosphorylation modulates its interaction with the HK CheA [118], the flagellar motor protein FliM [117], and a dephosphorylating protein CheZ [119]. Overlapping but nonidentical surfaces are used for these interactions [120].

REGULATION OF RESPONSE REGULATOR PHOSPHORYLATION

In all two-component systems, the output response of the system is dependent on the level of phosphorylation of the RR. Thus, regulation of RR phosphorylation must be tightly coupled to input signals. Many different, and often complex, strategies have been incorporated into two-component pathways to achieve such regulation. The mechanisms that regulate RR phosphorylation can be divided into two broad categories: those that are mediated directly by HKs and those that are not. It should be noted however, that because HKs typically function as sensors for input signals, even mechanisms that do not directly involve phosphorylation or dephosphorylation of RRs by HKs ultimately rely on the signaling state of the HK for regulation.

Histidine Kinase-Mediated Strategies

There are two ways in which HKs can directly regulate the phosphorylation level of RRs: through phosphorylation or dephosphorylation. In a few pathways, the rate-limiting step in phosphotransfer is autophosphorylation of the HK. In these systems, typically ones with short-lived phospho-RRs, such as the bacterial chemotaxis system [121], the level of RR phosphorylation is controlled by regulation of the autophosphorylation activity of the HK.

In most systems, however, the level of RR phosphorylation is regulated primarily by an RR phosphatase activity of the HK [122–126]. In many cases, ligand binding or other stimuli directly modulate the phosphatase activity of HKs. In some systems, the phosphatase activity of HKs is regulated by auxiliary proteins that are linked either directly or indirectly to sensing.

Examples of these more complex regulatory schemes include the cytoplasmic kinase NtrB that is regulated by an auxiliary cytoplasmic protein PII [122] and the transmembrane kinase PhoR that is regulated by a PstABCSPhoU complex at the cytoplasmic membrane [127, 128].

Alternative Strategies

As mentioned earlier, most RRs have autophosphatase activity, the level of which varies substantially from one RR to another. Half-lives of phosphorylated RRs range over greater than four orders of magnitude. Autophosphatase activity appears to be important, as it is a conserved feature of RRs. Even a *Thermotoga maritima* RR that has evolved special features to stabilize phospho-Asp at temperatures >80°C retains autophosphatase activity [129]. Whether the autophosphatase activities of RRs are regulated or whether they function solely to set the lifetime of the phosphoRR remains an open question. However, it seems likely that many RR "phosphatases" may function allosterically to enhance RR autophosphatase activity rather than directly catalyzing RR dephosphorylation.

A small number of systems utilize auxiliary phosphatases or proteins that accelerate RR dephosphorylation. The system that controls sporulation in *B. subtilis* contains four highly regulated phosphatases: RapA, RapB, and RapE that dephosphorylate Spo0F [130] and Spo0E that dephosphorylates Spo0A [131]. Additional phospho-Asp phosphatases have been identified by sequence analysis [132]. Chemotaxis systems of enteric bacteria utilize CheZ, a protein that oligomerizes with phospho-CheY and accelerates its dephosphorylation [20, 133, 134]. Characterization of RR mutants and metal ion specificity suggests a requirement for RR residues in auxiliary protein-assisted dephosphorylation [37]. In all cases, RR phosphatases appear to be extremely specific, further supporting the notion that they may work allosterically by affecting nucleophilic residues or bound water molecules of the RR rather than by directly catalyzing phosphate hydrolysis.

An additional mode of regulation is available to systems that contain more than one RR. In bacterial chemotaxis, both CheY and CheB compete for phosphoryl groups from the kinase CheA [118]. Another regulatory strategy has been postulated for chemotaxis systems that contain multiple CheY proteins. In the *Rhizobium meliloti* chemotaxis system, CheY2 is thought to function as a conventional RR, interacting with the flagellar motor. CheY1 appears to function as a "CheY2 phosphatase," providing a sink for phosphoryl groups that are passed from phospho-CheY2 through the HK CheA to CheY1. Undoubtedly, as more systems are characterized in greater detail, additional strategies for regulating RR phosphorylation will be identified.

INTERACTIONS OF RESPONSE REGULATORS WITH HISTIDINE KINASES AND HISTIDINE-CONTAINING PHOSPHOTRANSFER DOMAINS

PHOSPHOTRANSFER

In the cell, HKs and histidine-containing phosphotransfer (HPt) domains (hereafter referred to collectively as His-containing domains) serve as the primary phosphodonors to RR proteins. Kinetic studies have established that small molecule phosphodonors are far less efficient (several orders of magnitude slower) in providing phosphoryl groups to RRs than their cognate HKs or HPt domains [135–138]. These observations suggest that a phosphoimidazole presented on the surface of a His-containing domain provides an apparent mechanistic advantage in the phosphotransfer reaction catalyzed by RR domains. Furthermore, protein–protein interaction surfaces must be important in dictating the specificity of RR–HK interactions, thereby preventing undesirable "cross talk" *in vivo*.

The interaction of RRs with His-containing domains presumably occurs only transiently during the phosphotransfer reaction. The phosphorylated RR would then be expected to dissociate and interact with other downstream effectors, as observed with phospho-CheY [139]. There are several structures known for His-containing domains from HKs [140, 141], an HPt domain from a hybrid HK [142, 143] and independent HPt proteins [144–146], as well as numerous structures of conserved regulatory domains of RRs (see earlier discussion). However, structures of complexes have been difficult to obtain. To date, there has been only one structure reported of a complex between the single domain RR, Spo0F, and one of its phosphorelay partners, the HPt protein Spo0B [147]. These two proteins were cocrystallized in the presence of AlF_3, and even though electron density was not evident for this phosphate analog, it may have helped to promote or stabilize the complex.

Spo0B forms a dimer both in solution and in the crystalline state [145]. Two long antiparallel helices near the N terminus of one Spo0B monomer associate closely with the corresponding N-terminal helices in another monomer to form a central four-helix bundle core (Fig. 7a). The C-terminal α/β domains flank the four-helix bundle on each side. A key feature common to all the known structures of His-containing domains is that the imidazole side chain of the phosphorylatable His is almost completely solvent exposed and positioned prominently in the middle of a long α helix supported overall by a four-helix bundle scaffold. In contrast to monomeric HPt domains, the Spo0B dimer has two symmetrically located sites of phosphorylation, His30, which lie in the middle of the $\alpha 1$ helix.

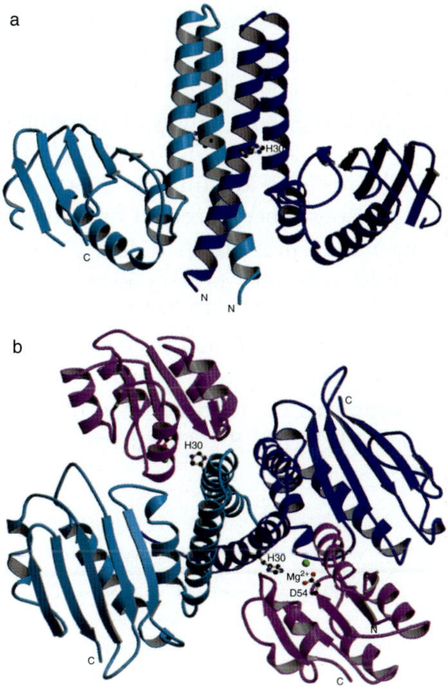

FIGURE 7 HK–RR interactions. (a) The structure of Spo0B [145], a dimeric histidine-containing phosphotransfer (HPt) protein, is illustrated in this ribbon diagram with the side chain for His30, the site of phosphorylation, shown as a ball-and-stick representation. Monomer subunits (light and dark blue) dimerize via association of the N-terminal α1 and α2 helices to form a central four-helix bundle, which is flanked on both sides by the C-terminal α/β domains. (b) The structure of the Spo0F–Spo0B complex [147] is shown with the HPt protein rotated approximately 90° from the view in (a). Two Spo0F RR proteins (magenta) bind independently on opposite faces of the four-helix bundle of Spo0B. The active site residues, His30 from Spo0B and Asp54 from Spo0F (ball-and-stick representation with carbon atoms colored black, nitrogen blue, and oxygen red) are in close proximity to each other and, together with the bound Mg^{2+} [illegible] ([illegible]), appear to be poised for phosphoryl transfer.

In the crystalline complex, two Spo0F monomers bind independently on opposite sides of the Spo0B dimer, making contacts with both the four-helix bundle and the C-terminal domain (Fig. 7b). No large changes were observed in backbone conformation for either protein on complex formation. Spo0F is structurally similar to the other regulatory domains of RRs for which structures are known (as described earlier) [25, 57]. The interaction surface, which covers an area of about 1200 Å^2 surrounding His30 on Spo0B, involves all five β-α loop regions around the active site of Spo0F. In addition, the complex buries a catalytically important Mg^{2+} ion (as discussed earlier) in the active site of Spo0F. Unexpectedly, the α1 helix of each Spo0F monomer makes extensive parallel contacts with α1 of Spo0B, effectively expanding the four-helix bundle to a six-helix bundle. Hydrophilic and hydrophobic residues from both proteins are involved in making specific contacts. Interestingly, many of the residues identified in Spo0F as contributing to the binding interface had previously been deemed functionally important *in vivo* based on Ala-scanning mutagenesis [61]. It had been predicted that these residues might be involved in protein–protein interactions.

Importantly, the positioning of the two proteins in the complex brings the active site residues, His30 on Spo0B and Asp54 on Spo0F, into reasonable alignment for phosphotransfer. A transition-state structure was proposed in which a phosphoryl group was modeled between the N^{ε} atom of His30 and the O^{δ} atom of Asp54 [147] (Fig. 8). The distances and geometry of the

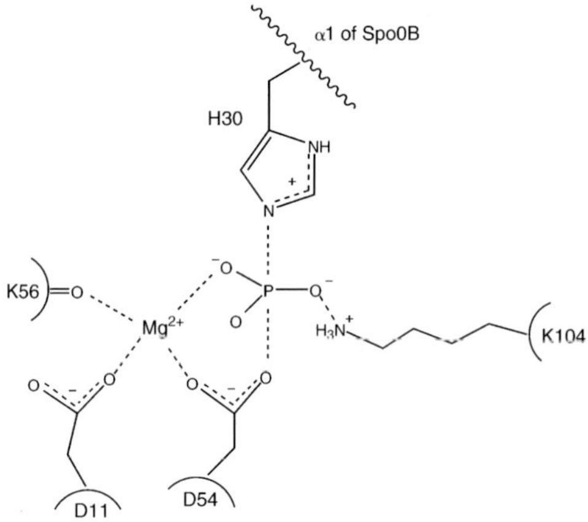

FIGURE 8 Transition state model for phosphotransfer. Based on the positioning of a planar phosphoryl group between Asp54 of Spo0F and His30 of Spo0B in the structure of the complex, a transition state structure for phosphoryl transfer between Spo0B and Spo0F was proposed [147]. In this schematic diagram, the only residue from Spo0B shown is the His30 side chain, which protrudes from helix α1; all other residues are from Spo0F. Phosphoryl transfer between Spo0B and Spo0F is freely reversible [61]; hence in the proposed trigonal bipyramidal transition state, either the N^{ε} of His30 or the O^{δ} of Asp54 could represent the attacking nucleophile in the reaction, whereas the opposite axial ligand would become part of the leaving group. The transition state is presumably stabilized by electrostatic interactions between the phosphoryl oxygens, the Mg^{2+} ion, and the highly conserved Lys104 side chain. The Mg^{2+} ion, which is essential for phosphoryl transfer, is coordinated by three ligands from the protein (carboxylate oxygens from Asp54 and Asp11, and the carbonyl oxygen from Lys56) and a phosphoryl oxygen. Presumably, two water molecules (not shown in the structure) complete the octahedral coordination geometry.

proposed pentavalent transition-state fall within an acceptable range for phosphoryl transfer reactions [148]. When bound to Spo0B, two loops around the active site of Spo0F become displaced slightly relative to the uncomplexed structure. The authors suggest that this subtle conformational change may be necessary for making the active site aspartate more accessible for phosphoryl transfer and possibly for accommodating the transition state structure [147].

Is the Spo0B–Spo0F complex a fair representative for how other RR domains may interact with His-containing domains? Of course, one can only answer this question speculatively, as no other structures of complexes are currently available. However, the four-helix bundle architecture from which a phosphoHis protrudes appears to be a conserved structural feature for His-containing domains that is recognized by RRs. Likewise, the highly conserved α/β fold for RR domains with the Asp phosphorylation site poised among the β-α loops regions strongly suggests a mode of bimolecular interaction at least grossly similar to the Spo0B–Spo0F example. Hydrophobic interactions are postulated to be the main attracting force in bringing the two proteins together, whereas nonconserved contact residues presumably confer specificity within signaling systems [147].

The NMR structure of the His-containing dimerization domain of EnvZ [140] revealed a very similar four-helix bundle architecture as seen in Spo0B. Intriguingly, NMR titration studies carried out using increasing amounts of the OmpR regulatory domain indicated that the region of EnvZ that underwent the most significant changes mapped from the His phosphorylation site all the way to the helical hairpin turn regions of the dimerization domain. It therefore appears possible that the molecular surface of HKs that is recognized by RRs may be quite extensive and is not necessarily limited to the near vicinity of the phosphorylated His.

It should be noted, however, that in the chemotaxis system, RRs CheY and CheB interact with their HK CheA in an entirely different manner than the Spo0B–Spo0F example. CheA has an unconventional domain organization relative to other HKs. CheA consists of five domains designated P1–P5 [149, 150]. The N-terminal domain (P1) contains the His phosphorylation site and forms a helical bundle characteristic of HPt domains [141, 151]. Although the phosphorylated P1 domain can readily transfer phosphoryl groups to RRs domains *in vitro*, in the context of the whole protein, the P2 domain connected to P1 through a flexible linker functions as a RR-binding domain [150]. The C-terminal domains, P3, P4, and P5, function as dimerization, kinase, and chemoreceptor-coupling domains, respectively. Several theories have been put forth to explain the evolution of a separate RR-binding domain in CheA [152, 153]. For instance, binding of CheY or CheB to the P2 domain may help properly orient the RR domain with respect to the P1 domain in order for phosphoryl transfer to proceed efficiently. In this respect, it is interesting to note that in structures of the CheY–P2 complex [152, 154], the active site of CheY is located on a face of the molecule opposite the P2 interaction surface and thus remains completely exposed for interaction with the His-containing P1 domain. A high-affinity binding site located adjacent to the phosphorylated P1 domain might also serve to increase the effective local concentration of RR available for phosphotransfer.

Dephosphorylation

As described earlier, many HKs exhibit phosphatase activity toward their cognate RRs. Thus, not all RR–HK interactions are designed for phosphotransfer. A notable example is the well-characterized EnvZ–OmpR osmoregulatory system in *E. coli* [155]. The opposing kinase/phosphatase activities of EnvZ control the level of phospho-OmpR in response to environmental cues. Both of these activities require the active site His as discussed in greater detail in Chapter 3. Other documented examples include NarX, NtrB (NRII), FixL, DegU, and KdpD kinase/phosphatases [156]. Mutations that affect kinase but not phosphatase activity, and vice versa, demonstrate that these activities are clearly distinguishable within HKs [157–159]. The RR dephosphorylation activity of HKs is Mg^{2+} dependent and requires the presence of ADP, ATP, or nonhydrolyzable analogs of ATP [160–163]. In some systems, the mechanism of HK-mediated RR dephosphorylation is clearly not a reversal of the phosphotransfer step, as evidenced by observations that mutation of the active site His in several HKs abolishes autokinase activity but still allows retention of RR phosphatase activity [164–168].

For EnvZ and NtrB, RR phosphatase activity was found to reside within the His-containing dimerization domains [161, 163]. Thus, phosphorylated RRs apparently retain some affinity for the His-containing four-helix bundle domain. Interestingly, in the case of EnvZ, activity was enhanced in the presence of the C-terminal kinase domain, but this enhancement required ADP, ATP, or nonhydrolyzable analogs of ATP. The authors suggest that the relative positioning of the dimerization and kinase domains, in response to environmental signals, may determine the extent of phosphatase activity [163]. A similar conclusion was drawn for the NtrB kinase/phosphatase [161].

We now have some clues about how a RR domain might recognize and bind His-containing domains (see earlier discussion). However, a big question remains. What is the molecular basis of the interaction between phosphorylated RRs and HKs that results in RR dephosphorylation? Mechanistically, one can envision a transition state for dephosphorylation in which a water molecule (or possibly a nucleophilic residue of the HK) occupies an axial position opposite the other axial ligand, the carboxylate oxygen of the active site aspartate on the RR. Where does this nucleophilic molecule come from and how does it become activated and oriented properly in the vicinity of the phospho-Asp of the RR? Several possibilities can be considered, none that are mutually exclusive. Binding of a HK to a phosphorylated RR could cause a conformational change that accelerates the intrinsic autophosphatase activity of the RR. In this manner, the HK is acting more like an allosteric regulator rather than participating directly in the hydrolysis reaction. Alternatively, the HK itself could provide the hydrolytic water molecule and/or the residue(s) that serves as a nucleophile or to position the water and enhance its nucleophilicity via abstraction of a proton. For a few individual systems, data that address the mechanism of hydrolysis are beginning to accumulate, but it seems unlikely that a single universal mechanism will emerge. In the end, we may find that HK-mediated dephosphorylation between different HK–RR pairs is modulated by different means and that, in essence, reflects the diversity of two-component systems.

PERSPECTIVES

Since the mid-1980s, when two-component systems first began to be recognized as a fundamental strategy for microbial signal transduction, the individual protein components have been characterized extensively with respect to both biochemical activities and structure. The modularity of the proteins has allowed an extension of knowledge gathered from representative examples to other proteins of specific interest. However, there are significant limits to such extrapolation. While fundamental mechanisms are conserved, two-component proteins are very versatile and have been specifically adapted to the pathways in which they function. As discussed within this chapter, even among members of the same RR subfamily, there is great diversity in mechanisms of function. Clearly, within any given system, numerous details remain to be elucidated, but there are also gaps in our understanding of several fundamental processes common to all systems.

Our understanding of interactions between HKs and RRs is significantly less complete than our knowledge of the activities of the individual proteins. A number of central questions remain to be answered. What determines the specificity between an HK and its cognate RR? Do HKs contribute side chains that chemically facilitate phosphotransfer or is their role limited to promoting the proper orientation of the substrate phospho-His side chains with respect to the active sites of RRs? Is there a single mode of interaction between HKs and RRs or do phosphotransfer and RR dephosphorylation involve distinctly different complexes? Do HKs actively catalyze dephosphorylation of RRs or do they work through allosteric means? How do stimuli alter HK–RR interactions to promote either phosphotransfer or RR dephosphorylation? Answers to several of these questions may perhaps be addressed most definitively through the structural analysis of HK–RR complexes. Advances in determining structures of HK domains and in producing analogs of phosphorylated RRs prophesy bright prospects for future studies.

REFERENCES

1. Stock, J. B., Ninfa, A. J., and Stock, A. M. (1989). Protein phosphorylation and regulation of adaptive responses in bacteria. *Microbiol. Rev.* **53**, 450–490.
2. Bourret, R. B., Borkovich, K. A., and Simon, M. I. (1991). Signal transduction pathways involving protein phosphorylation in prokaryotes. *Annu. Rev. Biochem.* **60**, 401–441.
3. Volz, K. (1993). Structural conservation in the CheY superfamily. *Biochemistry* **32**, 11741–11753.
4. Stock, J. B., Surette, M. G., Levit, M., and Park, P. (1995). Two-component signal transduction systems: Structure-function relationships and mechanisms of catalysis. *In* "Two-Component Signal Transduction" (J. A. Hoch and T. J. Silhavy, eds.), pp. 25–51. American Society for Microbiology, Washington, D.C.
5. Volz, K. (1995). Structural and functional conservation in response regulators. *In* "Two-Component Signal Transduction" (J. A. Hoch and T. J. Silhavy, eds.), pp. 53–64. American Society for Microbiology Press, Washington, DC.
6. Mizuno, T. (1997). Compilation of all genes encoding two-component phosphotransfer signal transducers in the genome of *Escherichia coli*. *DNA Res.* **4**, 161–168.
7. Chang, C., and Stewart, R. C. (1998). The two-component system. Regulation of diverse signaling pathways in prokaryotes and eukaryotes. *Plant Physiol.* **117**, 723–731.
8. Loomis, W. F., Kuspa, A., and Shaulsky, G. (1998). Two-component signal transduction systems in eukaryotic microorganisms. *Curr. Opin. Microbiol.* **1**, 643–648.
9. Mizuno, T. (1998). His-Asp phosphotransfer signal transduction. *J. Biochem.* **123**, 555–563.
10. D'Agostino, I. B., and Kieber, J. J. (1999). Phosphorelay signal transduction: The emerging family of plant response regulators. *Trends Biochem. Sci.* **24**, 452–456.
11. Perraud, A.-L., Weiss, V., and Gross, R. (1999). Signalling pathways in two-component phosphorelay systems. *Trends Microbiol.* **7**, 115–120.
12. Hoch, J. A. (2000). Two-component and phosphorelay signal transduction. *Curr. Opin. Microbiol.* **3**, 165–170.
13. Robinson, V. L., Buckler, D. R., and Stock, A. M. (2000). A tale of two components: A novel kinase and a regulatory switch. *Nature Struct. Biol.* **7**, 628–633.
14. Stock, A. M., Robinson, V. L., and Goudreau, P. N. (2000). Two-component signal transduction. *Annu. Rev. Biochem.* **69**, 183–215.
15. Stock, J., and Da Re, S. (2000). Signal transduction: Response regulators on and off. *Curr. Biol.* **10**, R420–R424.
16. Lukat, G. S., McCleary, W. R., Stock, A. M., and Stock, J. B. (1992). Phosphorylation of bacterial response regulator proteins by low molecular weight phospho-donors. *Proc. Natl. Acad. Sci. USA* **89**, 718–722.
17. Feng, J., Atkinson, M. R., McCleary, W., Stock, J. B., Wanner, B. L., and Ninfa, A. J. (1992). Role of phosphorylated metabolic intermediates in the regulation of glutamine synthetase synthesis in *Escherichia coli*. *J. Bacteriol.* **174**, 6061–6070.
18. McCleary, W. R., Stock, J. B., and Ninfa, A. J. (1993). Is acetyl phosphate a global signal in *Escherichia coli*? *J. Bacteriol.* **175**, 2793–2798.

19. McCleary, W. R., and Stock, J. B. (1994). Acetyl phosphate and the activation of two-component response regulators. *J. Biol. Chem.* **269**, 31567–31572.
20. Hess, J. F., Oosawa, K., Kaplan, N., and Simon, M. I. (1988). Phosphorylation of three proteins in the signaling pathway of bacterial chemotaxis. *Cell* **53**, 79–87.
21. Zapf, J., Madhusudan, M., Grimshaw, C. E., Hoch, J. A., Varughese, K. I., and Whiteley, J. M. (1998). A source of response regulator autophosphatase activity: The critical role of a residue adjacent to the Spo0F autophosphorylation active site. *Biochemistry* **37**, 7725–7732.
22. Stock, A. M., Mottonen, J. M., Stock, J. B., and Schutt, C. E. (1989). Three-dimensional structure of CheY, the response regulator of bacterial chemotaxis. *Nature* **337**, 745–749.
23. Stock, J. B., Stock, A. M., and Mottonen, J. M. (1990). Signal transduction in bacteria. *Nature* **344**, 395–400.
24. Volz, K., and Matsumura, P. (1991). Crystal structure of *Escherichia coli* CheY refined at 1.7 Å resolution. *J. Biol. Chem.* **266**, 15511–15519.
25. Madhusudan, Zapf, J., Whiteley, J. M., Hoch, J. A., Xuong, N. H., and Varughese, K. I. (1996). Crystal structure of a phosphatase-resistant mutant of sporulation response regulator Spo0F from *Bacillus subtilis*. *Structure* **4**, 679–690.
26. Djordjevic, S., Goudreau, P. N., Xu, Q., Stock, A. M., and West, A. H. (1998). Structural basis for methylesterase CheB regulation by a phosphorylation-activated domain. *Proc. Natl. Acad. Sci. USA* **95**, 1381–1386.
27. Solà, M., Gomis-Rüth, F. X., Serrano, L., González, A., and Coll, M. (1999). Three-dimensional crystal structure of the transcription factor PhoB receiver domain. *J. Mol. Biol.* **285**, 675–687.
28. Baikalov, I., Schröder, I., Kaczor-Grzeskowiak, M., Grzeskowiak, K., Gunsalus, R. P., and Dickerson, R. E. (1996). Structure of the *Escherichia coli* response regulator NarL. *Biochemistry* **35**, 11053–11061.
29. Volkman, B. F., Nohaile, M. J., Amy, N. K., Kustu, S., and Wemmer, D. E. (1995). Three-dimensional solution structure of the N-terminal receiver domain of NtrC. *Biochemistry* **34**, 1413–1424.
30. Lewis, R. J., Muchova, K., Brannigan, J. A., Barak, I., Leonard, G., and Wilkinson, A. J. (2000). Domain swapping in the sporulation response regulator Spo0A. *J. Mol. Biol.* **297**, 757–770.
31. Gouet, P., Fabry, B., Guillet, V., Birck, C., Mourey, L., Kahn, D., and Samama, J.-P. (1999). Structural transitions in the FixJ receiver domain. *Structure* **7**, 1517–1526.
32. Muller-Dieckmann, H. J., Grantz, A. A., and Kim, S. H. (2000). The structure of the signal receiver domain of the *Arabidopsis thaliana* ethylene receptor ETR1. *Structure Fold. Des.* **7**, 1547–1556
33. Birck, C., Liu, W., Cabantous, S., Hulett, M., and Samama, J.-P. (2001). X-ray crystal structure of the receiver domain of PhoP. *In* "VI Conference on Bacterial Locomotion and Signal Transduction." Cuernavaca, México.
34. Buckler, D. R., and Stock, A. M. (2001). Structural analysis of a full length OmpR homolog from *Thermotoga maritima*. *In* "VI Conference on Bacterial Locomotion and Signal Transduction." Cuernavaca, México.
35. Sanders, D. A., Gillece-Castro, B. L., Stock, A. M., Burlingame, A. L., and Koshland, D. E., Jr. (1989). Identification of the site of phosphorylation of the chemotaxis response regulator protein, CheY. *J. Biol. Chem.* **264**, 21770–21778.
36. Weiss, V., and Magasanik, B. (1988). Phosphorylation of nitrogen regulator I (NR_I) of *Escherichia coli*. *Proc. Natl. Acad. Sci. USA* **85**, 8919–8923.
37. Lukat, G. S., Stock, A. M., and Stock, J. B. (1990). Divalent metal ion binding to the CheY protein and its significance to phosphotransfer in bacterial chemotaxis. *Biochemistry* **29**, 5436–5442.

38. Needham, J. V., Chen, T. Y., and Falke, J. J. (1993). Novel ion specificity of a carboxylate cluster Mg(II) binding site: Strong charge slectivity and weak size selectivity. *Biochemistry* **32**, 3363–3367.
39. Stock, A. M., Martinez-Hackert, E., Rasmussen, B. F., West, A. H., Stock, J. B., Ringe, D., and Petsko, G. A. (1993). Structure of the Mg^{2+}-bound form of CheY and mechanism of phosphoryl transfer in bacterial chemotaxis. *Biochemistry* **32**, 13375–13380.
40. Herschlag, D., and Jencks, W. P. (1990). The effects of Mg^{2+}, hydrogen bonding, and steric factors on rate and equilibrium constants for phosphoryl transfer between carboxylate ions and pyridines. *J. Am. Chem. Soc.* **112**, 1942–1950.
41. Lewis, R. J., Brannigan, J. A., Muchová, K., Barák, I., and Wilkinson, A. J. (1999). Phosphorylated aspartate in the structure of a response regulator protein. *J. Mol. Biol.* **294**, 9–15.
42. Birck, C., Mourey, L., Gouet, P., Fabry, B., Schumacher, J., Rousseau, P., Kahn, D., and Samama, J.-P. (1999). Conformational changes induced by phosphorylation of the FixJ receiver domain. *Structure Fold. Des.* **7**, 1505–1515.
43. Kern, D., Volkman, B. F., Luginbuhl, P., Nohaile, M. J., Kustu, S., and Wemmer, D. E. (1999). Structure of a transiently phosphorylated switch in bacterial signal transduction. *Nature* **40**, 894–898.
44. Silversmith, R. E., and Bourret, R. B. (1998). Synthesis and characterization of a stable analog of the phosphorylated form of the chemotaxis protein CheY. *Protein Engin.* **11**, 205–212.
45. Halkides, C. J., Zhu, X., Phillion, D. P., Matsumura, P., and Dahlquist, F. W. (1998). Synthesis and biochemical characterization of an analogue of CheY-phosphate, a signal transduction protein in bacterial chemotaxis. *Biochemistry* **37**, 13674–13680.
46. Halkides, C. J., McEvoy, M. M., Casper, E., Matsumura, P., Volz, K., and Dahlquist, F. W. (2000). The 1.9 Å resolution crystal structure of phosphono-CheY, an analogue of the active form of the response regulator, CheY. *Biochemistry* **39**, 5280–5286.
47. Yan, D., Cho, H. S., Hastings, C. A., Igo, M. M., Lee, S. Y., Pelton, J. G., Stewart, V., Wemmer, D. E., and Kustu, S. (1999). Beryllofluoride mimics phosphorylation of NtrC and other bacterial response regulators. *Proc. Natl. Acad. Sci. USA* **96**, 14789–14794.
48. Lee, S.-Y., Cho, H. S., Pelton, J. G., Yan, D., Henderson, R. K., King, D. S., Huang, L.-S., Kustu, S., Berry, E. A., and Wemmer, D. E. (2001). Crystal structure of an activated response regulator bound to its target. *Nature Struct. Biol.* **8**, 52–56.
49. Wemmer, D. (2001). *In* "VI Conference on Bacterial Locomotion and Signal Transduction." Cuernavaca, México.
50. Welch, M., Oosawa, K., Aizawa, S.-I., and Eisenbach, M. (1994). Effects of phosphorylation, Mg^{2+}, and conformation of the chemotaxis protein CheY on its binding to the flagellar switch protein FliM^{+}. *Biochemistry* **33**, 10470–10476.
51. West, A. H., and Stock, A. M. (2001). Histidine kinases and response regulator proteins in two-component signaling systems. *Trends Biochem. Sci.* **26**.
52. Bellsolell, L., Prieto, J., Serrano, L., and Coll, M. (1994). Magnesium binding to the bacterial chemotaxis protein CheY results in large conformational changes involving its functional surface. *J. Mol. Biol.* **238**, 489–495.
53. Ganguli, S., Wang, H., Matsumura, P., and Volz, K. (1995). Uncoupled phosphorylation and activation in bacterial chemotaxis: The 2.1-Å structure of a threonine to isoleucine mutant at position 87 of CheY. *J. Biol. Chem.* **270**, 1–8.
54. Jiang, M., Bourret, R. B., Simon, M. I., and Volz, K. (1997). Uncoupled phosphorylation and activation in bacterial chemotaxis: The 2.3 Å structure of an aspartate to lysine mutant at position 13 of CheY. *J. Biol. Chem.* **272**, 11850–11855.
55. Zhu, X., Rebello, J., Matsumura, P., and Volz, K. (1997). Crystal structures of CheY mutants Y106W and T87I/Y106W: CheY activation correlates with movement of residue 106. *J. Biol.

Chem. **272**, 5000–5006.

56. Feher, V. A., and Cavanagh, J. (1999). Millisecond-timescale motions contribute to the function of the bacterial response regulator protein Spo0F. *Nature* **400**, 289–293.
57. Feher, V. A., Zapf, J. W., Hoch, J. A., Whiteley, J. M., McIntosh, L. P., Rance, M., Skelton, N. J., Dahlquist, F. W., and Cavanagh, J. (1997). High-resolution NMR structure and backbone dynamics of the *Bacillus subtilis* response regulator, Spo0F: Implications for phosphorylation and molecular recognition. *Biochemistry* **36**, 10015–10025.
58. Moy, F. J., Lowry, D. F., Matsumura, P., Dahlquist, F. W., Krywko, J. E., and Domaille, P. J. (1994). Assignments, secondary structure, global fold, and dynamics of chemotaxis Y protein using three- and four-dimensional heteronuclear (^{13}C, ^{15}N) NMR spectroscopy. *Biochemistry* **33**, 10731–10742.
59. Nohaile, M., Kern, D., Wemmer, D., Stedman, K., and Kustu, S. (1997). Structural and functional analyses of activating amino acid substitutions in the receiver domain of NtrC: Evidence for an activating surface. *J. Mol. Biol.* **273**, 299–316.
60. Volkman, B. F., Lipson, D., Wemmer, D. E., and Kern, D. (2001). Two-state allosteric behavior in a single domain signaling protein. *Science* **291**, 2429–2433.
61. Tzeng, Y.-L., and Hoch, J. A. (1997). Molecular recognition in signal transduction: The interaction surfaces of the Spo0F response regulator with its cognate phosphorelay proteins revealed by alanine scanning mutagenesis. *J. Mol. Biol.* **272**, 200–212.
62. Tzeng, Y.-L., Feher, V. A., Cavanagh, J., Perego, M., and Hoch, J. A. (1998). Characterization of interactions between a two-component response regulator, Spo0F, and its phosphatase, RapB. *Biochemistry* **37**, 16538–16545.
63. Aiba, H., Nakasai, F., Mizushima, S., and Mizuno, T. (1989). Phosphorylation of a bacterial activator protein, OmpR, by a protein kinase, EnvZ, results in a stimulation of its DNA-binding ability. *J. Biochem.* **106**, 5–7.
64. Huang, K.-E., Lan, C.-Y., and Igo, M. M. (1997). Phosphorylation stimulates the cooperative DNA-binding properties of the transcription factor OmpR. *Proc. Natl. Acad. Sci. USA* **94**, 2828–2832.
65. Head, C. G., Tardy, A., and Kenney, L. J. (1998). Relative binding affinities of OmpR and OmpR-phosphate at the *ompF* and *ompC* regulatory sites. *J. Mol. Biol.* **281**, 857–870.
66. Ames, S. K., Frankema, N., and Kenney, L. J. (1999). C-terminal DNA binding stimulates N-terminal phosphorylation of the outer membrane protein regulator OmpR from *Escherichia coli*. *Proc. Natl. Acad. Sci. USA* **96**, 11792–11797.
67. Qin, L., Yoshida, T., and Inouye, M. (2001). The critical role of DNA in the equilibrium between OmpR and phosphorylated OmpR mediated by EnvZ in *Escherichia coli*. *Proc. Natl. Acad. Sci. USA* **98**, 908–913.
68. Schuster, M., Silversmith, R. E., and Bourret, R. B. (2001). Conformational coupling between the signaling surface and the phosphorylation site in the chemotaxis response regulator CheY. *Proc. Natl. Acad. Sci. USA.*
69. Bren, A., and Eisenbach, M. (2000). How signals are heard during bacterial chemotaxis: Protein-protein interactions in sensory signal propagation. *J. Bacteriol.* **182**, 6865–6873.
70. Moreno, M., Audia, J. P., Bearson, S. M., Webb, C., and Foster, J. W. (2000). Regulation of sigma S degradation in *Salmonella enterica* var *typhimurium*: In vivo interactions between sigma S, the response regulator MviA(RssB) and ClpX. *J. Mol. Microbiol. Biotech.* **2**, 245–254.
71. Becker, G., Klauck, E., and Hengge-Aronis, R. (2000). The response regulator RssB, a recognition factor for sigmaS proteolysis in *Escherichia coli*, can act like an anti-sigmaS factor. *Mol. Microbiol.* **35**, 657–666.
72. Brown, J. L., North, S., and Bussey, H. (1993). SKN7, a yeast multicopy suppressor of a mutation affecting cell wall b-glucan assembly, encodes a product with domains homolo-

gous to prokaryotic two-component regulators and to heat shock transciption factors. *J. Bacteriol.* **175**, 6908–6915.

73. Brown, J. M., and Firtel, R. A. (1998). Phosphorelay signalling: New tricks for an ancient pathway. *Curr. Biol.* **8**, R662–R665.
74. Makino, K., Shinagawa, H., Amemura, M., Kimura, S., Nakata, A., and Ishihama, A. (1988). Regulation of the phosphate regulon of *Escherichia coli*: Activation of *pstS* transcription by PhoB protein *in vitro*. *J. Mol. Biol.* **203**, 85–95.
75. Winans, S. C. (1990). Transcriptional induction of an *Agrobacterium* regulatory gene at tandem promoters by plant-released phenolic compounds, phosphate starvation, and acidic growth media. *J. Bacteriol.* **172**, 2433–2438.
76. Holman, T. R., Wu, Z., Wanner, B. L., and Walsh, C. T. (1994). Identification of the DNA-binding site for the phosphorylated VanR protein required for vancomycin resistance in *Enterococcus faecium*. *Biochemistry* **33**, 4625–4631.
77. Mills, S. D., Lim, C. K., and Cooksey, D. A. (1994). Purification and characterization of CopR, a transcriptional activator protein that binds to a conserved domain (cop box) in copper-inducible promoters of *Pseudomonas syringae*. *Mol. Gen. Gene.* **244**, 341–351.
78. Higgins, D. E., and DiRita, V. J. (1996). Genetic analysis of the interaction between *Vibrio cholerae* transcription activator ToxR and *toxT* promoter DNA. *J. Bacteriol.* **178**, 1080–1087.
79. Aguirre, A., Lejona, S., Vescovi, E. G., and Soncini, F. C. (2000). Aguirre A. Lejona S. Vescovi EG. Soncini FC. Phosphorylated PmrA interacts with the promoter region of ugd in *Salmonella enterica* serovar *typhimurium*. *J. Bacteriol.* **182**, 3874–3876.
80. Makino, K., Amemura, M., Kim, S.-K., Nakata, A., and Shinagawa, H. (1993). Role of the s70 subunit of RNA polymerase in transcription activation by activator protein PhoB in *Escherichia coli*. *Genes Dev.* **7**, 149–160.
81. Kumar, A., Grimes, B., Fujita, N., Makino, K., Malloch, R. A., Hayward, R. S., and Ishihama, A. (1994). Role of sigma70 subunit of *Escherichia coli* RNA polymerase in transcription activation. *J. Mol. Biol.* **235**, 405–413.
82. Matsuyama, S., and Mizushima, S. (1987). Novel *rpoA* mutation that interferes with the function of OmpR and EnvZ, positive regulators of the *ompF* and *ompC* genes that code for the outer-membrane proteins in *Escherichia coli* K12. *J. Mol. Biol.* **195**, 847–853.
83. Igarashi, K., and Ishihama, A. (1991). Bipartite functional map of the *E. coli* RNA polymerase a subunit: Involvement of the C-terminal region in transcription activation by cAMP-CRP. *Cell* **65**, 1015–1022.
84. Slauch, J. M., Russo, F. D., and Silhavy, T. J. (1991). Suppressor mutations in *rpoA* suggest that OmpR controls transcription by direct interaction with the alpha subunit of RNA polymerase. *J. Bacteriol.* **173**, 7501–7510.
85. Bowrin, V., Brissette, R., Tsung, K., and Inouye, M. (1994). The a subunit of RNA polymerase specifically inhibits expression of the porin genes *ompF* and *ompC in vivo* and *in vitro* in *Escherichia coli*. *FEMS Microbiol. Lett.* **115**, 1–6.
86. Pratt, L. A., and Silhavy, T. J. (1994). OmpR mutants specifically defective for transcriptional activation. *J. Mol. Biol.* **243**, 579–594.
87. Kondo, H., Nakagawa, A., Nishihira, J., Nishimura, Y., Mizuno, T., and Tanaka, I. (1997). *Escherichia coli* positive regulator OmpR has a large loop structure at the putative RNA polymerase interaction site. *Nature Struct. Biol.* **4**, 28–31.
88. Martinez-Hackert, E., and Stock, A. M. (1997). The DNA-binding domain of OmpR: crystal structure of a winged-helix transcription factor. *Structure* **5**, 109–124.
89. Martinez-Hackert, E., and Stock, A. M. (1997). Structural relationships in the OmpR family of winged-helix transcription factors. *J. Mol. Biol.* **269**, 301–312.
90. Mizuno, T., and Tanaka, I. (1997). Structure of the DNA-binding domain of the OmpR family of response regulators. *Mol. Microbiol.* **24**, 665–667.

91. Okamura, H., Hanaoka, S., Nagadoi, A., Makino, K., and Nishimura, Y. (2000). Structural comparison of the PhoB and OmpR DNA-binding/transactivation domains and the arrangement of PhoB molecules on the phosphate box. *J. Mol. Biol.* **295**, 1225–1236.
92. Tyson, K. L., Bell, A. I., Cole, J. A., and Busby, J. W. (1993). Definition of nitrite and nitrate response elements at the anaerobically inducible *E. coli nirB* promoter: Interactions between FNR and NarL. *Mol. Microbiol.* 7, 151–157.
93. Li, J., Kustu, S., and Stewart, V. (1994). In vitro interaction of nitrate-responsive regulatory protein NarL with DNA target sequences in the *fdnG*, *narG*, *narK*, and *frdA* operon control regions of *Escherichia coli* K-12. *J. Bacteriol.* **241**, 150–165.
94. Walker, M. S., and DeMoss, J. A. (1991). Promoter sequence requirements for Fnr-dependent activation of transcription of the *narGHJI* operon. *Mol. Microbiol.* **aa5**, 353–360.
95. Rabin, R. S., Collins, L. A., and Stewart, V. (1992). In vivo requirement of integration host factor for nitrate recductase (*nar*) operon expression in *Escherichia coli* K-12. *Proc. Natl. Acad. Sci. USA* **89**, 8701–8705.
96. Merkel, T. J., Dahl, J. L., Ebright, R. H., and Kadner, R. J. (1995). Transcription activation at the *Escherichia coli uhpT* promoter by the catabolite gene activator protein. *J. Bacteriol.* **177**, 1712–1718.
97. Baikalov, I., Schröder, I., Kaczor-Grzeskowiak, M., Cascio, D., Gunsalus, R. P., and Dickerson, R. E. (1998). NarL dimerization? Suggestive evidence from a new crystal form. *Biochemistry* **37**, 3665–3676.
98. Kustu, S., North, A. K., and Weiss, D. S. (1991). Prokaryyotic transcriptional enhancers and enhancer-binding proteins. *Trends Biochem. Sci.* **16**, 397–402.
99. Weiss, V., Claverie-Martin, F., and Magasanik, B. (1992). Phosphorylation of nitrogen regulator I of *Escherichia coli* induces strong cooperative binding to DNA essential for activation of transcription. *Proc. Natl. Acad. Sci. USA* **89**, 5088–5092.
100. Wyman, C., Rombel, I., North, A. K., Bustamente, C., and Kustu, S. (1997). Unusual oligomerization required for activity of NtrC, a bacterial enhancer-binding protein. *Science* **275**, 1658–1661.
101. Weiss, D. S., Batut, J., Klose, K. E., Keener, J., and Kustu, S. (1991). The phosphorylated form of the enhancer-binding protein NTRC has an ATPase activity that is essential for activation of transcription. *Cell* **67**, 155–167.
102. Austin, S., and Dixon, R. (1992). The prokaryotic enhancer binding protein NtrC has an ATPase activity which is phosphorylation and DNA dependent. *EMBO J.* **11**, 2219–2228.
103. Wedel, A., and Kustu, S. (1995). The bacterial enhancer-binding protein NtrC is a molecular machine:ATP hydrolysis is coupled to transcriptional activation. *Genes Dev.* **9**, 2042–2052.
104. Osuna, J., Soberon, X., and Morett, E. (1997). A proposed architecture for the central domain of the bacterial enhancer-binding proteins based on secondary structure prediction and fold recognition. *Protein Sci.* **6**, 543–555.
105. Contreras, A., and Drummond, M. (1988). The effect on the function of the transcriptional activator NtrC from *K. pneumoniae* of mutations in the DNA-recognition helix. *Nucleic Acids Res.* **16**, 4025–4039.
106. Porter, S. C., North, A. K., Wedel, A. B., and Kustu, S. (1993). Oligomerization of NtrC at the *glnA* enhancer is required for transcritpional activation. *Genes Dev.* 7, 2258–2273.
107. Pelton, J. G., Kustu, S., and Wemmer, D. E. (1999). Solution structure of the DNA-binding domain of NtrC with three alanine substitutions. *J. Mol. Biol.* **292**, 1095–1110.
108. Brennan, R. G., and Matthews, B. W. (1989). The helix-turn-helix DNA-binding motif. *J. Biol. Chem.* **264**, 1903–1906.
109. Huala, E., Stigter, J., and Ausubel, F. M. (1992). The central domain of *Rhizobium leguminosarum* DctD functions independently to activate transcription. *J. Bacteriol.* **174**, 1428–1431.
110. Drummond, M. H., Contreras, A., and Mitchenall, L. A. (1990). The function of isolated

domains and chimaeric proteins constucted from the transcriptional activators NifA and NtrC of *Klebsiella pneumoniae*. *Mol. Microbiol.* 4, 29–37.

111. Tate, S., Kato, M., Nishimura, Y., Arata, Y., and Mizuno, T. (1988). Location of DNA-binding segment of a positive regulator, OmpR, involved in activation of the *ompF* and *ompC* genes of *Escherichia coli*. *FEBS Lett.* **242**, 27–30.
112. Simms, S. A., Keane, M. G., and Stock, J. (1985). Multiple forms of the CheB methylesterase in bacterial chemosensing. *J. Biol. Chem.* **260**, 10161–10168.
113. Anand, G. S., Goudreau, P. N., and Stock, A. M. (1998). Activation of methylesterase CheB: Evidence of a dual role for the regulatory domain. *Biochemistry* **37**, 14038–14047.
114. Fiedler, U., and Weiss, V. (1995). A common switch in activation of the response regulators NtrC and PhoB: Phosphorylation induces dimerization of the receiver modules. *EMBO J.* **14**, 3696–3705.
115. Da Re, S., Schumacher, J., Rousseau, P., Fourment, J., Ebel, C., and Kahn, D. (1999). Phosphorylation-induced dimerisation of the FixJ receiver domain. *Mol. Microbiol.* **34**, 504–511.
116. Harlocker, S. L., Bergstrom, L., and Inouye, M. (1995). Tandem binding of six OmpR proteins to the *ompF* upstream regulatory sequence of *Escherichia coli*. *J. Biol. Chem.* **270**, 26849–26856.
117. Welch, M., Oosawa, K., Aizawa, S.-I., and Eisenbach, M. (1993). Phosphorylation-dependent binding of a signal molecule to the flagellar switch of bacteria. *Proc. Natl. Acad. Sci. USA* **90**, 8787–8791.
118. Li, J., Swanson, R. V., Simon, M. I., and Weis, R. M. (1995). The response regulators CheB and CheY exhibit competitive binding to the kinase CheA. *Biochemistry* **34**, 14626–14636.
119. Blat, Y., and Eisenbach, M. (1994). Phosphorylation-dependent binding of the chemotaxis signal molecule CheY to its phosphatase, CheZ. *Biochemistry* **33**, 902–906.
120. Zhu, X., Volz, K., and Matsumura, P. (1997). The CheZ-binding surface of CheY overlaps the CheA- and FliM-binding surfaces. *J. Biol. Chem.* **272**, 23758–23764.
121. Borkovich, K. A., Kaplan, N., Hess, J. F., and Simon, M. I. (1989). Transmembrane signal transduction in bacterial chemotaxis involves ligand dependent activation of phosphate group transfer. *Proc. Natl. Acad. Sci. USA* **86**, 1208–1212.
122. Ninfa, A. J., and Magasanik, B. (1986). Covalent modification of the *glnG* product, NR_I, by the *glnL* product, NR_{II}, regulates the transcription of the *glnALG* operon in *Escherichia coli*. *Proc. Natl. Acad. Sci. USA* **83**, 5909–5913.
123. Aiba, H., Mizuno, T., and Mizushima, S. (1989). Transfer of phosphoryl group between two regulatory proteins involved in osmoregulatory expression of the *ompF* and *ompC* genes in *Escherichia coli*. *J. Biol. Chem.* **264**, 8563–8567.
124. Lois, A. F., Weinstein, M., Ditta, G. S., and Helinski, D. R. (1993). Autophosphorylation and phosphatase activities of the oxygen-sensing protein FixL of *Rhizobium meliloti* are coordinately regulated by oxygen. *J. Biol. Chem.* **268**, 4370–4375.
125. Dahl, M. K., Msadek, T., Kunst, F., and Rapoport, G. (1992). The phosphorylation state of the DegU response regulator acts as a molecular switch allowing either degradative enzyme synthesis or expression of genetic competence in *Bacillus subtilis*. *J. Biol. Chem.* **267**, 14509–14514.
126. Jung, K., Tjaden, B., and Altendorf, K. (1997). Purification, reconstitution, and characterization of KdpD, the turgor sensor of *Escherichia coli*. *J. Biol. Chem.* **272**, 10847–10852.
127. Steed, P. M., and Wanner, B. L. (1993). Use of the *rep* technique for allele replacement to constuct mutants with deletions of the *pstSCAB-phoU* operon: Evidence of a new role for the PhoU protein in the phosphate regulon. *J. Bacteriol.* **175**, 6797–6809.
128. Wanner, B. L. (1995). Signal transduction and cross regulation in the *Escherichia coli* phosphate regulon by PhoR, CreC, and acetyl phosphate. *In* "Two-Component Signal

Transduction" (J. A. Hoch and T. J. Silhavy, eds.), pp. 203–221. American Society for Microbiology Press, Washington, DC.

129. Goudreau, P. N., Lee, P.-J., and Stock, A. M. (1998). Stabilization of the phospho-aspartyl residue in a two-component signal transduction system in *Thermotoga maritima*. *Biochemistry* **37**, 14575–14584.

130. Perego, M., Hanstein, C., Welsh, K. M., Djavakhishvili, T., Glaser, P., and Hoch, J. A. (1994). Multiple protein-aspartate phosphatases provide a mechanism for the integration of diverse signals in the control of development in *B. subtilis*. *Cell* **79**, 1047–1055.

131. Ohlsen, K. L., Grimsley, J. K., and Hoch, J. A. (1994). Deactivation of the sporulation transcription factor Spo0A by the Spo0E protein phosphatase. *Proc. Natl. Acad. Sci. USA* **91**, 1756–1760.

132. Reizer, J., Reizer, A., Perego, M., and Saier, M. H., Jr. (1997). Characterization of a family of bacterial response regulator aspartyl-phosphate (RAP) phosphatases. *Microbial Comp. Genet.* **2**, 103–111.

133. Blat, Y., and Eisenbach, M. (1996). Mutants with defective phosphatase activity show no phosphorylation-dependent oligomerization of CheZ: The phosphatase of bacterial chemotaxis. *J. Biol. Chem.* **271**, 1232–1236.

134. Blat, Y., and Eisenbach, M. (1996). Oligomerization of the phosphatase CheZ upon interaction with the phosphorylated form of CheY. *J. Biol. Chem.* **271**, 1226–1231.

135. Da Re, S. S., Deville-Bonne.D., Tolstykh, T., Veron, M., and Stock, J. B. (1999). Kinetics of CheY phosphorylation by small molecule phosphodonors. *FEBS Lett.* **457**, 323–326.

136. Mayover, T. L., Halkides, C. J., and Stewart, R. C. (1999). Kinetic characterization of CheY phosphorylation reactions: Comparison of P-CheA and small-molecule phosphodonors. *Biochemistry* **38**, 2259–2271.

137. Silversmith, R. E., Appleby, J. L., and Bourret, R. B. (1997). Catalytic mechanism of phosphorylation and dephosphorylation of CheY: Kinetic characterization of imidazole phosphates as phosphodonors and the role of acid catalysis. *Biochemisty* **36**, 14965–14974.

138. Zapf, J. W., Hoch, J. A., and Whiteley, J. M. (1996). A phosphotransferase activity of the *Bacillus subtilis* sporulation protein Spo0F that employs phosphoramidate substrates. *Biochemistry* **35**, 2926–2933.

139. Schuster, S. C., Swanson, R. V., Alex, L. A., Bourret, R. B., and Simon, M. I. (1993). Assembly and function of a quaternary signal transduction complex monitored by surface plasmon resonance. *Nature* **365**, 343–347.

140. Tomomori, C., Tanaka, T., Dutta, R., Park, H., Saha, S. K., Zhu, Y., Ishima, R., Liu, D., Tong, K. I., Kurokawa, H., Qian, H., Inouye, M., and Ikura, M. (1999). Solution structure of the homodimeric core domain of *Escherichia coli* histidine kinase EnvZ. *Nature Struct. Biol.* **6**, 729–734.

141. Zhou, H., and Dahlquist, F. W. (1997). Phosphotransfer site of the chemotaxis-specific protein kinase CheA as revealed by NMR. *Biochemistry* **36**, 699–710.

142. Ikegami, T., Okada, T., Ohki, I., Hirayama, J., Mizuno, T., and Shirakawa, M. (2001). Solution structure and dynamic character of the histidine-containing phosphotransfer domain of anaerobic sensor kinase ArcB from *Escherichia coli*. *Biochemistry* **40**, 375–386.

143. Kato, M., Mizuno, T., Shimizu, T., and Hakoshima, T. (1997). Insights into multistep phosphorelay from the crystal structure of the C-terminal HPt domain of ArcB. *Cell* **88**, 717–723.

144. Song, H. K., Lee, J. Y., Lee, M. G., Moon, J., Min, K., Yang, J. K., and Suh, S. W. (1999). Insights into eukaryotic multistep phosphorelay signal transduction revealed by the crystal structure of Ypd1p from *Saccharomyces cerevisiae*. *J. Mol. Biol.* **293**, 753–761.

145. Varughese, K. I., Madhusudan, Zhou, X. Z., Whiteley, J. M., and Hoch, J. A. (1998). Formation of a novel four-helix bundle and molecular recognition sites by dimerization of a response regulator phosphotransferase. *Mol. Cell.* **2**, 485–493.

146. Xu, Q., and West, A. H. (1999). Conservation of structure and function among histidine-containing phosphotransfer (HPt) domains as revealed by the crystal structure of YPD1. *J. Mol. Biol.* **292**, 1039–1050.
147. Zapf, J., Sen, U., Madhusudan, Hoch, J. A., and Varughese, K. I. (2000). A transient interaction between two phosphorelay proteins trapped in a crystal lattice reveals the mechanism of molecular recognition and phosphotransfer in signal transduction. *Structure Fold Des.* **8**, 851–862.
148. Knowles, J. R. (1980). Enzyme-catalyzed phosphoryl transfer reactions. *Annu. Rev. Biochem.* **49**, 877–919.
149. Hess, J. F., Bourret, R. B., and Simon, M. I. (1988). Histidine phosphorylation and phosphoryl group transfer in bacterial chemotaxis. *Nature* **336**, 139–143.
150. Swanson, R. V., Schuster, S. C., and Simon, M. I. (1993). Expression of CheA fragments which define domains encoding kinase, phosphotransfer, and CheY binding activities. *Biochemistry* **32**, 7623–7629.
151. Zhou, H., Lowry, D. F., Swanson, R. V., Simon, M. I., and Dahlquist, F. W. (1995). NMR studies of the phosphotransfer domain of the histidine kinase CheA from *Escherichia coli*: Assignments, secondary structure, general fold, and backbone dynamics. *Biochemistry* **34**, 13858–13870.
152. McEvoy, M. M., Hausrath, A. C., Randolph, G. B., Remington, S. J., and Dahlquist, F. W. (1998). Two binding modes reveal flexibility in kinase/response regulator interactions in the bacterial chemotaxis pathway. *Proc. Natl. Acad. Sci. USA* **95**, 7333–7338.
153. Stewart, R. C., Jahreis, K., and Parkinson, J. S. (2000). Rapid phosphotransfer to CheY from a CheA protein lacking the CheY-binding domain. *Biochemistry* **39**, 13157–13165.
154. Welch, M., Chinardet, N., Mourey, L., Birck, C., and Samama, J.-P. (1998). Structure of the CheY-binding domain of histidine kinase CheA in complex with CheY. *Nature Struct. Biol.* **5**, 25–29.
155. Pratt, L. A., and Silhavy, T. J. (1995). Porin regulon of *Escherichia coli*. *In* "Two-Component Signal Transduction" (J. A. Hoch and T. J. Silhavy, eds.), pp. 105–127. Am. Soc. Microbiol. Press, Washington, DC.
156. Hoch, J. A., and Silhavy, T. J. (1995). *In* "Two-Component Signal Transduction." American Society for Microbiology Press, Washington, DC.
157. Aiba, H., Nakasai, F., Mizushima, S., and Mizuno, T. (1989). Evidence for the physiological importance of the phosphotransfer between the two regulatory components, EnvZ and OmpR, in osmoregulation of *Escherichia coli*. *J. Biol. Chem.* **264**, 14090–14094.
158. Hsing, W., Russo, F. D., Bernd, K. K., and Silhavy, T. J. (1998). Mutations that alter the kinase and phosphatase activities of the two-component sensor EnvZ. *J. Bacteriol.* **180**, 4538–4546.
159. Jung, K., and Altendorf, K. (1998). Truncation of amino acids 12-128 causes deregulation of the phosphatase activity of the sensor kinase KdpD of *Escherichia coli*. *J. Biol. Chem.* **273**, 17406–17410.
160. Igo, M. M., Ninfa, A. J., Stock, J. B., and Silhavy, T. J. (1989). Phosphorylation and dephosphorylation of a bacterial transcriptional activator by a transmembrane receptor. *Genes Dev.* **3**, 1725–1734.
161. Kramer, G., and Weiss, V. (1999). Functional dissection of the transmitter module of the histidine kinase NtrB in *Escherichia coli*. *Proc. Natl. Acad. Sci. USA* **96**, 604–609.
162. Walker, M. S., and DeMoss, J. D. (1993). Phosphorylation and dephosphorylation catalyzed in vitro by purified components of the nitrate sensing system, NarX and NarL. *J.Biol.Chem.* **268**, 8391–8393.
163. Zhu, Y., Qin, L., Yoshida, T., and Inouye, M. (2000). Phosphatase activity of histidine kinase EnvZ without kinase catalytic domain. *Proc. Natl. Acad. Sci. USA* **97**, 7808–7813.

164. Atkinson, M. R., and Ninfa, A. J. (1993). Mutational analysis of the bacterial signal-transducing protein kinase/phosphatase nitrogen egulator II (NR_{II} or NtrB). *J. Bacteriol.* **175**, 7016–7023.
165. Cavicchioli, R., Schroder, I., Constanti, M., and Gunsalus, R. P. (1995). The NarX and NarQ sensor-transmitter proteins of *Escherichia coli* each require two conserved histidines for nitrate-dependent signal transduction to NarL. *J. Bacteriol.* **177**, 2416–2424.
166. Hsing, W., and Silhavy, T. J. (1997). Function of conserved histidine-243 in phosphatase activity of EnvZ, the sensor for porin osmoregulation in *Escherichia coli*. *J. Bacteriol.* **179**, 3729–3735.
167. Kamberov, E. S., Atkinson, M. R., Chandran, P., and Ninfa, A. J. (1994). Effect of mutations in *Escherichia coli glnL* (*ntrB*), encoding nitrogen regulator II (NR_{II} or NtrB), on the phosphatase activity involved in bacterial nitrogen regulation. *J. Biol. Chem.* **269**, 28294–28299.
168. Skarphol, K., Waukau, J., and Forst, S. A. (1997). Role of His243 in the phosphatase activity of EnvZ in *Escherichia coli*. *J. Bacteriol.* **179**, 1413–1416.
169. West, A. H., Martinez-Hackert, E., and Stock, A. M. (1995). Crystal structure of the catalytic domain of the chemotaxis receptor methylesterase, CheB. *J. Mol. Biol.* **250**, 276–290.

CHAPTER 13

Cyanophytochromes, Bacteriophytochromes, and Plant Phytochromes: Light-Regulated Kinases Related to Bacterial Two-Component Regulators

RICHARD DAVID VIERSTRA

Cellular and Molecular Biology Program and the Department of Horticulture, University of Wisconsin–Madison, Wisconsin 53706

Histidine Kinases in Signal Transduction

Phytochromes (phys) are a family of photoreceptors used by higher plants to coordinate their growth and development to the ambient light environment. Through the unique photochromic properties of their bilin chromophore, phys behave as light-modulated switches sensitive to red and far-red light. Despite their agricultural importance, we still do not understand how phys function upon photoactivation. Previous biochemical studies suggested that phys are light-regulated protein kinases. This idea has been strongly supported recently with the discovery of phy-like proteins in cyanobacteria and eubacteria called cyanophytochromes (CphPs) and bacteriophytochromes (BphPs), respectively. CphPs and BphPs both contain an N-terminal sensor module homologous to higher plant phys that binds the bilin chromophore followed by a C-terminal transmitter module with sequence similarity to two-component histidine kinases common among other bacterial signaling systems. Whereas phys use phytochromobilin as the chromophore, CphPs and BphPs likely use phycocyanobilin and biliverdin, respectively. Several of these CphPs/BphPs behave *in vitro* as histidine kinases using an associated response regulator as the phosphoacceptor. In the few cases studied, the CphP/BphP sensory chain ultimately affects motility or pigmentation, presumably as a way to enhance photosynthetic light capture or to protect the bacterium from light damage. *In vitro* studies with recombinant phys suggest that they are also light-regulated kinases. However, phys appear more related biochemically to serine/threonine kinases, despite their evolutionary ancestory. This kinase activity could have multiple functions in plants that include initiating signal transduction as well as affecting localization, activity, and/or degradation of the photoreceptor.

INTRODUCTION TO PHYTOCHROMES (Phys)

Light is a critical environmental factor for plants. It provides not only the necessary radiant energy for photosynthesis, but also the positional information that plants use to adapt and optimize their growth and development to the ambient light environment [1]. Perception and interpretation of light signals are accomplished by an array of photoreceptors, including the ultraviolet (UV)-A and UV-B photoreceptors that absorb UV light, cryptochromes that absorb blue light, and phytochromes (phys) that primarily absorb red (R) and far-red (FR) light [1–4]. Each of these photoreceptors is associated with signal transduction pathways that work either alone or in concert with others to provide an integrated view of the prevailing light conditions. Through their combined action, plants can measure the intensity, direction, duration, and spectral quality of the light, with the latter imparting a crude form of color vision. The output of these signaling pathways can be as short as seconds, thus reflecting adaptation to rapid changes in light conditions, and as long as months, thus entraining the life cycle of plants to the seasonal cycles. In most cases, the transduction chains end with alterations in gene expression.

Among the receptors, the most influential are the phys, a family of chromoproteins universally present in all plants from algae to angiosperms [4, 5]. They exist as soluble homodimers with each subunit containing the linear tetrapyrrole (bilin) chromophore phytochromobilin (PΦB) covalently coupled to a ~120-kDa polypeptide (Fig. 1). Phys sense R and FR through the photointerconversion between two stable conformations: a R-absorbing Pr form (λ_{max} 660 nm) that is biologically inactive and a FR-absorbing Pfr form (λ_{max} 730 nm) that is biologically active (Fig. 1). R converts Pr to Pfr whereas FR converts Pfr back to Pr. By this unique photochromic behavior, phys function as light-regulated switches for a number of essential processes, including seed germination, chloroplast development, pigmentation, shade avoidance, entrainment of circadian rhythms, flowering time, and senescence [1, 2, 4].

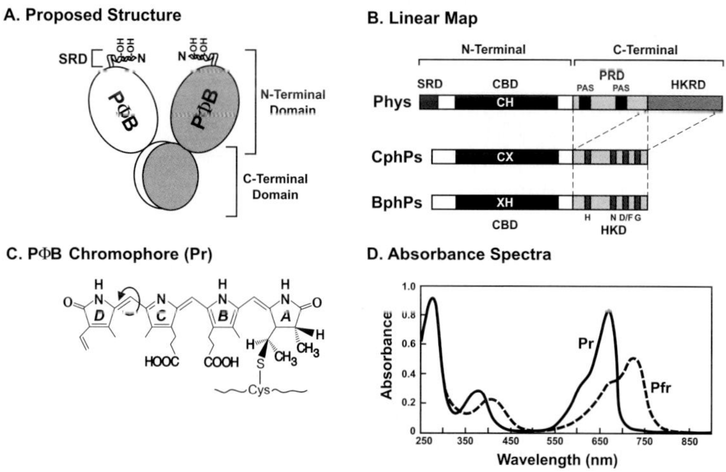

FIGURE 1 Structure and organization of plant phys. (A) Proposed structure of a phy homodimer showing N- and C-terminal domains and the exposed serine-rich domain (SRD) at the N terminus. (B) Linear map of a higher plant phy aligned with that of a typical CphP and BphP from bacteria. CBD, chromophore-binding domain that uses either a cysteine (C) in phys and CphPs or a histidine (H) in BphPs to covalently attach the bilin chromophore (X = other residue). PAS, Per/Arndt/Sim motif [25]; PRD, PAS-related domain; HKRD, histidine kinase-related domain; HKD, histidine kinase domain. (C) Structure of the PΦB chromophore attached to phys from higher plants. The arrow shows the C15 double bond that undergoes a *cis*-to-*trans* isomerization during photoconversion between Pr and Pfr. (D) Absorbance spectra of phyA following saturating FR (Pr) and R (mainly Pfr) [64].

Molecular cloning has established that phy apoproteins are synthesized from a family of nuclear genes; five exist in *Arabidopsis thaliana* (designated *PHYA–E*) with counterparts present in many other plant species [6]. Upon translation in the cytosol, apo-phy proteins autocatalytically bind PΦB via a thiol-ether linkage to a unique cysteine, using a bilin lyase activity intrinsic to the polypeptide [7]. This prosthetic group is synthesized in the plastid in a pathway that branches off that used to produce heme and chlorophylls [8, 9]. Presumably, PΦB is exported to the cytoplasm before ligation. Once assembled, chromoproteins acquire their novel photochromic spectral properties and high extinction coefficient (e.g., 1.2×10^5 l mol^{-1} cm^{-1} for oat phyA). Photoconversion of Pr to Pfr involves a *cis*-to-*trans* isomerization of the C15 double bond between the C and the D rings of PΦB (Fig. 1), a reorientation of the chromophore in its binding pocket, and subsequent conformation changes within the polypeptide. Genetic studies indicate that each phy isoform has unique and overlapping roles in controlling photomorphogenesis [4, 10]. For example, phyA is the predominant isoform in etiolated seedlings and detects FR-enriched environments, whereas phyB is more influential in adult green plants and detects R-rich environments.

In an effort to determine the mode of action of phys, their biochemical properties have been analyzed in considerable detail [4, 5]). Phy holoproteins can be divided into two parts separated by a flexible hinge (Fig. 1). The N-terminal half contains the chromophore-binding domain (CBD), an overlapping region responsible for PΦB ligation, and a flexible and exposed serine-rich domain (SRD) near the N terminus that can be phosphorylated *in vivo* [11, 12]. The CBD contains all the residues necessary for the unique photochromic spectral characteristics of the molecule. The C-terminal half contains a site(s) required for dimerization and one or more regions necessary for signal output. Presumably these signaling regions interact with downstream components of the transduction cascade. Two regions show sequence similarity to the Per/Arndt/Sim (PAS) domain. This 40 amino acid module has been found in other signaling proteins that detect small ligands, light, oxygen levels, or redox potential in a variety of organisms [13]. Of special interest are domains that change conformation on photoconversion from Pr to Pfr, as these may be relevant to phy action. Thus far, several have been detected, including a 6-kDa region surrounding the SRD and a region bracketing the hinge [4, 12].

Phys AS PROTEIN KINASES?

Since their discovery in the early 1960s, it has been assumed that phys function as enzymes that initiate signal transduction upon activation by light.

First analyses of purified phyA in the 1970s revealed that these photoreceptors are phosphoproteins, suggesting an integral role for this posttranslational modification [11, 14]. The notion that phys could be protein kinases arose in the 1980s from biochemical studies showing that the phyA holoprotein could bind ATP and that highly enriched preparations contained kinase activity [15–17]. Both autophosphorylation and phosphorylation of artificial substrates were observed. However, controversy persisted regarding the relevance of this activity because it appeared to be independent of the spectral form (e.g., both Pr and Pfr were active) and because several reports suggested that the kinase activity was not intrinsic to the chromoprotein but originated from a contaminating protein [18, 19].

Schneider-Poetsch and co-workers refueled the notion that phys are protein kinases upon inspection of the amino acid sequences of several that became available in the early 1990s. They detected weak but potentially significant amino acid sequence similarity between a segment of the C-terminal domain of phys and the transmitter module of two-component histidine kinases [20]. These kinase are prevalent in bacteria where they serve to initiate phosphorelay cascades responsible for sensing a number of environmental stimuli, including osmolarity, nitrogen levels, and nutrients, and for detecting hosts by pathogens [21, 22]. While originally thought to be restricted to bacteria, they now have been found in a number of eukaryotes, including plants where some have been implicated in hormone signaling [23, 24] (see Chapters 19–21).

Typically, two-component histidine kinases receive environmental stimuli through an N-terminal sensor domain [21, 22]. Following stimulation, the sensor domain activates a histidine kinase domain (HKD) in a C-terminal transmitter module. The signal relay proceeds with phosphorylation of a conserved histidine in the transmitter module, typically by cross-phosphorylation of the sensor/transmitter dimer. The bound phosphate is then transferred to an aspartate residue in a response regulator (RR) module, which directs an output module to initiate an appropriate response (e.g., changes in gene expression or locomotion). Often the RR and output modules reside in separate polypeptides from those bearing the sensor/transmitter modules. However, a number of other permutations are possible, including fusions bearing both the sensor/transmitter and the RR, and situations were a second histidine phosphotransferase (HPT)/RR pair is positioned between the input and output domains, thus creating a four-step His-Asp-His-Asp phosphorelay [21]. Well-described sensory processes regulated by two-component regulators include chemotaxis (*Escherichia coli* CheA/CheY), sporulation (*Bacillis subtilis* KimA-B/SpoOF), osmosensing (*Saccharomyces cerevisae* Sln1/Ydp1), nitrogen fixation (*Rhizobium meliloti* FixL/FixJ), and pathogenesis (*Agrobacterium tumefaciens* VirA/VirG) [21, 22].

As intriguing as this possibility was that phys act as histidine kinases, several studies questioned its validity. The essential histidine residue that becomes phosphorylated to initiate the phosphorelay cascade in *bona fide* two-component histidine kinases is notably absent in most higher plant phys (see later) [20, 25]. Moreover, several other residues that are typically essential in other histidine kinases were shown to be functionally unimportant in phyA and phyB [26, 27]. So while the sequences showed potential relatedness, it appeared clear that phys are not typical two-component regulators.

DISCOVERY OF CYANOPHYTOCHROMES (CphPs) AND BACTERIOPHYTOCHROMES (BphPs)

Until recently, the prevailing notion was that phys are present only in higher and lower plants and a few alga [6, 28]. However, this idea changed dramatically with the pioneering work of Kehoe and Grossman, studying complementary chromatic adaptation (CCA) in the cyanobacterium *Fremyella diplosiphon* [29]. During a genetic screen to isolate mutants defective in CCA, they identified *rcaE* for *response to chromatic adaption E*, a mutant that was unable to respond to R or green light. The encoded 74-kDa protein has a 150 amino acid N-terminal region with limited similarity to the CBD of higher plant phys attached to a prototypical HKD common among two-component transmitter kinases. In contrast to phys, the HKD in RcaE has all the residues expected of such transmitter kinases including the active-site histidine. Both its proposed position in the CCA signal-transduction chain and its relation to phys suggested that RcaE controls CCA by acting as a R photoreceptor [29, 30].

Kehoe and Grossman also noted that other cyanobacteria contain phy-like sequences; in the complete genome of *Synechocystis* sp. PCC6803 alone, four genes were detected with varying degrees of relatedness to *RcaE* and higher plant *PHYs* [29, 31–33]. Several laboratories subsequently showed that one of these *Synechocystis* sequences (*Cph1*) does indeed encode a phy-like photoreceptor capable of binding bilins such as phycocyanobilin (PCB) and PΦB and becoming R/FR photochromic [32–34]. These observations encouraged a number of investigators to perform similar genomic searches to define the limits of the prokaryotic phy kingdom. Of the 11 cyanobacterial species surveyed, including *Anabena*, *Calothrix*, and *Nostoc*, all were found to contain one or more related phy protein sequences, indicating that these photoreceptors may be common to this phylogenetic group [7, 35]. *Synechocystis elongatus* CikA [36] and another CphP from *Synechocystis* PCC6803 (TaxD1 [37]) were subsequently shown via genetics to be functional, required to reset the circadian clock, and for motility, respectively. Multiple phy-like sequences were also detected in the genomes of the purple bacteria, *Rhodospirillum centenum*, *Rhodopseudomonas palustris*, and *Rhodobacter sphaeroides*, thus extending the distribution of prokaryotic phys to photosynthetic eubacteria [7, 38–40]. More recently, we expanded the distribution even beyond phototrophs with the discovery of phy-like sequences in nonphotosynthetic eubacteria such as *Deinococcus radiodurans*, *Pseudomonomas aeruginosa*, *Pseudomonas syringae*, *Pseudomononas putida*, and *Rhizobium leguminosarium* and several fungi including *Neurospora crassa* [39–41]. These nonphotosynthetic eubacterial examples may be particularly useful because they provided the first opportunity to studies phys without the photosynthetic light reactions.

Unfortunately, these prokaryotic phys have a confusing array of designation. To help simplify the nomenclature, we have collectively called them cynanophytochrome photoreceptors or CphPs and bacteriophytochrome photoreceptors or BphPs, using rhodopsin and bacteriorhodopsin as the prototype. These two separate classes reflect subtle but important structural differences between the cyanobacterial and the eubacterial versions that appear to be functionally relevant (see later) [40]. All the CphPs and BphPs have a core structure that is similar to higher plant phys. As shown in Figs. 1 and 2A, N-terminal regions contain a domain homologous to the CBD of higher plant phys whereas the C-terminal ends bear a HKD. However, from the examples shown, it is obvious that a number of variations exist (Fig. 2A). *R. centenum* Ppr has the binding domain for an additional pigment [38], and *S. elongatus* CikA and *R. leguminosarium* BphP have a RR fused directly to the C terminus of HKD ([36] A. Johnston and M. Wexler, unpublished observations).

PHOTOCHEMICAL PROPERTIES OF CphPs AND BphPs

The CBD of CphPs are highly related to phys, including the presence of the positionally conserved cysteine that binds the chromophore. Where tested, the CBD region of CphPs, like those from phys, has the capacity to covalently bind bilins to generate photochromic photoreceptors. For example, recombinant *Synechocystis* Cph1 and Cph2 and *F. diplosiphon* RcaE will bind autocatalytically both PCB and PΦB *in vitro* in the absence of other factors to generate the R/FR photochromic chromoproteins [7, 32–34], (D. Kehoe, unpublished results). The absorbance spectra of both Pr and Pfr forms of Cph1 are similar to those of higher plant phys with absorbance maxima at 668 and 718 nm for Pr and Pfr [32], respectively, as compared to 666 and

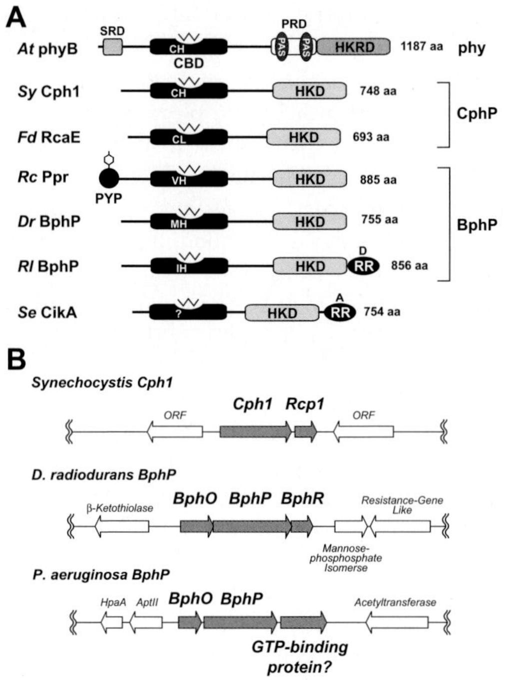

FIGURE 2 Organization of representative prokaryotic phys. (A) Domain structure of CphPs and BphPs in relation to the higher plant phy, *A. thaliana* phyB. Sequences include *Synechocystis* Cph1 (*Sy* Cph1), *F. diplosiphon* (*Fd*) RcaE, *R. centenum* (*Rc*) Ppr, *D. radiodurans* (*Dr*) BphP, *R. leguminosarium* (*Rl*) BphP, and *S. elongatus* (*Se*) CikA. CBD, chromophore-bearing domain; HKRD, histidine kinase-related domain; HKD, histidine kinase domain; PAS, Per/Arndt/Sim repeats; PYP, photoactive yellow protein domain; RR, response regulator; SRD, serine-rich domain. The D in the RR denotes the active-site aspartate that accepts the phosphoryl group from the HKD; this residue is alanine (A) in CikA [36]. Numbers on the right indicate the amino acid length of each polypeptide. Brackets denote the potential inclusion of each protein in the CphP or BphP groups. (B) Genomic organization of the *Cph1* operon in *Synechocystis*, as compared to *BphP* operons in *D. radiodurans* and *P. aeruginosa*, respectively. The *Synechocystis Cph1* operon is proposed to be a two member transcriptional unit encoding Cph1 and its cognate RR, Rcp1. The *D. radiodurans BphP* operon encodes three proteins: a heme oxygenase BphO, BphP, and its cognate RR, BphR. The *P. aeruginosa BphP* operon encodes three proteins, a heme oxygenase BphO, BphP, and a putative GTP-binding protein.

730 nm for higher plant phyA [4] (see Fig. 1). Analysis of Cph1 by a variety of spectroscopic techniques indicates that it has physicochemical properties remarkably similar to those of higher plant phys [42–44], despite substantial differences in the CBD sequence (Fig. 3). This conservation implies that the apoprotein–chromophore contacts involved in chromophore ligation and the unique photochromic properties of CphPs and phys have been maintained

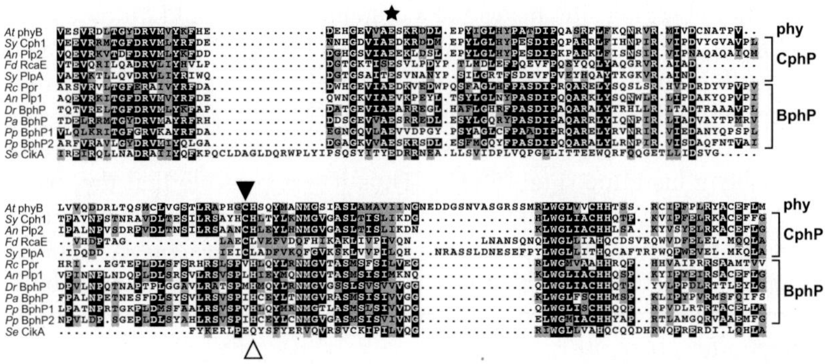

FIGURE 3 Amino acid sequence alignments of the chromophore-bearing domain (CBD) among phy-related proteins: *A. thaliana* (*At*) phyB (GenBank X17342), *Synechocystis* sp. PCC6803 (*Sy*) Cph1 (GenBank AB001339), *Anabaena* sp. PCC7120 (*An*) Plp2 (GenBank AB034952), *F. diplosiphon* (*Fd*) RcaE (GenBank U59741), *Synechocystis* sp. PCC6803 (*Sy*) plpA (GenBank U67397), *R. centenum* (*Rc*) Ppr (GenBank AF064527), *Anabaena* sp. PCC7120 (*An*) Plp1 (GenBank AB028873), *D. radiodurans* (*Dr*) BphP (GenBank AE001862), *P. aeruginosa* (*Pa*) BphP, *P. putida* (*Pp*) BphP1 and Bph2, and *S. elongatus* (*Se*) CikA (GenBank AF258464). Closed and open arrowheads identify the positionally conserved cysteine and histidine residues that serve as the chromophore-binding site in *Dr* BphP and higher plant phys, respectively [12, 41]. The star indicate the conserved glutamate residue essential for bilin attachment [7]. Sequences proposed to be included in the phy, CphP, and BphP groups are indicated on the right. Reverse type and gray boxes denote identical and similar amino acids, respectively.

during evolution. By identifying residues conserved among the CBDs, Wu and Lagarias [7] have detected several potentially critical amino acids, including a positionally conserved glutamate which is required for bilin ligation (Fig. 3).

What are the identities of the native CphP chromophore(s)? Unfortunately, addressing this question has been particularly difficult because the photoreceptors are present at very low levels [45]. Given the efficiency with which several of the recombinant apoproteins bind a variety of bilins *in vitro*, including PΦB, PCB, and phycoerythrobilin (PEB), it is likely that these or related molecules are the prosthetic group(s). Whereas PΦB–CphPs and PCB–CphPs are photochromic, PEB–CphPs are not, consistent with the absence of the C15 double bond between C and D rings [32, 45]. In most cases, the chromophore is likely to be PCB, as these cyanobacteria make copious amounts of this bilin as a photosynthetic accessory pigment [46]. In support, Hübschmann *et al.* [47] purified *Synechocystis* Cph1 directly from the cyanobacterium and found the spectral properties to be indistinguishable to that of Cph1 assembled *in vitro* with PCB.

In contrast to phys and CphPs, BphPs have a distinct CBD. Although most of this ~150 amino acid region is similar, a major change is evident at the chromophore-binding site. Instead of a cysteine, which could link the bilin by a thiol-ether bond, a small hydrophobic residue (e.g., leucine, isoleucine, valine, methionine) is most often in this position, thus implying that a unique chromophore linkage is used by BphPs (Figs. 2A and 3). Sequence comparisons of a number of BphPs revealed that the residue distal to this cysteine is invariably a histidine (Fig. 3) [40]. Thus it was possible that the histidine is the binding site, potentially using a Schiff base-type linkage to attach the bilin. This possibility was confirmed for *D. radiodurans* BphP; mutational analysis showed that this histidine is essential for chromophore binding with the linkage confirmed directly by mass spectrometric analysis of chromopeptides [41].

Differences in the chromophore ligation site between BphPs and phys/CphPs appear to translate into differences in the nature of the chromophore. Instead of using PCB or PΦB, BphPs appear to prefer the immediate precursor of these bilins, biliverdin (BV) [40]. In fact, whereas both CphPs and phys cannot bind BV (at least *in vitro*), a number of BphPs can bind BV with high efficiency both *in vitro* and *in vivo* [40]. The spectral properties of BphPs, coupled with BV, are more similar to phys and CphPs than when PCB and PΦB are bound, suggesting that BV is the preferred chromophore. The only significant difference is a substantial red shift of the absorbance spectra as compared to phys and CphPs; absorbance maxima for the Pr and Pfr forms of BV–BphPs are at 698 and 750 nm, respectively [40]. This shift is likely explained by the increased double bond conjugation of BV as compared to PCB and PΦB.

Further support for using BV as the chromophore came from the analysis of various *BphP* operons. Under the assumption that a bilin is the chromophore, BphP-containing species should contain a heme oxygenase (HO) to convert heme to BV [8]. (In photosynthetic organisms, BV is converted to PCB, PEB, or PΦB by a synthase that reduces the C3 double bond in the A ring [9].) Surprisingly, when known HO sequences were used as queries, we detected putative *HO* genes within *BphP* operons from *D. radiodurans*, *P. aerginosa*, and *R. leguminosarium*, just upstream of the *BphP* coding sequence (Figs. 2B) [40]. Thus, some BphPs have developed a transcriptional connection between chromophore and apoprotein synthesis that may have interesting regulatory implications (Fig. 2B).

Several novel photoreceptors are evident among the prokaryotic phy families. For example, *R. centenum* Ppr represents a unique BphP in that it contains, in addition to a CBD, an N-terminal extension similar to the chromophore pocket in PYP (Fig. 2A). Like PYP, Ppr binds the *p*-hydroxycinnamic acid chromophore via a thiol-ether linkage to a conserved cysteine to generate a blue light-absorbing chromoprotein [38]. This organization raises the interesting possibility that Ppr has two chromophores, which in combination can absorb both blue and R. *S elongatus* CikA is unique in that the CBD has neither the positionally conserved cysteine present in phys/CphPs nor histidine present in BphPs to bind the chromophore [36]. Whether this means that a third type of linkage is used by CikA or that CikA binds bilins noncovalently remains to be addressed. It is also of interest that some cyanobacteria contain BphP-like sequences in addition to proteins typical of CphPs (e.g., *Anabena* Plp1 and Plp2) (Fig. 3). Thus it is possible that some species simultaneously use BV- and PCB-containing photoreceptors. *Synechocystis* TaxD1 may be an extreme case of this duality; it has a putative BphP CBD and a CphP CBD in the same polypeptide that could bind both chromophores [37].

HISTIDINE KINASE DOMAINS AND KINASE ACTIVITY FOR CphPs AND BphPs

Unlike higher plant phys, where sequence analysis has not provided clear clues to their mode of action, the sequence of CphPs and BphPs has unambiguously supported the notion that these receptors are two-component histidine kinases [39, 45]. Attached to the C-terminal end of the CBD in all prokaryotic sequences discovered to date is a 250 amino acid domain homologous to the consensus HKD (Fig. 2A). All four kinase motifs necessary for catalysis (H, N, D/F, and G) are present, including the histidine required for phosphotransfer and the nucleotide-binding GXGXG motif (Fig. 4).

Given this homology to two-component regulators, a mechanism for CphPs and BphP action can be proposed (Fig. 5). Like other histidine kinases and higher plant phys, it is likely that they function as dimers [12, 21, 22]. CphP/BphP dimers would receive the light signal through the N-terminal sensor module (CBD) and then activate the HKD in a C-terminal transmitter module by appropriate conformational changes. The signal relay would begin by phosphorylation of a conserved histidine in the HKD, typically by *trans*-phosphorylation of the BphP dimer. This signal would then be relayed to a RR module by transfer of the phosphate from the histidyl residue to a conserved aspartic acid in the RR. For photoreceptors such as *R. leguminosarium* BphP or *S. elongatus* CikA, which contain an appended RR, this step could be an intramolecular transfer (Fig. 2A). For the others, a separate RR would be necessary, thus requiring an intermolecular phosphotransfer.

For *F. diplosiphon RcaE*, genetic evidence indicates that the immediately downstream RR is encoded by *RcaF* [30]. It follows *RcaE* in a two-membered open reading frame operon and encodes an archetypal RR domain bearing the conserved aspartate phosphoacceptor site. Several associated RRs have been identified in other bacteria by analysis of *CphP* and *BphP* genes/operons. For

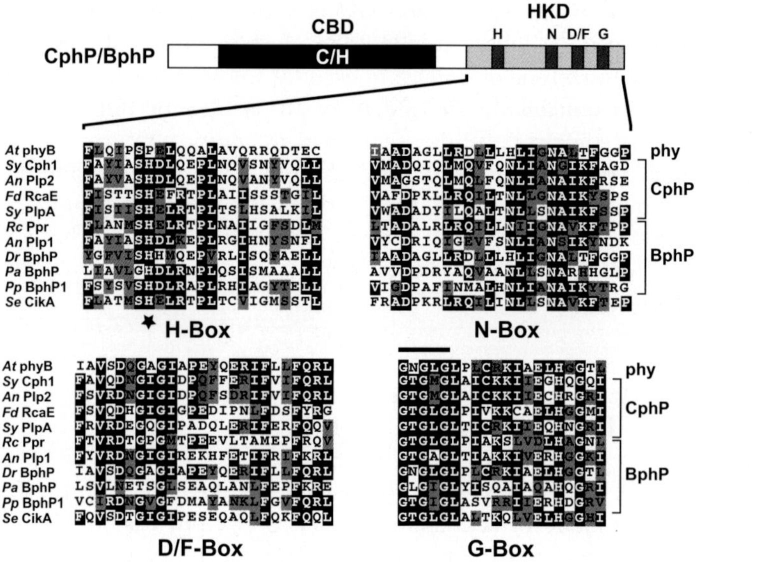

FIGURE 4 Amino acid sequence alignments of the conserved H, N, G/F, and G boxes present within the histidine kinase domain (HKD) of various prokaryotic phys with those of the HKRD in *A. thaliana* phyB [39]. Diagram of a typical CphP/BphP shows the location of the four boxes within the polypeptide. Descriptions of the various CphPs and BphPs are given in Fig. 3. CBD, chromophore-binding domain. The star indicates the position of the consensus histidine residue that serves as the autophosphorylation site for two-component histidine kinases. The bar locates the consensus nucleotide-binding motif GXGXG in the G box.

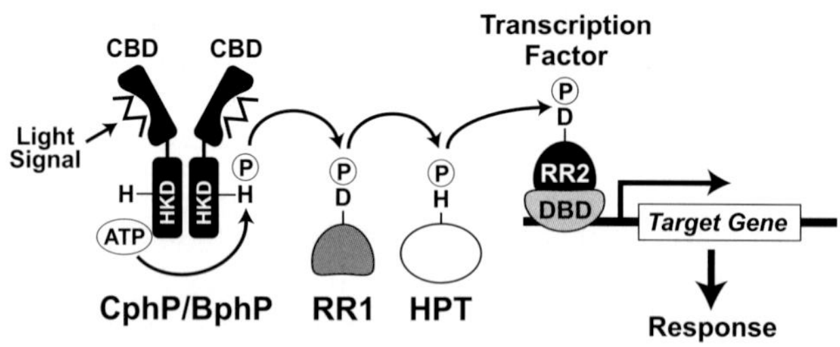

FIGURE 5 Proposed scheme for the function of prokaryotic phys in light perception. Absorption of light by the CphP/BphP homodimer triggers a conformation change in the sensor domain of the photoreceptor that activates (or inactivates) histidine kinase activity in the HKD. Active BphP cross-phosphorylates the conserved histidine (H) in the dimer and then transfers this phosphate to an aspartate residue (D) on an associated response regulator (RR1). The RR1 then transfers the phosphate to a conserved histidine in a histidine phosphotransferase (HPT), which then donates the phosphate to a second RR (RR2). RR2 contains an output domain that initiates the response. In this example, RR2 is fused to a DNA-binding domain (DBD) and serves as a transcription factor to affect gene expression.

example, the *Synechocystis Cph1* [32] and the *D. radiodurans BphP* operons [40], like that of *RcaE*, also contain an open reading frame for a separate RR (*Rcp1* and *BphR*, respectively), 3′ to that for the photoreceptor gene (Fig. 2B). In fact, the closest sequence relative to *D. radiodurans* BphP in the entire bacterial DNA database is Rcp1. The *Synechocystis* and *D. radiodurans* sensors and RRs were subsequently linked functionally by biochemical and genetic approaches; *Synechocystis* Rcp1 was shown to be a kinase substrate for Cph1 *in vitro* [32] and the mutant phenotype of *D. radiodurans* Δ*bphR* was found to be phenotypically similar to that of Δ*bphP* [41].

Whereas many RRs are fused to output domains that elicit the response, known RR domains associated with CphPs/BphPs contain only the 120 amino acid catalytic domain without such recognizable output motifs [30, 32, 36, 39, 40]. Consequently, additional components must follow the RR domain in the phosphorelay. In other two-component phosphorelay cascades, these often include a HPT that accepts the aspartyl phosphate and a second RR with an output domain that accepts the bound phosphate from the HPT, thus creating a four-step phosphorelay of input→HKD→RR1→HPT→RR2→output [21, 22]. Candidates for HPT and RR2 activities are known only for the *F. diplosiphon* RcaE pathway. Both domains are included in a single polypeptide encoded by the *RcaC* gene, also identified genetically as required for CCA [30]. RcaC protein has four modules, an HPT domain, and two RRs, one of which is connected to a DNA-binding domain (DBD). Presumably, the phosphate transferred to histidyl residue within the HPT of RcaC is then donated to the aspartate residues in either of the two RRs, one of which activates the DBD. As with similar two-component histidine kinase cascades, the output of the BphP signal likely affects the expression of specific genes (Fig. 5). This is true for *F. diplosiphon* RcaE, *S. elongatus* CikA, and *R. centenum* Ppr [30, 36, 38] and is likely true for *D. radiodurans* BphP [41].

One notable exception to the finding that many *BphP* genes are physically associated with cognate *RR* genes is *P. aeruginosa* BphP [40]. Instead of having a RR grouped in the same operon, the *BphP* gene is associated with coding regions for a putative GTP-binding protein (Fig. 2B). Whether this organization implies a distinct mode of action for *P. aeruginosa* BphP is unclear.

Experimental studies are now underway to prove this four-step phosphorelay pathway (Fig. 5). Using recombinant holoprotein assembled with BV, we have shown that *P. syringae* BphP has histidine-kinase activity *in vitro* and will phosphorylate itself at the expected histidine and transfer the phosphate a RR [40]. In agreement with the expectation that Pfr is the physiologically active form [4, 5], this kinase activity is stimulated substantially upon photoconversion of the chromoprotein from Pr to Pfr. Yeh and co-workers [32] have demonstrated similarly that *Synechocystis* Cph1 assembled with PCB is a histidine kinase. In contrast to *P. syringae* BphP, Cph1 had less kinase activity as Pfr than Pr. Likewise, *R. centenum* Ppr apoprotein and holoprotein assembled with *p*-hydroxycinnamic acid have kinase activity *in vitro* [38]. For the holoprotein, the activity is repressed by blue light, suggesting that the ground state of the photoreceptor (Pr) is also the active form. Why *D. radiodurans* BphP is activated by light whereas *Synechocystis* Cph1 and *R. centenum* Ppr are inactivated by light remains to be determined. Based on other prokaryotic examples (*e.g.*, *Rhizobium* FixL), Hughes and Lamparter [45] have suggested that BphPs could be bifunctional, capable of acting as a protein kinase in one form (Pr) and as a protein phosphatase in the other (Pfr). If this scenario is true, then the long-held belief that the phy family of photoreceptors has a single "active form" may need revising.

BIOLOGICAL FUNCTIONS OF PROKARYOTIC Phys

As predicted, CphPs and BphP families likely play a critical role in the response of bacteria to their light environment. Genetic analyses have uncovered the functions of several. For others, their functions remains unclear, in some cases because these bacteria, like higher plants, contain multiple BphPs that likely overlap functionally. We predict that most functions will reiterate what is known about phys from higher plants, i.e., CphPs/BphPs are important sensors for regulating photosynthetic potential, movement toward or away from favorable/unfavorable light conditions, and/or pigmentation [37, 39, 45]. At present, little is know about the sensory cascade associated with each CphP/BphP except for *F. diplosiphon* RcaE. However, given the simplicity of bacterial genetics in some species, uncovering the components of each pathway should be reasonably straightforward.

As described earlier, the RcaE system in *F. diplosiphon* is critical for CCA. This filamentous cyanobacterium responds to changing light intensity and quality by varying the bili-protein composition of the phycobilisomes that comprise the light-harvesting antenna system [29]. This change in protein composition is controlled in part by transcription. In R-enriched light, the expression of operons encoding the phycocyanin polypeptides is enhanced, whereas in green-enriched light, the expression of operons encoding phycoerythrin polypeptides is enhanced. Mutants in the *RcaE* pathway fail to show CCA and have unregulated expression of both phycocyanin and phycoerythrin operons, indicating that RcaE is one important light sensor [29]. As CCA responds to both R and green light, an intriguing possibility is that RcaE detects both colors, using its photochromicity to distinguish between green- and R-rich environments much the same way that higher plant phys can detect R- and FR-rich environments [4].

For other cyanobacteria, the function(s) of phy-type photoreceptors remains unclear. In *S. elongatus,* numerous photosynthetic genes are expressed in a circadian periodicity, with higher expression in the light phase and lower expression in the dark phase. Mutant analysis of *CikA* suggests that the encoded protein is required for entraining the expression of these gene to the diurnal light conditions [36]. However, because CikA protein lacks both the conserved cysteine and histidine residues that could bind a bilin and because the associated RR domain lacks the consensus aspartate that serves as the phosphoacceptor of the HKD (Fig. 2A), its role as a light-regulated histidine kinase is not assured [36]. Several CphPs exist in *Synechocystis* PCC 6803. TaxD1 appears important for motility and may be used to phototax toward favorable light conditions [37]. Disruptions of either *Cph1* or *Rcp1* are without an observable phenotypic consequence [45, 48]. *Synechocystic* mutants disrupted in *PlpA* are normal when grown photoheterotrophically (with light plus glucose), but when grown photoautotropically (light without glucose), the mutants are normal under white and FR but are unable to grow under blue light [31]. Although it appears that PlpA is a photoreceptor that controls growth, why this expected R/FR-absorbing pigment would regulate a blue light response is unclear.

The purple bacterium *R. centenum* is remarkable for its group-coordinated phototaxis. Although the cells are autonomous, they move synchronously as a single colony toward unidirectional blue light. Ppr was first isolated based on the potential role of PYP-related chromoproteins in this phototactic behavior [38]. However, subsequent disruption of the *Ppr* gene showed that this photoreceptor does not control this movement. *ppr* mutants do show reduced expression of the *chalcone synthase* (*Csh*) gene [38], which encodes a rate-limiting enzyme in the biosynthesis of the photoprotective anthocyanins pigments. In higher plants, the accumulation of anthocyanins is regulated by phys and other photoreceptors through the photoregulation of *Chs* transcription [3, 4].

Most of the BphP systems have not yet been investigated with regard to function [39–41]. Searches of the *D. radiodurans* genome using known chromoproteins (e.g., rhodopsin, cryptochromes, phys) indicate that BphP may be the sole photoreceptor in this eubacterium. As a result, this BphP offers the first opportunity to study a single phy system, not only without the complication of photosynthesis, but also in the possible absence of other light-sensing pathways. The *D. radiodurans* BphP pathway appears to help regulate carotenoid biosynthesis [41]. Whereas dark-grown cells accumulate carotenoids, the most abundant being a novel form called deinoxanthin, light increases these levels substantially. In fact, light-grown *D. radiodurans* colonies are bright red. This increase is attenuated markedly in strains disrupted in *BphP* and/or *BphR*. The Δ*bphP* mutant is also slow growing in the high-light conditions [41]. Thus, BphP-stimulated carotenogenesis may protect the bacterium when subjected to intense light, much the same way that phy-simulated production of carotenoids photoprotects higher plants. A likely mechanism is transcriptional activation of the *phytoene synthase* (*PSY*) gene that encodes the rate-limiting enzyme in carotenoid production [39].

DO HIGHER PLANT Phys FUNCTION AS TWO-COMPONENT HISTIDINE KINASES?

The discovery of CphPs and BphPs has further renewed attempts to determine if phys are light-regulated kinases. By developing recombinant systems to produce holo-phys and thus eliminate the potential for contamination by other plant kinases, Yeh and Lagarias [25] showed conclusively that an algal phy and higher plant phyA have intrinsic kinase activities. These phys could direct autophosphorylation and phosphorylation of other proteins but still showed only modest activation (or repression) by light. Biochemical analysis of autophosphorylation indicated that a serine/threonine was the acceptor site, suggesting that phys are actually serine/threonine protein kinases and not the expected histidine kinases. However, more refined sequence analysis still suggests that phys arose from a two-component histidine kinase and then diverged following duplication of the HKD [25]. The distal histidine kinase-related domain (HKRD) remained similar to the HKD but lost the essential histidine in the H box and presumably its histidine kinase activity (Figs. 1 and 4). The proximal PRD diverged substantially and acquired the two PAS repeats. Collectively, data suggest that the first phys were histidine kinases but have since developed a new phosphotransferase activity that favors serines/theonines. Which domain (HKRD or PRD) contains the kinase activity remains to be determined.

What are the downstream components of the proposed phy kinase cascade? One possibility is that phys, despite their divergence, still initiate a two-component phosphorelay cascade similar to that of BphPs. Higher plants contain a number of RR proteins. In fact, one RR in *A. thaliana* encoded by the *TOC1* gene is involved in the circadian clock; coincidentally, this clock can be entrained by phys [49]. Alternatively, other types of phosphorelays could be employed, involving possible downstream interacting components, such as *Arabidopsis* nucleotide diphosphate kinase (NDPK)-2, phytochrome kinase substrate (PSK)-1, and phytochrome-interacting factor (PIF)-3. PKS1 and NDPK2 were isolated by a yeast two-hybrid screen using the C-terminal domain of phyB as bait [50, 51], whereas PIF3 was identified both by two-hybrid screen and by cloning of a FR-insensitive mutation *poc1* [52, 53]. PKS1 can be a substrate of the *in vitro* protein kinase activity of phyA [50].

Of interest is that both NDPK2 and PIF3 prefer binding to the Pfr form of phyB [51, 54]. The function(s) of NDPKs in plants is unknown; in other organisms they play a role in various developmental processes. PIF3 is a member of the helix-loop-helix DNA-binding protein family and will bind to the promoter regions of known light-regulated genes [55]. Thus PIF3 could function in an analogous manner to the RR-DBD components of two-component signal relays (Fig. 5). In this case, phy could directly affect transcription by phosphorylation of this transcription factor.

FUNCTIONS OF THE KINASE ACTIVITY OF Phys

What is the function(s) of the protein kinase activity of phys? Although no definitive data are currently available, the kinase activity could have several roles (Fig. 6). Certainly the most intriguing would be a role in initiating the phy signal transduction cascade. Similar to BphPs, phys could *trans*-phosphorylate one or more downstream components as Pfr (or Pr) and thus begin (or stop) a light-regulated phosphorelay cascade that ultimately affects gene expression and development [18, 19].

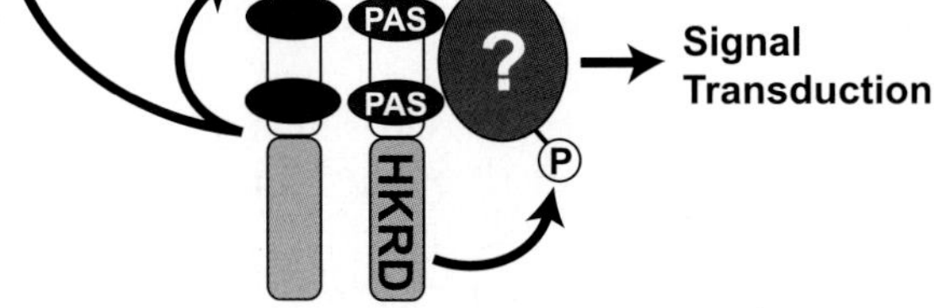

FIGURE 6 Possible functions of the protein kinase activity of higher plant phys. These include autophosphorylation of serines within the hinge region and the serine-rich domain (SRD) at the N terminus and the phosphorylation of associated proteins in the signal transduction cascade. The outcome of such phosphoryation could (1) induce a phosphorelay within the sensory transduction chain that could begin with or without an intramolecular phosphotransfer, (2) changes in the location of the photoreceptor, (3) changes in the activity of the photoreceptor, and/or (4) changes in the degradation rate of the photoreceptor. The location of the kinase active site in phys is not yet known but is presumed to be in the C-terminal half.

It is also possible that phosphorylation is responsible for modulating phys themselves to enhance or repress their action. For example, autophosphorylation could affect the subcellular localization of phys. It has been shown for several phy isoforms that the majority of phy pool is in the cytoplasm as Pr but slowly relocalizes to the nucleus as Pfr [56, 57]. The light-activated/repressed kinase activity could phosphorylate the phy homodimer itself by either *cis* or *trans*-phosphorylation, which in turn could direct nuclear import. This import could occur by enhancing the release of a cytoplasmic retention factor or by directing association with a nuclear import factor.

Alternatively, autophosphorylation of phys could provide a mechanism to modify the activity of Pfr. Similar to the repression of rhodopsin signaling by phosphorylation-induced binding by arrestin [58], photoconversion to Pfr could simulate autophosphorylation of the holoprotein, thereby promoting interaction with an inhibitor. In support, we and others have previously found that removal of the serines within the SRD of phyA hyperactivates the photoreceptor *in vivo*, indicating that the SRD serves to attenuate phy function [59, 60]. Because one or more serines in the SRD are known to be phosphorylated *in vivo* [11], the repression of phy activity could be initiated by this modification. Finally, autophosphorylation could direct phy degradation. For phyA in particular, it has been shown that photoconversion to Pfr stimulates the rapid degradation of the photoreceptor by the ubiquitin/26S proteasome proteolytic pathway [61, 62]. Many examples now exist where the substrates must first be phosphorylated before they are selectively recognized by the ubiquitination machinery [63].

BphP, CphP, AND Phy EVOLUTION

The discovery of CphPs and BphPs now provides a potential route for the evolution of phy-type photoreceptors. Given their widespread distribution in a number of distantly related bacteria and simple organization [40], BphPs likely represent the progenitors of all of the bilin-containing photochromic photoreceptors. They likely use BV as the chromophore, thus exploiting the simplest way to obtain a linear bilin in which just one enzymatic step is required from the heme precursor. However, it should be mentioned that BphPs are not universally present among eubacteria. Of the 30 or so other eubacterial genomes completed to date (e.g., *Escherichia coli*), only a handful have *BphP*-related genes, some of which include other *Psuedomonas* strains and eubacteria related to *D. radiodurans* (e.g., *Thermus aquaticus*) [40]. Whether these species lost their BphPs over time or descended from lineages that arose before this receptor is unknown.

Upon the appearance of cyanobacteria, CphP type photoreceptors emerged that replaced BV with PCB as the chromophore. Why was this change made? One reason may be reflected by a substantial blue absorbance shift of the PCB–CphP chromoproteins as compared to BV–BphPs [32, 40]. Whereas the absorbance spectra of PCB–CphPs are perfectly adapted for measuring shade avoidance by photosynthetic organisms because the Pr spectrum closely overlaps that of chlorophylls [4], BV–BphPs are not given their Pr maximum at 700 nm [40]. As a consequence, cyanobacteria could have adopted more complex blue-shifted bilins to better detect competition and thus enhance light capture. This switch would require the ability to make and bind bilins such as PCB and the ability to prevent coupling of BV, an obligate precursor of these chromophores. Such a conversion may have been facilitated in cyanobacteria by their exploitation of PCB in photosynthetic light harvesting [46] and then fixed by the use of a cysteine instead of a histidine to bind the chromophore.

Once in plants, BphPs evolved further to help control the myriad of developmental responses in these complex species. In particular, SRD was added and HKD was duplicated and altered to create HKRD and PRD (Fig. 1). PRD then acquired/evolved PAS repeats. Mutant analysis suggests that PAS domains are critical for output of the light signal and may interact with a number of components of the phy transduction cascade [50, 51, 54]. Additional changes also occurred in the HKRD, especially the loss of the histidine necessary in BphP for kinase activity.

Another evolutionary consideration is the extent to which phy-like photoreceptors are distributed in biology. We have not detected phy sequences in the genomes of *Drosophila melanogaster, Caenorhabditis elegans*, or humans, suggesting that these receptors did not invade the animal kingdom (S. Davis and R.D. Vierstra, unpublished results). We also predict that phys are not widely distributed in the archaebacterial kingdom, especially in anerobic species, given that the heme oxygenase activity essential for creating linear bilins from heme obligatorily requires oxygen [8]. Taken together, it appears that phy-like photoreceptors have continued to evolve from their BphP progenitors to generate the complex photoreceptors now present in plants that are capable of highly precise measurements of light quantity, quality, and duration.

CONCLUSIONS

Although it was proposed almost 30 years ago that phys are light-regulated kinases, only now are definitive data appearing that support this view unambiguously. Of particular help has been the identification of numerous phy-like photoreceptors in bacteria that show substantial homology to two-component histidine kinases. Clearly, given the ease of bacterial genetics, these CphPs and BphPs should serve as useful paradigms to study phy structure and function. They may be particularly useful in attempts to solve the three-dimensional structures of these unique photoreceptors and to understand the conformational changes that occur during the transformation of Pr to Pfr, as methods are now available to synthesize the photochemically active holoproteins via recombinant approaches [40]. Ultimately, it is hoped that CphPs and BphPs will help resolve questions on the mode of action of higher plant phys. Do phys really function as light-regulated protein kinases? If so, which form is more active, Pr or Pfr? What is the next step in the signal transduction chain? Despite the recent wealth of information, it is clear that many questions remain concerning the organization, structure, and function of the phy family of chromoproteins. To date, the signal transduction cascade for only one of these photoreceptors is even remotely clear (*F. diplosiphon* RcaE [29, 30]). However, given the accelerating interest in these pigments, we expect that a general understanding of several will soon emerge.

ACKNOWLEDGMENTS

I thank S.-H. Bhoo, S.J. Davis, B. Karniol, D. Kehoe, A. Johnston, and M. Wexler for providing information prior to publication. Our work was supported by grants from the U.S. Department of Energy and the National Science Foundation.

REFERENCES

1. Neff, M. M., Fankhauser, C., and Chory, J. (2000). Light: An indicator of time and place. *Genes Dev.* **14**, 257–271.
2. Fankhauser, C. and Chory, J. (1997). Light control of plant development. *Annu. Rev. Cell. Dev. Biol.* **13**, 203–229.
3. Kendrick, R. E., and Kronenberg, G. H. M. (1994). "Photomorphogenesis in Plants," 2d Ed. Martinus Nijhoff, The Netherlands.
4. Smith, H. (2000). Phytochromes and light signal perception by plants: An emerging synthesis. *Nature* **407**, 585–591.
5. Quail, P. H., Boylan, M. T., Parks, B. M., Short, T. W., Xu, Y., and Wagner, D. (1995). Phytochromes: Photosensory perception and signal transduction. *Science* **268**, 675–680.
6. Mathews, S., and Sharrock, R.A. (1997). Phytochrome gene diversity. *Plant Cell Environ.* **20**, 666–671.
7. Wu, S. H. and Lagarias, J. C. (2000). Defining the bilin lyase domain: Lessons from the extended phytochrome superfamily. *Biochemistry* **39**, 13487–13495.
8. Davis, S. J., Kurepa, J., and Vierstra, R. D. (1999). The *Arabidopsis thaliana HY1* locus, required for phytochrome-chromophore biosynthesis, encodes a protein related to heme oxygenases. *Proc. Natl. Acad. Sci. USA* **96**, 6541–6546.
9. Terry, M. L. (1997). Phytochrome deficient mutants. *Plant Cell Environ.* **20**, 740–745.

10. Whitelam, G. C., and Devlin, P. F. (1997). Roles of different phytochromes in *Arabidopsis* morphogenesis. *Plant Cell Environ.* **20**, 752–758.
11. Lapko, V. N., Jiang, X. Y., Smith, D. L., and Song, P. S. (1999). Mass spectrometric characterization of oat phytochrome A: Isoforms and post-translational modifications. *Protein Sci.* **8**, 1032–1044.
12. Quail, P. H. (1997). An emerging molecular map of phytochromes. *Plant Cell Environ.* **20**, 657–665.
13. Taylor, B. L., and Zhulin, I. B. (1999). PAS domains: Internal sensors of oxygen, redox potential, and light. *Microbiol. Mol. Biol. Rev.* **63**, 1051–1058.
14. Hunt, R. E. and Pratt, L. H. (1980). Partial characterization of undegraded oat phytochrome. *Biochemistry* **19**, 390–394.
15. Wong, Y. S., Cheng, H. C., Walsh, D. A., and Lagarias, J. C. (1986). Phosphorylation of *Avena* phytochrome *in vitro* as a probe of light-induced conformational changes. *J. Biol. Chem.* **261**, 12089–12097.
16. Wong, Y. S. and Lagarias, J. C. (1989). Affinity labeling of *Avena* phytochrome with ATP analogs. *Proc. Natl. Acad. Sci. USA* **86**, 3469–3473.
17. Biermann, B. J., Pao, L. I., and Feldman, L. J. (1994). Pr-specific phytochrome phosphorylation *in vitro* by a protein kinase present in anti-phytochrome maize immunoprecipitates. *Plant Physiol.* **105**, 243–251.
18. Elich, T. D., and Chory, J. (1997). Phytochrome: If it looks and smells like a histidine kinase, is it a histidine kinase? *Cell* **91**, 713–716.
19. Reed, J. W. (1998). Phytochrome autophosphorylation: No longer a red/far red herring? *Trends Plant Sci.* **3**, 43–44.
20. Schneider-Poetsch, H. A. (1992). Signal transduction by phytochrome: Phytochromes have a module related to the transmitter modules of bacterial sensor proteins. *Photochem. Photobiol.* **56**, 839–846.
21. Appleby, J. L., Parkinson, J. S., and Bourret, R. B. (1996). Signal transduction via the multistep phosphorelay: Not necessarily the road less traveled. *Cell* **86**, 845–848.
22. Hoch, J. A. and Silhavy, T. J. (1995). Two component signal transduction. Amer. Soc. Microbiol. Press, Washington, DC.
23. Chang, C., Kwok, S. F., Bleecker, A. B., and Meyerowitz, E. M. (1993). *Arabidopsis* ethylene-response gene *ETR1*: Similarity of product to two-component regulators. *Science* **262**, 539–544.
24. Kakimoto, T. (1996). CKI1, a Histidine kinase homolog implicated in cytokinin signal transduction. *Science* **274**, 982–985.
25. Yeh, K. C., and Lagarias, J. C. (1998). Eukaryotic phytochromes: light-regulated serine/threonine protein kinases with histidine kinase ancestry. *Proc. Natl. Acad. Sci. USA* **95**, 13976–13981.
26. Boylan, M. T., and Quail, P. H. (1996). Are phytochromes protein kinases? *Protoplasma* **195**, 59–67.
27. Krall, L., and Reed, J. W. (2000). The histidine kinase-related domain participates in phytochrome B function but is dispensable. *Proc. Natl. Acad. Sci. USA* **97**, 8169–8174.
28. Kolukisaoglu, H. U., Marx, S., Wiegmann, C., Hanelt, S., and Schneider-Poetsch, H. A. (1995). Divergence of the phytochrome gene Family predates angiosperm evolution and suggests that *Selaginella* and *Equisetum* arose prior to *Psilotum*. *J. Mol. Evol.* **41**, 329–337.
29. Kehoe, D. M., and Grossman, A. R. (1996). Similarity of a chromatic adaptation sensor to phytochrome and ethylene receptors. *Science* **273**, 1409–1412.
30. Kehoe, D. M., and Grossman, A. R. (1997). New classes of mutants in complementary chromatic adaptation provide evidence for a novel four-step phosphorelay system. *J. Bacteriol.* **179**, 3914–3921.

31. Wilde, A., Churin, Y., Schubert, H., and Borner, T. (1997). Disruption of a *Synechocystis* sp. PCC 6803 gene with partial similarity to phytochrome genes alters growth under changing light qualities. *FEBS Lett.* **406**, 89–92.
32. Yeh, K. C., Wu, S. H., Murphy, J. T., and Lagarias, J. C. (1997). A Cyanobacterial phytochrome two-component light sensory system. *Science* **277**, 1505–1508.
33. Hughes, J., Lamparter, T., Mittmann, F., Hartmann, E., Gartner, W., Wilde, A., and Borner, T. (1997). A prokaryotic phytochrome. *Nature* **386**, 663.
34. Park, C. M., Shim, J. Y., Yang, S. S., Kang, J. G., Kim, J. I., Luka, Z., and Song, P .S. (2000). Chromophore-apoprotein interactions in *Synechocystis* sp. PCC6803 phytochrome Cph1. *Biochemistry* **39**, 6349–6356.
35. Herdman, M., Coursin, T., Rippka, R., Houmard, J., and Tandeau de Marsac, N. (2000). A new appraisal of the prokaryotic origin of eukaryotic phytochromes. *J. Mol. Evol.* **51**, 205–213.
36. Schmitz, O., Katayama, M., Williams, S. B., Kondo, T., and Golden, S. S. (2000). CikA, a bacteriophytochrome that resets the cyanobacterial circadian clock. *Science* **289**, 765–768.
37. Bhaya, D., Takahashi, A., and Grossman, A. R. (2001) Light regulation of type IV pilus-dependent motility by chemosenor-like elements in *Synechocystis* PCC6803. *Proc. Natl. Acad. Sci. USA* **98**, 7540–7545.
38. Jiang, Z., Swem, L. R., Rushing, B. G., Devanathan, S., Tollin, G., and Bauer, C. E. (1999). Bacterial photoreceptor with similarity to photoactive yellow protein and plant phytochromes. *Science* **285**, 406–409.
39. Vierstra, R. D., and Davis, S. J. (2000). Bacteriophytochromes: New tools for understanding phytochrome signal transduction. *Sem. Cell. Dev. Biol.* **11**, 511–521.
40. Bhoo, S.-H. Davis, S. J., Walker, J. M., Karniol, B., and Vierstra, R. D. (2001). Bacteriophytochromes are photochromic histidine kinases that use a biliverdin chromophore. *Nature* **414**, 776–779.
41. Davis, S. J., Vener, A. V., and Vierstra, R. D. (1999). Bacteriophytochromes: Phytochrome-like photoreceptors from nonphotosynthetic eubacteria. *Science* **286**, 2517–2520.
42. Remberg, A., Lindner, I., Lamparter, T., Hughes, J., Kneip, C., Hildebrandt, P., Braslavsky, S. E., Gartner, W., and Schaffner, K. (1997). Raman spectroscopic and light-induced kinetic characterization of a recombinant phytochrome of the cyanobacterium *Synechocystis*. *Biochemistry* **36**, 13389–13395.
43. Sineshchekov, V., Hughes, J., Hartmann, E., and Lamparter, T. (1998). Fluorescence and photochemistry of recombinant phytochrome from the cyanobacterium *Synechocystis*. *Photochem. Photobiol.* **67**, 263–267.
44. Lamparter, T., Mittmann, F., Gartner, W., Borner, T., Hartmann, E., and Hughes, J. (1997). Characterization of recombinant phytochrome from the cyanobacterium *Synechocystis*. *Proc. Natl. Acad. Sci. USA* **94**, 11792–11797.
45. Hughes, J., and Lamparter, T. (1999). Prokaryotes and phytochrome. The connection to chromophores and signaling. *Plant Physiol.* **121**, 1059–1068.
46. Beale, S. I. (1993) Biosynthesis of phycobilins. *Chem. Rev.* **93**, 785–802.
47. Hübschmann, T., Börner, T., Hartmann, E., and Lampartner, T. (2001). Characterization of the Cph1 holo-protein from *Synechocystis* sp. PCC6803. *Eur J. Biochem.* **268**, 2005–2063.
48. Garcia-Dominguez, M., Muro-Pastor, M. I., Reyes, J. C., and Florencio, F .J. (2000). Light-dependent regulation of cyanobacterial phytochrome expression. *J. Bacteriol.* **182**, 38–44.
49. Strayer, C., Oyama, T., Schultz, T. F., Raman, R., Somers, D. E., Mas, P., Panda, S., Kreps, J. A., and Kay, S. A. (2000). Cloning of the *Arabidopsis* clock gene *TOC1*, an autoregulatory response regulator homolog. *Science* **289**, 768–771.
50. Fankhauser, C., Yeh, K. C., Lagarias, J. C., Zhang, H., Elich, T. D., and Chory, J. (1999). PKS1, a substrate phosphorylated by phytochrome that modulates light signaling in

Arabidopsis. *Science* **284**, 1539–1541.

51. Choi, G., Yi, H., Lee, J., Kwon, Y. K., Soh, M. S., Shin, B., Luka, Z., Hahn, T. R., and Song, P. S. (1999). Phytochrome signaling is mediated through nucleoside diphosphate kinase 2. *Nature* **401**, 610–613.
52. Halliday, K. J., Hudson, M., Ni, M., Qin, M., and Quail, P. H. (1999). *poc1*: An *Arabidopsis* mutant perturbed in phytochrome signaling because of a T-DNA insertion in the promoter of PIF3, a gene encoding a phytochrome-interacting BHLH protein. *Proc. Natl. Acad. Sci. USA* **96**, 5832–5837.
53. Ni, M., Tepperman, J. M., and Quail, P. H. (1998). PIF3, a Phytochrome-interacting factor necessary for normal photoinduced signal transduction, is a novel basic helix-loop-helix protein. *Cell* **95**, 657–667.
54. Ni, M., Tepperman, J. M., and Quail, P. H. (1999). Binding of phytochrome B to its nuclear sSignaling partner PIF3 is reversibly induced by light. *Nature* **400**, 781–784.
55. Martinez-Garcia, J. F., Huq, E., and Quail, P. H. (2000). Direct targeting of light signals to a promoter element-bound transcription factor. *Science* **288**, 859–863.
56. Kircher, S., Kozma-Bognar, L., Kim, L., Adam, E., Harter, K., Schafer, E., and Nagy, F. (1999). Light quality-dependent nuclear import of the plant photoreceptors phytochrome A and B. *Plant Cell.* **11**, 1445–1456.
57. Hisada, A., Hanzawa, H., Weller, J. L., Nagatani, A., Reid, J. B., and Furuya, M. (2000). Light-Induced nuclear translocation of endogenous Pea phytochrome A visualized by immunocytochemical procedures. *Plant Cell.* **12**, 1063–1078.
58. Yarfitz, S., and Hurley, J. B. (1994). Transduction mechanisms of vertebrate and invertebrate photoreceptors. *J. Biol. Chem.* **269**, 14329–14332.
59. Jordan, E. T., Marita, J. M., Clough, R. C., and Vierstra, R. D. (1997). Characterization of regions within the N-terminal 6-kilodalton domain of phytochrome A that modulate its biological activity. *Plant Physiol.* **115**, 693–704.
60. Stockhaus, J., Nagatani, A., Halfter, U., Kay, S., Furuya, M., and Chua, N. H. (1992). Serine-to-alanine substitutions at the amino-terminal region of phytochrome A result in an increase in biological activity. *Genes Dev.* **6**, 2364–2372.
61. Clough, R. C., Jordan-Beebe, E. T., Lohman, K. N., Marita, J. M., Walker, J. M., Gatz, C., and Vierstra, R. D. (1999). Sequences within both the N- and C-terminal domains of phytochrome A are required for Pfr ubiquitination and degradation. *Plant J.* **17**, 155–167.
62. Vierstra, R. D. (1996). Proteolysis in Plants: Mechanisms and functions. *Plant Mol. Biol.* **32**, 275–302.
63. Hershko, A., and Ciechanover, A. (1998). The ubiquitin system. *Annu. Rev. Biochem.* **67**, 425–479.
64. Vierstra, R. D., and Quail, P. H. (1983). Purification and initial characterization of 124-kilodalton phytochrome from *Avena*. *Biochemistry* **22**, 2498–2505.

CHAPTER 14

Histidine Kinases in the Cyanobacterial Circadian System

HIDEO IWASAKI AND TAKAO KONDO

Division of Biological Science, Graduate School of Science, Nagoya University and CREST, JST, Furo-cho, Chikusa, Aichi 464-8602, Japan

Histidine Kinases in Signal Transduction

Circadian rhythms, endogenous biological oscillations with a period of about a day, have been observed in innumerable physiological processes in most of organisms from cyanobacteria to higher plants and mammals. These rhythms are thought to be adaptive to daily environmental changes on the earth. The circadian system is composed schematically of a central oscillator (circadian clock) and inputs to and outputs from the clock. The cyanobacterium is the simplest organism known to harbor circadian clocks, and now becomes one of most successful model organisms for comprehensive understanding the oscillatory system at the molecular level. Molecular genetic studies on circadian clocks in the cyanobacterium *Synechococcus* identified two histidine kinases, SasA and CikA. SasA interacts with a well-established clock protein KaiC and is necessary to sustain robust circadian oscillation, whereas CikA functions in a photic input pathway to the clock. Thus, multiple His-to-Asp signaling pathways are likely to play important roles in the *Synechococcus* circadian system.

INTRODUCTION

Circadian rhythms are daily fluctuations in cellular, physiological, and behavioral parameters, which are observed ubiquitously in many organisms. In many cases, organisms coordinate such a diurnal pattern of biological activities by their endogenous pacemakers, called circadian clocks [1]. Circadian rhythms are characterized by three physiological features. First, the rhythms persist even in the absence of external cues with a period length (free-running period) that is close to, but not always exactly equal to, the 24-h period of the earth's rotation. Second, circadian rhythms can be entrained to the external earth's cycles, such as the predictable alternation of light and dark. Third, the period length is relatively stable at different constant temperatures, i.e., the free-running period is temperature compensated. All of these features support the idea that circadian clocks are physiologically significant for organisms as adaptive systems to daily changes in environmental conditions on the earth, such as light, temperature, and humidity [1]. Furthermore, the circadian clock is suggested as a fundamental basis of photoperiodic behavior in plants, insects, and mammals. In humans, circadian clocks are involved in many physiological regulations, such as the wake/sleep cycle, hormone secretions, and body temperature. Such a human circadian control has become a medical consequence because a loss of clock function causes the genesis of depression, jet lag, or insomnia [2].

Schematically, circadian organization can be considered to be organized by three elements: the circadian oscillator that generates basic oscillation, input pathways that transfer external environmental information to the oscillator, and output pathways from the oscillator that regulate many temporal events (Fig. 1A). During the past decades, studies on circadian rhythms have led to considerable amounts of information about the ubiquity, general features of the oscillation, and a variety set of rhythmic phenomena at the physiological levels. At the cellular and molecular levels, however, it is still far from comprehensive understanding of the circadian system. Among many model organisms, cyanobacteria, the simplest organisms to exhibit circadian rhythms, have become analyzed intensively to address fundamental questions in circadian biology.

CYANOBACTERIAL CIRCADIAN RHYTHMS

Until the beginning of the 1990s, a dogma that clocks exist exclusively in eukaryotes was influential among circadian clock researchers. However, several observations in the late 1980s established circadian regulation in cyanobacteria [3]. Initially, daily rhythms in nitrogen-fixation in some

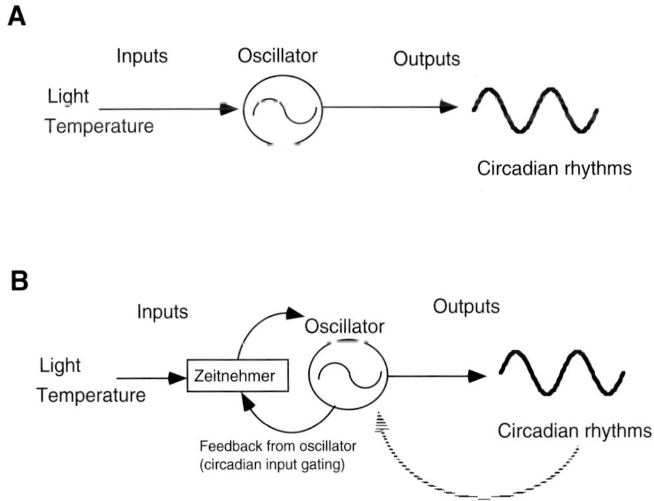

FIGURE 1 Heuristic models of circadian system components. (A) The traditional "unidirectional cascade" model of a circadian system. This model includes a central oscillator that generates circadian rhythms, input factors that receive environmental stimuli to synchronize the oscillator to environmental cycles, and output pathways through which the oscillator regulates cellular and behavioral rhythmic activities. (B) The "multiple feedback loop" model (or the "zeitnehmer" model). A clock-controlled input factor (*zeitnehmer*) regulates photic (or/and temperature) input pathways to the oscillator in a day-time-dependent manner to adjust period length, amplitude, and phase angle of circadian rhythms. In some organisms, circadian output activities have been shown to feed back to the oscillator function to additonally modify physiological rhythms (dotted line), further representing complicated multiple feedback loops in circadian systems.

diazotrophic (nitrogen-fixing) cyanobacteria were characterized as circadian phenomena by Huang and colleagues [4]. Currently, the cyanobacterial circadian clock is known to regulate many physiological processes, such as photosynthesis, amino acid uptake, cell division cycle, and carbohydrate synthesis [3]. Interestingly, these rhythms are able to be driven and sustained with a circadian period even in cells dividing faster than once every 24 h [5, 6]. Moreover, Mori and co-workers [5] demonstrated circadian gating of the cell division cycle in *Synechococcus*.

The significance of the circadian program in cyanobacteria has been examined in the nondiazotrophic cyanobacterium *Synechococcus elongatus* PCC 7942 (renamed from *Synechococcus* sp. strain PCC 7942) using wild-type and mutant strains whose free-running periods differ from each other [7]. Although growth rates of these strains are quite similar [8], the strain with a period most similar to that of the external light/dark cycles had a selective advantage and eventually dominated the population in competitive situation [7]. This is the best demonstration of the selective advantage of circadian systems for fitness in any organism.

MOLECULAR GENETICS OF CYANOBACTERIAL CIRCADIAN SYSTEM: *kai* GENES

KAI GENES AS MAIN CIRCADIAN REGULATORS IN *SYNECHOCOCCUS*

The genetic approach to the dissection of the circadian timing mechanism initially arose from isolation of period mutants in *Drosophila* and *Neurospora* in the early 1970s, followed by identification and intensive analyses of the causal genes, *period* (*per*) and *frequency* (*frq*), respectively [9].

In cyanobacteria, Kondo and co-workers [10] chose a genetically tractable, unicellular nondiazotrophic freshwater strain, *S. elongatus* PCC 7942, as a model organism for comprehensive analyses of the circadidan clock. Studies in which a bioluminescence reporter (the *luxAB* bacterial luciferase gene set from *Vibrio harveyi*) was driven by a photosynthetic gene promoter demonstrated robust circadian luminescence rhythms in *S. elongatus* [10]. Using a newly developed automonitoring system, more than a hundred *S. elongatus* mutants were isolated that exhibited an altered period length ranging from 16 to 60 h, abnormal shape of the rhythm, and even arhythmia [8]. By genetic complementation of the mutants, a gene cluster *kai* composed of three genes, *kaiA*, *kaiB*, and *kaiC*, was cloned [11]. These genes share no homology to known genes, including any clock-related genes of eukaryotes. Moreover, Kai proteins do not contain any functional motifs except for several amino acyl clusters in KaiC that are conserved in many ATP/GTP-binding proteins. Strikingly, most of the rhythm mutants are complemented by the *kai* gene cluster, suggesting that these genes reside at the center of the oscillatory mechanism. Disruption of any *kai* genes abolishes circadian rhythms, indicating that *kai* genes are all essential for circadian function. Altered circadian phenotypes can arise from mutation in any of the *kai* genes [11].

Transcriptional Feedback Model for the KaiABC-Based Circadian Clock

kaiB and *kaiC* are organized into an operon to be cotranscribed, whereas *kaiA* is transcribed independently [11]. All three genes exhibit circadian expression profiles. This is not surprising because most of the gene promoters have been demonstrated to exhibit circadian rhythms in *Synechococcus* [12].

However, based on genetic analyses on the transcriptional regulation of *kai* genes, Ishiura and co-workers [11] proposed a KaiABC-based circadian oscillatory mechanism: KaiC negatively regulates its own transcription to generate a basic feedback, whereas KaiA enhances *kaiBC* expression to make the loop oscillate. This transcription/translation-based oscillator (TTO) model is schematically similar as proposed for other circadian clocks in *Neurospora*, *Drosophila*, and mammals [9, 13]. Circadian fluctuation was also observed in accumulation levels of the Kai proteins, and a transient increase in the KaiC protein at the physiological range sets the phase of the clock during a certain period of the day [14].

Alternative models have been proposed for the *Neurospora* clock protein FRQ, suggesting that it may be an autoregulating photic input factor controlled by an unknown clock mechanism [15–17], although the FRQ-less oscillator (FLO) impairs typical circadian charcteristics, such as a period length of ca. 24 h, photic entrainment, and temperature compensation [13]. Potential problems in circadian studies in cyanobacteria have been discussed by Iwasaki and Kondo [18]. For example, it has not yet been strictly examined whether the transcriptional regulation of *kai* genes is indeed necessary for circadian rhythm generation or not essential but important for sustaining robust oscillation of the *S. elongatus* clock.

Biochemical Information of Kai Proteins

The biochemical mechanism by which Kai proteins contribute to circadian rhythm generation is still wrapped in mystery, whereas fragmentary information on biochemical properties of the Kai proteins has been obtained. KaiA, KaiB, and KaiC proteins physically associate in all possible combinations in yeast cells by the two-hybrid system, *in vitro* and in cyanobacterial cells [19]. The three Kai proteins are able to form a heteromultimeric complex. A long period mutation *kaiA1* dramatically enhances the KaiA–KaiB interaction *in vitro*. Thus, direct protein–protein association among the Kai proteins is likely to be a critical process in determining the phenotype of circadian rhythms in the cyanobacterium.

We also demonstrated ATP-binding and autophosphorylating activities of KaiC *in vitro*. Their importance was strongly supported by genetic disruption of a nucleotide-binding site (P-loop1) of KaiC that severely impaired circadian *kaiBC* expression profiles [20].

SasA, A KaiC-BINDING HISTIDINE KINASE AS A CIRCADIAN AMPLIFIER

SasA Interacts with KaiC

A yeast two-hybrid screen for KaiC-associating proteins identified a histidine protein kinase, SasA [21]. The *sasA* gene had been identified by Nagaya and co-workers [22] to encode a *S. elongatus* histidine kinase that complemented *E. coli EnvZ* mutants, although its legitimate function was unknown. Interestingly, the presumed input domain of SasA shares sequence homology to KaiB and appears sufficient for interaction with KaiC *in vitro* and in yeast. Immunoprecipitation experiments confirmed an *in vivo* association of SasA with KaiC [21].

SasA Is Necessary for Robust Circadian Oscillation

The functional relevance of SasA to the cyanobacterial circadian clock was analyzed by testing *S. elongatus* in which the *sasA* gene was disrupted or overexpressed [21]. *sasA*-disrupted cells grew as well as wild-type cells under continuous light condition (LL), as observed previously [22]. Strikingly, *sasA* disruption lowered the magnitude of *kaiBC* expression, reduced the amplitude of the rhythm dramatically, and shortened the period length by 3 h in LL (Fig. 2). Interestingly, robustness of the attenuated oscillation is dependent of light fluence. At the higher light intensity (50 μE m^{-2} s^{-1}), amplitudes of expression rhythms driven from both *kaiA* and *kaiBC* pro-

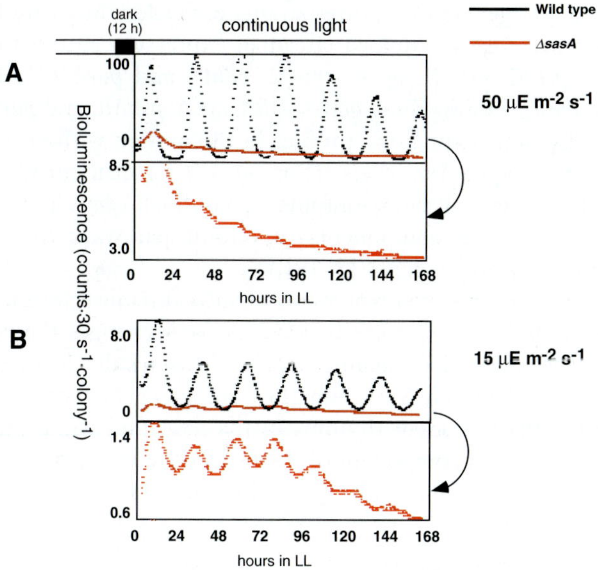

FIGURE 2 Attenuation of circadian kai gene expression by disruption of the *sasA* gene. Temporal profiles of *kaiBC* expression were monitored using a *luxAB*-based *in vivo* realtime luciferase reporter. Wild type (black trace) and *sasA*-disrupted (Δ*sasA*, red trace) cells that carried the *kaiBC* reporter cassette were grown for 3–4 days on agar medium in continuous light (LL) at 50 μE m^{-2} s^{-1} (A) or 15 μE m^{-2} s^{-1} (B), and the bioluminescence was monitored using a photomultiplier tube after a 12-h dark break to synchronize the oscillator. For clarity, bioluminescence profiles from *sasA*-disrupted cells are also shown in magnified scales (modified from [21]).

moters (P_{kaiA} and P_{kaiBC}, respectively) were reduced dramatically and damped out to be arhythmic within a few days. Moreover, the magnitude of P_{kaiA} and P_{kaiBC} activities was lowered in the *sasA*-disruptant cells to 40–60% and 5–10% of the peak levels, respectively. At the lower light intensity (15 μE m^{-2} s^{-1}), both the magnitude of expression and the amplitude of the P_{kaiBC} rhythm were still diminished by the inactivation of *sasA*, whereas the magnitude of expression from P_{kaiA} was less affected. Importantly, the residual rhythmicity was more robust than at 50 μE m^{-2} s^{-1}.

Attenuation of circadian oscillation and reduction in *kaiBC* promoter activity were also confirmed in a mutant *S. elongatus* in which a putative autophosphorylation site expected to be essential for phosphoryl transfer activity was substituted with glutamine (H162Q). Thus, it is likely that the histidine kinase function of SasA is indeed necessary for normal robust rhythmicity in *S. elongatus*, consistent with a requirement for autophosphorylation.

Monitoring a larger set of reporter strains under low-intensity LL revealed that disruption of *sasA* attenuated circadian expression patterns of all other tested genes: *psbAI*, *psbAII*, *opcA*, *rpoD2*, *ndhD*, and *purF* [21; H. Iwasaki, unpublished data]. Among these genes, *cikA*, *opcA*, *psbAII*, and *purF* exhibited completely arhythmic expression profiles in the mutant without reduction in the magnitude of expression levels. Thus, *sasA* is a global circadian regulator presumably to regulate all clock outputs. In contrast, *cpmA* and *rpoD2* genes have been proposed as components of output pathways that affect only subsets of clock-controlled genes [3, 18].

Moreover, continuous overexpression of *sasA* disrupts circadian rhythms, whereas a temporal and transient increase in SasA changes the phase of the clock in a phase-dependent manner [21]. These results further support a close relationship between SasA and the timing mechanism. However, the presence of residual circadian rhythms in the *sasA*-null strain indicates that SasA is not an essential component of the oscillatory mechanism. Rather, it is necessary to sustain robust circadian oscillation in *S. elongatus*.

Perspective of the Function of SasA in *Synechococcus*

How does SasA contribute to the circadian clock? Effects of *sasA*-disruption on the magnitude of various promoter activities suggest that SasA may act as a positive regulator of a set of genes, including the *kaiBC* operon. Interaction of KaiC with the presumed sensory domain of SasA strongly suggests KaiC-dependent modification of the enzymatic activity of SasA. Thus, SasA is proposed to be a "circadian amplifier" to form a secondary loop [21] (Fig. 3).

SasA may function as the first output of the oscillatory mechanism to regulate all downstream clock-controlled processes or as an element of redundant input pathways. Clock-regulated sensory components that regulate (photic) inputs to the clock have been called *zeitnehmers* (time takers): the clock-regulated *zeitnehmer* loop stabilizes the central oscillator to sustain robust rhythmicity with an appropriate circadian period length [15, 23] (Fig. 1B). Attenuation of rhythmicity in a light-dependent manner with a slight change in the period length caused by *sasA* inactivation would support the presumed KaiC-regulated function of SasA as a *zeitnehmer*.

Additional evidence exists for the roles of SasA in light-dependent metabolism. Although inactivation of *sasA* does not affect the growth rate of *S. elongatus* under various LL conditions, *sasA*-disrupted cells grow much slower than the wild type or a *kaiABC*-deficient strain under light/dark (LD) cycles [21]. Thus, SasA has a "two fold significance" for growth in a diurnal environment: it is necessary to sustain robust circadian rhythms and to adapt to LD alterations.

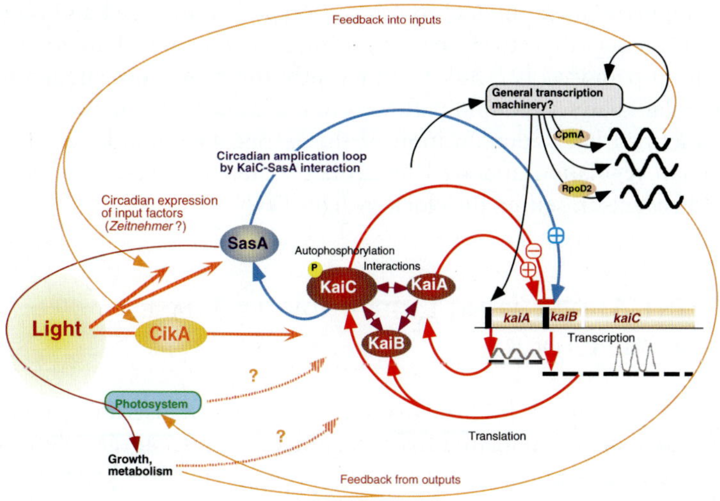

FIGURE 3 Possible multiple feedback regulatory loops in the cyanobacterial circadian system. In a classical model, a KaiABC-based autoregulatory loop (red) forms a primary molecular oscillatory circuit, while newly identified SasA-KaiC loop (blue) functions as a secondary loop to amplify the basic oscillation. SasA links not only to the oscillator but also to a photic input pathway and light-dependent metabolic growth regulation. The second histidine kinase, CikA, is a phytochrome-related circadian input factor that is important for setting the phase angle of the oscillator. Importantly, almost all genes including input factor genes express in a circadian fashion in *Synechococcus*. Thus, the resulting circadian modulation of input pathways (*zeitnehmers*) may contribute to robust circadian oscillation. The global circadian genome regulation could be due to *kai*-dependent regulation of general transcription/translation machineries. Some factors such as RpoD2 and CpmA have been identified as output factors that affect subsets of clock-controlled genes [18].

CikA, A BACTERIOPHYTOCHROME FAMILY HISTIDINE KINASE AS A CIRCADIAN PHOTIC INPUT FACTOR

Circadian Photoreception in Cyanobacteria

Light is the dominant factor for setting the clock in many organisms. In *Synechococcus* sp. RF-1, light/dark alteration appears more efficient for shifting the phase of circadian rhythms than temperature change [24], in contrast to the case for the *Neurospora* clock input system [13]. Based on several physiological and pharmacological experiments in *Synechococcus* sp. RF-1, Huang and colleagues proposed that a break of photosynthesis by darkness is responsible for the photic-entraining mechanism in cyanobacteria [25, 26]. In contrast, in higher plants, phtytochromes and cryptochromes, red and blue light receptor proteins, respectively, have been suggested to contribute to circadian photic input pathways [27–30]. Until recently, there were not enough reports on molecular analyses of input pathways to the cyanobacterial clock.

As described later, identification of the second histidine kinase involved in circadian regulation appeared immediately after the report on SasA: this kinase likely acts in setting the clock by light [31].

A Mutant Attenuating Light-Induced Phase Shift of the Clock

Schmitz and co-workers [31] isolated a Tn5 transposon insertion mutant of *S. elongatus* that slightly impaired light-responsive expression of a photosystem II gene, *psbAI*, and tested this mutant in terms of circadian gene expression and a light-resetting property. The mutant somehow reduced the amplitude of circadian expression rhythms of all tested genes and shortened the period length by 2 h.

In wild-type cells, a 5-h dark pulse treatment after cells are returned to continuous light (LL) conditions changes the phase of the rhythm during certain periods of the day [31]. For example, dark treatment from hours 5 to 10 in LL shifted the phase by 10–12 h, whereas that from hours 15 to 20 affected the phase less so. Strikingly, the Tn-5 insertion mutant showed little phase resetting throughout the day [31].

CikA Histidine Kinase as a Bacteriophytochrome

The causal gene of the mutant encodes a histidine kinase, and the amino-terminal sequence has striking similarity to the chromophore-binding domain of bacteriophytochromes (see Chapter 15), similar to *Synechocystis* sp. PCC 6803 Cph1, *Fremyella diplosiphon* RcaE, and *Arabidopsis thaliana* PhyE [31]. From the phenotype and sequence information, the gene was named *cikA* for circadian input kinase [31]. Its similarity to bacteriophytochromes suggests a possibility that CikA acts as a circadian photoreceptor. Because it lacks the conserved Cys or His redisue for bilin-binding, CikA may bind a chromophore by a novel attachment or may function without bilins [31].

Disruption of *cikA* enhances the magnitude of *kaiBC* expression, while it lowered amplitude of the rhythm of gene expression [31]. Considering the robust circadian profile of *cikA* promoter activity [21], circadian *cikA* expression would somehow modulate the clock in a day-phase-specific manner to contribute the appropriate period length and amplitude of the clock. If this is the case, the role of CikA would also be fit for the *zeitnehmer* concept.

PERSPECTIVES: TOWARD FURTHER UNDERSTANDING OF His-TO-Asp SIGNALING PATHWAYS IN THE CIRCADIAN NETWORK IN CYANOBACTERIA

As described earlier, two histidine kinases, SasA and CikA, have been found to have distinct roles in the *S. elongatus* circadian system. However, biochemical mechanisms by which SasA and CikA function contribute to the circadian system are largely unknown. For example, KaiC-dependent modification of the enzymatic activity of SasA has not been yet examined, while it is also possible that SasA senses some environmental signals. The relationship between two functions of SasA in the stabilization of circadian oscillatation and in LD responsive growth control remains to be solved. As for CikA, its chromophore-binding activity needs to be examined. Moreover, cognate response regulators for both kinases have not been identified: *cikA* and *sasA* genes do not form gene clusters with their cognate partners. Genetic or biochemical identification and further characterization of these regulators should provide better understanding of the biochemical properties of SasA and CikA.

More globally, in addition to classical forward genetics, functional genomics is an important approach to the comprehensive understanding of the signaling network in circadian systems. Although the complete genomic sequence of our standard model strain *S. elongatus* PCC 7942 is not available, that of a different cyanobacterial strain, *Synechocystis* sp. PCC 6803, has been reported [32]. In *Synechocystis* genomic DNA, at least 80 genes are found to encode His-to-Asp regulatory elements, 42 of which encode histidine kinases [33]. Splendid serial disruption of 41 histidine kinase genes by Suzuki and co-workers [34] identified two genes involved in gene expression responsive to cold acclimation (see also Chapter 19). Circadian gene expression in *Synechocystis* has been demonstrated using the bioluminescence reporter [35], and tests for each histidine kinase function in the *Synechocystis* circadian system are currently underway. For example, disruption of a *sasA* homologue (sll0750) has also been confirmed to impair circadian *dnaK* gene expression in *Synechocystis* (R. Kito, I. Suzuki, and H. Iwasaki, unpublished data). Further analyses would find novel histidine kinases functioning at interesting aspects of the circadian system in cyanobacteria.

His-to-Asp Signaling-Related Molecules in Plant Circadian Systems

Cyanobacteria are considered as phylogenically ancestral to chloroplasts in plants in the cellular symbiosis hypothesis [36]. Many genes encoded by chloroplast genomes in *Chlamydomonas* [37] and in higher plants [38] express under the control of circadian clocks. Although *kai* genes are not found in plants, identification of the sensory histidine kinase genes *sasA* and *cikA* as clock-related genes has opened a possibility that some domains of organisms use a similar signal transduction scheme for circadian regulation. Finally, we summarize briefly current knowledge on atypical two-component regulatory factors involved in plant circadian systems.

Phytochromes for Circadian Photoreception in Both Plants and Cyanobacteria

CikA is the first potential evolutionary parallel between cyanobacterial and eukaryotic circadian systems [31]. Phytochromes in higher plants are also involved in circadian photoreception [28, 29]. Although phytochromes in higher plants harbor the PAS domain, which are found in several eukaryotic clock-related proteins, WC-1 and WC-2 in *Neurospora*, dClock, Cycle (dBmal1), and Period proteins in *Drosophila*, and their orthologs in mammals [9], CikA lacks the PAS domain.

The C-terminal domains in plant phytochromes are similar to the transmitter domain of the histidine kinase, whereas the His residue, a target for potential autophosphorylation, is poorly conserved. Rather, studies have suggested that plant phytochromes function as Ser/Thr kinases in a light-dependent manner [39, 40]. However, there is no doubt that the plant phytochrome is derived from bacterial histidine kinase family proteins.

Pseudo-Response Regulator TOC1 as a Circadian Factor in *Arabidopsis*

Based on a molecular genetic approach, Strayer and co-workers [41] cloned a novel *Arabidopsis* clock gene, *toc1*, that encoded a response–regulator-like protein and expressed in a robust circadian fashion. Mutations in *toc1* alter the period length of circadian rhythm in gene expression and photoperiodic flowering. Interestingly, despite the similarity of TOC1 to response regulators, it has a Glu residue instead of the essential Asp residue.

Makino and co-workers [42] independently cloned this gene (named APRR1) for an *Arabidopsis* pseudo-response regulator. Interestingly, additional four pseudo-response regulator genes (APRR3, 5, 7, and 9) similar to TOC1/APRR1 were identified and found to be expressed in robust circadian fashion with different phase relationship to each other [43]. Although this finding may suggest that coordinated TOC1/APRR1 family functions are important for circadian timing, biochemical characteristics of them remain a mystery. However, it is now clear that both TOC1 and phytochromes that must originate from the bacterial response regulator and histidine kinase, respectively, function at different aspects of circadian systems in higher plants.

ACKNOWLEDGMENTS

We thank Susan S. Golden and Stan B. Williams (Texas A&M Univ.) and Yohko Kitayama (Nagoya Univ.) for very "positive feedback" collaboration on the SasA project, Masahiro Ishiura (Nagoya Univ.) and Shinsuke Kutsuna (Yokohama City Univ.) for the early phase of the *kai* gene analysis, Taeko Nishiwaki (Nagoya Univ.) for the KaiC biochemistry project, and S. Golden and Mitsunori Katayama (Nagoya Univ.) for collaboration on the CikA studies. This work was partly supported by grants-in-aid from the Ministry of Education, Culture, Sports, Science and Technology (2051585935), from the Kurata Memorial Hitachi Foundation, and the Inoue Foundation to H. I.

REFERENCES

1. Bünning, E. (1973). "The Physiological Clock," 3rd Ed. Springer Verlag, Heidelberg.
2. Moore-Ede, M. C., Sulzman, F. M., and Fuller, C. A. (1982) "The Clocks That Time Us: Physiology of the Circadian Timing System." Harvard University Press, Cambridge.
3. Golden, S. S., Ishiura, M., Johnson, C. H., and Kondo, T. (1997). Cyanobacterial circadian rhythms. *Annu. Rev. Plant Physiol. Mol. Biol.* **48**, 327–354.
4. Grobbelaar, N., Huang, T.-C., Lin, H.-Y., and Chow, T.-J. (1986). Dinitrogen-fixing endogenous rhythm in *Synechococcus* RF-1. *FEMS Microbiol. Lett.* **37**, 173–177.
5. Mori, T., Binder, B., and Johnson, C. H. (1996). Circadian gating of cell division in cyanobacteria growing with average doubling times of less than 24 hours. *Proc. Natl. Acad. Sci. USA* **93**, 10183–18186.
6. Kondo, T., Mori, T., Lebedeva, N. V., Aoki, S., Ishiura, M., and Golden, S. S. (1997). Circadian rhythms in rapidly dividing cyanobacteria. *Science* **275**, 224–227.
7. Ouyang, Y., Andersson, C. R., Kondo, T., Golden, S. S., and Johnson. C. H. (1998). Resonating circadian clocks enhance fitness in cyanobacteria. *Proc. Natl. Acad. Sci. USA* **95**, 8660–8664.
8. Kondo, T., Tsinoremas, N. F., Golden, S. S., Johnson, C. H., Kutsuna, S., and Ishiura, M. (1994). Circadian clock mutants of cyanobacteria. *Science* **266**, 1233–1236.
9. Dunlap, J. C. (1999). Molecular bases for circadian clocks. *Cell* **96**, 271–290.
10. Kondo, T., Strayer, C. A., Kulkarni, R. D., Taylor, W., Ishiura, M., Golden, S. S., and Johnson,

C. H. (1993). Circadian rhythms in prokaryotes: Luciferase as a reporter of circadian gene expression in cyanobacteria. *Proc. Natl. Acad. Sci. USA* **90**, 5672–5676.
11. Ishiura, M., Kutsuna, K., Aoki, S., Iwasaki, H., Andersson, C. R., Tanabe, A., Golden, S. S., Johnson, C. H., and Kondo, T. (1998). Expression of a gene cluster *kaiABC* as a circadian feedback process in cyanobacteria. *Science* **281**, 1519–1523.
12. Liu, Y., Tsinoremas, N. F., Johnson, C.H., Lebedeva, N. V., Golden, S. S., Ishiura, M., and Kondo, T. (1995). Circadian orchestration of gene expression in cyanobacteria. *Genes Dev.* **9**, 1469–1478.
13. Iwasaki, H., and Dunlap, J. C. (2000). Microbial circadian oscillatory systems in *Neurospora* and *Synechococcus*: Models for cellular clocks. *Curr. Opin. Microbiol.* **3**, 189–196.
14. Xu, Y., Mori, T., and Johnson, C. H. (2000). Circadian clock-protein expression in cyanobacteria: Rhythms and phase setting. *EMBO J.* **19**, 3349–3357.
15. Roenneberg, T., and Merrow, M. (1998). Molecular circadian oscillators: An alternative hypothesis. *J. Biol. Rhythms* **13**, 167–179.
16. Merrow, M., Brunner, M., and Roenneberg, T. (1999). Assignment of circadian function for the *Neurospora* clock gene *frequency*. *Nature* **399**, 584–586.
17. Lakin-Thomas, P. (2000). Circadian rhythms: New functions for old clock genes. *Trends Genet.* **16**, 135–142.
18. Iwasaki, H., and Kondo, T. (2000). The current state and problems of circadian clock studies in cyanobacteria. *Plant Cell Physiol.* **41**, 1013–1020.
19. Iwasaki, H., Taniguchi, Y., Ishiura, M., and Kondo, T. (1999). Physical interactions among circadian clock proteins, KaiA, KaiB and KaiC, in cyanobacteria. *EMBO J.* **18**, 1137–1145.
20. Nishiwaki, T., Iwasaki, H., Ishiura, M., and Kondo, T. (2000). Nucleotide binding and autophosphorylation of clock protein KaiC as a circadian timing process of cyanobacteria. *Proc. Natl. Acad. Sci. USA* **97**, 495–499.
21. Iwasaki, H., Williams, S. B., Kitayama, Y., Ishiura, M. Golden, S. S., and Kondo, T. (2000). A KaiC-interacting sensory histidine kinase, SasA, necessary to sustain robust circadian oscillation in cyanobacteria. *Cell* **101**, 223–233.
22. Nagaya, M., Aiba, H., and Mizuno, T. (1993). Cloning of a sensory-kinase-encoding gene that belongs to the two-component regulatory family from the cyanobacterium *Synechococcus* sp. strain PCC7942. *Gene* **131**, 119–124.
23. Roenneberg, T., and Merrow, M. (1999). Circadian systems and metabolism. *J. Biol. Rhythms* **14**, 449–459.
24. Lin, R. F., Chou, H. M., and Huang, T.-C. (1999). Priority of light/dark entrainment over temperature in setting the circadian rhythms of the prokaryote *Synechococcus* RF-1. *Planta* **209**, 202–206.
25. Huang, T.-C., and Chow, T.-J. (1990). Characterization of the rhythmic nitrogen-fixing activity of *Synechococcus* sp. RF-1 at the transcriptional level. *Curr. Microbiol.* **20**, 23–26
26. Chen, T.-H., Pen, S.-Y., and Huang , T.-C. (1993). Induction of nitrogen-fixing circadian rhythm in *Synechococcus* sp. RF-1 by light signals. *Plant Sci.* **92**, 179–183.
27. Nagy, F., Kay, S. A., and Chua, N.-H. (1988). A circadian clock regulates transcription of the wheat *Cab-1* gene. *Genes Dev.* **2**, 376–382.
28. Millar, A. J., Straume, M., Chory, J., Chua, N.-H., and Kay, S. A. (1995). The regulation of circadian period by phototransduction pathways in *Arabidopsis*. *Science* **267**, 1163–1166.
29. Somers, D. E., Devlin, P. F., and Kay, S. A.(1998). Phytochromes and cryptochromes in the entrainment of the *Arabidopsis* circadian clock. *Science* **282**, 1488–1490.
30. Yanovsky, M. J., Izaguirre, M., Wagmaister, J. A., Gatz, C., Jackson, S. D., Thomas, B., and Casal, J. J. (2000). Phytochrome A resets the circadian clock and delays tuber formation under long days in potato. *Plant J.* **23**, 223–232.
31. Schmitz, O., Williams, S. B., Katayama, M., Kondo, T., and Golden, S. S. (2000). CikA, a bac-

teriophytochrome that resets the cyanobacterial circadian clock. *Science* **289**, 765–768.
32. Kaneko, T., Sato, S., Kotani, H., Tanaka, A., Asamizu, E., Nakamura, Y., Miyajima, N., Hirosawa, M., Sugiura, M., Sasamoto, S., Kimura, T., Hosouchi, T., Matsuno, A., Muraki, A., Nakazaki, N., Naruo, K., Okumura, S., Shimpo, S., Takeuchi, C., Wada, T., Watanabe, A., Yamada, M., Yasuda, M., and Tabata, S. (1996). Sequence analysis of the genome of the unicellular cyanobacterium *Synechocystis* sp. strain PCC6803. II. sequence determination of the entire genome and assignment of potential protein-coding regions. *DNA Res.* **3**, 109–136.
33. Mizuno, T., Kaneko, T., and Tabata, S. (1996). Compilation of all genes encoding bacterial two-component signal transducers in the genome of the cyanobacterium, *Synechocystis* sp. strain PCC6803. *DNA Res.* **3**, 407–414.
34. Suzuki, I., Los, D. A., Kanesaki, Y., Mikami, K., and Murata, N. (2000). The pathway for perception and transduction of low-temperature signals in Synechocystis. *EMBO J.* **19**, 1327–1334.
35. Aoki, S., Kondo T., and Ishiura, M. (1995). Circadian expression of the *dnaK* gene in the cyanobacterium *Synechocystis* sp. strain PCC 6803. *J. Bacteriol.* **177**, 5606–5611.
36. Margulis, L. (1993). "Symbiosis in Cell Evolution," 2nd Ed. Freeman, New York.
37. Hwang, S., Kawazoe, R., and Herrin, D. L. (1996). Transcription of *tufA* and other chloroplast-encoded genes is controlled by a circadian clock in *Chlamydomonas*. *Proc. Natl. Acad. Sci. USA* **93**, 996–1000.
38. Nakahira, Y., Baba, K., Yoneda, A., Shiina, T., and Toyoshima, Y. (1998). Circadian-regulated transcription of the *psbD* light-responsive promoter in wheat chloroplasts. *Plant Physiol.* **118**, 1079–1088.
39. Yeh, K. C., and Lagarias, J. C. (1998). Eukaryotic phytochromes: Light-regulated serine/threonine protein kinases with histidine kinase ancestry. *Proc. Natl. Sci. Acad. USA* **95**, 13976–13981
40. Fankhauser, C., Yeh, K. C., Lagarias, J. C., Zhang, H., Elich, T., and Chory, J. (1999). PKS1, a substrate phosphorulated by phytochrome that modulate light signaling in *Arabidopsis*. *Science* 284, 1539–1541.
41. Strayer, C. A., Oyama, T., Schultz, T. F., Raman, R., Somers, D. E., Mas, P., Panda, S., Kreps, J. A., and Kay, S. A. (2000). Cloning of the *Arabidopsis* clock gene TOC1, an autoregulatory response regulator homolog. *Science* **289**, 768–771.
42. Makino, S., Kiba, T., Imamura, A., Hanaki, N., Nakamura, A., Suzuki, T., Taniguchi, M., Ueguchi, C., Sugiyama, T., and Mizuno, T. (2000). Genes encoding pseudo-response regulators: Insight into His-to-Asp phosphorelay and circadian rhythm in *Arabidopsis thaliana*. *Plant Cell Physiol.* **41**, 791–803.
43. Matsushika, A., Makino, S., Kojima, M., and Mizuno, T. (2000). Circadian waves of expression of the APRR1/TOC1 family of pseudo-response regulators in *Arabidopsis thaliana*: Insight into the plant circadian clock. *Plant Cell Physiol.* **41**, 1002–1012.

CHAPTER 15

Two-Component Control of Quorum Sensing in Gram-Negative Bacteria

KENNY C. MOK AND BONNIE L. BASSLER
Department of Molecular Biology, Princeton University, Princeton, New Jersey 08544

INTRODUCTION

Bacteria living in communities convey their presence to each other by producing, releasing, and subsequently responding to the accumulation of a minimal threshold concentration of chemical signaling molecules termed autoinducers. This method of cell–cell communication is called quorum sensing and it allows a population of bacteria to coordinate behavior, and thus acquire some of the characteristics of multicellular organisms.

Quorum sensing was first identified in a luminous marine bacterium called *Vibrio fischeri* [1, 2]. In *V. fischeri*, quorum sensing controls the cell density-dependent production of light. *V. fischeri* lives in mutualistic associations with several marine animal hosts. In each of these specific symbioses, the animal uses the light produced by *V. fischeri* for tactics such as attracting prey, escaping from predators, or appealing to mates. The advantage gained by *V. fischeri* in producing light is that the bacterium obtains a nutrient-rich environment in which to exist [3, 4].

Engebrecht and Silverman discovered and characterized the regulatory components that govern density-dependent light production (Lux) in *V. fischeri* [5–7]. They showed that quorum sensing in *V. fischeri* is accomplished through regulation of the luciferase enzyme complex encoded by the *luxCDABE* operon. Two regulatory proteins comprise the *V. fischeri* quorum sensing control apparatus. The LuxI protein is the autoinducer synthase. LuxI produces the acylated homoserine lactone (HSL) signaling molecule *N*-(3-oxohexanoyl)-L-homoserine lactone, and the concentration of the HSL increases in proportion to increasing cell density [5, 8]. The second regulatory protein is called LuxR. LuxR is a transcriptional activator and the autoinducer-binding protein. Only when bound to the HSL autoinducer can LuxR bind to the *luxCDABE* promoter and activate transcription of the luciferase operon to allow light to be produced [5]. A positive feedback regulatory loop operates in the system because an increase in autoinducer concentration leads to an increase in expression of the *luxI* gene encoding the HSL synthase. The increased autoinducer, in turn, results in the increased transcription of luciferase. A negative feedback loop compensates for the positive feedback loop. Specifically, the LuxR–autoinducer complex represses transcription of *luxR*, which decreases the intracellular concentration of LuxR. This quorum sensing circuit enables light production to be tightly coupled to the bacterial population density [5, 6]. Quorum sensing enables *V. fischeri* to emit light only when it resides inside the specialized light organ of a host but not when it exists free-living in the ocean. Presumably, only in the nutrient-rich environment of a host light organ can *V. fischeri* grow to high population densities, and additionally, only in the light organ can the diffusible autoinducer molecule accumulate to a sufficient concentration that *V. fischeri* is able to detect it [3, 4].

For over 10 years the *V. fischeri* LuxI/LuxR quorum sensing system was considered an isolated example of bacterial communication that had presumably evolved for the explicit purpose of bacterial colonization of a symbiotic host. However, homologues of the LuxI/LuxR system of *V. fischeri* have now been found in over 50 other Gram-negative bacterial species, many of which are important clinical or agricultural pathogens. LuxI/LuxR-like quorum sensing systems control processes that necessitate the cooperation of a large number of bacteria in order to be effective, and individuals in the group benefit from the activity of the entire assembly. In every case, an HSL autoinducer and a cognate LuxR-like transcriptional activator regulate quorum sensing. Figure 1 shows a general model for a LuxI/LuxR-type quorum sensing circuit. Many LuxI/LuxR circuits contain additional quorum sensing regulatory components. For a review of LuxI/LuxR quorum sensing, see Miller and Bassler [9].

Processes controlled by LuxI/LuxR-like quorum sensing systems in Gram-negative bacteria include bioluminescence, virulence, antibiotic production, sporulation, and biofilm formation [9–12]. In every case, the specificity inherent in autoinducer detection by a cognate LuxR protein stems from a precise interaction between a given LuxR-type protein and the acyl side chain of each homoserine lactone. The acyl chains vary in length and in the location of attached functional groups, such as hydroxyl and carbonyl groups. Rather exquisite signaling specificity exists in these quorum sensing circuits,

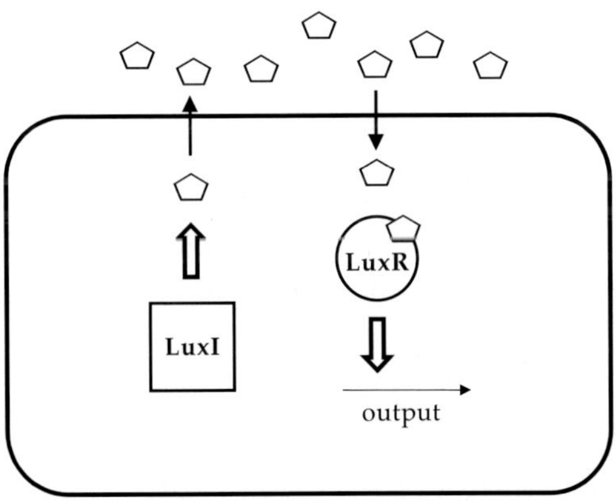

FIGURE 1 LuxI/LuxR-type quorum sensing in Gram-negative bacteria. In most Gram-negative bacteria, LuxI/LuxR-type regulatory circuits control quorum sensing. LuxI-type autoinducer synthases produce specific acylated homoserine lactone (HSL) signaling molecules termed autoinducers (pentagons). This type of molecule diffuses freely into and out of the cell [86]. Therefore, the concentration of HSL increases in proportion to increasing bacterial cell density. As the cells grow to high cell density, the autoinducer molecule reaches a critical threshold concentration and can be bound by a cognate LuxR-type transcriptional activator. The LuxR–autoinducer complex induces the expression of specific quorum sensing regulated genes (output).

which is quite remarkable given the structural similarity of the family of HSL signal molecules.

Gram-positive bacteria also use quorum sensing systems to communicate. Similar to Gram-negative bacteria, a variety of processes are under quorum sensing control in Gram-positive bacteria. For example, virulence, sporulation, and competence are a few of the processes regulated by quorum sensing in Gram-positive bacterial species. Gram-positive bacteria have developed a mechanism for communication that differs fundamentally from that of Gram-negative bacteria. Unlike Gram-negative bacteria that use HSL autoinducers and LuxI/LuxR-like regulatory circuits, Gram-positive bacteria communicate using secreted oligopeptides as autoinducers, and two-component signaling circuits are used for autoinducer detection and signal transduction [9, 13]. Two-component regulation of quorum sensing in Gram-positive bacteria is covered in Chapter 16.

In contrast to all other Gram-negative bacteria that regulate quorum sensing with LuxI/LuxR-type systems, *Vibrio harveyi*, a free-living bioluminescent marine bacterium, and *Myxococcus xanthus*, a sporulating soil bacterium, control quorum sensing with complex two-component signaling circuits. Homologues of *V. harveyi* quorum sensing regulatory proteins have been identified in many other marine *Vibrios* sp. These findings indicate that the two-component regulation of quorum sensing could be a common feature in Gram-negative bacteria. This present chapter focuses exclusively on the *V. harveyi* and *M. xanthus* two-component regulation of quorum sensing.

QUORUM SENSING IN *Vibrio harveyi*

V. harveyi is closely related to *V. fischeri*. However, unlike *V. fischeri*, *V. harveyi* is not known to exist in symbiotic associations with marine eukaryotes. Rather, *V. harveyi* is ubiquitous in the marine environment, residing in the water column, in shallow sediments, and on the surfaces and in the gut tracts of various marine animals [1]. *V. harveyi* is also a pathogen of black tiger prawns [14, 15]. As in *V. fischeri*, *V. harveyi* light production requires the luciferase operon *luxCDABE*. The ecological significance of light production in *V. harveyi* is not understood.

Unlike *V. fischeri* and other Gram-negative quorum sensing bacteria, *V. harveyi* does not employ a LuxI/LuxR-type system to regulate quorum sensing. *V. harveyi* controls density-dependent expression of luciferase genes using a hybrid quorum sensing system with components common to both Gram-negative and Gram-positive quorum sensing systems [9, 10]. Similar to other Gram-negative quorum sensing bacteria, *V. harveyi* employs an acyl-homoserine lactone autoinducer (called AI-1) as a signal molecule. In this case the autoinducer is *N*-(3-hydroxybutanoyl)-L-homoserine lactone [16, 17]. Similar to Gram-positive quorum sensing bacteria, *V. harveyi* employs a two-component signaling circuit for autoinducer detection and signal transduction [16, 18–23]. In addition, a second, novel autoinducer, called AI-2, is also involved in the regulation of quorum sensing in *V. harveyi* [16, 18, 24, 25]. AI-1 is proposed to be used for intraspecies cell–cell signaling, and AI-2 is hypothesized to be used for interspecies cell–cell communication [24–26]. *V. harveyi* exists in mixed species communities in the ocean, and the two different autoinducer signals are presumed to enable *V. harveyi* to vary gene expression in response to changes in total cell number and, additionally, in response to fluctuations in the species composition of the consortium.

A Multichannel Two-Component Circuit Regulates Quorum Sensing in *V. harveyi*

Genetic analysis in *V. harveyi* reveals that it utilizes two two-component signaling systems that operate in parallel to regulate the density-dependent expression of the *lux* operon [16, 18, 19]. The two circuits are called system 1 and system 2. System 1 is composed of AI-1 and its sensor, LuxN. As mentioned, AI-1 is *N*-(3-hydroxybutanoyl)-L-homoserine lactone [17]. The *luxLM* locus encodes the AI-1 synthase [16]. Although AI-1 is an HSL with structural similarity to other quorum sensing HSL signaling molecules, the LuxLM synthase does not resemble the LuxI type of autoinducer synthase. However, both the LuxLM and the LuxI synthases likely use an identical biochemical mechanism for autoinducer production [27, 28]. The AI-1 sensor, LuxN, is a hybrid two-component kinase containing both a histidine kinase domain and a response regulator domain [16]. The second *V. harveyi* quorum sensing circuit (system 2) is composed of AI-2 and its sensor, LuxPQ [18]. The AI-2 synthase is called LuxS [25]. AI-2 is derived from the ribosyl moiety of *S*-ribosyl homocysteine [28, 29]. The detector for AI-2 is composed of two proteins, LuxP and LuxQ. LuxP is a periplasmic protein that has homology to the ribose-binding proteins of *Escherichia coli and Salmonella typhimurium.* LuxP is proposed to be the primary sensor for AI-2, and the LuxP–AI-2 complex is hypothesized to interact with LuxQ. LuxQ, like LuxN, is a hybrid kinase containing both a sensor kinase and a response regulator domain [18]. A shared histidine phosphotransfer (HPt) protein called LuxU receives both system 1 and system 2 sensory inputs [21]. LuxU, in turn, transmits this information to the response regulator protein called LuxO, and LuxO regulates the expression of the *luxCDABE* operon [20, 21, 23]. A protein called LuxR is also required for the expression of *luxCDABE* in *V. harveyi*. The *V. harveyi* LuxR protein is a transcriptional activator, but shares no homology to the LuxR-type proteins found in other Gram-negative bacterial quorum sensing systems [30, 31]. The *V. harveyi* Lux quorum sensing circuit is shown in Fig. 2.

The LuxO Response Regulator

Data obtained in early *V. harveyi* mutant analyses indicated that the response regulator LuxO functioned in both autoinducer pathways [19]. Therefore,

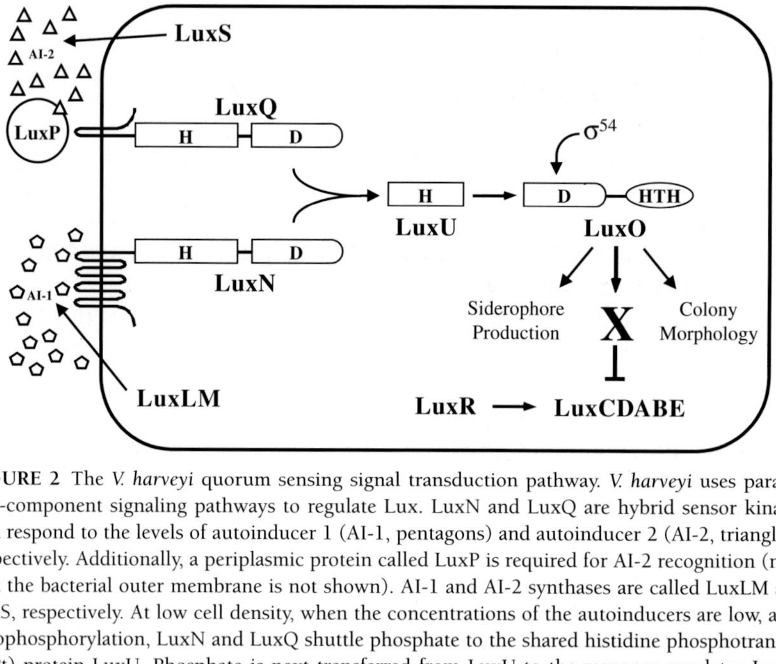

FIGURE 2 The *V. harveyi* quorum sensing signal transduction pathway. *V. harveyi* uses parallel two-component signaling pathways to regulate Lux. LuxN and LuxQ are hybrid sensor kinases that respond to the levels of autoinducer 1 (AI-1, pentagons) and autoinducer 2 (AI-2, triangles), respectively. Additionally, a periplasmic protein called LuxP is required for AI-2 recognition (note that the bacterial outer membrane is not shown). AI-1 and AI-2 synthases are called LuxLM and LuxS, respectively. At low cell density, when the concentrations of the autoinducers are low, after autophosphorylation, LuxN and LuxQ shuttle phosphate to the shared histidine phosphotransfer (HPt) protein LuxU. Phosphate is next transferred from LuxU to the response regulator LuxO. Phospho–LuxO acts to repress the expression of the luciferase structural operon (*luxCDABE*). However, this repression is indirect, as phospho-LuxO and σ^{54} activate the expression of a putative repressor protein called X. At high cell density, in the presence of autoinducers, LuxN and LuxQ convert from kinases to phosphatases. The phosphatase activities of the sensors lead to the dephosphorylation and inactivation of LuxO. This step also requires LuxU. Inactivation of LuxO terminates the expression of the gene encoding the repressor X, which relieves the repression of *luxCDABE*, resulting in light production. LuxR is a required transcriptional activator of the *luxCDABE* operon. Phospho-LuxO regulates other *V. harveyi* genes, including those involved in siderophore production and rugose colony morphology. This model comes from work reported elsewhere [16, 18–23, 25, 28, 30, 31]. H, D, and HTH denote histidine, aspartate, and helix-turn-helix, respectively.

LuxO was suggested to have a role in integrating sensory information from both system 1 and system 2. To investigate how the two autoinducer signals are integrated in *V. harveyi*, specific mutations in *luxO* were engineered and recombined onto the *V. harveyi* chromosome for *in vivo* phenotypic analysis. The Lux phenotypes of some of these mutant strains are shown in Fig. 3 [20].

Initially, a *luxO* null mutant was constructed and its phenotype compared to the wild-type parent. In this experiment, dense, overnight cultures of *V. harveyi* were diluted 1:5000 into fresh growth medium, and the light production of each culture was measured thereafter as a function of increasing cell density. Figure 3 shows that, in the wild-type strain, following dilution into fresh medium, light production declines rapidly over 1000-fold (squares). The decrease in luminescence happens in response to the decrease in autoinducer concentration that occurs upon dilution of the overnight culture. Specifically, autoinducers present in the culture medium following overnight growth are diluted out, and in response to the lack of autoinducer signal, *V. harveyi* terminates light production. Therefore, luminescence declines in the initial phase of the experiment. However, as the wild-type culture grows, endogenously produced AI-1 and AI-2 accumulate in the medium. At a critical concentration of autoinducer, which corresponds to a critical cell density, *V. harveyi* responds to the autoinducers, and an increase in light output of

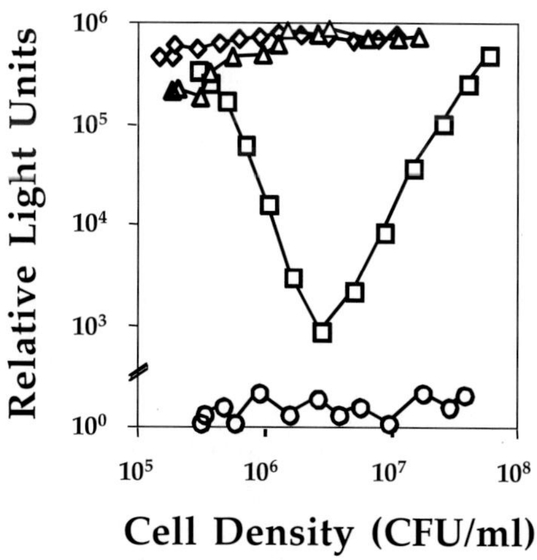

FIGURE 3 Phenotypes of *V. harveyi luxO* mutants. Density-dependent bioluminescence assays are shown for various *V. harveyi* strains. The strains and their symbols are as follows: wild-type (□), *luxO* (△), *luxO* D47A (◇), and *luxO* D47E (○). The assay is described in the text. Relative light units are defined as light emission per cell (i.e., counts min^{-1} ml^{-1} × 10^3/cfu ml^{-1}) [20].

1000-fold per cell occurs. At this point, the light output of the cells has again reached the predilution level. This is the characteristic response of the wild-type *V. harveyi* strain to the autoinducers. Figure 3 also shows that inactivation of *luxO* results in a constitutively bright *V. harveyi* strain (triangles), indicating that LuxO is required by both autoinducer systems for density-dependent *lux* expression. Furthermore, this result shows that the wild-type function of LuxO is to cause repression of the expression of luciferase at low cell density [19, 20].

To determine the mechanism of repression by LuxO, a series of *luxO* missense alleles were constructed and incorporated at the *luxO* locus in the *V. harveyi* chromosome. Among these alleles were ones in which the conserved aspartate residue (Asp-47), presumed to be the site of phosphorylation in LuxO, was mutated. Analysis of mutants with various alterations at Asp-47 was used to determine whether phospho-LuxO or dephospho-LuxO acts to repress Lux expression at low cell density. Specifically, the alteration of Asp-47 to Ala was used to lock the protein in a form mimicking dephospho-LuxO, whereas the mutation of Asp-47 to Glu was used to convert the LuxO protein into a state mimicking the phospho-LuxO form [20].

Figure 3 shows some representative density-dependent bioluminescence assays used to assess the activities of phospho- and dephospho-LuxO. Replacement of the *luxO* D47A mutation onto the chromosome of *V. harveyi* results in a strain that constitutively expresses luminescence (diamonds). The *V. harveyi luxO* D47A phenotype resembles the high cell density phenotype of wild-type *V. harveyi*. This result suggests that dephospho-LuxO is the form of LuxO that predominates at high cell density when no repression of Lux occurs. Therefore, dephospho-LuxO is inactive in Lux repression, indicating that phospho-LuxO is the active species and is responsible for the repression of light production at low cell density. This hypothesis was verified by demonstrating that the *V. harveyi luxO* D47E mutant, which locks LuxO into a form mimicking the phospho-LuxO state, exhibits a constitutively dark phenotype (Fig. 3, circles). Together these results show that, at low cell densities, LuxO is phosphorylated and active and Lux is repressed. However, at high cell density, in the presence of autoinducer, the inactive dephospho-LuxO protein exists in *V. harveyi*. Inactivation of LuxO allows the *lux* operon to be expressed, and light is produced by *V. harveyi* [20].

The aforementioned results demonstrate that, in *V. harveyi*, the transition from the low cell density, low light-producing state to the high cell density, maximal light-producing state requires inactivation (i.e., dephosphorylation) of LuxO. Furthermore, in wild-type *V. harveyi*, this transition requires autoinducers and their cognate sensors. Therefore, LuxN and LuxQ must act as LuxO phosphatases at high cell density in the presence of their autoinducing ligands [20].

LuxO Activates a Lux Repressor

LuxO is a homologue of the NtrC protein that is involved in regulating nitrogen assimilation in *E. coli* and other bacteria. Both LuxO and NtrC proteins contain response regulator domains, σ^{54} activation domains, and helix-turn-helix DNA-binding domains [23, 32, 33]. These domains are conserved in the NtrC family of proteins and are hallmarks of proteins that interact with σ^{54} to activate gene expression [34]. However, the initial LuxO studies indicated that LuxO was a repressor of *lux* expression. Proteins exist in the NtrC family that act as both activators and repressors. Two such cases are FlbD of *Caulobacter crescentus* and NtrC itself. Phospho-FlbD activates the expression of class III flagellar genes and also represses expression from the *fliF* operon [35–37]. Phospho-NtrC activates transcription of the *glnA* gene and also represses transcription at a minor σ^{70} promoter upstream of the major *glnA* σ^{54} promoter [38, 39]. In both cases, activation of gene expression is dependent on phosphorylation and on σ^{54}, whereas repression is independent of σ^{54}. Therefore, it was conceivable that LuxO could be either a direct repressor of Lux or an activator of a repressor of Lux. To distinguish between these two possibilities, the dependence of phospho-LuxO on σ^{54} was tested. A *V. harveyi* σ^{54} mutant strain was constructed, and its phenotype is presented in Fig. 4. Figure 4 shows that a null mutation in *rpoN* (encoding σ^{54}) results in a constitutively bright *V. harveyi* strain (circles). This phenotype is identical to the *luxO* null phenotype (triangles). This experiment confirmed that σ^{54} is necessary for the repression of luminescence expression in *V. harveyi*. The requirement of σ^{54} in the repression of Lux is consistent with a role for phospho-LuxO as an activator, not a repressor. Therefore, it is hypothesized that phospho-LuxO, together with σ^{54}, activates a downstream Lux repressor protein [23]. This putative repressor protein has not yet been identified, and it is denoted X in the *V. harveyi* quorum sensing model presented in Fig. 2. While we favor this hypothesis, it remains formally possible that at high concentrations of phospho-LuxO, the latter, together with σ^{54}, could act directly as a repressor of *luxCDABE*.

LuxN and LuxQ Hybrid Sensors

The LuxO studies just described demonstrated that the function of the LuxN and LuxQ sensors is to cause dephosphorylation of LuxO at high cell density. However, it was not clear what *V. harveyi* cellular component(s) was responsible for the phosphorylation of LuxO at low cell density. To examine whether LuxN and LuxQ played a role in the conversion of LuxO from the inactive form to the phosphorylated, active form, *luxN* null, *luxQ* null, and double

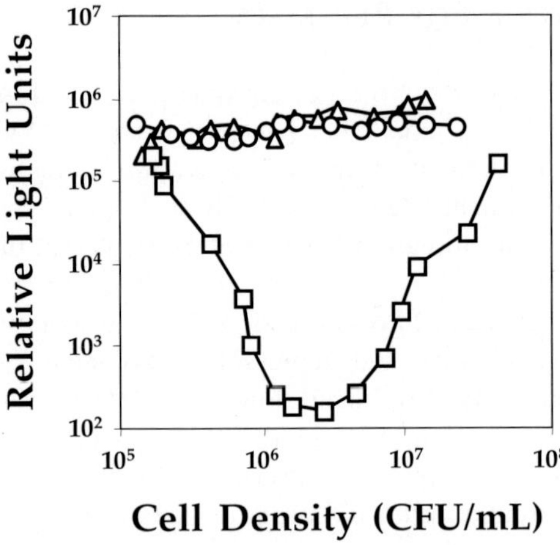

FIGURE 4 LuxO requires σ^{54} to regulate quorum sensing in *V. harveyi*. Results of density-dependent bioluminescence assays are presented for *V. harveyi* wild-type (□), *luxO* (△), and *rpoN* strains (○) (23).

luxN, *luxQ* null mutant *V. harveyi* strains were constructed. Subsequently, the density-dependent light production in these strains was measured and compared to that of the wild-type *V. harveyi* strain [22]. Figure 5 shows the results of this study.

Figure 5 shows the characteristic wild-type *V. harveyi* density-dependent Lux profile (squares). Additionally, both *V. harveyi luxN* and *luxQ* single mutants maintain density-dependent expression of luminescence (diamonds and circles, respectively). This result shows that the two sensory systems are redundant, and operation of either system alone is sufficient for quorum sensing in *V. harveyi*. However, the *luxN*, *luxQ* double mutant strain does not display density-dependent regulation of *lux* expression (triangles). Rather, this strain expresses bioluminescence constitutively. The phenotype of the $LuxN^-$, $LuxQ^-$ strain is identical to a *luxO* null mutant. First, this result indicates that functional LuxN and LuxQ sensors are required for the low cell density repression of Lux. Second, this phenotype shows that in the absence of LuxN and LuxQ, no active (phosphorylated) LuxO exists in *V. harveyi*, and Lux is not repressed. Therefore, LuxN and LuxQ are inferred to have kinase activities that are required for the phosphorylation of LuxO at low cell density [22]. Together with the aforementioned results suggesting that LuxN and LuxQ are LuxO phosphatases, these findings imply that the autoinducer sensors LuxN and LuxQ act like switches to control the activity of LuxO in a density-dependent manner. At low cell density, when the cognate autoinducers are at low levels, the LuxN and LuxQ sensors act as autophosphory-

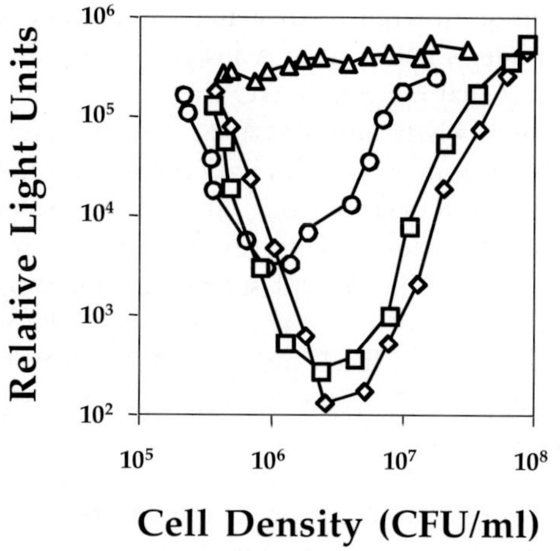

FIGURE 5 LuxN and LuxQ have different effects on *V. harveyi* quorum sensing. Density-dependent bioluminescence phenotypes of *V. harveyi* wild-type (□), *luxN* (◇), *luxQ* (○), and *luxN*, *luxQ* (△) strains are shown [20].

lating kinases. Ultimately, phosphate is transferred to the conserved aspartate of the response regulator protein LuxO. Phospho LuxO shuts off the expression of luminescence. In contrast, at high cell density, when the autoinducers are at high levels, the sensors change from being kinases to phosphatases. In this mode, LuxN and LuxQ promote the dephosphorylation of LuxO. Dephosphorylation of LuxO inactivates it and leads to light production [20, 22]. Both phosphorylation and dephosphorylation of LuxO were shown to be dependent on the HPt protein LuxU (Fig. 2) [21, 22].

Signal Integration in *V. harveyi*

The phenotypes of the *luxN* and *luxQ* single mutants presented in Fig. 5 demonstrate that although the activities of LuxN and LuxQ are redundant, differences in the effectiveness of the two circuits exist. Compared to wild-type *V. harveyi* (squares), the $LuxN^+$, $LuxQ^-$ strain shows a much more shallow dip in the curve and begins to increase its light production at a lower cell density (circles). In contrast, the $LuxN^-$, $LuxQ^+$ mutant has a slightly deeper dip in the curve and begins to induce light production at a higher cell density than the wild-type strain (diamonds). The switch from repression of Lux to derepression occurs much earlier in the $LuxN^+$, $LuxQ^-$ strain than in the $LuxN^-$, $LuxQ^+$ strain, implying that LuxN converts from the kinase mode to the phosphatase mode at a lower cell density than LuxQ. Therefore, the critical AI-1 concentration required to switch LuxN from the kinase mode to the phosphatase mode is achieved at a lower cell density compared to the concentration of AI-2 that converts LuxQ from a kinase to a phosphatase. In the wild-type situation, when both sensors are present, LuxN is a LuxO phosphatase for a period of time in which LuxQ remains a LuxO kinase. Because the wild-type *V. harveyi* quorum sensing circuit receives sensory inputs from both LuxN and LuxQ sensor kinases, the wild-type response to the autoinducers is due to the combined, and sometimes opposing, activities of the two sensors. Figure 5 bears this out as it shows that the wild-type response curve (squares) declines to a level intermediate between that of the LuxN$^+$, LuxQ$^-$ and the LuxN$^-$, LuxQ$^+$ strains. Finally, in the Lux signaling circuit, the interplay between the relative kinase and the phosphatase levels of the two sensors dictates the exact level of phosphorylation of LuxO and allows precise modulation of the *lux* structural operon. It appears that the AI-1/LuxN system has a greater influence on light production than the AI-2/LuxPQ system [20].

LuxN Sensor Protein

The LuxN protein was studied extensively to probe the requirements for kinase and phosphatase activity [22]. This hybrid sensor kinase protein contains a conserved His at residue 471 in the sensor kinase domain and a conserved Asp at residue 771 in the response regulator domain that are predicted to be involved in the autophosphorylation, phosphorelay, and phosphatase activities of the protein. His-471 was mutagenized to a number of different residues in *V. harveyi*, and the effect of mutation of the invariant His residue was examined using density-dependent Lux assays. Figure 6A shows the phenotype of the wild-type strain (squares), a *luxN* null mutant (diamonds), and the *V. harveyi luxN* H471Q mutant strain as a representative for these experiments (circles). The LuxN H471Q mutation renders *V. harveyi* constitutively bright (circles). This phenotype indicates that, in the absence of His 471, LuxN has no kinase activity because *V. harveyi* is locked into the high cell density mode. In this strain, LuxQ is wild-type, and at low cell density, LuxQ is in kinase mode and transfers the phosphate (*via* the HPt protein LuxU) to LuxO. LuxQ kinase activity should result in the repression of Lux expression. However, Fig. 6 shows that LuxQ causes only a modest repression of luminescence in the early phase of the experiment. Because the phenotype of the LuxQ$^+$, LuxN H471Q mutant is nearly constitutively bright, it shows that LuxN phosphatase activity remains in the mutant protein and that the LuxN phosphatase activity is dominant to the LuxQ kinase activity [22].

Mutagenesis of the LuxN Asp-771 to Ala was used to determine the role of this critical Asp residue in each of the activities of LuxN. Figure 6B shows

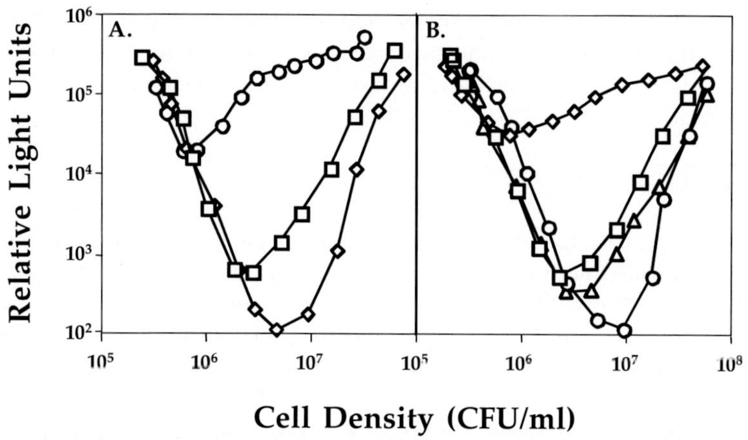

FIGURE 6 Analysis of kinase and phosphatase activities of the *V. harveyi* LuxN protein. (A) The histidine residue (H471) is required for kinase but not phosphatase activity in *V. harveyi*. Density-dependent bioluminescence was measured for the *V. harveyi luxN* H471Q strain (○) and compared to that of the wild-type (□) and a *luxN* null (◇) strain. (B) The aspartate residue (D771) is necessary for LuxN kinase and phosphatase activity. Bioluminescence was assayed in *V. harveyi* wild-type and *V. harveyi luxN* D771A mutant strains as a function of cell density in the presence and absence of exogenously supplied AI-1. Symbols denote phenotypes for wild-type (□), wild-type + AI-1 (◇), *luxN* D771A (○), and *luxN* D771A + AI-1 (△) [22].

that the LuxN D771A mutant displays a phenotype similar to that of a LuxN null mutant. Specifically, the Lux curve for the LuxN D771A mutant (circles) declines to a deeper level than the wild-type (squares), as was shown for the *LuxN* null mutant in Fig. 5. The response to AI-1 of the D771A mutant is also shown in Fig. 6B. Unlike the wild-type strain, in which light production is stimulated upon the exogenous addition of AI-1 (diamonds), the LuxN D771A mutant is incapable of a response to AI-1 (triangles). The phenotype of the LuxN D771A mutant suggests that the Asp-771 residue is required for both LuxN kinase and phosphatase activities [22].

To further characterize LuxN phosphatase activity, a *V. harveyi* strain was constructed that was wild-type for LuxQ and expressed only the response regulator domain of LuxN. This strain has a constitutively bright phenotype; a phenotype identical to that conferred by the full-length LuxN protein carrying the LuxN H471Q mutation. This result shows that the response regulator module of LuxN possesses a dominant phosphatase activity, indicating that the amino acids required for phosphatase activity are localized exclusively to this domain of LuxN (not shown). Similar to the description for Fig. 6A, a constitutively bright phenotype indicates that LuxN phosphatase activity in this construction retains its dominance over LuxQ kinase activity. Expression of the LuxN response regulator construction containing the D771A alteration resulted in a *luxN* null phenotype (not shown). Therefore, Asp-771 is absolutely required for both LuxN kinase and LuxN phosphatase activities [22].

LuxQ Sensor Protein

The LuxQ protein has not been examined in an in-depth manner similar to that performed with the LuxN protein. However, from the analysis of the phenotypes of the wild-type and the *luxQ* null mutant, LuxQ is known to possess both kinase and phosphatase activity [20]. Furthermore, at least with respect to Lux expression, the AI-2/LuxPQ system is recessive to the activities of the AI-1/LuxN system. An examination of the phenotypes of *V. harveyi* mutants containing missense mutations in *luxQ*, analogous to those described earlier for *luxN* is required in order to define the modular domain organization of the LuxQ enzymatic functions. Additionally, because LuxP is also involved in AI-2 signal transduction, mutations in LuxQ (and LuxP) that promote or eliminate LuxP-LuxQ interactions, as well as mutations that facilitate or interfere with AI-2 detection, will be required to begin to understand the initial events in AI-2 signal recognition and relay.

Quorum Sensing Signal Transduction in *V. harveyi*

As mentioned, Fig. 2 shows the current model for quorum sensing regulation in *V. harveyi*. The circuit functions as follows. At low cell densities, when AI-1 and AI-2 are absent, LuxN and LuxQ are kinases. Under these conditions, the sensors autophosphorylate on their conserved histidine residues. The phosphoryl group is next transferred intramolecularly to the conserved aspartate in the attached response regulator modules. Subsequently, intermolecular phosphotransfer occurs to the conserved histidine in the histidine phosphotransfer protein LuxU. Finally, the phosphate is transferred to the conserved aspartate of the response regulator protein LuxO. Phospho-LuxO, in conjunction with σ^{54}, shuts off the expression of luminescence by activating the transcription of a putative repressor protein (denoted X). In contrast, at high cell density, when the autoinducers are at high levels, LuxN and LuxQ switch their activities and change from being kinases to being phosphatases. In this mode, the sensors drain phosphate out of the system. The phosphatase activities of the sensors result in rapid elimination of phospho-LuxO, and dephospho-LuxO is inactive. It is hypothesized that the phosphate flows from LuxO to LuxU to the critical Asp on the sensors, and this is the site of phosphotransfer and phosphate hydrolysis. The putative repressor protein X is not transcribed, the activator protein LuxR functions to promote the transcription of *luxCDABE*, and the bacteria emit light [9, 20–23].

In addition to light production, LuxO and σ^{54} regulate other cellular processes in *V. harveyi*. These results suggest that a complex quorum sensing regulon exists in *V. harveyi*. Specifically, phospho-LuxO, in conjunction with σ^{54}, was shown to activate siderophore production and to control the transition from a smooth to a rugose colony morphology in *V. harveyi* (Fig. 2) [23]. Preliminary evidence suggests that additional targets exist in the *V. harveyi* quorum sensing regulon, some of which are activated and some of which are repressed. Taken together, all these studies show that the *V. harveyi* quorum sensing circuit is designed to facilitate both positive and negative regulation of genes in response to fluctuations in population density.

Intra- and Interspecies Quorum Sensing in *V. harveyi*

As mentioned, it is hypothesized that *V. harveyi* uses AI-1 for intraspecies cell–cell communication and AI-2 for interspecies cell–cell signaling [24–26, 40]. Presumably, *V. harveyi* responds to changes in total cell number and also to changes in the species of bacteria that make up the community. Consistent with a role for AI-2 as a universal signal used for bacterial interspecies communication, over 30 species of Gram-negative and Gram-positive bacteria have now been shown to produce AI-2 [9, 10, 25]. In every case, an AI-2 synthase that is highly homologous to *V. harveyi* LuxS AI-2 synthase is required for signal production [25, 28, 29].

The biosynthetic pathway for AI-2 synthesis has been reported [28, 29]. AI-2 is produced from *S*-ribosylhomocysteine (SRH), a product in the *S*-adenosylmethionine (SAM) consumption pathway. LuxS hydrolyzes SRH to produce homocysteine and AI-2. AI-2 is proposed to have structural similarity to ribose [28]. As opposed to the variable nature of acyl-homoserine lactone and oligo-peptide autoinducers, the structures of AI-2 molecules from various species of bacteria appear to be identical. If AI-2 is employed for interspecies bacterial communication, a common signal structure could be useful in order for AI-2 to be recognized by some or all of the members of a multispecies consortium [29].

Again, as mentioned, the two *V. harveyi* signaling circuits are not precisely redundant because the activities of LuxN are dominant to those of LuxQ. The difference in the strengths of the two inputs into the circuit could allow *V. harveyi* to fine-tune the expression of specific quorum sensing controlled genes for periods when it comprises a majority versus periods when it makes up only a minority of a given mixed species community. Although this may not seem useful in the regulation of light production, it could be critical for other quorum sensing regulated targets.

Finally, this complex quorum sensing circuit possesses a number of features that could grant *V. harveyi* the flexibility to adapt rapidly and precisely to its changing ocean environment. The two sensors, LuxN and LuxQ, allow the input of species-specific and species-nonspecific signals. The HPt protein LuxU enables the integration of various pieces of environmental information. LuxO, together with X, allows the precise regulation of multiple target genes, and each gene can be controlled either positively or negatively. Finally, if regulators other than LuxO exist downstream of LuxN and LuxQ, additional opportunities remain to build complexity into the signal transduction pathway.

V. harveyi-like Quorum Sensing Systems

Genes that are regulated by AI-2 and the mechanism of AI-2 detection in other *luxS*-containing bacteria remain unknown for the most part. A few preliminary reports show that virulence is regulated by *luxS* in some bacteria [41–45]. Given the widespread distribution of LuxS, AI-2, and the conserved AI-2 biosynthetic pathway, it is very possible that other two-component systems are involved in regulating Gram-negative quorum sensing systems. Perhaps numerous species of bacteria have a quorum sensing architecture resembling that of *V. harveyi* in that they make a species-specific signal and AI-2 is used as a species-nonspecific signal. If so, the *V. harveyi* two-component quorum sensing circuit may not be an isolated exception to the LuxI/LuxR rule, but may well be a common method that Gram-negative bacteria have evolved to regulate cell density-dependent gene expression.

Homologues of many of the components of the *V. harveyi* circuit have been identified in other bacteria. For example, the *V. harveyi* LuxR transcriptional activator has been identified in several other marine *Vibrio* sp. [46–48]. *Vibrio cholerae* has been shown to contain a *luxS* homologue and to produce AI-2. Completion of the sequencing of the *V. cholerae* genome shows that the entire *V. harveyi* system 2 (*luxS*, *luxP*, *luxQ*, *luxU*, *luxO*, and *luxR*) exists in this bacterium. However, no obvious candidates for *V. harveyi*-like system 1 components exist in *V. cholerae*, and the targets of the putative *V. cholerae* system 2 circuit remain to be identified. In *V. fischeri*, a gene called *ainS* is similar to the *luxM* gene of *V. harveyi* [49]. The AinS protein has been shown to direct the synthesis of a second acyl-homoserine lactone autoinducer in *V. fischeri*. The primary role of this HSL in *V. fischeri* is unknown, as this autoinducer has only a minor influence on the regulation of bioluminescence. In *V. fischeri*, downstream of *ainS* is the *ainR* gene. AinR shows homology to LuxN in *V.*

harveyi. AinR is likely the sensor for the AinS-dependent autoinducer. However, the quorum sensing target genes that this sensor–autoinducer pair regulates in *V. fischeri* remain unknown [49]. Homologues of *luxO* and *luxU* have also been identified in *V. fischeri* [50]. The role of LuxO in *V. fischeri* luminescence regulation was examined by constructing a *luxO* null mutant strain. Inactivation of *V. fischeri luxO* resulted in a phenotype similar to that of a *V. harveyi luxO* mutant. Specifically, light production in the *V. fischeri luxO* mutant was largely independent of cell density. In addition, the *V. harveyi luxO* gene could complement the *V. fischeri luxO* mutant [50]. The *V. fischeri luxO* null mutant remains responsive to stimulation by the LuxI-dependent HSL. This result indicates that, in *V. fischeri*, the LuxI/LuxR circuit and the *V. harveyi*-like quorum sensing circuit function independently [50]. A *luxQ* homologue was identified in *V. fischeri* that is required for proper colonization of a squid-host light organ [51]. Finally, *V. anguillarum*, a fish pathogen, appears to possess both *V. harveyi* circuits as well as a *V. fischeri*-like LuxI/LuxR system [52, 53]. The emergence of *V. harveyi*-like quorum sensing components in several marine *Vibrio* sp. suggests that, at least in *Vibrionaciae*, two-component control of quorum sensing could be a conserved trait. This could well be the case in other Gram-negative bacteria.

QUORUM SENSING IN *Myxococcus xanthus*

The Gram-negative bacterium *M. xanthus* is a soil dwelling organism with a complex life cycle and an interesting mode of motility. The bacteria glide over surfaces, traveling and hunting for food in swarms, and releasing antibiotics that kill other species of bacteria. *M. xanthus* subsequently scavenges the nutrients released by the dead cells. This complex social behavior allows the individual *M. xanthus* cells to profit from the hydrolytic enzymes secreted by other cells in the swarm [54]. In the presence of nutrients, *M. xanthus* undergoes vegetative growth. However, as nutrients become scarce, *M. xanthus* initiates a developmental program that results in the formation of fruiting bodies containing myxospores. The myxospores are dispersed and remain capable of survival for long periods of time in the absence of nutrients.

To begin the formation of fruiting bodies, high cell density, a lack of nutrients, and a solid surface are required. Under these conditions, bacteria swarm together and aggregate into mounds. These mounds of cells develop into fruiting bodies, and within these structures, vegetative bacteria are converted from rod-shaped cells into spherically shaped, heat- and desiccation-resistant myxospores. This developmental process is partially dependent on quorum sensing because the early events in fruiting body formation require high cell density and cell-to-cell signaling [54–56].

Density Sensing and the A Signal

At least five extracellular signals are produced by *M. xanthus* during the process of myxospore formation. These signals are named A, B, C, D, and E signals [57, 58]. *M. xanthus* uses the A signal, which is produced early in the developmental time line, as the quorum sensing signal to measure cell population density. This section focuses exclusively on A signaling. For a more complete review of *M. xanthus* extracellular signaling, see Shimkets [59]. The A signal functions during the developmental stage prior to aggregation [60]. The A signal is not the only signal required at this time, because, as mentioned, *M. xanthus* cells must also be starving. Considerable evidence suggests that the starvation signal is guanosine tetra- or pentaphosphate ([p]ppGpp) [61, 62].

Biochemical analysis of the A signal shows that it is composed of two distinct activities: one that is heat labile and one that is heat stable [63]. The heat-labile activity consists of at least two proteases that possess different substrate specificities. The heat-stable activity was determined to be a mixture of amino acids and peptides. Fifteen amino acids and a variety of small peptides were shown to have A signal activity. This mixture of amino acids is produced as a consequence of the activity of the secreted proteases [64]. The mixture of amino acids and peptides is now considered to be the primary A signal, and proteases are responsible for the production of the signal.

M. xanthus A signal activity has been studied extensively using an A signal-dependent Tn*5lac* reporter insertion in the *4521* gene [65–67]. The *4521* gene is of unknown function and is expressed early in the *M. xanthus* developmental cycle. Expression of the *4521* gene requires the A signal [60]. In wild-type *M. xanthus*, the *4521* gene is only expressed at high cell densities [68]. However, under starvation conditions, the addition of amino acids to *M. xanthus* at low cell density results in stimulation of the expression of *4521*. Furthermore, in an A signal mutant strain that is only capable of producing very low levels of the A signal, increasing the cell density 10- to 20-fold over that normally required for development restores *4521* gene expression. Finally, the extracellular concentration of the amino acids that compose the A signal is directly proportional to the cell density [68]. All of these findings led to the idea that the A signal is a quorum sensing signal. Use of the A signal amino acid mixture as the "autoinducer" for assessing cell density is unique among quorum sensing systems. In all other known Gram-negative and Gram-positive quorum sensing systems, the signaling molecules are each a single, unique compound. Apparently, some aspect(s) of the intricate social lifestyle of *M. xanthus* necessitates the use of a mixture of molecules for the signaling process.

AsgA, AsgB, and AsgC Control A Signaling

Four of the five different extracellular signals produced by *M. xanthus* were identified in experiments in which different *M. xanthus* developmental mutants were mixed together and screened for the restoration of normal sporulation [57, 69]. This analysis allowed the various *M. xanthus* mutants to be classified according to the signal they were missing. Four complementation groups were established, named A, B, C, and D. The E signal was identified in later experiments [58]. Approximately one-third of the mutants belonged to group A, and they were defective in A signaling. Examination of the A signaling mutants revealed three unlinked loci that were subsequently named *asgA*, *asgB*, and *asgC* [70].

The *asgA* gene encodes a two-component hybrid kinase protein [66]. AsgA possesses both a sensor kinase domain and a response regulator domain. Interestingly, AsgA contains the response regulator domain N-terminal to the histidine kinase domain. Additionally, AsgA does not contain any predicted transmembrane regions as are common to two-component sensors [71]. The AsgA protein has been shown to be capable of autophosphorylation, and it is likely that the conserved His residue is the site of phosphorylation [71]. AsgA is currently hypothesized to function in a phosphorelay signaling cascade that is required for A signal production in response to nutrient deprivation.

AsgB is hypothesized to be a transcription factor that may recognize specific upstream sequences in the promoters of AsgB-dependent target genes. Evidence for this supposition comes from identification of a predicted helix-turn-helix (HTH) DNA-binding domain near the C-terminus of AsgB [72]. This putative HTH domain is highly similar to the HTH domains present in other bacterial σ factors that contact the -35 regions of bacterial promoters [73–77]. However, AsgB does not contain any motifs predicted to be involved in interacting with RNA polymerase, suggesting that AsgB does not function as a σ factor. AsgB-dependent genes are predicted to encode proteins involved in early developmental steps, such as A signaling.

M. xanthus asgC mutations map to a gene called *sigA* [78]. The *sigA* gene encodes the major σ factor in *M. xanthus*. SigA is a homologue of the *E. coli* RpoD (i.e., σ^{70}) protein [78, 79]. The link between *asgC* and A signaling could be that SigA is required for the expression of genes encoding proteins necessary for A signal production. Possibly, SigA interacts with specific transcriptional regulators, such as AsgB, to activate expression of the genes required for production of the A signal. Developmental genes activated by this signaling pathway presumably include those encoding the proteases required to generate the A signal or regulators required for the expression of these proteases. Proteases, in turn, are hypothesized to act on cell wall proteins to produce the mixture of amino acids that constitutes the A signal.

A model showing the *M. xanthus* two-component regulated quorum sensing system is presented in Fig. 7. An unknown protein that is the detector for the starvation signal is hypothesized to function upstream of AsgA in the phosphorelay cascade. This putative sensor is proposed to be activated upon nutrient limitation. The starvation sensor signals to AsgA by an unknown mechanism, and AsgA autophosphorylates. Phospho-AsgA subsequently transfers this phosphoryl group to an unknown downstream protein or proteins (i.e., to a cognate phosphotransferase and response regulator). Ultimately, modulation of the activity of AsgB leads to the expression of genes required for A signal production. Transcription of these target genes could additionally require the major σ factor SigA (σ^{70} in Fig. 7). If so, each of the proteins originally identified in the A signaling mutants would have a required role in the same signal transduction pathway.

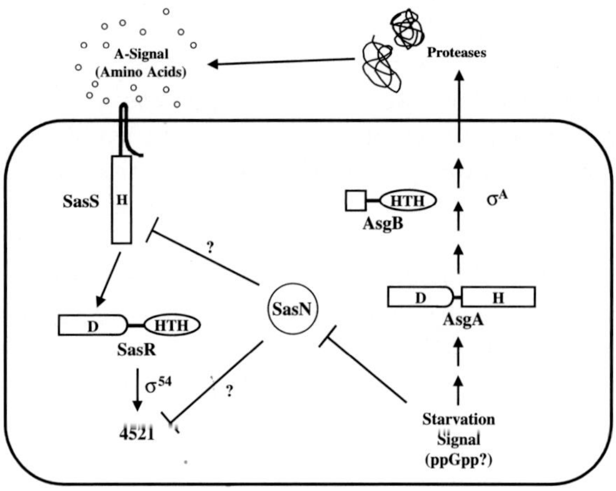

FIGURE 7 The *M. xanthus* quorum sensing circuit. The *M. xanthus* quorum sensing signal is called the A signal, and is a mixture of amino acids. Generation of the A signal requires the action of extracellular proteases that most likely degrade bacterial cell wall components. Synthesis of these proteases is dependent on AsgA, a hybrid two-component sensor kinase, AsgB, a proposed transcriptional regulator, and AsgC (SigA), the major σ factor of *M. xanthus* (designated σ^A). Under starvation conditions, the AsgA, B, and C sensory circuit leads to the expression and secretion of the extracellular proteases and the subsequent formation of the A signal. Multiple arrows are drawn between components of the Asg system because the exact number of regulatory steps in this process is not known. Response to the A signal requires the two-component SasS/SasR, sensor kinase/response regulator pair. In concert with σ^{54}, SasR induces the expression of A signal-dependent genes, including the *4521* gene. SasN is a negative regulator of *4521*. Exactly where SasN acts in the circuit remains unknown. SasN activity is inhibited under starvation conditions. Information for this model comes primarily from Plamann and Kaplan [83].

AsgD is a Novel Member of the A Signaling Pathway

An additional protein involved in A signaling has been identified [80]. A newly isolated *M. xanthus* developmental mutant was classified as an *asg* mutant because it could be complemented to sporulate in the presence of *bsgA*, *csgA*, *dsgA*, and *esgA* mutants (i.e., mutants defective in the B, C, D, or E signal), but not by *asgA*, *asgB*, or *asgC* mutant strains. Furthermore, sporulation in this mutant could be restored following the addition of several amino acids present in the A signal. The mutation in this *M. xanthus* strain was mapped to a previously unknown gene and was subsequently named *asgD*. Sequence analysis of *asgD* shows that, similar to AsgA, the AsgD protein contains regions homologous to both two-component histidine kinases and response regulators. Again, similar to AsgA, AsgD is organized with the response regulator domain at the N-terminus and the histidine kinase domain in the C-terminus. However, unlike AsgA, AsgD contains an ~380 residue intermediate region of unknown function that separates the response regulator domain from the histidine kinase domain.

Although *asgD* was classified as an A signaling mutant, this mutant is distinct from other *asg* mutants. As mentioned earlier, the A signal is required for the expression of the *4521* developmental reporter fusion. Therefore, transcription of *4521* is abolished in *asgA*, *asgB*, and *asgC* mutants. However, the expression of *4521* is not affected in the *asgD* mutant, suggesting that the level of A signal produced by this strain is wild-type. Furthermore, although some of the amino acids present in the A signal complement the *asgD* developmental defect, other A signal amino acids inhibit the development of the *asgD* mutant. Another difference between the *asgD* mutant and the other *asg* mutants is in their developmental response to nutrient levels. Specifically, *asgD* mutants require more stringent starvation conditions than wild-type to initiate sporulation. AsgD is therefore hypothesized to reduce the threshold for *M. xanthus* to perceive the starvation signal. Thus, the AsgD protein is inferred to have a role in sensing conditions of starvation [80].

The Sas Two-Component System Controls A Signaling

In order to identify regulators of quorum sensing, UV mutagenesis was carried out in the A signal-dependent *4521-lacZ* reporter fusion strain [65]. Under starvation conditions, when the A signal is supplied, the mutant expresses the reporter. To identify regulators of *4521*, suppressors were isolated that bypassed the requirement for both the A signal and the starvation signal, as well as suppressors that only bypassed the A signal requirement. The bypass suppressors of the first class, those that bypassed both starvation and the need for the A signal, were localized to one region of the chromosome, which was called the *sasB* locus. The *sasB* locus encodes two regulators of *4521*: *sasN* and *sasS* [81, 82].

SasN Regulator

SasN shows no homology to any known protein. However, it does contain a hydrophobic region and a leucine zipper motif in its N-terminus. SasN is a negative regulator of *4521* because null mutations generated in *sasN* result in *4521* expression in the absence of starvation and A signaling. SasN likely regulates additional genes because the *sasN* null mutant forms defective fruiting bodies and sporulates at much lower levels than wild-type [81].

SasS Sensor Kinase

The *sasS* gene encodes a protein with homology to two-component sensor kinases. SasS is predicted to contain two transmembrane domains and a histidine kinase domain. The suppressor mutation isolated in *sasS* that activated the expression of *4521* was a gain-of-function allele because a null mutation in *sasS* eliminates the expression of *4521* [82]. The identification of SasS as a two-component sensor kinase required for the regulation of A signal dependent genes indicates that SasS could be the A signal sensor. Interestingly, the two transmembrane regions and the N-terminal predicted input domain of SasS show limited homology to the corresponding domains in *E. coli* chemotaxis receptors (MCPs) [83]. Many of the chemotaxis receptors are known to interact directly with amino acids to convey the chemotaxis signal. Given that amino acids are the main component of the A signal, it is hypothesized that they interact with SasS directly.

SasR Response Regulator

Identification of mutations resulting in the second class of *4521* suppressors, those that only bypassed the A signal requirement, indicates a role for lipopolysaccharide (LPS) O-antigen biosynthesis in quorum sensing because LPS biosynthesis was disrupted in this class of suppressor mutants [65]. The mechanism by which a defect in O-antigen biosynthesis activates the expression of *4521* is unknown. These suppressor mutants proved useful because one was used successfully in a subsequent insertion mutagenesis screen to identify additional positive regulators of *4521* [84]. In this experiment, among other things, another gene within the *sasB* locus was identified. This gene, called *sasR*, is located ~3.5 kb downstream of *sasS* and directly upstream of *sasN*. SasR is an NtrC-like response regulator protein [84]. Epistasis analysis demonstrated that SasR acts in the signaling circuit with SasS and, as expected, SasR functions downstream of SasS in the signaling cascade. Apparently, the SasS sensor kinase and the SasR response regulator form a cognate two-component pair that regulates the expression of the *4521* gene. SasR is predicted to directly control the transcription of *4521*. Evidence for this assertion comes from the fact that SasR is an NtrC homologue. NtrC, as mentioned earlier, is a σ^{54}-dependent transcriptional activator protein. Specifically, phospho-NtrC oligomerizes on DNA far upstream of the target promoter. DNA between the NtrC complex and the target promoter is looped out, allowing phospho-NtrC to interact with σ^{54}, which is situated at the target promoter. Subsequent ATP hydrolysis by NtrC leads to open complex formation and transcription of the target gene [34]. Previous work on the promoter of *4521* identified it as a σ^{54}-dependent promoter [85]. Additionally, DNA sequences far upstream from the promoter (125 to 146 bp upstream of the transcriptional start site) are required for expression of *4521* [85].

The current model for *4521* regulation places SasS as the A signal sensor. When the quorum sensing A signal reaches a critical concentration, SasS autophosphorylates and transfers the phosphoryl group to SasR. Phospho-SasR interacts with σ^{54} to activate the expression of downstream genes, one of which is the *4521* gene (Fig. 7). The SasN negative regulator could be responsible for integrating the starvation information into the signaling system. It is not known where SasN exerts its negative effect in the SasS/SasR quorum sensing pathway. Presumably, depletion of nutrients leads to inactivation of the SasN protein, allowing SasS/SasR signaling to occur in the presence of a sufficient level of the density-dependent A signal.

It should be emphasized that Fig. 7 and the discussion presented in this section describe the few known components of the quorum sensing signaling pathway in *M. xanthus*. Many components of this quorum sensing circuit remain to be identified. In addition, besides the quorum sensing signal, at least four other extracellular signals and the sensory apparatuses required for their detection and response are also necessary for development in *M. xanthus*. Therefore, cell–cell signaling in *M. xanthus* is a much more complicated process than shown in Fig. 7.

CONCLUSIONS

We now understand that most bacteria communicate using extracellular signaling molecules to coordinately control the behavior of the group. We also understand that a considerable array of molecules are employed as signals, that individual species of bacteria have the capability to simultaneously produce, detect, and respond to multiple types of chemical signals, and that the signal detection systems are highly variable and are exquisitely adapted for optimized cell-to-cell communication in specialized niches. The *V. harveyi* and *M. xanthus* systems covered in this chapter are two of the most complex quorum sensing circuits known today. These two systems serve to exemplify the diversity of signals and the range of activities that are regulated by cell–cell communication in bacteria. Additionally, the *V. harveyi* and *M. xanthus* multichannel two-component circuits enable these bacteria to collect, integrate, and process multiple sensory inputs. The use of two-component circuits that enable the influx and outflow of phosphate at multiple locations in the signal transduction circuit presumably allows very precise modulation of the output responses. Other simpler quorum sensing circuits, such as LuxI–LuxR circuits, may not possess the inherent plasticity to integrate multiple signaling inputs nor the fine control that can be incorporated into multichannel two-component relays. Finally, investigation of these two particular quorum sensing systems is only at the beginning. Continued study of quorum sensing in bacteria such as *V. harveyi* and *M. xanthus* could allow us to uncover novel mechanisms for intracellular, intercellular, intraspecies, and interspecies signal transduction.

REFERENCES

1. Hastings, J. W., and Nealson, K. H. (1977). Bacterial bioluminescence *Annu. Rev. Microbiol.* **31**, 549–595.
2. Nealson, K. H., and Hastings, J. W. (1979). Bacterial bioluminescence: Its control and ecological significance *Microbiol. Rev.* **43**, 496–518.
3. Ruby, E. G. (1996). Lessons from a cooperative, bacterial-animal association: The *Vibrio fischeri-Euprymna scolopes* light organ symbiosis *Annu Rev. Microbiol.* **50**, 591–624.
4. Visick, K. L., and McFall-Ngai, M. J. (2000). An exclusive contract: Specificity in the *Vibrio fischeri-Euprymna scolopes* partnership. *J. Bacteriol.* **182**, 1779–1787.
5. Engebrecht, J., Nealson, K., and Silverman, M. (1983). Bacterial bioluminescence: Isolation and genetic analysis of functions from *Vibrio fischeri*. *Cell* **32**, 773–781.
6. Engebrecht, J., and Silverman, M. (1984). Identification of genes and gene products necessary for bacterial bioluminescence. *Proc. Natl. Acad. Sci. USA* **81**, 4154–4158.
7. Engebrecht, J., and Silverman, M. (1987). Nucleotide sequence of the regulatory locus controlling expression of bacterial genes for bioluminescence. *Nucleic Acids Res.* **15**, 10455–10467.

8. Eberhard, A., Burlingame, A. L., Eberhard, C., Kenyon, G. L., Nealson, K. H., and Oppenheimer, N. J. (1981). Structural identification of autoinducer of *Photobacterium fischeri* luciferase. *Biochemistry* **20**, 2444–2449.
9. Miller, M. B., and Bassler, B. L. (2001). Quorum sensing in bacteria. *Annu. Rev. Microbiol.* **55**, 165–199.
10. Bassler, B. L. (1999). How bacteria talk to each other: Regulation of gene expression by quorum sensing. *Curr. Opin. Microbiol.* **2**, 582–587.
11. de Kievit, T. R., and Iglewski, B. H. (2000). Bacterial quorum sensing in pathogenic relationships. *Infect. Immun.* **68**, 4839–4849.
12. Fuqua, C., Winans, S. C., and Greenberg, E. P. (1996). Census and consensus in bacterial ecosystems: The LuxR-LuxI family of quorum sensing transcriptional regulators. *Annu. Rev. Microbiol.* **50**, 727–751.
13. Lazazzera, B. A., and Grossman, A. D. (1998). The ins and outs of peptide signaling. *Trends Microbiol.* **6**, 288–294.
14. Lavilla-Pitogo, C. R., Baticados, M. C. L., Cruz-Lacierda, E. R., and de la Pena, L. D. (1990). Occurrence of the luminous bacterial disease of *Penaeus monodon* larvae in the Phillippines. *Aquaculture* **91**, 1–13.
15. Song, Y.-L., and Lee, S. P. (1993). Characterization of ecological implication of luminous *Vibrio harveyi* isolated from tiger shrimp (*Penaeus monodon*). *Bull. Inst. Zool. Acad. Sin.* **32**, 217–220.
16. Bassler, B. L., Wright, M., Showalter, R. E., and Silverman, M. R. (1993). Intercellular signaling in *Vibrio harveyi*: Sequence and function of genes regulating expression of luminescence. *Mol. Microbiol.* **9**, 773–786.
17. Cao, J. G., and Meighen, E. A. (1989). Purification and structural identification of an autoinducer for the luminescence system of *Vibrio harveyi J. Biol. Chem.* **264**, 21670–21676
18. Bassler, B. L., Wright, M., and Silverman, M. R. (1994). Multiple signaling systems controlling expression of luminescence in *Vibrio harveyi*: Sequence and function of genes encoding a second sensory pathway. *Mol. Microbiol.* **13**, 273–286.
19. Bassler, B. L., Wright, M., and Silverman, M. R. (1994). Sequence and function of LuxO, a negative regulator of luminescence in *Vibrio harveyi*. *Mol. Microbiol.* **12**, 403–412.
20. Freeman, J. A., and Bassler, B. L. (1999). A genetic analysis of the function of LuxO, a two-component response regulator involved in quorum sensing in *Vibrio harveyi*. *Mol. Microbiol.* **31**, 665–677.
21. Freeman, J. A., and Bassler, B. L. (1999). Sequence and function of LuxU: A two-component phosphorelay protein that regulates quorum sensing in *Vibrio harveyi*. *J. Bacteriol.* **181**, 899–906.
22. Freeman, J. A., Lilley, B. N., and Bassler, B. L. (2000). A genetic analysis of the functions of LuxN: A two-component hybrid sensor kinase that regulates quorum sensing in *Vibrio harveyi*. *Mol. Microbiol.* **35**, 139–149.
23. Lilley, B. N., and Bassler, B. L. (2000). Regulation of quorum sensing in *Vibrio harveyi* by LuxO and sigma-54. *Mol. Microbiol.* **36**, 940–954.
24. Surette, M. G., and Bassler, B. L. (1998). Quorum sensing in *Escherichia coli* and *Salmonella typhimurium*. *Proc. Natl. Acad. Sci. USA* **95**, 7046–7050.
25. Surette, M. G., Miller, M. B., and Bassler, B. L. (1999). Quorum sensing in *Escherichia coli*, *Salmonella typhimurium*, and *Vibrio harveyi*: A new family of genes responsible for autoinducer production. *Proc. Natl. Acad. Sci. USA* **96**, 1639–1644.
26. Bassler, B. L., Greenberg, E. P., and Stevens, A. M. (1997). Cross-species induction of luminescence in the quorum sensing bacterium *Vibrio harveyi*. *J. Bacteriol.* **179**, 4043–4045.
27. Hanzelka, B. L., and Greenberg, E. P. (1996). Quorum sensing in *Vibrio fischeri*: Evidence that S-adenosylmethionine is the amino acid substrate for autoinducer synthesis. *J. Bacteriol.*

178, 5291–5294.

28. Schauder, S., Shokat, K., Surette, M. G., and Bassler, B. L. (2001). The LuxS family of bacterial autoinducers: Biosynthesis of a novel quorum sensing signal molecule. *Mol. Microbiol.*
29. Schauder, S., and Bassler, B. L. (2001). The languages of bacteria. *Genes Dev.* **15**, 1468–1480.
30. Martin, M., Showalter, R., and Silverman, M. (1989). Identification of a locus controlling expression of luminescence genes in *Vibrio harveyi*. *J. Bacteriol.* **171**, 2406–2414.
31. Showalter, R. E., Martin, M. O., and Silverman, M. R. (1990). Cloning and nucleotide sequence of *luxR*, a regulatory gene controlling bioluminescence in *Vibrio harveyi*. *J. Bacteriol.* **172**, 2946–2954.
32. Popham, D. L., Szeto, D., Keener, J., and Kustu, S. (1989). Function of a bacterial activator protein that binds to transcriptional enhancers. *Science* **243**, 629–635.
33. Weiss, D. S., Batut, J., Klose, K. E., Keener, J., and Kustu, S. (1991). The phosphorylated form of the enhancer-binding protein NTRC has an ATPase activity that is essential for activation of transcription. *Cell* **67**, 155–167.
34. Kustu, S., Santero, E., Keener, J., Popham, D., and Weiss, D. (1989). Expression of sigma 54 (*ntrA*)-dependent genes is probably united by a common mechanism. *Microbiol. Rev.* **53**, 367–376.
35. Wu, J., and Newton, A. (1997). Regulation of the *Caulobacter flagellar* gene hierarchy; not just for motility. *Mol. Microbiol.* **24**, 233–239.
36. Benson, A. K., Ramakrishnan, G., Ohta, N., Feng, J., Ninfa, A. J., and Newton, A. (1994). The *Caulobacter crescentus* FlbD protein acts at ftr sequence elements both to activate and to repress transcription of cell cycle-regulated flagellar genes. *Proc. Natl. Acad. Sci. USA* **91**, 4989–4993.
37. Wingrove, J. A., and Gober, J. W. (1994). A sigma 54 transcriptional activator also functions as a pole-specific repressor in *Caulobacter. Genes Dev.* **8**, 1839–1852.
38. North, A. K., Weiss, D. S., Suzuki, H., Flashner, Y., and Kustu, S. (1996). Repressor forms of the enhancer-binding protein NrtC: Some fail in coupling ATP hydrolysis to open complex formation by sigma 54- holoenzyme. *J. Mol. Biol.* **260**, 317–331.
39. Reitzer, L. J., and Magasanik, B. (1985). Expression of *glnA* in *Escherichia coli* is regulated at tandem promoters. *Proc. Natl. Acad. Sci. USA* **82**, 1979–1983.
40. Surette, M. G., and Bassler, B. L. (1999). Regulation of autoinducer production in *Salmonella typhimurium*. *Mol. Microbiol.* **31**, 585–595.
41. Joyce, E. A., Bassler, B. L., and Wright, A. (2000). Evidence for a signaling system in *Helicobacter pylori*: Detection of a *luxS*-encoded autoinducer. *J. Bacteriol.* **182**, 3638–3643.
42. Kim, S. Y., Lee, S.E., Kim, Y. R., Kim, J. H., Ryu, P. Y., Chung, S. S., and Rhee, J. H. (2000). "Virulence Regulatory Role of *luxS* Quorum Sensing System in *Vibrio vulmificus*," pp. 97–98. ASM General Meeting.
43. Lyon, W. R., Madden, J. C., Stein, J., and Caparon, M. G. (2001). Mutation of *luxS* affects growth and virulence factor expression in *Streptococcus pyogenes*. *Mol. Microbiol.*
44. Sperandio, V., Mellies, J. L., Nguyen, W., Shin, S., and Kaper, J. B. (1999). Quorum sensing controls expression of the type III secretion gene transcription and protein secretion in enterohemorrhagic and enteropathogenic *Escherichia coli*. *Proc. Natl. Acad. Sci. USA* **96**, 15196–15201.
45. Day, W. A., Jr., and Maurelli, A. T. (2001). *Shigella flexneri* LuxS quorum sensing system modulates *virB* expression but is not essential for virulence. *Infect. Immun.* **69**, 15–23.
46. McCarter, L. L. (1998). OpaR, a homolog of *Vibrio harveyi* LuxR, controls opacity of *Vibrio parahaemolyticus*. *J. Bacteriol.* **180**, 3166–3173.
47. Jobling, M. G., and Holmes, R. K. (1997). Characterization of *hapR*, a positive regulator of the *Vibrio cholerae* HA/protease gene hap, and its identification as a functional homologue of the *Vibrio harveyi luxR* gene. *Mol. Microbiol.* **26**, 1023–1034.

48. McDougald, D., Rice, S. A., and Kjelleberg, S. (2000). The marine pathogen *Vibrio vulnificus* encodes a putative homologue of the *Vibrio harveyi* regulatory gene, *luxR*: A genetic and phylogenetic comparison. *Gene* **248**, 213–221.
49. Gilson, L., Kuo, A., and Dunlap, P. V. (1995). AinS and a new family of autoinducer synthesis proteins. *J. Bacteriol.* **177**, 6946–6951.
50. Miyamoto, C. M., Lin, Y.H., and Meighen, E. A. (2000). Control of bioluminescence in *Vibrio fischeri* by the LuxO signal response regulator. *Mol. Microbiol.* **36**, 594–607.
51. Visick, K. L., Foster, J., Doino, J., McFall-Ngai, M., and Ruby, E. G. (2000). *Vibrio fischeri lux* genes play an important role in colonization and development of the host light organ. *J. Bacteriol.* **182**, 4578–4586.
52. Milton, D. L., Hardman, A., Camara, M., Chhabra, S. R., Bycroft, B. W., Stewart, G. S., and Williams, P. (1997). Quorum sensing in *Vibrio anguillarum*: Characterization of the *vanI/vanR* locus and identification of the autoinducer *N*-(3-oxodecanoyl)-L-homoserine lactone. *J. Bacteriol.* **179**, 3004–3012.
53. Milton, D. L., Chalker, V .J., Kirke, D., Hardman, A., Camara, M., and Williams, P. (2001). The LuxM homologue VanM from *Vibrio anguillarum* directs the synthesis of N-(3-hydroxyhexanoyl)homoserine lactone and N-hexanoylhomoserine lactone. *J. Bacteriol.* **183**, 3537–3547.
54. Dworkin, M. (1973). Cell-cell interactions in the myxobacteria. *Symp. Soc. Gen. Microbiol.* **23**, 125–147.
55. Dworkin, M., and Kaiser, D. (1985). Cell interactions in myxobacterial growth and development. *Science* **230**, 18–24.
56. Kaiser, D. (1984). Regulation of multicellular development in myxobacteria. *In* "Microbial Development" (R. Losick and L. Shapiro, eds.) pp. 197–218. Cold Spring Harbor Laboratory Press, New York, Cold Spring Harbor, NY.
57. Hagen, D. C., Bretscher, A. P., and Kaiser, D. (1978). Synergism between morphogenetic mutants of *Myxococcus xanthus*. *Dev. Biol.* **64**, 284–296.
58. Downard, J., Ramaswamy, S. V., and Kil, K. S. (1993). Identification of *esg*, a genetic locus involved in cell-cell signaling during *Myxococcus xanthus* development. *J. Bacteriol.* **175**, 7762–7770.
59. Shimkets, L. J. (1999). Intercellular signaling during fruiting-body development of *Myxococcus xanthus*. *Annu. Rev. Microbiol.* **53**, 525–549.
60. Kuspa, A., Kroos, L., and Kaiser, D. (1986). Intercellular signaling is required for developmental gene expression in *Myxococcus xanthus*. *Dev. Biol.* **117**, 267–276.
61. Singer, M., and Kaiser, D. (1995). Ectopic production of guanosine penta- and tetraphosphate can initiate early developmental gene expression in *Myxococcus xanthus*. *Genes Dev.* **9**, 1633–1644.
62. Harris, B. Z., Kaiser, D., and Singer, M. (1998). The guanosine nucleotide (p)ppGpp initiates development and A-factor production in *Myxococcus xanthus*. *Genes Dev.* **12**, 1022–1035.
63. Plamann, L., Kuspa, A., and Kaiser, D. (1992). Proteins that rescue A signal-defective mutants of *Myxococcus xanthus*. *J. Bacteriol.* **174**, 3311–3318.
64. Kuspa, A., Plamann, L., and Kaiser, D. (1992). Identification of heat-stable A-factor from *Myxococcus xanthus*. *J. Bacteriol.* **174**, 3319–3326.
65. Kaplan, H. B., Kuspa, A., and Kaiser, D. (1991). Suppressors that permit A signal-independent developmental gene expression in *Myxococcus xanthus*. *J. Bacteriol.* **173**, 1460–1470.
66. Plamann, L., Li, Y., Cantwell, B., and Mayor, J. (1995). The *Myxococcus xanthus asgA* gene encodes a novel signal transduction protein required for multicellular development. *J. Bacteriol.* **177**, 2014–2020.
67. Bowden, M. G., and Kaplan, H. B. (1996). The *Myxococcus xanthus* developmentally expressed *asgB*-dependent genes can be targets of the A signal-generating or A signal-

responding pathway. *J. Bacteriol.* **178**, 6628–6631.
68. Kuspa, A., Plamann, L., and Kaiser, D. (1992). A signaling and the cell density requirement for *Myxococcus xanthus* development. *J. Bacteriol.* **174**, 7360–7369.
69. McVittie, A., Messik, F., and Zahler, S. A. (1962). Developmental biology of *Myxococcus*. *J. Bacteriol.* **84**, 546–551.
70. Kuspa, A., and Kaiser, D. (1989). Genes required for developmental signaling in *Myxococcus xanthus*: Three *asg* loci. *J. Bacteriol.* **171**, 2762–2772.
71. Li, Y., and Plamann, L. (1996). Purification and *in vitro* phosphorylation of *Myxococcus xanthus* AsgA protein. *J. Bacteriol.* **178**, 289–292.
72. Plamann, L., Davis, J. M., Cantwell, B., and Mayor, J. (1994). Evidence that *asgB* encodes a DNA-binding protein essential for growth and development of *Myxococcus xanthus*. *J Bacteriol.* **176**, 2013–2020.
73. Dombroski, A. J., Walter, W. A., Record, M. T., Jr., Siegele, D. A., and Gross, C. A. (1992). Polypeptides containing highly conserved regions of transcription initiation factor sigma 70 exhibit specificity of binding to promoter DNA. *Cell* **70**, 501–512.
74. Gardella, T., Moyle, H., and Susskind, M. M. (1989). A mutant *Escherichia coli* sigma 70 subunit of RNA polymerase with altered promoter specificity. *J. Mol Biol.* **206**, 579–590.
75. Lonetto, M., Gribskov, M., and Gross, C. A. (1992). The sigma 70 family: Sequence conservation and evolutionary relationships. *J. Bacteriol.* **174**, 3843–3849.
76. Margolis, P., Driks, A., and Losick, R. (1991). Establishment of cell type by compartmentalized activation of a transcription factor. *Science* **254**, 562–565.
77. Siegele, D. A., Hu, J. C., Walter, W. A., and Gross, C. A. (1989). Altered promoter recognition by mutant forms of the sigma 70 subunit of *Escherichia coli* RNA polymerase. *J. Mol. Biol.* **206**, 591–603.
78. Davis, J. M., Mayor, J., and Plamann, L. (1995). A missense mutation in *rpoD* results in an A signaling defect in *Myxococcus xanthus*. *Mol. Microbiol.* **18**, 943–952.
79. Hernandez, V. J., and Cashel, M. (1995). Changes in conserved region 3 of *Escherichia coli* sigma 70 mediate ppGpp-dependent functions *in vivo*. *J. Mol. Biol.* **252**, 536–549.
80. Cho, K., and Zusman, D. R. (1999). AsgD, a new two-component regulator required for A signaling and nutrient sensing during early development of *Myxococcus xanthus*. *Mol. Microbiol.* **34**, 268–281.
81. Xu, D., Yang, C., and Kaplan, H. B. (1998). *Myxococcus xanthus sasN* encodes a regulator that prevents developmental gene expression during growth. *J. Bacteriol.* **180**, 6215–6223.
82. Yang, C., and Kaplan, H. B. (1997). *Myxococcus xanthus sasS* encodes a sensor histidine kinase required for early developmental gene expression. *J. Bacteriol.* **179**, 7759–7767.
83. Plamann, L., and Kaplan, H. B. (1999). Cell-density sensing during early development in *Myxococcus xanthus*, *In* "Cell–Cell Signaling in Bacteria" (G. M. Dunny and S. C. Winans, eds), pp. 67–82. ASM Press, Washington DC.
84. Guo, D., Wu, Y., and Kaplan, H. B. (2000). Identification and characterization of genes required for early *Myxococcus xanthus* developmental gene expression. *J. Bacteriol.* **182**, 4564–4571.
85. Keseler, I. M., and Kaiser, D. (1995). An early A signal dependent gene in *Myxococcus xanthus* has a sigma 54- like promoter. *J. Bacteriol.* **177**, 4638–4644.
86. Kaplan, H. B., and Greenberg, E. P. (1985). Diffusion of autoinducer is involved in regulation of the *Vibrio fischeri* luminescence system. *J. Bacteriol.* **163**, 1210–1214.

CHAPTER 16

Intercellular Communication in Gram-Positive Bacteria Depends on Peptide Pheromones and Their Histidine Kinase Receptors

LEIV SIGVE HÅVARSTEIN
Department of Chemistry and Biotechnology, Agricultural University of Norway, N-1432 Ås, Norway

Intercellular communication mediated by peptide pheromones controls several different biological processes in gram-positive bacteria, such as virulence and horizontal gene transfer. In the majority of cases, the pheromones are sensed by a membrane-localized histidine kinase, which is part of a two-component signal transduction pathway. Pheromone binding results in autophosphorylation of the histidine kinase receptor, followed by transfer of the phosphoryl group to the cognate cytoplasmic response regulator. This regulator in turn activates transcription of target genes. Gram-positive bacteria use this kind of cell-to-cell signaling to monitor population density and to coordinate the activities of the population. Communication seems to be

Histidine Kinases in Signal Transduction

restricted to closely related bacteria, as different species and strains produce pheromones with different primary structures and specificity. Interestingly, most peptide pheromone receptors belong to a newly defined subfamily of histidine kinases termed HPK_{10}, indicating that they have evolved from a common ancestor. The most prominent feature of members of this subfamily is an unusual sensor domain with five to eight transmembrane segments.

INTRODUCTION

To survive physiochemical changes in their environment, bacteria must constantly monitor external conditions and adjust their pattern of gene expression and physiology accordingly. Perhaps the most important type of sensor mechanism bacteria have developed for this purpose is the two-component regulatory systems [1–4, 4a]. These systems, which in their simplest form consist of a histidine kinase and a response regulator, are also used for intercellular communication [5]. In this case the signal perceived by the sensor histidine kinase is synthesized and secreted by the bacteria themselves. Bacteria use this kind of cell-to-cell signaling to monitor population density and to coordinate the activity of the bacterial population. This group behavior, often called quorum sensing, regulates many different processes in bacteria, such as bacteriocin production, virulence, natural genetic transformation, bioluminescence, secondary metabolite production, and sporulation [5]. One common type of cell-to-cell signaling is acyl-homoserine lactone quorum sensing in gram-negative bacteria [6]. In all known cases, except one, this type of quorum sensing does not involve a membrane-bound histidine kinase but a LuxI–LuxR quorum-sensing circuit where LuxI is the homoserine lactone synthase and LuxR is a transcription factor activated by the homoserine lactone autoinducer. Bacterial cells seem to be freely permeable to homoserine lactones, and internal receptors instead of membrane-bound histidine kinase sensors are therefore used in these systems. An exception to this rule is quorum sensing in *Vibrio harveyi* where two parallel two-component systems regulate bioluminescence [7, 8]. In gram-positive bacteria peptide pheromones, rather than acyl-homoserine lactones, seem to be the preferred signaling molecules used for intercellular communication [9, 10]. The first evidence that a proteinaceous extracellular factor was involved in signaling between bacterial cells came from work on natural genetic transformation in streptococci [11–15]. The signaling factor was originally believed to be a protein, and it was not until the mid-90s that the pneumococcal competence pheromone was purified and shown to be a peptide consisting of only 17 amino acid residues [16]. At about the same time, several other peptide pheromones, regulating competence in *Bacillus subtilis*, virulence in *Staphylococcus aureus*, and production of bacteriocins in lactic acid bacteria, were identified [17–21]. In all of these systems, signaling peptides accumulate in the external environment as the cells grow, and after reaching a threshold concentration they trigger a phosphotransfer pathway consisting of a sensor histidine kinase and a cytoplasmic response regulator. Phosphorylation of the response regulator, which in all cases functions as a transcription factor, directly or indirectly activates the transcription of genes involved in the different biological processes mentioned earlier [5]. Conjugative plasmid transfer in the gram-positive bacterium *Enterococcus faecalis* also depends on peptide pheromones [22], but detection of the pheromones takes place by a mechanism that does not involve a membrane histidine kinase receptor. Instead, an oligopeptide permease actively transports the peptide pheromones into the bacterial cells where they presumably interact with intracellular receptors to initiate conjugation. The same mechanism is used for internalization of the competence and sporulation factor (CSF), a pheromone, which together with another peptide pheromone (ComX), regulates competence development in *Bacillus subtilis* [9]. Biological processes regulated by peptide pheromones with intracellular targets are not reviewed here. This chapter focuses on processes in gram-positive bacteria that depend on peptide pheromones with membrane-bound histidine kinase receptors.

INTERCELLULAR COMMUNICATION BY UNMODIFIED PEPTIDES

In lactic acid bacteria, which include genera such as *Streptococcus*, *Lactococcus*, *Enterococcus*, and *Lactobacillus*, unmodified peptides consisting of 14–27 amino acid residues are used for intercellular communication [16,19, 23–27]. So far, unmodified peptide pheromones have been found to regulate two different biological processes in these genera: competence for natural genetic transformation and production of class II bacteriocins [16,19]. Class II bacteriocins are a group of unmodified antimicrobial peptides ranging in size from 30 to about 60 amino acids. Most bacteriocins seem to be directed against closely related species and act by inducing leakage through the cell membrane of sensitive cells [30–32]. Bacteria that are naturally competent have the ability to take up naked DNA from the surrounding medium and incorporate this DNA into their own genome by homologous recombination. It is most likely that natural transformation is a DNA uptake and recombination system for the generation of genetic diversity in related species [33]. Among the genera mentioned earlier, only some of the species belonging to the genus *Streptococcus* are known to be naturally competent [23]. In contrast, the production of class II bacteriocins seems to be ubiquitous among lactic acid bacteria.

Practically, all unmodified peptide pheromones are synthesized ribosomally as precursor peptides with a characteristic Gly-Gly leader at the N-terminal end [28]. The double glycine leader is removed concomitant with export and probably remains inside the cell while the biologically active pheromone is transported across the cytoplasmic membrane. ABC transporters involved in the secretion of gene products with Gly-Gly leaders have a unique proteolytic domain at their N-terminal end, which distinguish them from all other ABC transporters [29]. Many peptide bacteriocins (class I and class II) are also synthesized with double glycine leaders and use the same type of ABC transporters for their export. Whereas pheromones with Gly-Gly leaders have only been found in lactic acid bacteria, bacteriocins with this type of leader are produced by gram-negative bacteria as well [32, 34].

Natural competence in streptococci is not a constitutive state, but a strictly regulated property that depends on the expression of a unique set of gene products. Five of these (ComAB and ComCDE) make up a quorum-sensing circuit that regulates competence development by monitoring the concentration of a peptide pheromone in the environment [31]. The rest, which are encoded by at least 20 late genes, are involved in binding, uptake, and integration of extracellular DNA [35,36]. The quorum-sensing circuit consists of a peptide pheromone, its secretion apparatus (ComAB) [37], and a two-component regulatory system (ComDE) [38]. The *comC* gene encodes the precursor of the competence pheromone, also called the competence-stimulating peptide (CSP) [16]. As described earlier, this precursor contains a Gly-Gly leader peptide, which is removed concomitantly with export by the proteolytic N-terminal domain of its dedicated ABC transporter (ComA). In a vigorously growing culture, CSP accumulates slowly in the medium and triggers competence induction when it reaches a concentration of 0.2–1 ng/ml. At this stage the cell density of the population is in the range of 10^5–5×10^8 cells/ml [39]. The signaling cascade, which leads to transcriptional activation of the late genes, begins with the interaction of CSP with its membrane-bound histidine kinase receptor ComD [40], which responds by transferring a phosphoryl group to its cognate response regulator ComE. Upon phosphorylation, ComE activates transcription from promoters containing an imperfect direct repeat motif [41]. The presence of this motif in the promoters of the *comAB* and *comCDE* operons generates the autoinducing behavior typical for quorum-sensing systems. The autoinduction mechanism amplifies the response to the pheromone, resulting in higher amounts of CSP in the environment and an increased level of phosphorylated ComE inside the cells. Autoinduction increases the amount of pheromone in the medium 30-fold or more and presumably serves to coordinate competence induction throughout the bacterial population [31]. The late genes, which together constitute a competence regulon, are expressed through a common regulatory element called the cinbox or combox [42]. This conserved – 10 sequence (TACGAATA) is recognized by an alternative σ factor, ComX , which replaces the primary σ factor from the RNA polymerase holoenzyme [43]. It is still not known how the expression of ComX is turned on, but close examination of the promoter region of the *comX* gene has revealed a direct repeat motif similar to the binding site of ComE. However, in one of the repeated sequence motifs, two con-

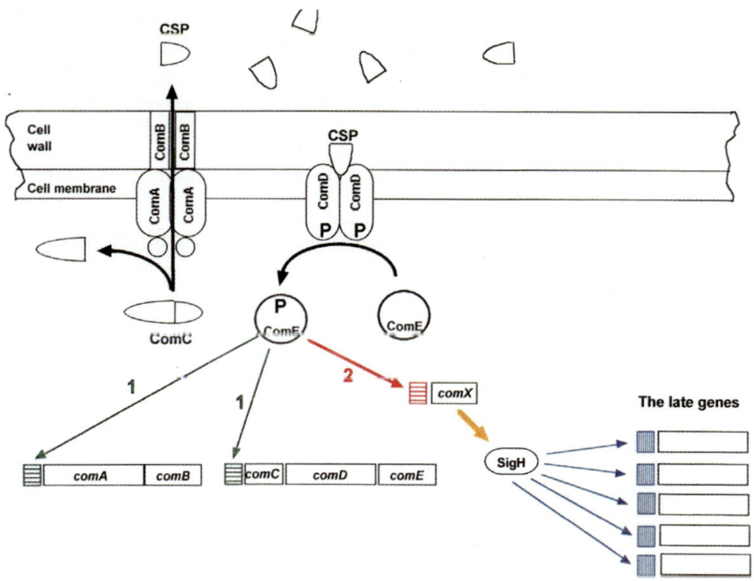

FIGURE 1 Schematic representation of signal transduction and information flow in the quorum sensing circuit regulating competence development in *Streptococcus pneumoniae* and related streptococci. ABC transporter proteins (ComAB), the competence stimulating peptide precursor (ComC), and two-component regulatory proteins (ComDE) are synthesized at a low rate in vigorously growing cultures of streptococci. Concomitant with export, the leader peptide of ComC is removed, and mature competence stimulating peptides (CSP) accumulate slowly in the medium. At the CSP concentration triggering competence induction (0.2–1 ng/ml), sufficient ComE-P is present in the cytoplasm to initiate transcription from the *comAB* and *comCDE* promoters (arrow marked 1). This results in a rapid increase in the synthesis of the ComABCDE gene products, followed by a 30- to 100-fold increase in the CSP concentration in the medium, which in turn will lead to a significant increase in the ComE-P concentration in the cytoplasm. At this high-level concentration of ComE-P, the response regulator presumably binds to a low-affinity ComE-binding site in the *comX* promoter region and activates synthesis of the alternative σ factor SigH (arrow marked 2). SigH will replace the primary σ factor from the RNA polymerase holoenzyme and direct expression of the cinbox containing late genes. The gene products of the late genes are involved in binding, processing, uptake, and incorporation of extracellular DNA.

served thymines are replaced by adenines, suggesting that this potential ComE-binding site will bind ComE-P with low affinity. If this is the case, transcriptional activation of the *comX* gene will require high levels of ComE-P and will therefore take place only after the autoinduction mechanism has amplified the level of the phosphorylated response regulator in the cell (see Fig. 1).

The production of class II bacteriocins in lactic acid bacteria is regulated by quorum sensing circuits consisting of five proteins that are highly homologous to and have the same function as the corresponding ComABCDE proteins regulating competence development in streptococci [28–30, 44–46]. Motifs of the direct repeat type are also found in bacteriocin quorum sensing circuits, where they are recognized by homologues of ComE [47]. The major difference between the two systems is that ComE activates expression of the late genes through an alternative σ factor, whereas the corresponding bacteriocin response regulators directly activate transcription of the bacteriocin structural genes by binding to the direct repeat motif in their promoter regions (Fig. 2).

In naturally transformable pneumococci, which also produce class II bacteriocins, competence and bacteriocin production appear to be regulated inde-

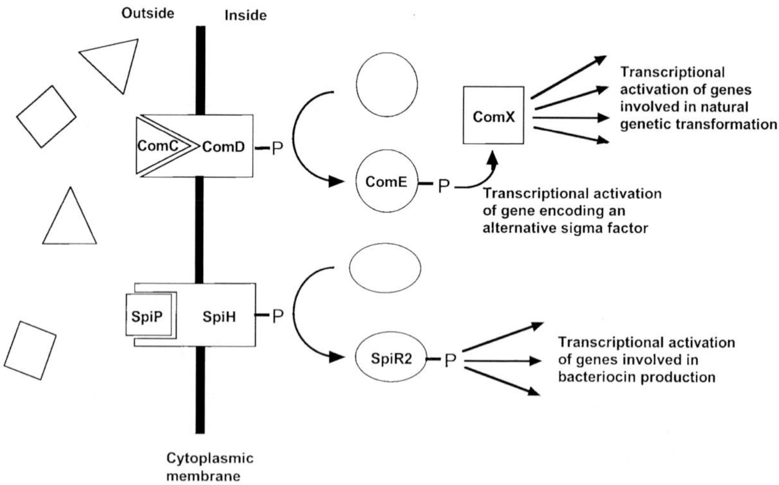

FIGURE 2 Schematic diagram showing the signal transduction pathways regulating competence development and bacteriocin production in *Streptococcus pneumoniae*. CSP, competence pheromone; ComD, CSP receptor; ComE, response regulator; ComX, alternative σ factor; SpiP, pheromone regulating induction of bacteriocin production; SpiH, SpiP receptor; and SpiR2, response regulator.

pendently, even though both processes are controlled by population density. Sequencing of the complete genome of the pneumococcus has shown that this species possesses two distinct but closely related quorum-sensing systems, one for the control of DNA uptake and the other for regulating bacteriocin production [48–51]. The primary structures of the two unmodified peptide pheromones regulating competence and bacteriocin production in *Streptococcus pneumoniae* strain Rx are completely unrelated. They are also different in size, consisting of 17 and 27 amino acid residues, respectively [50, 51]. Addition of the pheromone inducing bacteriocin production to a culture of strain Rx will not induce the competent state (unpublished results), suggesting that cross communication between the two signaling pathways does not take place (Fig. 2).

INTERCELLULAR COMMUNICATION BY MODIFIED PEPTIDES

Many gram-positive bacteria produce ribosomally synthesized antimicrobial peptides termed class I bacteriocins or lantibiotics. In contrast to class II bacteriocins, class I bacteriocins are modified posttranslationally and contain unusual amino acids, such as didehydroalanine, didehydrobutyrine, lanthionine, and β-methyllanthionine [46, 52, 53]. The structural gene of class I bacteriocins encodes a precursor protein consisting of an N-terminal leader peptide and a C-terminal prolantibiotic region, which are modified posttranslationally to form the mature lantibiotic. In addition to the enzymes involved in modification of the prolantibiotic peptide, the production of class I bacteriocins requires a processing protease for removal of the leader peptide, a translocator of the ABC transporter family, regulatory proteins, and a self-protection mechanism [52]. Many class I bacteriocins are only expressed during the late exponential or early stationary phase, showing that their synthesis is subjected to regulation [54]. In the biosynthetic gene cluster of nisin, a lantibiotic consisting of 34 amino acid residues, genes encoding a histidine kinase (NisK) and a response regulator (NisR) have been identified and shown to be essential for nisin production [54]. It follows from this observation that some environmental stimulus sensed by NisK controls the induction of nisin production. Unexpectedly, Kuipers *et al.* [55] dicovered that a mutant with a 4-bp deletion in the nisin structural gene (*nisA*) had lost its capacity to produce the *nisA* transcript. This discovery led to the finding that the addition of subinhibitory amounts of extracellular nisin restored the transcription of Δ*nisA* [21]. In sum, these and other data show that nisin is the environmental signal sensed by NisK and that nisin stimulates its own synthesis by triggering the NisKR signal transduction pathway in a quorum-sensing-like manner [56]. Based on the high similarity between the biosynthesis systems of nisin and a related lantibiotic called subtilin, it is likely that the latter also acts as a peptide pheromone stimulating its own production [54].

The synthesis of virulence factors and other exoproteins in *Staphylococcus aureus* is regulated globally by a cell density-sensing system called *agr* (accessory gene regulator) [57,58], which depends on a modified octapeptide pheromone produced by the bacterium itself [20]. The *agr* locus consists of two divergent transcription units, termed RNAII and RNAIII, transcribed from neighboring but separate promoters, P2 and P3. The RNAII operon contains four genes (*agr*BDCA), which together make up a quorum-sensing circuit. *agrD* encodes the precursor of the autoinducing octapeptide pheromone (AIP), which probably is processed and secreted by the gene product of *agrB*, whereas *agrC* and *agrA* encode a membrane histidine kinase and its cognate response regulator, respectively [59, 60]. Several lines of evidence strongly indicate that AgrC is the receptor of the AIP pheromone [61]. By analogy with other two component systems, binding of the AIP ligand to AgrC is supposed to induce autophosporylation of the receptor followed by transfer of the phosphoryl group to AgrA. The phosphorylated form of AgrA will then, in conjunction with another transcription factor (SarA), upregulate transcription from its own promoter (P2) [62–64]. Triggering of this autoinducing circuit brings about a rapid increase in the level of phosphorylated AgrA, resulting in activation of transcription from the P3 promoter. The P3 operon specifies a 0.5-kb regulatory RNA transcript (RNAIII), the actual effector of exoprotein gene regulation [59,60]. The exact mechanism by which this transcript regulates the synthesis of virulence factors and other exoproteins has yet to be determined.

Purification and sequencing of the AIP pheromone from *S. aureus* showed that it is a short peptide consisting of eight amino acid residues. It was also found that the sequence of this octapeptide corresponds to an internal sequence of AgrD, indicating that AIP is processed from within the 46 amino acid long *agrD* gene product [20]. A synthetic version of the peptide lacked biological activity, indicating that the native peptide is modified posttranslationally. In accordance with this theory, it was observed that the molecular mass of the native peptide was about 18 Da smaller than predicted from its amino acid sequence [20]. From this observation, and from the fact that the pheromone contains an internal cysteine residue, it was deduced that the native octapeptide contains an internal anhydride bond, presumably involving the cysteine thiol group. This hypothesis has been confirmed through total chemical synthesis of a biologically active thiolactone peptide [65].

In contrast to *S. pneumoniae*, where competence for natural genetic transformation develops in 100% of the population during exponential growth, competence in *B. subtilis* develops in a subpopulation of the bacteria in the beginning of the stationary phase [66]. Under conditions like these, where nutrients are becoming limiting for growth, *B. subtilis* can choose between two separate developmental pathways: sporulation or competence development [9]. Natural competence, which is focused on here, is controlled by a quorum-sensing mechanism [67]. Two different peptide pheromones, ComX and CSF, are released in the medium by growing cells and increase in concentration concomitantly with cell density [17, 68]. ComX is part of a quorum-sensing circuit, ComQXPA, which resembles the AgrBDCA cell density-sensing system of *S. aureus*. The genetic organization of the two gene clusters is the same with respect to gene order, and in both systems, two of the genes (*agrBD* and *comQX*) are involved in pheromone production and maturation, whereas the other two (*agrCA* and *comPA*) encode two-component regulatory systems [9, 67]. Reminiscent of AgrD, the ComX pheromone is synthesized ribosomally as a 55 residue precursor protein that is processed and modified before it is released to the extracellular medium as a biologically active pheromone [17]. The peptide moiety of the pheromone is derived from the 10 C-terminal amino acid residues of the ComX precursor protein. Because a synthetic version of this pheromone lacks biological activity and has less mass (206 Da) than the purified pheromone, the native pheromone must be modified. The exact nature of this modification is not known, but circumstantial evidence suggests that ComX contains a hydrophobic modification linked to the tryptophan residue three amino acids from the C-terminal end of the pheromone [17]. As in the *agr* system, production of the active pheromone depends on the gene located upstream of the pheromone structural gene. A precise role for this protein (ComQ) is not known, but it shares some amino acid similarity to isoprenyl transferases, suggesting that it may be used to modify the ComX pheromone [9, 67]. The CSF pheromone is an unmodified peptide, which corresponds to the C-terminal 5 amino acids of the *phrC* gene product [69]. ComX and CSF act on two different quorum sensing pathways that converge to regulate the level of phosphorylation of the response reulator ComA. ComX is thought to bind to the membrane domain of the histidine kinase ComP, initiating a phosphorylation cascade that culminates in the accumulation of phosphorylated ComA. CFS, however, is internalized via an oligopeptide permease and probably acts by inhibiting the dephosphorylation of ComA [9, 68, 69]. The phosphorylated form of ComA turns on the transcription of *comS*, a small gene embedded within the *srfA* operon.The synthesis of ComS in response to the converging quorum-sensing pathways described earlier, releases ComK from a ternary ComK/MecA/ClpC complex in which ComK is held inactive. This key competence transcription factor is then free to activate transcription of the late competence genes encoding the DNA binding and uptake machinery [9, 67, 72].

BACTERIA SPEAK DIFFERENT LANGUAGES

Many species of streptococci, such as *S. pneumoniae*, *S. mitis*, *S. oralis*, *S. gordonii*, *S. crista*, and *S. sanguis*, are naturally competent. Sequencing of the *comC* gene from different isolates of these species has shown that they produce a great variety of competence pheromones [23,24,73]. Preliminary studies indicate that the diversity is most extensive among strains belonging to the species *S. mitis* and *S. oralis* (unpublished results). Similarly, the amino acid sequences of pheromones regulating bacteriocin production in lactic acid bacteria vary among different species and strains [29, 51]. This extensive polymorphism is reflected in membrane-bound histidine kinase receptors, which are highly divergent in the N-terminal half of their membrane domains [23, 24]. Most likely, this part of the receptor, encompassing about 100 amino acid residues, determines the specificity of the receptor–ligand interaction [23, 40].

At least four variants of the thiolactone peptide are produced by different strains of *S. aureus*, and two additional variants are produced by *S. epidermidis* and *S. lugdunensis* [64, 74]. A comparison of the *agrBDCA* sequences from each strain revealed a continuous region of extensive variability, which includes most of the *agrB* gene, *agrD*, and the region of *agrC* encoding the sensor domain of the receptor [74–76]. Interestingly, a corresponding sequence comparison of *comQXPA* genes from several different *Bacillus* isolates revealed the same pattern of hypervariability as detected within the *agrBDCA* locus [77]. One aspect unique to the *agr* system is that the peptide pheromone produced by one strain of *S. aureus* will inhibit virulence expression in a strain producing another variant of the thiolactone peptide [75]. In other words, the thiolactone peptide pheromone activates *agr* expression in strains belonging to the same pheromone group as the producer, but inhibits *agr* expression in strains belonging to other pheromone groups. Studies have shown that both intragroup activation and intergroup inhibition are mediated by the same histidine kinase receptor, AgrC [64].

The biological significance of this diversity in pheromone types (pherotypes) is not fully understood, but it must have been generated by a strong selection pressure. The existence of different pherotypes strongly influences intercellular communication, which becomes restricted to those cells that are able to recognize each others' pheromones. This restriction ensures that bacteriocin production, competence development, or virulence will be induced only if a minimal number of bacteria with the same pherotype, i.e., closely related bacteria, are present in a certain volume of space. In the case of bacteriocin production, this makes sense. Bacteria having the same pherotype must act as a group in order to make enough bacteriocin to eradicate other bacteria competing for the same ecological niche. For self-protection, a specific immunity protein is usually coexpressed with its cognate bacteriocin. Consequently, it is important that the initiation of bacteriocin production is coordinated by a quorum sensing mechanism. If the bacteria did not act in concert, they could kill each other instead of their competitors.

The production of pathogenicity factors in *S. aureus* is a carefully orchestrated temporal program. Several regulatory elements are involved, but the best characterized and probably the most important is the *agr* system. Mutants lacking this qourum-sensing system are attenuated greatly, but are not completely avirulent [60]. Why is cell density used to regulate the production of pathogenicity factors in *S. aureus*? Many of these factors, such as toxins and enzymes, are used during an infection to degrade tissue components and ward off the defense system of the host. Presumably, the bacterium will succeed only if the concentration of pathogenicity factors at the focus of infection is high. This can only be obtained by cooperative behavior of a bacterial population consisting of a sufficient number of cells. As pointed out by Novick [60], a cell density-dependent expression of pathogenicity determinants could be useful if bacteria have become trapped in a fibrin clot or in an abscess. In a confined environment like this, diffusion of the secreted peptide phromone will be restricted, and its concentration will therefore reflect the cell density within the abscess. When the critical population density is reached, the high-level expression of secreted virulence factors could enable the bacteria to escape from the enclosure and disseminate to initiate new foci of infection. Remarkably, this kind of group behavior does not take place unless the bacteria belong to the same pherotype. On the contrary, staphylococci producing one type of pheromone will inhibit *agr* expression in strains belonging to any other pherotype [75]. This chemical warfare may serve the same function as production of bacteriocins in lactic acid bacteria, namely as a weapon to eradicate competitors within the same ecological niche.

Even though researchers have studied natural transformation in streptococci, bacilli, and other bacteria for a considerable period of time, the ultimate advantage for these organisms to be naturally competent is still not clearly understood. It has been proposed that the biological role of natural competence is to aid in the repair of damaged chromosomes. Alternatively, competence could be advantageous to the bacterium because it allows the acquisition of new traits from genetically distinct organisms. Another possibility is that natural competence is a parasexual mechanism, which allows the assembly of new combinations of genes and thereby increases diversity and speeds up evolution. All of these possible biological roles would require both a donor and a recipient organism. Up untill now, research has focused on the recipient and how it develops competence when grown in a laboratory environment [31]. It is therefore far from clear how gene exchange takes place under natural conditions. A question arising in this connection is whether release and uptake of DNA are separate events. Do competent bacteria depend on DNA, which is released to the environment at random when other closely related bacteria die and fall apart or is release of DNA from the donor coordinated in time and space with uptake by the recipient? In 1960, Ottolenghi and Hotchkiss [78] reported *in vitro* gene exchange between mixed growing populations of two mutants of *S. pneumoniae* strain R_6, resistant to sulfonamid/amethopterin and streptomycin/micrococcin, respectively. They were not successful, however, in determining whether DNA released came from dead cells as a result of normal population turnover or from active release triggered by induction of the competent state [78].

The effectiveness of bacteriocins and pathogenicity factors is compromised by dilution. Consequently, bacteria have evolved quorum-sensing systems to coordinate the release of these products from a large number of bacteria. In the case of natural genetic transformation, however, uptake and incorporation of naked DNA do not seem to rely on any secreted product. There must therefore be a different reason why the uptake of DNA should be regulated by a cell density-dependent mechanism. Most likely the purpose of quorum sensing in this context is to monitor the concentration of potential gene donors in the immediate neighborhood. Transformation is most effective with chromosomal DNA from closely related bacteria. This is due to a RecA-based homologous recombination system whose efficiency depends on the degree of homology between the donor DNA and the corresponding sequences in the genome of the recipients [79]. Thus, it appears that competence primarily has developed to promote the exchange of genetic material within the species. Naturally competent Gram-positive bacteria seem to have evolved a system of highly divergent pheromone types to distinguish between closely related donors and other bacteria in the environment. It is conceivable that gene exchange in nature takes place primarily between bacteria that share the same pherotype. If this is true, acquisition of a new pherotype will lead to sexual isolation and eventually to the development of a new bacterial species.

PEPTIDE PHEROMONES DEPEND ON HISTIDINE KINASE RECEPTORS

The unmodified and modified peptide pheromones discussed earlier depend on membrane-bound histidine kinase receptors for their signaling activity [40, 45, 54, 61, 80]. All of these receptors belong to the so-called orthodox kinases, which are composed of a highly variable, membrane-spanning, N-terminal sensor region and a more conserved C-terminal kinase core domain that features the phosphoaccepting histidine as well as the various homology boxes. Within the C-terminal kinase domain are four to five such boxes (H, N, D, F, and G), amino acid motifs that have been named after the most conserved residue in each box [49]. In the majority of membrane-localized histidine kinases, the N-terminal membrane-associated domain consists of two transmembrane segments located on either side of a periplasmic loop believed to interact with the external stimulus [81]. This simple organization of the membrane domain is found in NisK [54], but not in the other peptide pheromone receptors. Instead, they belong to a small group of histidine kinases that possess a sensor domain consisting of five to eight membrane-spanning segments [40, 45, 61, 80]. In an article by Grebe and Stock [49], a multisequence alignment of 348 kinase domains revealed that histidine kinases can be divided into subgroups based on their degree of amino acid sequence homology. According to the criteria used, most of the histidine kinases fell into one of 11 different subfamilies. Interestingly, all known peptide pheromone receptors, except ComP and NisK, fell into the same subfamily (HPK_{10}) and have probably evolved from a common ancestral histidine kinase. The HPK_{10} subfamily includes the recepetors of (i) streptococcal competence pheromones (ComD), (ii) pheromones inducing the production of class II bacteriocins, and (iii) pheromones regulating the production of pathogenicity factors in staphylococci (AgrC) [49]. Other histidine kinases that clearly belong to the HPK_{10} subfamily are VirS and FsrC. Together with their cognate response regulators, VirS and FsrC control the production of several pathogenicity factors in *Clostridium perfringens* and *E. faecalis*, respectively [82, 83]. The stimulus sensed by VirS has been suggested to be a small molecule termed substance A, the molecular mass of which has been estimated to be less than 2000 Da [84, 85]. The *fsr* genes of *E. faecalis* were discovered very recently, and the nature of the signal detected by FsrC is therefore still unknown. Nevertheless, as VirS and FsrC both belong to the HPK_{10} subfamily, chances are good that their ligands will turn out to be peptide pheromones.

ComP, the histidine kinase receptor of the ComX pheromone, possesses an N-terminal domain with six or eight membrane-spanning segments [80]. In this respect, ComP resembles sensor kinases belonging to the HPK_{10} subfamily, but sequence comparison shows that its kinase domain is related more closely to members of the HPK_7 subfamily. This mosaic-like structure suggests that shuffling of domains may have taken place between the HPK_{10} and the HPK_7 subfamilies. However, a comparison of the N-terminal sensor domains of ComP and members of the HPK_{10} subfamily does not lend much support to this hypothesis. The membrane domain of ComP is almost twice as large and has two prominent extracellular loops that seem to be involved in the binding of ComX. These loops are lacking in the HPK_{10} subfamily, suggesting that members of this subfamily interact with their ligands in a different way.

In contrast to the other peptide pheromone receptors, NisK and SpaK, the sensor histidine kinases of the lantibiotics nisin and subtilin, have only two predicted membrane-spanning segments in their N-terminal ends [54]. The two transmembrane segments are separated by an extracellular domain of approximately 110 amino acids, which presumably contains the binding site of the ligand. In an attempt to map the regions important for ligand binding, various chimeric *nisK/spaK* genes were constructed. Bacterial cells expressing hybrid sensor proteins were then induced with either nisin or subtilin, and the ability of the hybrid sensor to phosphorylate NisR was assayed by means of a *nisA::gusA* transcriptional fusion. Results obtained from these experiments showed that the entire sensor domain could be exchanged between NisK and SpaK without loss of function, whereas all hybrid sensors with fusion points within the sensor domain were inactive [54].

THE HPK_{10} SUBFAMILY OF HISTIDINE KINASES

Members of the HPK_{10} subfamily differ from most membrane-localized histidine kinases in that their kinase core domains lack a D box and contain only one asparagine in the N box. In addition, they possess a characteristic sensor domain with five to eight transmembrane segments [40, 49, 61]. This subfamily is of special interest, as all of its members seem to have the same type of ligand, namely peptide pheromones. If this turns out to be the case, it will be possible to identify new quorum-sensing systems, which employ peptide pheromones for intercellular communication, just by searching sequence databases for the presence of HPK_{10}-type histidine kinases. So far, these kinases have only been found in association with gram-positive bacteria, suggesting that they have evolved within this bacterial line of descent.

The exact topology of the sensor domain is not known for any HPK_{10} kinases, but judging from amino acid sequence alignments, it is not unreasonable to suppose that all members of this subfamily have membrane domains with very similar or identical two-dimensional topologies. Because membrane-embedded proteins are tricky to handle biochemically, are notoriously difficult to overexpress, and seldom form high-quality crystals, other methods than three-dimensional structure determination by crystallography and X-ray diffraction must also be employed to get structural information on these proteins. Fortunately, the theoretical prediction of structural aspects is simpler for membrane proteins than for globular proteins because the lipid layer imposes strong constraints on the degrees of freedom for the three-dimensional structure. Computer programs have been introduced that predict the two-dimensional topology of membrane proteins very accurately. When using the so-called Dense Alignment Surface (DAS) method [86], seven membrane-

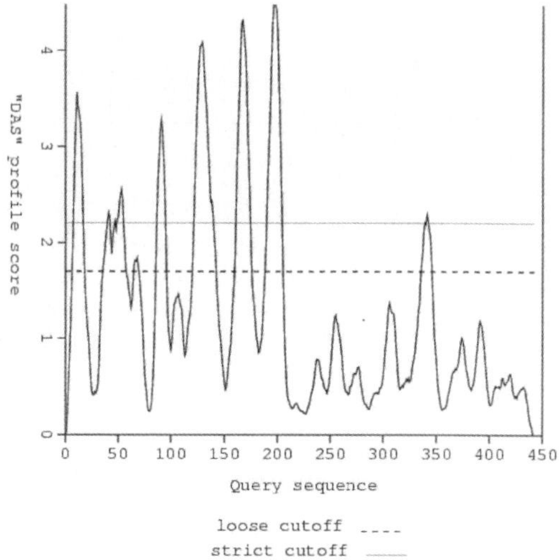

FIGURE 3 Prediction of transmembrane segments in histidine kinase ComD using the Dense Alignment Surface (DAS) method.

spanning regions were predicted in the N-terminal domain of ComD from *S. peumoniae* strain Rx (Fig. 3). The same result was obtained with TMAP, a method for the prediction of transmembrane segments from multiply aligned amino acid sequences [87]. An amino acid sequence alignment of different ComD proteins from *S. pneumoniae*, *S. mits* , *S. oralis*, *S. gordonii*, *S. sanguis*, and *S. anginosus* was used as input for this analysis. The most strongly preferred model predicted by the TMAP program placed the N terminus outside the cytoplasmic membrane, the kinase domain inside the cell, and seven transmembrane segments in between (Fig. 4). Although computer programs predicting two-dimensional topology of membrane proteins have become very reliable, experimental evidence is still required to verify the predicted models. One commonly used approach is to construct fusions of the protein of interest to alkaline phosphatase (PhoA), which will be active only when located in the periplasm. This method was used by Lina *et al.* [61] to study the topology of AgrC, the receptor of the thiolactone peptide pheromone from *S. aureus*. Similar to ComD, computer-aided analysis of AgrC suggested that its N-terminal half contains six or seven membrane-spanning segments. This predicted topology of AgrC was partially confirmed by the construction of PhoA fusions at various points in the pheromone receptor. However, the topology of the 60–100 amino acid residues at the extreme N-terminal end, which most likely contains three or four transmembrane segments, could not

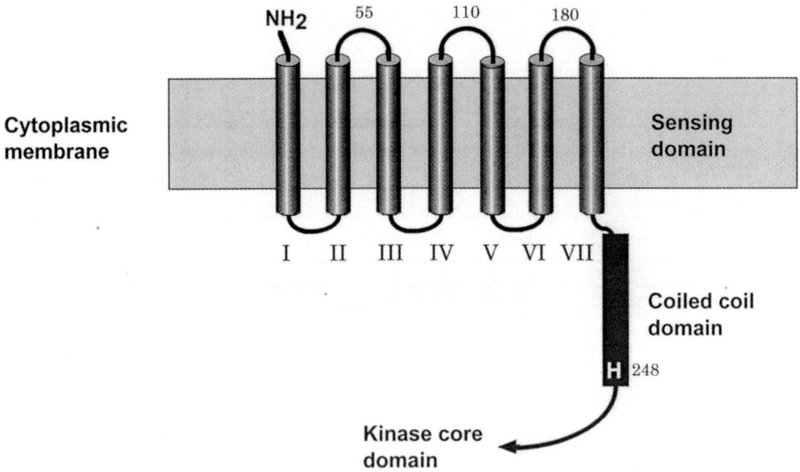

FIGURE 4 Predicted model for ComD topology and position of a coiled coil-like motif in the histidine kinase linker domain. Arabic numbers indicate approximate amino acid positions.

be resolved [61]. This region is also the most difficult to predict in ComD, where transmembrane segments II and III may be too short to span the membran if they have an α-helical configuration. Interestingly, highly homologous ComD receptors, which are specific for different but closely related pheromone types, are most divergent in the region containing transmembrane segments I, II, and III, suggesting that this part of the receptor determines the specificity of the receptor–ligand interaction [23, 40]. A comparison of the two dominating pherotypes found within the species *S. pneumoniae* clearly demonstrates the importance of this region with respect to ligand recognition. CSP-1 (EMRLSKFFRDFILQRKK) and CSP-2 (EMRISRIILDFLFLRKK) are approximately 50% identical, but the CSP-2 pheromone is unable to induce competence in *S. pneumoniae* strain Rx, which belongs to the CSP-1 pherotype. The amino acid sequences of the corresponding pheromone receptors, ComD-1 (strain Rx) and ComD-2 (strain A66), are identical except for 12 amino acid substitutions within the 60 N-terminal amino acids. Seven of the substitutions involve hydrophobic amino acid positions, suggesting that hydrophobic interactions between receptor and ligand may be important. Experiments addressing the effect of various changes in the primary structure of the CSP-1 pheromone support this view. An alanine scan carried out on CSP-1 [31] and analyses of the biological activity of different synthetic CSP-1/CSP-2 hybrid pheromones (unpublished results) show that several hydrophobic amino acid positions are very important for pheromone activity and specificy. The charged amino acids in positions 1 (E or D) and 3 (R) from the N-terminal end are also essential for pheromone activity, but do not seem to be involved in specificity as they are highly conserved in all streptococcal competence pheromones [23, 24].

An alanine scan carried out on the modified peptide pheromone produced by *S. aureus* group II (AgrDII) revealed that amino acids in both the ring and the tail region were critical for activation of the *agr* response [65]. The thiolactone bond between the sulfhydryl group of the conserved cysteine and the C terminus was also shown to be essential for the biological activity of the pheromone. Homologs containing a lactone or a lactame in place of the more reactive thiolactone bond had no activity, suggesting that *agr* activation may involve formation of a covalent bond between the pheromone and its AgrC receptor. Interestingly, neither the tail region nor the thiolactone bond was necessary for inhibition of the *agr* response in *S. aureus* strains belonging to other pherotypes, suggesting that activation and inhibition occur through different mechanisms [65]. In contrast to ComD, where the N-terminal 60–100 amino acid residues are important for ligand binding, deletion mutagenesis on AgrC seems to indicate that the most C-terminal extracellular loop in the membrane domain is sufficient for ligand binding and receptor activation [40, 61].

It is now generally accepted that dimerization is a prerequisite for signaling in most or all bacterial histidine kinases. It has also been established for several histidine kinases (i.e., EnvZ, CheA, and NtrB) that sensing of the external stimulus is followed by autophosphorylation in *trans* between the monomers [4, 4a]. How sensing of the input signal brings about the conformational change required for transducing the signal through the dimer is not understood, but structural information from EnvZ [88] has revealed that the dimerization domain is located in the cytoplasm between the most C-terminal transmembrane segment and the catalytic domain. However this region, also frequently called the linker domain is not present in CheA [89]. A coiled-coil motif within the linker domains of many histidine kinases has been identified [90]. Coiled-coils are an oligomerization motif consisting of two or more α-helices that wrap around each other with a slight left-handed superhelical twist. The appearance of this structural motif in a functionally important part of the receptors suggests that it plays a critical role in kinase regulation and signal transduction. By using the LEARNCOIL program [91] a coiled-coil motif, encompassing most of the H box and the region directly preceding this box, was also predicted in all members of the HPK_{10} subfamily (Figs. 4 and 5). It is reasonable to assume that this coiled-coil domain is responsible for receptor dimerization in HPK_{10} kinases. Presumably, structural changes in the sensor domain resulting from ligand binding will be relayed to the coiled-coil

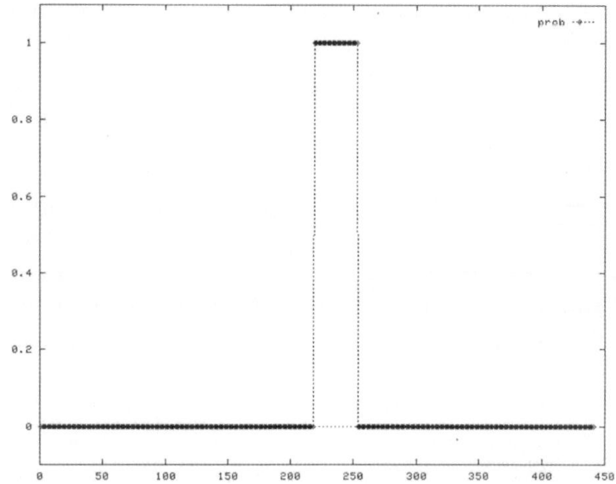

FIGURE 5 LEARNCOIL score using the amino acid sequence of histidine kinase ComD. Ordinate: probability. Abscissa: amino acid position. The H box histidine is located at position 248.

domain and give rise to small conformational shifts in this region, which stimulates autophosphorylation.

It is an open question whether signal transmission in HPK_{10} type kinases takes place by the same mechanism as in histidine kinases that have only two membrane-spanning segments in their sensor domains. Available data strongly indicate that dimerization and autophosporylation steps are common for all histidine kinases [4, 4a, 49]. However, this does not necessarily mean that sensing of the input signal and transfer of this signal to the dimerization domain take place by the same mechanisms in all membrane-localized histidine kinases. Considering the predicted topology of HPK_{10}-type receptors, it is likely that the extracellular parts of the five to eight transmembrane segments, and the relatively short loops connecting them, form the peptide binding pocket on the surface of the cytoplasmic membrane. Presumably, binding of the peptide ligand will be relayed to the dimerization domain by conformational rearrangements that involve all five to eight transmembrane segments. Because the sensor domains of most histidine kinase have a different architecture, consisting of only two transmembrane segments bordering a large extracellular loop, it is to be expected that they do not communicate with the cytoplasmic dimerization and catalytic domains in quite the same way as HPK_{10}-type sensor domains.

REFERENCES

1. Stock, J. B., Ninfa, A. J., and Stock, A. M. (1989). Protein phosphorylation and regulation of adaptive responses in bacteria. *Microbiol. Rev.* **53**, 450–490.
2. Parkinson, J. S., and Kofoid, E. C. (1992). Communication modules in bacterial signaling proteins. *Annu. Rev. Genet.* **26**, 71–112.
3. Hakenbeck, R., and Stock, J. B. (1996). Analysis of two-component signal transduction systems involved in transcriptional regulation. *Methods Enzymol.* **273**, 281–300.
4. Robinson, V. L., Buckler, D. R., and Stock, A. M. (2000). A tale of two components: A novel kinase and a regulatory switch. *Nature Struct. Biol.* **7**, 626–633.

4a. Stock, A. M., Robinson, V. L., and Goudreau, P. N. (2000). Two-component signal transduction. *Annu. Rev. Biochem.* **69**, 183–215.

5. Dunny, G. M., and Winans, S. C. (1999). "Cell–Cell Signaling in Bacteria." ASM Press, Washington, DC.
6. Fuqua,C., and Eberhard, A. (1999). Signal generation in autoinduction systems: Synthesis of acylated homoserine lactones by LuxI-type proteins. *In* "Cell–Cell Signaling in Bacteria" (G. M. Dunny and S. C. Winans, eds.), pp. 211–230. ASM Press, Washington, DC.
7. Surette, M. G., Miller, M. B., and Bassler, B. L. (1999). Quorum sensing in *Escherichia coli*, *Salmonella typhimurium*, and *Vibrio harveyi*: A new family of genes responsible for autoinducer production. *Proc. Natl. Acad. Sci. USA* **96**, 1639–1644.
8. Lilley, B. N., and Brassier, B. L. (2000). Regulation of quorum sensing in *Vibrio harveyi* by LuxO and Sigma-54. *Mol. Microbiol.* **36**, 940–954.
9. Lazazzera, B. A., and Grossman, A. D. (1998) The ins and out of peptide signaling. *Trends Microbiol.* **6**, 288–294.
10. Bassler, B. L. (1999). How bacteria talk to each other: Regulation of gene expression by quorum sensing. *Curr. Opin. Microbiol.* **2**, 582–587.
11. Dawson, M. H., and Sia, R. H. (1931). In vitro transformation of pneumococcal types. I. A technique for inducing transformation of pneumococcal types in vitro. *J. Exp. Med.* **54**, 681–699.
12. Pakula, R., and Walczak, W. (1963) On the nature of competence of transformable streptococci. *J. Gen. Microbiol.* **31**, 125–133.
13. Tomasz, A., and Hotchkiss, R. D. (1964). Regulation of the transformability of pneumococcal cultures by macromolecular cell products. *Proc. Natl. Acad. Sci. USA* **51**, 480–487.
14. Tomasz, A. (1965). Control of the competent state in pneumococcus by a hormone-like cell product: An example for a new type of regulatory mechanism in bacteria. *Nature* **208**, 155–159.
15. Leonard, C. G., and Cole, R. M. (1972). Purification and properties of streptococcal competance factor isolated from chemically defined medium. *J. Bacteriol.* **110**, 273–280.
16. Håvarstein, L. S., Coomaraswamy, G., and Morrison, D. A. (1995). An unmodified heptadecapeptide pheromone induces competence for genetic transformation in *Streptococcus pneumoniae*. *Proc. Natl. Acad. Sci. USA* **92**, 11140–11144.
17. Magnuson, R., Solomon, J., and Grossman, A. D. (1994). Biochemical and genetic characterization of a competence pheromone from *B. subtilis*. *Cell* **77**, 207–216.
18. Diep, D. B., Håvarstein, L. S., Nissen-Meyer, J., and Nes, I. F. (1994). The gene encoding plantaricin A, a bacteriocin from *Lactobacillus plantarum* C11, is located on the same transcription unit as an *agr*-like regulatory system. *Appl. Environ. Microbiol.* **60**, 160–166.
19. Diep, D. B., Håvarstein, L. S., and Nes, I. F. (1995). A bacteriocin-like peptide induces bacteriocin synthesis in *Lactobacillus plantarum* C11. *Mol. Microbiol.* **18**, 631–639.
20. Guangyong, J., Beavis, R. C., Novick, R. P. (1995). Cell density control of staphylococcal virulence mediated by an octapeptide pheromone. *Proc. Natl. Acad. Sci. USA* **92**, 12055–12059.

21. Kuipers, O. P., Beerthuyzen, M. M., de Ruyter, P. G. G. A., Luesink, E. J., and de Vos, W. M. (1995). Autoregulation of nisin biosynthesis in *Lactococcus lactis* by signal transduction. *J. Biol. Chem.* **270**, 27299–27304.
22. Clewell, D. B. (1999). Sex pheromone systems in enterococci. *In* "Cell–Cell Signaling in Bacteria" (G. M. Dunny and S. C. Winans, eds.), pp. 47–65. ASM Press, Washington, DC.
23. Håvarstein, L. S., Hakenbeck, R., and Gaustad, P. (1997). Natural competence in the genus *Streptococcus*: Evidence that streptococci can change pherotype by interspecies recombinational exchanges. *J. Bacteriol.* **179**, 6589–6594.
24. Whatmore, A. M., Barcus, V. A., and Dowson, C. G. (1999). Genetic diversity of the streptococcal competence (*com*) gene locus. *J. Bacteriol.* **181**, 3144–3154.
25. Eijsink, V. G. H., Brurberg, M. B., Middelhoven, P. H., and Nes, I. F. (1996). Induction of bacteriocin production in Lactobacillus sake by a secreted peptide. *J. Bacteriol.* **178**, 2232–2237.
26. Quadri, L. E. N., Kleerebezem, M., Kuipers, O. P., de Vos, W. M., Roy, K. L., Vederas, J. C., and Stiles, M. E. (1997). Characterization of a locus from *Carnobacterium piscicola* LV17B involved in bacteriocin production and immunity: Evidence for global inducer-mediated transcriptional regulation. *J. Bacteriol.* **179**, 6163–6171.
27. Nilsen, T., Nes, I. F., and Holo, H. (1998). An exported inducer peptide regulates bacteriocin production in *Enterococcus faecium* CTC492. *J. Bacteriol.* **180**, 1848–1854.
28. Nes, I. F., Diep, D. B., Håvarstein, L. S., Brurberg, M. B., Eijsink, V. G. H., and Holo, H. (1996). Biosynthesis of bacteriocins in lactic acid bacteria. *Antonie van Leeuwehoek* **70**, 113–128.
29. Nes, I. F., and Eijsink, V. G. H. (1999). Regulation of group II peptide bacteriocin synthesis by quorum-sensing mechanisms. *In* "Cell–Cell Signaling in Bacteria" (G. M. Dunny and S. C. Winans, eds.), pp. 175–192. ASM Press, Washington, DC.
30. Ennahar, S., Sashihara, T., Sonomoto, K., and Ishizaki, A. (2000). Class IIa bacteriocins: Biosynthesis, structure and activity. *FEMS Microbiol. Rev.* **24**, 85–106.
31. Håvarstein, L. S., and Morrison, D. A. (1999). Quorum-sensing and peptide pheromones in streptococcal competence for genetic transformation. *In* "Cell–Cell Signaling in Bacteria" (G. M. Dunny and S. C. Winans, eds.), pp. 9–26. ASM Press, Washington, DC.
32. Håvarstein, L. S., Holo, H., and Nes, I. F. (1994). The leader peptide of colicin V shares consensus sequences with leader peptides that are common among peptide bacteriocins produced by Gram-positive bacteria. *Microbiology* **140**, 2383–2389.
33. Håvarstein, L. S., Diep, D. B., and Nes, I. F. (1995). A family of bacteriocin ABC transporters carry out proteolytic processing of their substrates concomitant with export. *Mol. Microbiol.* **16**, 229–240.
34. Lagos, R., Villaneuva, J. E., and Monasterio, O. (1999).Identification and properties of the genes encoding microcin E492 and its immunity protein. *J. Bacteriol.* **181**, 212–217.
35. Rimini, R., Jansson, B., Feger, G., Roberts, T. C., de Francesco, M., Gozzi, A., Faggioni, F., Domenici, E., Wallace, D. M., Frandsen, N., and Polissi, A. (2000). Global analysis of transcription kinetics during competence development in *Streptococcus pneumoniae* using high density DNA arrays. *Mol. Microbiol.* **36**, 1279–1292.
36. Peterson, S., Cline, R. T., Tettelin, H., Sharov, V., and Morrison, D. A. (2000). Gene expression analysis of the *Streptococcus pneumoniae* competence reulons by use of DNA microarrays. *J. Bacteriol.* **182**, 6192–6202.
37. Hui, F. M., and Morrison, D. A. (1991). Genetic transformation in *Streptococcus pneumoniae*: Nucleotide sequence analysis shows *comA*, a gene required for competence induction, to be a member of the bacterial ATP-dependent transport protein family. *J. Bacteriol.* **173**, 372–381.
38. Pestova, E. V., Håvarstein, L. S., and Morrison, D. A. (1996). Regulation of competence for genetic transformation in *Streptococcus pneumoniae* by an auto-induced peptide pheromone and a two-component regulatory system. *Mol. Microbiol.* **21**, 853–862.

39. Morrison, D. A. (1997). Streptococcal competence for genetic transformation: Regulation by peptide pheromones. *Microbial Drug Resist.* **3**, 27–37.
40. Håvarstein, L. S., Gaustad, P., Nes, I. F., and Morrison, D. A. (1996). Identification of the streptococcal competence-pheromone receptor. *Mol. Microbiol.* **21**, 863–869.
41. Ween, O., Gaustad, P., and Håvarstein, L. S. (1999). Identification of DNA binding sites for ComE, a key regulator of natural competence in *Streptococcus pneumoniae. Mol. Microbiol.* **33**, 817–827.
42. Campbell, E. A., Choi, S. Y., and Masure, H. R. (1998) A competence regulon in *Streptococcus pneumoniae* revealed by genomic analysis. *Mol. Microbiol.* **27**, 929–939.
43. Lee, M. S., and Morrison, D. A. (1999). Identification of a new regulator in *Streptococcus pneumoniae* linking quorum sensing to competence for genetic transformation. *J. Bacteriol.* **181**, 5004–5016.
44. Diep, D. B., Håvarstein, L. S., and Nes, I. F. (1996). Characterization of the locus responsible for the bacteriocin production in *Lactobacillus plantarum* C11. *J. Bacteriol.* **178**, 4472–4483.
45. Brurberg, M. B. Nes, I. F., and Eijsink, V. G. H. (1997). Pheromone-induced production of antimicrobial peptides in *Lactobacillus. Mol. Microbiol.* **26**, 347–360.
46. Kleerebezem, M., Quadri, L. E. N., Kuipers, O. P., and de Vos, W. M. (1997). Quorum sensing by peptide pheromones and two-component signal transduction systems in Gram-positive bacteria. *Mol. Microbiol.* **24**, 895–904.
47. Risøen, P. A., Håvarstein, L. S., Diep, D. B., and Nes, I. F. (1998). Identification of the DNA-binding sites for two response regulators involved in control of bacteriocin synthesis in *Lactobacillus plantarum* C11. *Mol. Gen. Genet.* **259**, 224–232.
48. Lange, R., Wagner, C., de Saizieu, A., Flint, N., Molnos, J., Stieger, M., Caspers, P., Kamber, M., Keck, W., and Amrein, K. E. (1999). Domain organization and molecular characterization of 13 two-component systems identified by genome sequencing of *Streptococcus pneumoniae. Gene* **237**, 223–234.
49. Grebe, T. W., and Stock, J. B. (1999). The histidine protein kinase superfamily. *In* "Advances in Microbial Physiology" (R. K. Poole, ed.), Vol. 41, pp. 139–227. Academic Press, London.
50. de Saizieu, A., Gardès, C., Flint, N., Wagner, C., Kamber, M., Mitchell, T. J., Keck, W., Amrein, K. E., and Lange, R. (2000). Microarray-based identification of a novel *Streptococcus pneumoniae* regulon controlled by an autoinduced peptide. *J. Bacteriol.* **182**, 4696–4703.
51. Reichmann, P., and Hakenbeck, R. (2000). Allelic variation in a peptide-inducible two component system of *Streptococcus pneumoniae. FEMS Microbiol. Lett.* **190**, 231–236.
52. Sahl, H. G., and Bierbaum, G. (1998). Lantibiotics: Biosynthesis and biological activities of uniquely modified peptides from Gram-positive bacteria. *Annu. Rev. Microbiol.* **52**, 41–79.
53. Baba, T., and Schneewind, O. (1998). Instruments of microbial warfare: Bacteriocin synthesis, toxicity and immunity. *Trends Microbiol.* **6**, 66–71
54. Kleerebezem, M., de Vos, W. M., and Kuipers, O. P. (1999). The lantibiotics nisin and subtilin act as extracellular regulators of their own biosynthesis. *In* "Cell–Cell Signaling in Bacteria" (G. M. Dunny and S. C. Winans, eds.), pp. 159–174. ASM Press, Washington, DC.
55. Kuipers, O. P., Beerthuizen, M. M., Siezen, R. J., and de Vos, W. M. (1993). Characterization of the nisin gene cluster *nisABTCIPR* of *Lactococcus lactis,* requirement of expression of the *nisA* and *nisI* genes for development of immunity. *Eur. J. Biochem.* **216**, 281–291.
56. Kuipers, O. P., de Ruyter, P. G. G. A., Kleerebezem, M., and de Vos, W. M. (1998). Quorum sensing-controlled gene expression in lactic acid bacteria. *J. Biotech.* **64**, 15–21.
57. Morfeldt, E., Janzon, L., Arvidson, S., and Löfdahl, S. (1988). Cloning of a chromosomal locus *(exp)* which regulates the expression of several exoprotein genes in *Staphylococcus aureus. Mol. Gen. Genet.* **211**, 435–440.
58. Peng, H. L., Novick, R. P., Kreiswirth, B., Kornblum, J., and Schlievert, P. (1988).Cloning, characterization and sequencing of an accessory gene regulator *(agr)* in *Staphylococcus*

aureus. *J. Bacteriol.* **170**, 4365–4372.

59. Novick, R. P. (1999). Regulation of pathogenicity in *Staphylococcus aureus* by a peptide-based density-sensing system. *In* "Cell–Cell Signaling in Bacteria" (G. M. Dunny and S. C. Winans, eds.), pp. 129–146. ASM Press, Washington, DC.
60. Novick, R. P. (2000). Pathogenicity factors and their regulation. *In* "Gram-Positive Pathogens" (V. A. Fischetti, R. P. Novick, J. J. Ferretti, D. A. Portnoy, and J. I. Rood, eds.), pp. 392–407. ASM Press, Washington, DC.
61. Lina, G., Jarraud, S., Ji, G., Greenland, T., Pedraza, A., Etienne, J., Novick, R. P., and Vandenesch, F. (1998). Transmembrane topology and histidine protein kinase activity of AgrC, the *agr* signal receptor in *Staphylococcus aureus*. *Mol. Microbiol.* **28**, 655–662.
62. Cheung, A. L. and Projan, S. J. (1994). Cloning and sequencing of *sarA* of *Staphylococcus aureus*, a gene required for the expression of *agr*. *J. Bacteriol.* **176**, 4168–4172.
63. Chien, Y., and Cheung, A. L. (1998). Molecular interactions between two global regulators, *sar* and *agr*, in *Staphylococcus aureus*. *J. Biol. Chem.* **273**, 2645–2652.
64. Lyon, G. J., Mayville, P., Muir, T. W., and Novick, R. P. (2000). Rational design of a global inhibitor of the virulence response in *Staphylococcus aureus*, based in part on localization of the site of inhibition to the receptor-histidine kinase, AgrC. *Proc. Natl. Acad. Sci. USA* **97**, 13330–13335.
65. Mayville, P., Ji, G., Beavis, R., Yang, H., Goger, M., Novick, R. P., and Muir, T. W. (1999). Structure-activity analysis of synthetic autoinducing thiolactone peptides from *Staphylococcus aureus* responsible for virulence. *Proc. Natl. Acad. Sci. USA* **96**, 1218–1223.
66. Dubnau, D. (1991). Genetic competence in *Bacillus subtilis*. *Microbiol. Rev.* **55**, 395–424.
67. Lazazzera, B. A., Palmer, T., Quisel, J., and Grossman A. D. (1999). Cell density control of gene expression and development in *Bacillus subtilis*. *In* "Cell–Cell Signaling in Bacteria" (G. M. Dunny and S. C. Winans, eds.), pp. 27–46. ASM Press, Washington, DC.
68. Solomon, J. M., Magnuson, R., Srivastava, A., and Grossman, A. D. (1995). Convergent sensing pathways mediate response to two extracellular competence factors in *Bacillus subtilis*. *Genes Dev.* **9**, 547–558.
69. Solomon, J. M., Lazazzera, B. A., and Grossman, A. D. (1996). Purification and characterization of an extracellular peptide factor that effects two different developmental pathways in *Bacillus subtilis*. *Genes Dev.* **10**, 2014–2024.
70. Roggiani, M., and Dubnau, D. (1993). ComA, a phosphorylated response regulator protein of *Bacillus subtilis*, binds to the promoter region of *srfA*. *J. Bacteriol.* **175**, 3182–3187.
71. Hamoen, L. W., Eshuis, H., Jongbloed, J., Venema, G., and Van Sinderen, D. (1995). A small gene, designated *comS*, located within the coding region of the fourth amino acid-activation domain of *srfA*, is required for competence development in *Bacillus subtilis*. *Mol. Microbiol.* **15**, 55–63.
72. Van Sinderen, D., Luttinger, A., Kong, L., Dubnau, D., Venema, G., and Hamoen, L. (1995). *ComK* encodes the competence transcription factor, the key regulatory protein for competence development in *Bacillus subtilis*. *Mol. Microbiol.* **15**, 455–462.
73. Pozzi, G., Masala, L., Iannelli, F., Manganelli, R., Håvarstein, L.S., Piccoli, L., Simon, D., and Morrison, D. A. (1996). Competence for genetic transformation in encapsulated strains of *Streptococcus pneumoniae*: Two allelic variants of the peptide pheromone. *J. Bacteriol.* **178**, 6087–6090.
74. Jarraud, S., Lyon, G. J., Figueiredo, A. M. S., Gérard, L., Vandenesch, F., Etienne, J., Muir, T. W., and Novick, R. P. (2000). Exfoliatin-producing strains define a fourth *agr* specificity group in *Staphylococcus aureus*. *J. Bacteriol.* **182**, 6517–6522.
75. Ji, G., Beavis, R., and Novick, R. P. (1997). Bacterial interference caused by autoinducing peptide variants. *Science* **276**, 2027–2030.
76. Novick, R. P., and Muir, T. W. (1999). Virulence gene regulation by peptides in staphylococci

and other Gram-positive bacteria. *Curr. Opin. Microbiol.* **2**, 40–45.
77. Phan Tran, L. S., Nagai, T., and Itoh, Y. (2000). Divergent structure of the ComQXPA quorum-sensing components: Molecular basis of strain-specific communication mechanism in *Bacillus subtilis. Mol. Microbiol.* **37**, 1159–1171.
78. Ottolenghi, E., and Hotchkiss, R. D. (1960). Appearance of genetic transforming activity in pneumococcal cultures. *Science* **132**, 1257–1258.
79. Lorenz, M. G., and Wackernagel, W. (1994). Bacterial gene transfer by natural genetic transformation in the environment. *Microbiol. Rev.* **58**, 563–602.
80. Piazza, F., Tortosa, P., and Dubnau, D. (1999). Mutational analysis and membrane topology of ComP, a quorum-sensing histidine kinase of *Bacillus subtilis* controlling competence development. *J. Bacteriol.* **181**, 4540–4548.
81. Egger, L. A., Park, H., and Inouye. M. Signal transduction via the histidyl-aspartyl phosphorelay. *Genes Cells* **2**, 167–84.
82. Ba-Thein, W., Lyristis, M., Ohtani, K., Nisbet, I. T., Hayashi, H., Rood, J. I., and Shimizu, T. (1996). The *virR/virS* locus regulates the transcription of genes encoding extracellular toxin production in *Clostridium perfringens. J. Bacteriol.* **178**, 2514–2520.
83. Qin, X., Singh, K. V., Weinstock, G. M., and Murray, B. E. (2000). Effects of *Enterococcus faecalis fsr* genes on production of gelatinase and a serine protease and virulence. *Infect. Immun.* **68**, 2579–2586.
84. Imagawa, T., Tatsuki, T., Higashi, Y., and Amano, T. (1981). Complementation characteristics of newly isolated mutants from two groups of strains of *Clostridium perfringens. Biken J.* **24**, 13–21.
85. Imagawa, T., and Higashi, Y. (1992). An activity which restores theta toxin activity in some theta toxin-deficient mutants of *Clostridium perfringens. Microbiol. Immunol.* **36**, 523–527.
86. Cserzo, M., Wallin, E., Simon, I., von Heijne, G., and Elofsson, A. (1997). Prediction of transmembrane alpha-helices in prokaryotic membrane proteins: The dense alignment surface method. *Protein Eng.* **10**, 673–676.
87. Persson, B., and Argos, P. (1997). Prediction of membrane protein topology utilizing multiple sequence alignments. *J. Protein. Chem.* **16**, 453–457.
88. Tomomori, C., Tanaka, T., Dutta, R., Park, H. Y., Saha, S. K., Zhu, Y., Ishima, R., Liu, D. J., Tong, K. I., Kurokawa, H., Qian, H., Inouye, M., and Ikura, M. (1999). Solution structure of the homodimeric core domain of *Escherichia coli* histidine kinase EnvZ. *Nature Struct. Biol.* **6**, 729–734.
89. Bilwes, A. M., Alex, L. A., Crane, B. R., and Simon, M. I. (1999). Structure of CheA, a signal-transducing histidine kinase. *Cell* **96**, 131–141.
90. Singh, M., Berger, B., Kim, P. S., Berger, J. M., and Cochran, A. G. (1998). Computational learning reveals coiled coil-like motifs in histidine kinase linker domains. *Proc. Natl. Acad. Sci. USA* **95**, 2738–2743.
91. Berger, B., and Singh, M. (1997). An interative method for improved protein structural motif recognition. *J. Comput. Biol.* **4**, 261–273.

CHAPTER 17

Initiation of Bacterial Killing by Two-Component Sensing of a "Death Peptide": Development of Antibiotic Tolerance in *Streptococcus pneumoniae*

RODGER NOVAK
Institute of Microbiology and Genetics, Vienna Biocenter, Vienna A-1030, Austria

ELAINE TUOMANEN
Department of Infectious Diseases, St. Jude Children's Research Hospital, Memphis, Tennessee 38105

Classically, the term apoptosis has referred to programmed cell death in eukaryotic cells. In today's more expanded view, programmed cell death describes an active process that results in any form of cell death triggered by an intracellular death program in eukaryotic or prokaryotic organisms. In bacteria, the most obvious form of cell death is initiated by the action of bactericidal antibiotics. A broad scientific audience has acknowledged that antibiotic-induced bacterial death results from an active, bacterial-driven triggering of an innate death program. This concept is illustrated by the finding that the binding of β-lactam antibiotics to their bacterial targets stops bacterial growth but does not lead directly to cell death. Rather, a second process arising from the bacteria itself is necessary to trigger endogenous suicidal enzymes that dissolve the cell wall during autolysis, an event leading ultimately to cell death. The discovery in *Streptococcus pneumoniae* of a bacterial-derived peptide signal and a histidine kinase-mediated signal transduction apparatus that recognizes and delivers the death signal provides strong evidence for a link between an extracellular trigger and an intracellular death program.

INTRODUCTION

Bacterial Cell Death

Addiction Modules and Other Cell Death Mechanisms

The simplest death system would involve a toxin, sufficient to kill on its own, and a mechanism for protection of the toxin-producing cell. In many gram-negative bacteria, programmed cell death is mediated through so-called addiction modules that consist of two such genes (for reviews, see Engelberg-Kulka and Glaser [1]). This pair of genes encodes a stable toxin and a less stable antitoxin that prevents the lethal action of the toxin [2, 3]. These systems are usually plasmid encoded, and plasmid-cured cells are killed selectively because the unstable antitoxin is degraded faster than the more stable toxin. Thus, survival depends on the presence of the plasmid, i.e., addiction. These primitive systems involve little "signaling" for their behavior.

A more complex approach is used by organisms that produce antibiotics. Some of these systems are rather simple, such as colicins and microcins , but others involve true signal cascades embedded in normal bacterial physiology. Colicins and microcins act only after release from producer cells and subsequent penetration of their nonproducer target cells [4]. Although these diverse lethal components may target similar sites, their modes of action remain complex. ColicinE and related substances kill by membrane depolarization, similar to the intracellular toxins of the Gef family [5, 6]. DNA is the target of restriction-modification systems, such as Pae R7 and *Eco*RI, and of toxins, such as microcin B17 or CcdB [6]. Restriction enzymes cleave DNA directly, whereas microcin B17 and CcdB convert DNA gyrase into a DNA-damaging agent, leading to DNA degradation similar to that observed in eukaryotic apoptosis.

Although the physiological role of components such as polyketides and nonribosomal peptides in normal bacterial metabolism is often unclear, some of these substances have a clearly defined antibacterial activity, leading to cell death solely experienced by the target cells. Some of these products, produced primarily by actinomycetes, bacilli, and filamentous fungi, are widely used antibiotics, including rifamycin (inhibitor of bacterial RNA polymerase), penicillin, and vancomycin (cell wall biosynthesis inhibitor). Most interestingly, although the use of penicillin dates back to the beginning of the 1940s, the biological mode of action remains obscure. There is now growing evidence that penicillins and other cell wall synthesis inhibitors trigger bacterial programmed death pathways that are normally employed during cell death in the stationary phase. These newly recognized pathways seem to be regulated by classical two-component regulatory systems [7]. It is these latter signaling systems that are discussed in this chapter.

The Mode of Action of β-Lactams and the Concept of Antibiotic Tolerance

Bacteria are stabilized by the peptidoglycan that completely encloses the cell in a cytoskeleton [8]. Maintenance of this covalently closed network requires enzymes capable of cleaving and reorganizing the cell wall structure during bacterial growth and cell separation. However, in some bacteria, such as *S. pneumoniae*, cell wall hydrolases can also act as autolysins, completely degrading the cell wall and thereby representing suicidal enzymes [9]. β-Lactams, like penicillin, require these autolytic enzymes to induce bacterial death. Pneumococci lacking the major cell wall hydrolase (amidase, LytA) stop growing but do not undergo lysis or cell death upon exposure to penicillin, a phenomenon called antibiotic tolerance [10]. Thus, after β-lactams inhibit growth directly, irreversible lethal effects are caused by a set of secondary events that activate the autolytic system [11]. Thus, bacteria actively cooperate using their own enzymatic death machinery to execute the final killing event. Until recently, the only known element of this machinery was the set of cell wall hydrolases. Identification of a trigger signal involved in the regulation of pneumococcal hydrolases/autolysins has provided the first insights into the role of two-component systems in a complex bacterial cell death program [7].

CELL DEATH AND SIGNAL TRANSDUCTION

THE *VEX-PEP*^{27}*-VNCR/S* SYSTEM

Identification of the Two-Component System VncR/S

The bactericidal activity of antibiotics that inhibit cell wall synthesis relies on the activation of bacterial-encoded cell wall hydrolases. In pneumococcus, three autolytic enzymes serve this hydrolytic function. To detect genes that could trigger the pneumococcal autolysins, a library of loss of function mutants [12] was screened for survival upon treatment with penicillin (i.e., tolerance) [13]. The screen identified over 10 mutants that contained an active autolysin but were tolerant to antibiotics. One of the mutants harbored an interruption in a gene encoding a putative histidine kinase, VncS, of a two-component regulatory system. Sequence analysis of the gene locus revealed an upstream contiguous response regulator gene, *vncR* [14] (Fig. 1). Based on sequence alignments, the histidine kinase VncS and the response regulator VncR were assigned to the OmpR/Pho subfamily [15, 16]. The C terminus of the VncS protein contains four of five blocks of conserved amino acids characteristic for transmitter modules of histidine kinases [17]. The D box, which is likely to be involved in the binding of Mg, was missing, a finding similar to the *Enterococcus faecium* histidine kinase $VanS_B$ [18]. The predicted site for autophosphorylation was found at amino acid positions 239 to 246, $X_2HEX_{3\text{-}4}P$, and was identical to the site in $VanS_B$. Because the first 170 amino acids of VncS contain two clusters of hydrophobic amino acids, which could correspond to membrane-spanning regions, VncS is most likely a membrane protein.

The ABC Transporter *vex* and Identification of the Putative Signaling Peptide Pep^{27}

Analysis of the publicly available genome of *S. pneumoniae* revealed a locus encoding a putative ABC transporter directly upstream of *vncR* [7] (Fig. 1). The putative ABC transporter was named Vex and was predicted to have a

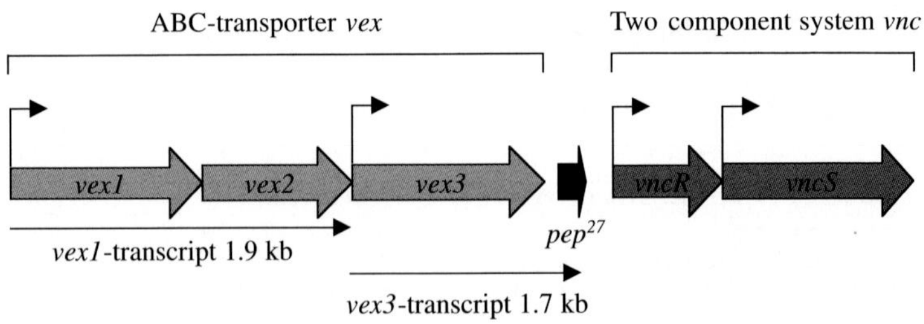

FIGURE 1 Schematic diagram of the locus signaling death in pneumococci. Vex is the peptide transporter. Pep^{27} is the death peptide signal. VncS is the histidine kinase proposed to sense Pep^{27}, whereas VncR is the DNA binding protein paired with VncS. Transcripts detected by Northern analysis are indicated by arrows.

fused ABC–ABC organization with heterodimeric transmembrane domains (TMDs). The genes *vex1* and *vex3* encode putative hydrophobic transmembrane proteins each consisting of four TMDs. *vex1* and *vex3* flank the gene *vex2* that encodes an ATP-binding cassette (ABC) protein, including the Walker A motif $GX_4GK(S/T)$ [19] at amino acid positions 41 to 46 and the Walker B motif $(R/K)X_{6\text{-}8}hyd_4D$ (hyd, hydrophobic residues) [20, 21], at amino acid positions 142 to 156. The majority of ABC transporters are associated with periplasmic-binding proteins, which provide the primary substrate-binding site for the uptake of solute into cells [22]. The absence of such a protein in the putative Vex ABC transporter indicates that it is expected to be involved in substrate export.

Northern blot analysis of the *vex* locus suggested that *vex1* and *vex2* are transcribed from one promoter located upstream of *vex1*. Transcription of *vex3* is initiated from another promoter upstream of *vex3*. Unexpectedly, the *vex3*-specific mRNA hybrid was 1.7 kb, 0.3 kb larger than the predicted size of the open reading frame. This suggested the possibility of a large untranslated 5' messenger RNA or the cotranscription of an 84-bp large open reading frame located directly downstream of *vex3*. Northern blot analysis confirmed the latter option. The identification of two putative stem loop structures directly downstream of the small open reading frame, designated pep^{27}, suggested the termination of transcription directly downstream of pep^{27}. To confirm transcription/translation of pep^{27}, the linear peptide — MRKEFHNVLSSGQLLADKRPARDYNRK — was synthesized and used to raise specific antibodies. Western blot analysis of the cytoplasmic fraction and the supernatant of wild-type pneumococci demonstrated a 3-kDa product reactive with the Pep^{27}-specific antiserum.

The *vex*-pep^{27}-*vncR/S* System: A Classical Quorum-Sensing System?

Peptide signaling as a form of cell density-dependent gene expression is widespread in bacteria. The activation of peptide expression is often characterized by autoinduction, a process where accumulation of a low level of peptide acts as a stimulus to further induce peptide expression. Autoinduction usually engenders a population-wide response. This molecular mechanism employed by bacteria to monitor cell density is termed quorum sensing [23]. These quorum-sensing-dependent regulatory mechanisms appear to follow a common theme in gram-positive bacteria in which the peptide is secreted by a dedicated ABC transporter and it subsequently functions as a signal for a specific sensor component of a two-component signal-transduction system [24].

The chromosomal arrangement of the two component system VncR/S and the ABC transporter Vex shows characteristic features of a gene locus involved in gram-positive cell density-dependent peptide signaling [24], in that genes involved in the export of peptides are usually found adjacent to the structural gene they transport [25–27]. Northern blot analysis demonstrated that transcription of *vncR* was cell density dependent, peaking during the stationary phase. In contrast, transcriptional levels of the *vex* locus and the *pep*^{27} gene seemed to be constitutive, and Western blot analysis suggested an extracellular accumulation of the peptide during logarithmic growth. One might therefore speculate that successful signal transduction would involve extracellular accumulation of the signal peptide and increased expression of the receptor two-component system.

The *vex-pep*^{27}-*vncR/S* System and Cell Death

The Role of VncR/S in Mediating Antibiotic Induced Cell Death

Because loss of function of the histidine kinase conveyed the phenotype of antibiotic tolerance, the role of the kinase VasS and its cognate partner, the response regulator VncR, in promoting death were investigated further. *vncS*-deficient mutants survived β-lactam antibiotics, even though growth was inhibited as effectively as in the wild-type strain. This indicated that the antibiotics were binding readily to their bacterial targets in the *vncS* mutants but that the autolytic enzymes were not subsequently triggered. Interestingly, mutation of the response regulator *vncR* did not lead to an antibiotic-tolerant phenotype and an overexpression of VncS suppressed autolysis. Based on these findings, a proposed model of VncS/R activity suggested that the system functioned to suppress autolysis (Fig. 2). When VncR is phosphorylated (active), genes that would be turned on in response to antibiotics — and kill the bacteria — are switched off. Mutations in the VncS component mean that VncR cannot be switched off, with the result that the death promoting genes remain inactive [14, 28].

Induction of Cell Death by Pep^{27}

Cultures of wild-type pneumococci exposed to synthetic Pep^{27} undergo a potent, dose-dependent loss of viability. At a concentration of 0.05 μM, the inhibition of growth was marginal, whereas at >50 μM the loss of viability was ≥2 log bacteria over 4 h. Analysis of different peptide structural variants indicated that the entire molecule was required for activity.

Several mechanisms of action can be considered for the biological effect of Pep^{27}. The net charge of +4 raised the possibility that Pep^{27} might be a cationic peptide. These peptides, which are widely spread among gram-

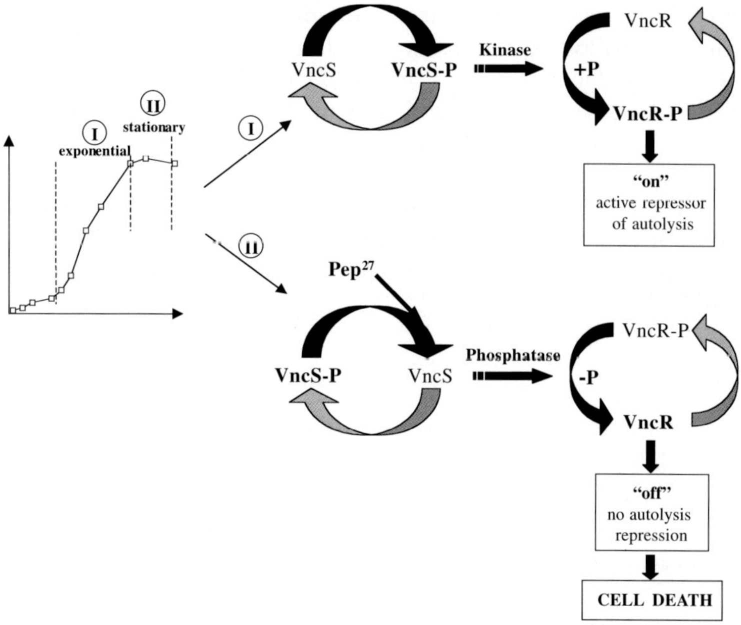

FIGURE 2 Model of regulation of autolytic activity in *S. pneumoniae*. Environmental signals regulate the addition of a phosphoryl group (P) to the sensor kinase (VncS). This, in turn, controls whether the response regulator (VncR) is on (phosphorylated) or off (dephosphorylated). When VncR is phosphorylated, genes that are turned on in response to antibiotics or stationary phase — and induce activation of autolysin, killing the bacteria — are switched off. One of the trigger signals for bacterial lysis seems to be the peptide Pep^{27}, which acts in a quorum-sensing manner. It is probably sensed by the two-component system VncS/R and determines with that the dephosphorylation of VncR, leading to cell death. It remains to be established how and where inhibition of cell wall synthesis by antibiotics feed into the death peptide pathway.

positive bacteria, interact directly with the bacterial cell membrane and do not require any receptor structure [29]. However, Pep^{27} failed to kill a *vncS*-deficient mutant consistent with the need for VncS to act as a receptor. The synthetic peptide was also not antibacterial against different streptococcal species and *Staphylococcus aureus*, further suggesting the need for a receptor.

Pep^{27} Triggers Different Death Pathways through VncS

Cell death associated with cell lysis is the most obvious and dramatic form of death in *S. pneumoniae*. Most pneumococcal strains employ this form of death predominantly, and mutation in the autolysin LytA leads to an antibiotic-tolerant phenotype. Although less prominent, pneumococci also possess nonlytic death pathways (referred to globally as *cid* loci) [30, 31].

To identify the ability of Pep^{27} to act independently of LytA, the effect of the peptide on the pneumococcal mutant Lyt-4-4, deficient in the major autolysin, was investigated. The mutant Lyt-4-4 was killed by Pep^{27} but not as efficiently as wild-type R6. The difference in the rate of killing of R6 versus Lyt-4-4 indicates that LytA participates in peptide-induced killing. However, these data also support the contention that in addition to LytA, alternative nonlytic death pathways exist [30] and that these pathways might also be activated by Pep^{27}.

Pep^{27} and the Stringent Response Mechanism

The stringent response is a very powerful mechanism to physiologicaly down-regulate autolysis during deprivation of an essential nutrient [32]. In this setting, antibiotic-induced lysis is blocked by an as yet uncharacterized defect in autolysin activation. Why β-lactam antibiotics bind readily to penicillin-binding proteins of starved bacteria but fail to activate hydrolytic enzymes remains undetermined.

Under starvation conditions, neither the addition of ten times the MIC of penicillin or vancomycin nor the addition of 200 μ*M* Pep^{27} alone resulted in lysis or death. However, a combination of antibiotic with Pep^{27} resulted in significant lysis of the cells. The minimum concentration of penicillin required to cause autolysis was equal to its MIC (0.1 μg/ml). This suggests that the peptide Pep^{27}might be able to bypass or override the block to lysis engendered by the stringent response.

Model

The analysis of genetic loci adjacent to the VncS/VncR system revealed an upstream ABC transporter, Vex, and a small orf encoding a 27 amino acid peptide, Pep^{27} [7]. The peptide appears to be exported by the dedicated ABC transporter, Vex. The model indicates that Pep^{27} functions as a signal necessary for the triggering of death pathways in *S. pneumoniae*. Cell death is triggered by the peptide in wild-type and autolysin-deficient strains, suggesting that the two-component system initiates several death pathways in pneumococcus.

The following model integrates the biology of the peptide Pep^{27} with the signal transduction pathway initiated by the two-component regulatory system VncR/S. During the logarithmic phase, it was suggested that the response regulator VncR is phosphorylated and binds chromosomal DNA, repressing autolytic and other death systems [14] (Fig. 2). Under these conditions, the *pep*^{27} gene is transcribed constitutively at low levels and Pep^{27} accumulates gradually in the extracellular compartment. During the stationary phase, which is characterized by autolysis, the concentration of Pep^{27} reaches a critical concentration sensed by the histidine kinase VncS. Under these conditions, VncS works predominantly as a phosphatase, leading to dephosphorylation of the response regulator VncR. The dephosphorylated response regulator detaches from its DNA binding site, initiating a change in target gene transcription, finally triggering the autolytic event. While these data suggest that the bactericidal effects of β-lactam antibiotics are mediated via the signal transduction pathway initiated by the two-component system VncR/S, it remains to be established how and where inhibition of cell wall synthesis by antibiotics feed into the death peptide pathway.

SUMMARY AND PERSPECTIVES

The detailed study of the VncS/R two-component system in pneumococcus has revealed a signal transduction cascade impacting on the control of an endogenous death pathway. As is true for signaling between gram-positive bacteria for other purposes, such as DNA transformation or virulence induction, the components of the death system are a signaling peptide, a peptide transporter, and a two-component system. This paradigm may be more broadly applicable in that the regulation of autolysis in *Staphylococcus aureus* has also revealed a role for a two-component system LytS/R, suggesting that autolytic activity responds to external signals in another gram-positive bacteria [33]. In contrast to the pneumococcus, the *lytS/R* regulatory system regulates *lrgA* and *lrgB* transcriptionally [34]. The function of the gene *lrgAB* is thought to be similar to that of a bacteriophage-encoded antiholin, which inhibits the formation of murein hydrolase transport channels in the bacterial membrane [35]. Mutations of the *lytS* kinase gene lead to a decreased expression of the *lrgAB* operon, resulting in reduced extracellular murein hydrolase activity. One result of a system involving "inside out" signaling is that an entire bacterial population can respond at once to the stimulus appearing in the medium. Thus, the bacterial population exhibits control of community behavior through two-component signal transduction.

REFERENCES

1. Engelberg-Kulka, H., and Glaser, G. (1999). Addiction modules and programmed cell death and antideath in bacterial cultures. *Annu. Rev. Microbiol.* 53, 43–70.

2. Jensen, R. B., and Gerdes, K. (1995). Programmed cell death in bacteria: Proteic plasmid stabilization systems. *Mol. Microbiol.* **17**, 205–210.
3. Lehnherr, H., Maguin, E., Jafri, S., and Yarmolinsky, M. B. (1993). Plasmid addiction genes of bacteriophage P1: doc, which causes cell death on curing of prophage, and phd, which prevents host death when prophage is retained. *J. Mol. Biol.* **233**, 414–428.
4. Kolter, R., and Moreno, F. (1992). Genetics of ribosomally synthesized peptide antibiotics. *Annu. Rev. Microbiol.* **46**, 141–163.
5. Jaffe, A., Ogura, T., and Hiraga, S. (1985). Effects of the ccd function of the F plasmid on bacterial growth. *J. Bacteriol.* **163**, 841–849.
6. Thisted, T., Nielsen, A. K., and Gerdes, K. (1994). Mechanism of post-segregational killing: Translation of Hok, SrnB and Pnd mRNAs of plasmids R1, F and R483 is activated by 3'-end processing. *EMBO J.* **13**, 1950–1959.
7. Novak, R., Charpentier, E., Braun, J. S., and Tuomanen, E. (2000). Signal transduction by a death signal peptide: Uncovering the mechanism of bacterial killing by penicillin. *Mol. Cell* **5**, 49–57.
8. Weidel, J., and Pelzer, H. (1964). Bag shaped macromolecules: A new look on bacterial cell walls. *Adv. Enzymol.* **26**, 193–232.
9. Tomasz, A. (1983). Murein hydrolases: Enzymes in search of a physiological function. *In* "The Target of Penicillin" (R. Hakenbeck, J. Holtje, and L. H., eds.) pp. 155–172. Walter de Gruyter, Berlin.
10. Tomasz, A., Albino, A., and Zanati, E. (1970). Multiple antibiotic resistance in a bacterium with suppressed autolytic system. *Nature* **227**, 138–140.
11. Tomasz, A., and Holtje, J. V. (1977). Murein hydrolases and the lytic and killing action of penicillin. *In* "Microbiology" (D. Schlesinger, ed.) pp. 209–215. American Society for Microbiology, Washington, DC.
12. Pearce, B. J., Yin, Y. B., and Masure, H. R. (1993). Genetic identification of exported proteins in *Streptococcus pneumoniae*. *Mol. Microbiol.* **9**, 1037–1050.
13. Williamson, R., and Tomasz, A. (1980). Antibiotic-tolerant mutants of *Streptococcus pneumoniae* that are not deficient in autolytic activity. *J. Bacteriol.* **144**, 105–113.
14. Novak, R., Henriques, B., Charpentier, E., Normark, S., and Tuomanen, E. (1999). Emergence of vancomycin tolerance in *Streptococcus pneumoniae*. *Nature* **399**, 590–593
15. Stock, J. B., Ninfa, A. J., and Stock, A. M. (1989). Protein phosphorylation and regulation of adaptive responses in bacteria. *Microbiol. Rev.* **53**, 450–490.
16. Throup, J. P., Koretke, K. K., Bryant, A. P., Ingraham, K. A., Chalker, A. F., Ge, Y., Marra, A., Wallis, N. G., Brown, J. R., Holmes, D. J., Rosenberg, M., and Burnham, M. K. (2000). A genomic analysis of two-component signal transduction in *Streptococcus pneumoniae*. *Mol. Microbiol.* **35**, 566–576.
17. Parkinson, J. S., and Kofoid, E. C. (1992). Communication modules in bacterial signaling proteins. *Annu. Rev. Genet.* **26**, 71–112.
18. Holman, T. R., Wu, Z., Wanner, B. L., and Walsh, C. T. (1994). Identification of the DNA-binding site for the phosphorylated VanR protein required for vancomycin resistance in *Enterococcus faecium*. *Biochemistry* **33**, 4625–4631.
19. Walker, J. E., Saraste, M., Runswich, M. J., and Gay, N. J. (1982). Distantly related sequences in the alpha- and beta-subunits of ATP synthase, myosin, kinases, and other ATP-requiring enzymes and a common nucleotide binding fold. *EMBO J* **1**, 945–951.
20. Ames, G. F., Mimura, C. S., and Shyamala, V. (1990). Bacterial periplasmic permeases belong to a family of transport proteins operating from *Escherichia coli* to human: Traffic ATPases. *FEMS Microbiol. Rev.* **6**, 429–446.
21. Hyde, S. C., Emsley, P., Hartshorn, M. J., Mimmack, M. M., Gileadi, U., Pearce, S. R., Gallagher, M. P., Gill, D. R., Hubbard, R. E., and Higgins, C. F. (1990). Structural model of

ATP-binding proteins associated with cystic fibrosis, multidrug resistance and bacterial transport. *Nature* **346**, 362–365.
22. Linton, K. J., and Higgins, C. F. (1998). The *Escherichia coli* ATP-binding cassette (ABC) proteins. *Mol. Microbiol.* **28**, 5–13.
23. Fuqua, W. C., Winans, S. C., and Greenberg, E. P. (1994). Quorum sensing in bacteria: The LuxR-LuxI family of cell density-responsive transcriptional regulators. *J. Bacteriol.* **176**, 269–275.
24. Kleerebezem, M., Quadri, L. E., Kuipers, O. P., and de Vos, W. M. (1997). Quorum sensing by peptide pheromones and two component signal-transduction systems in Gram-positive bacteria. *Mol. Microbiol.* **24**, 895–904.
25. Hui, F. M., Zhou, L., and Morrison, D. A. (1995). Competence for genetic transformation in *Streptococcus pneumoniae*: Organization of a regulatory locus with homology to two lactococcin A secretion genes. *Gene* **153**, 25–31.
26. Kolter, R., and Moreno, F. (1992). Genetics of ribosomally synthesized peptide antibiotics. *Annu. Rev. Microbiol.* **46**, 141–165.
27. Kornblum, J., Kreiswirth, B., Projan, S. J., Ross, H., and Novick, R. P. (1990). *agr*: A polycistronic locus regulating exoprotein synthesis in *Staphylococcus aureus*. *In* "Molecular Biology of the Staphylococci" (R. P. Novick, ed.) pp. 373–402. VCH, New York.
28. Gilmore, M. S., and Hoch, J. A. (1999). Antibiotic resistance: A vancomycin surprise. *Nature* **399**, 524–525.
29. Hancock, R. E. W. (1997). Peptide antibiotics. *Lancet* **349**, 418–422.
30. Moreillon, P., Markiewicz, Z., Nachman, S., and Tomasz, A. (1990). Two bactericidal targets for penicillin in pneumococci: Autolysis-dependent and autolysis-independent killing mechanisms. *Antimicrob. Agents Chemother.* **34**, 33–39.
31. Moreillon, P., and Tomasz, A. (1988). Penicillin resistance and defective lysis in clinical isolates of pneumococci: Evidence for two kinds of antibiotic pressure operating in the clinical environment. *J. Infect. Dis.* **157**, 1150–1157.
32. Cashel, M., Gentry, D. R., Hernandez, V. J., and Vinella, D. (1996). The stringent response. *In* "*Escherichia coli* and *Salmonella*: Cellular and Molecular Biology" (F. C. Neidhardt, R. Curtis III, J. L. Ingraham, E. C. C. Lin, K. B. Low, B. Magasanik, W. S. Reznikoff, M. Riley, M. Schaechter, and H. E. Umbarger, eds.). ASM Press, Washington, DC.
33. Brunskill, E. W., and Bayles, K. W. (1996). Identification of LytSR-regulated genes from *Staphylococcus aureus*. *J. Bacteriol.* **178**, 5810–5812.
34. Groicher, K. H., Firek, B. A., Fujimoto, D. F., and Bayles, K. W. (2000). The *Staphylococcus aureus* lrgAB operon modulates murein hydrolase activity and penicillin tolerance. *J. Bacteriol.* **182**, 1794–1801.
35. Bayles, K. W. (2000). The bactericidal action of penicillin: New clues to an unsolved mystery. *Trends Microbiol.* **8**, 274–278.
36. Garcia, P., Gonzalez, M. P., Garcia, E., Rubens, L., and Garcia, J. L. (1999). Lyt B, a novel pneumococcal murein hydrolase essential for cell separation. *Mol. Microbiol.* **31**, 1275–1277.

CHAPTER 18

Role of Multiple Sensor Kinases in Cell Cycle Progression and Differentiation in *Caulobacter crescentus*

AUSTIN NEWTON AND NORIKO OHTA
Department of Molecular Biology, Princeton University, Princeton, New Jersey 08544

Caulobacter crescentus, an asymmetrically dividing, gram-negative, α-proteobacterium, has been a fruitful system for studying the mechanism of cell differentiation. Internal cell cycle cues drive morphogenesis in these cells, with stages of the cell division cycle acting as checkpoints for successive developmental events. Experiments now show that signal transduction pathways mediated by essential two-component system proteins regulate the cell cycle and couple the developmental program to cell cycle progression. Pathways initiated by the histidine protein kinases DivJ, CckA, DivL, and PleC regulate the initiation of DNA synthesis, flagellum biosynthesis, cell division, and motility, among other cell cycle-regulated events. Results of genetic and biochemical experiments argue strongly that an essential transcription factor, CtrA, is the common downstream target regulated by these sensor kinases. Three of the kinases are localized dynamically either to the flagellated cell pole or to the stalked cell pole, apparently at times corresponding to their function in the cell cycle. The detailed organization of signal transduction networks and the regulation of sensor protein kinases in response to cell cycle checkpoints are promising future areas of investigation.

INTRODUCTION

Caulobacter crescentus is a gram-negative, α-proteobacterium whose genome sequence has been reported [1]. This bacterium has been a fruitful system for studying the mechanism of differentiation in an organism that is amenable to biochemical and genetic analysis. Internal cell cycle cues drive morphogenesis in these cells, with stages of the cell division cycle acting as checkpoints for successive developmental events. Experiments now show that signal transduction pathways mediated by essential His-Asp proteins regulate the cell cycle and couple the developmental program to cell cycle progression. This chapter considers the organization of these pathways with particular attention to the multiple histidine protein kinases and how they regulate signal transduction.

TEMPORAL AND SPATIAL CONTROL OF CELL CYCLE EVENTS

The repeated division of the nonmotile, "mother" stalked cell produces the mother cell plus a new daughter cell, which is the motile swarmer cell. The A → A + B pattern of asymmetric cell division in *C. crescentus* is reminiscent of the division pattern during stem cell differentiation observed in multicellular eukaryotes [2]. Although studies of DNA segregation at division suggest that the stalked and swarmer cells inherit identical chromosomes [3, 4], DNA synthesis is initiated asymmetrically in the two progeny cells [5]. The two genomes are programmed to direct very different sequences of developmental events (Fig. 1). Thus, after cell separation, the mother stalked cell immediately initiates DNA replication and then begins polar morphogenesis at the stalk-distal pole. The ensuing developmental events, including synthesis of the flagellum, adhesive holdfast, and DNA phage receptors, are required to build the new swarmer cell. The flagellum begins to rotate immediately before cell separation, and the motile swarmer cell is released once the mother and daughter cells separate. In contrast to the stalked cell, the newly divided swarmer cell is silent for DNA replication (Fig. 1) [5]. During the presynthetic gap, or G1 period, the swarmer cell remodels itself into a stalked cell that is competent for the initiation of DNA replication. Attendant developmental events include pili formation, loss of motility, ejection of the flagellum, and, finally, formation of a membranous stalk at the same cell pole [6–8].

LEVELS OF DEVELOPMENTAL REGULATION

Differential transcription is a major mechanism controlling the cell cycle and differentiation in *C. crescentus*. As shown by early experiments in synchronous cell cultures, *de novo* RNA synthesis is required for the loss of motility, stalk formation, cell division, and DNA initiation [9]. The best-studied example of transcriptional regulation is the flagellar gene cascade, which is largely responsible for sequential assembly of the polar flagellum [10, 11]. A microarray analysis of 2966 genes of the predicted 3767 genes in this bacterium revealed the extent of gene regulation in synchronous cells [12]. RNA levels change as a function of the cell cycle for 553 (*ca.* 19%) of the genes

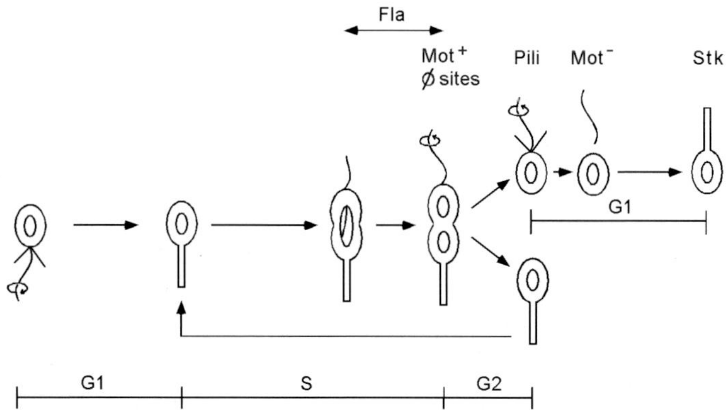

FIGURE 1 The *Caulobacter* cell cycle. Developmental events in wild-type strain CB15 include flagellum formation (Fla), activation of flagellum rotation (Mot⁺), appearance of polar bacteriophage receptors (ϕ sites), pili formation (Pili), flagellum ejection and loss of motility (Mot⁻), and stalk formation (Stk). Periods corresponding to presynthetic gap (G1), DNA synthesis (S), and postsynthetic gap (G2) are indicated.

examined. Examples of posttranslational control in *C. crescentus* that are crucial for cell differentiation include the cell cycle-regulated proteolysis of the cell division protein FtsZ [13] and the global transcription regulator CtrA (see later) [14]. In addition to the temporal control of developmental events, the localization of the flagellum, pili, stalk, and other differentiated surface structures in the asymmetric cells illustrates the spatial dimension of developmental control. As discussed later, the interplay of temporal and spatial mechanisms controlling differentiation is also reflected in the dynamic localization of His-Asp regulatory proteins within the developing swarmer and stalked cell compartments (see Fig. 3).

CONTROL OF DIFFERENTIATION BY CELL CYCLE CHECKPOINTS

Conditional *C. crescentus* mutants defective in either DNA replication (DNA) or cell division (DIV) form long, filamentous cells at the nonpermissive temperature (Fig. 2) [15, 16]. These mutants also arrest polar morphogenesis at specific stages, depending on where the mutant is blocked in the cell cycle [17]. This observation led to the conclusion that cell cycle and developmental events are not only coordinated, but that successive steps in the cell division cycle provide cues or checkpoints that determine the order of developmental events [17]. This independence of development from known environmental cues sets *C. crescentus* apart from most prokaryotes that undergo cell differentiation [18]. Cell cycle checkpoints (Fig. 2) include the requirement of (i) DNA synthesis for cell division [19] and flagellum biosynthesis [20, 21], (ii) completion of an early stage of division for motility and stalk formation [22, 17], and (iii) cell separation for pili formation [23, 24]. As outlined later, signal transduction pathways mediated by histidine protein kinases (HPKs), along with their cognate response regulators (RRs), are now known to couple development to cell cycle progression. More recently, Wortinger *et al.* [25] presented evidence that the DNA-dependent cell division checkpoint [19] is mediated by central RR CtrA and does not depend on SOS repair function. It had been shown previously that the DNA-dependent checkpoint for flagellum biosynthesis is also independent of the SOS system [26].

TWO-COMPONENT SIGNAL TRANSDUCTION AND CELL CYCLE REGULATION

Two-component systems in bacteria, eukaryotic microorganisms (e.g., *Candida albicans*, *Dictyostelium discoideum*, *Neurospora crassa*, *Saccharomyces*

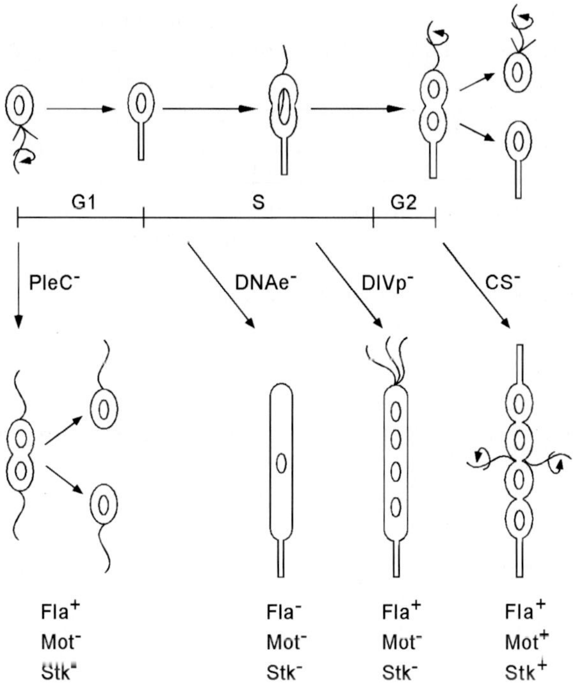

FIGURE 2 Cell cycle checkpoints. Diagrams depict the wild-type cell cycle and morphology of the temperature-sensitive cell cycle and developmental mutants at nonpermissive temperature. *pleC*⁻, *pleC319* mutant blocked in the swarmer to stalked cell transition; DNA⁻, defect in DNA chain elongation [26a]; DIVp⁻, *ftsA* mutation blocked in cell division progression [16]; CS⁻, *parE* mutation blocked in cell separation [24].

cerevisiae, and *Schizosacchromycyes pombe*), and plants (*Arabidopsis thaliana*) control many cellular responses to fluxes in growth conditions, stress, osmolarity, and changes in other environmental conditions [27, 28]. *C. crescentus* contains an unusually large number of HPKs (61) and RRs (71) [1]. In contrast, *Escherichia coli* and *Bacillus subtilis* contain *ca.* 29 and 35 HPKs, respectively [28].

Bacterial histidine kinases have been assigned to 11 subfamilies [29] based on the highly conserved H, N, D, F and G box sequence motifs [27, 30]. In *C. crescentus*, 32 of the HPKs belong to subfamily 1 [8], which contains the largest number of bacterial HPKs and all reported eukaryotic HPKs [29]. The remaining *C. crescentus* kinases are members of subfamilies 2–4 (15 HPKs), subfamily 9 (2 HPKs), which contains exclusively chemotaxis HPKs (CheA), and subfamily 11 (12 HPKs), a subfamily not represented in *E. coli* or *B. sub-*

tilis. Subfamily 11 was originally defined by *Methanobacter* kinases [29]; based on phylogenetic data, it seems probable that this subfamily originated in *C. crescentus* as a result of lateral gene transfer and radiation [8]. Compared to other reported genomes, an unusually large number of *C. crescentus* HPKs (27) are hybrid kinases that contain only a transmitter and receiver domain (H1-D1 protein). The largest number of these hybrid kinases is in subfamily 1 (22), with the remainder in subfamilies 4 (3) and 11 (2).

Identification of His-Asp Proteins in Cell Cycle Regulation

Although many of the *C. crescentus* His-Asp proteins mediate responses to environmental conditions [1], others are central players in the regulation of cell cycle progression, and several HPKs and RRs are essential for cell viability. The involvement of His-Asp proteins in cell cycle control first became apparent from a pseudoreversion analysis of the developmental gene *pleC* [31]. *pleC* mutants divide normally, but are blocked in the swarmer-to-stalked cell transition (see developmental events in Fig. 1). As a result, *pleC* mutants form stalkless, biflagellated cells that are nonmotile (Fig. 2). These defects in polar morphology are similar to those in filamentous mutants blocked early in cell division (see DIVp$^-$ mutants in Fig. 2) [18]. This similarity of the polar developmental phenotypes of these mutants suggested that PleC and cell cycle progression regulate development by a common mechanism [31]. To test this proposition, Mot$^+$ revertants isolated at 37°C from a nonmotile, temperature-sensitive *pleC* mutant were screened to identify second-site suppressor mutations that simultaneously conferred a cold-sensitive cell division phenotype at 24°C. These conditional mutations mapped to the cell cycle control genes *divK*, *divJ*, and *divL*. Another outside suppressor mutation in *pleD* suppressed the Mot$^-$ phenotype, but did not confer a cell division defect [32].

Of the genes examined in this study, *pleC*, *divJ*, and *divL* encode HPKs, whereas *divK* and *pleD* encode RRs. PleC [33], DivJ [34], and DivL [35], which are HPKs with N-terminal hydrophobic regions, belong to HPK subfamily 1 [8]. The DivL kinase is unusual in that the canonical His residue of the H box is substituted by a Tyr. The Tyr residue of DivL is autophosphorylated by ATP *in vitro* and is required for full biological function of this kinase *in vivo*. Only DivL among these three kinases is essential for cell viability [35].

DivK is a single domain RR [36] similar in structure to chemotaxis protein CheY of *Salmonella typhimurium* [37] and the phosphotransfer protein Spo0F of *B. subtilis* [38, 39]. Unlike other bacterial response regulators described at the time, however, DivK is essential for normal cell cycle progression and viability [36]. PleD is an unusual RR with two N-terminal receiver domains (D and D′) and a highly conserved C-terminal output domain containing the so-called GGDEF motif [40]. *pleD* null mutants are motile, but do not lose motility or form stalks. This phenotype and the observation that a *pleD* null mutation is recessive to the *pleD*$^+$ allele [40] imply that PleD is a negative regulator of motility and a positive regulator of flagellum loss and stalk formation (see Fig. 4). As shown by Aldridge and Jenal [41], the GGDEF sequence is required for loss of motility and stalk formation during the swarmer-to-stalked cell transition [41].

Defining Signal Transduction Pathways

Roles of DivJ and PleC Kinases

The phenotype of conditional mutations in DivK indicates that the RR affects the regulation of two different cell cycle functions: polar morphogenesis and cell division. The ability of a cold-sensitive mutation in *divK* (*divK341*) to suppress the Mot$^-$, Stk$^-$ phenotype of *pleC* strains at the permissive temperature suggests that PleC is upstream of DivK and regulates DivK activity [36].

The function of DivK in cell cycle regulation is supported by the observation that the *divK341* allele also blocks an early cell division step at the non-permissive temperature [36]. The resulting filamentous phenotype is similar to that displayed by *divJ* mutants [34], which are defective in an early stage of cell cycle progression [23, 42]. Moreover, the effects of changing *divJ* and *divK* expression levels have provided evidence for a DivJ → DivK pathway regulating cell division. While overexpression of *divK* produces severe filamentation accompanied by cell lysis, increasing the level of *divJ* expression reverses this phenotype [43] (N. Ohta and A. Newton, unpublished results).

Studies of DivK phosphorylation *in vitro* also indicate that DivK functions downstream of DivJ and PleC kinases. The purified cytoplasmic domains of DivJ and PleC efficiently catalyze the phosphorylation of DivK and the dephosphorylation of phospho-DivK [36]. Assays of the full-length PleC and DivJ kinases in membrane preparations from wild-type, *divJ*, and *pleC* mutant cells showed, however, that phosphorylation of purified DivK in the presence of ATP depended almost entirely on the presence of DivJ. Moreover, the presence of PleC in membrane preparations had little effect on the levels of DivK phosphorylation (T. Hofmeister and A. Newton, unpublished results). Among the questions raised by these results are whether protein kinase specificity is affected in assays using truncated HPK fragments and the possible effect of subcellular HPK localization on kinase activity in membrane preparations (see later).

HPK activity has also been measured *in vivo* by assaying DivK phosphorylation in intact cells labeled with [^{32}P]orthophosphate. The activity was only partially dependent on DivJ kinase, which suggests that an additional kinase(s) may be required for DivK phosphorylation *in vivo* [42]. The elevated levels of DivK phosphorylation measured in a *pleC*::Tn5 strain were attributed to the unregulated phosphotransfer from DivJ to DivK [42]. The authors concluded that DivJ is responsible for the phosphoryl transfer to DivK and that PleC is responsible, directly or indirectly, for the regulation of DivJ. It is, however, difficult to assess the degree to which PleC activity contributes directly to DivK phosphorylation based on *in vivo* measurements in *pleC* mutant cells. A loss of PleC kinase activity might well be obscured by increased activity of the mislocalized DivJ kinase. Results of a yeast two-hybrid screen, which is described later, demonstrate the direct interaction of the PleC and DivJ kinases with DivK.

Function of CtrA in Cell Cycle Regulation

Two findings led to a better understanding of how PleC and DivJ regulate polar morphogenesis and cell division. The first of these was the discovery and characterization of the global response regulator CtrA, which is essential for regulating multiple cell division cycle events [44, 14]. The second finding was the identification of CtrA as a downstream target of PleC and DivJ kinases [43].

CtrA was identified as a RR necessary for the regulation of class II flagellar gene transcription *in vivo* [44]. Examination of class II flagellar gene expression in a defined *in vitro* system showed directly that transcription initiation requires phosphorylation of CtrA and promoter recognition by the principal *C. crescentus* σ factor, σ^{73} [43]. Phospho-CtrA controls these and other cell cycle-regulated genes by binding to CtrA box sequences of the promoters [7]. The differential affinity of phospho-CtrA for CtrA-dependent promoters may be a factor in the cell cycle regulation of transcription from these promoters [45].

Asymmetric initiation of DNA synthesis in the new swarmer and mother stalked cell (Fig. 1) [5, 4] is controlled by the localization of CtrA to the swarmer cell compartment of the predivisional cell (Fig. 3A). At cell division, CtrA segregates preferentially to the new swarmer cell, where it silences initiation of DNA synthesis by binding to multiple sites near the replication origin [14]. CtrA is subsequently degraded at the G1 to S transition to permit initiation of DNA synthesis (Fig. 3A). In the stalked cell, which does not inherit CtrA, DNA synthesis is initiated immediately after division. CtrA is synthesized and activated by phosphorylation soon after the initiation of DNA synthesis and thus prevents premature reinitiation of DNA synthesis. The localization of CtrA to the swarmer cell compartment late in the cell cycle is achieved by CtrA proteolysis in the stalked cell compartment [14].

DivJ and PleC Kinases Regulate CtrA Activity

Pseudoreversion analysis of a conditional *divK* mutation was used to screen for downstream components of signal transduction pathways regulated by the

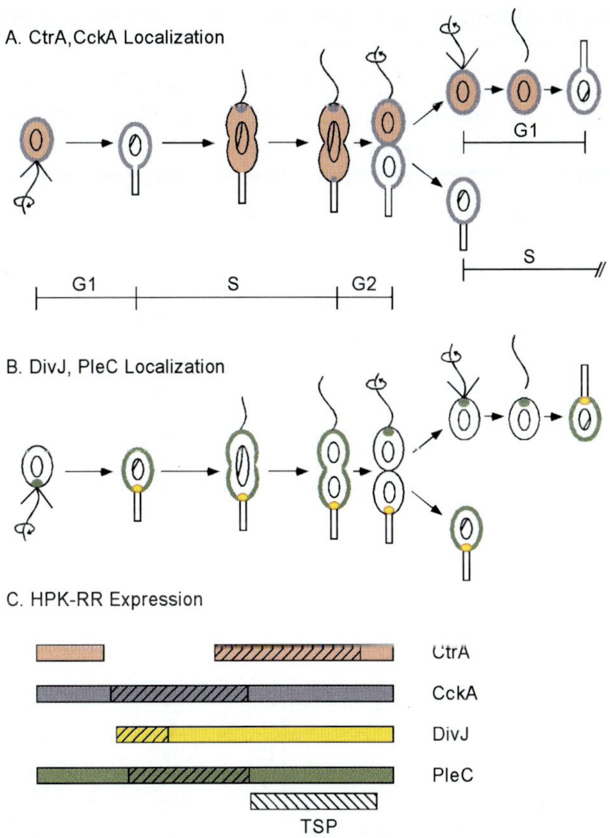

FIGURE 3 Localization of HPK and RR proteins during *Caulobacter* cell cycles. (A) CtrA [14] and CckA [49] localization. (B) DivJ and PleC [42] localization. (C) Cell cycle expression of CtrA [44], CckA [49], DviJ [34, 42], and PleC [42]. Protein localization was determined using protein fusions to Gfp (CckA, DvJ, PleC), immunogold electron microscopy (DivJ), and immunofluorescence microscopy of synchronized cells (CtrA). Solid bars represent steady-state protein levels, and cross-hatching indicates the transcription of corresponding genes. Colors are keyed to the proteins whose localization is shown in A and B. The cross-hatched bar under PleC is the temperature-sensitive period (TSP) for the activity of the protein in a conditional mutant [32].

PleC and DivJ kinases. Temperature-sensitive *sokA* (suppressors of *divK*) alleles that suppress the cell division defects of the cold-sensitive *divK* mutation were isolated. One of the *sokA* alleles maps in the C-terminal DNA-binding domain of CtrA [43] and also suppresses the lethal phenotype of *divK* gene disruptions, the cell division defect of *divJ* mutations [43], and the Mot$^-$ phenotype of *pleC* mutations (N. Ohta and A. Newton, unpublished results). These genetic results indicate that the signal transduction pathways initiated by DivJ (DivJ → DivK) in the regulation of cell division and by PleC (PleC → DivK) in the regulation of motility control these cell cycle events *via* CtrA (Fig. 4).

It seems unlikely that the aspartyl-phosphate of phospho-DivK transfers phosphate directly to the Asp residue of CtrA. Consistent with this view, purified DivJ phosphorylates DivK much more efficiently than CtrA. Moreover, DivJ exclusively phosphorylates DivK when both CtrA and DivK are present,

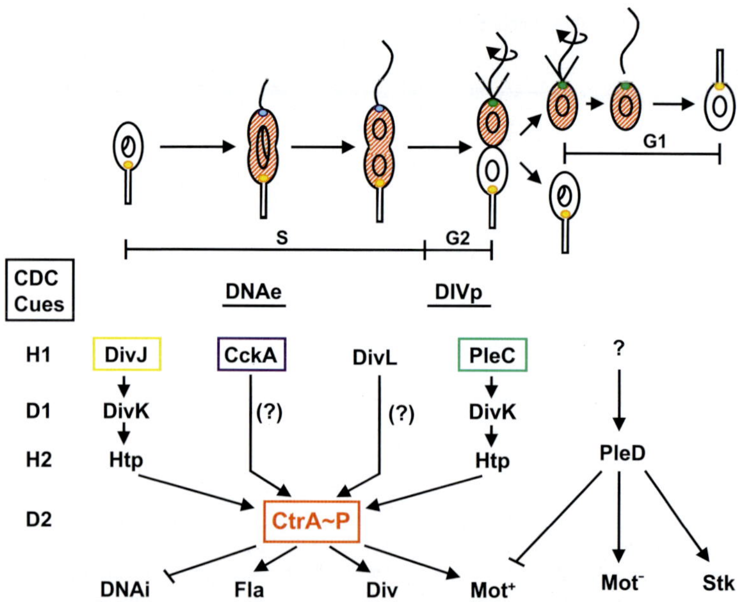

FIGURE 4 Proposed organization of signal transduction pathways regulating the cell cycle and polar morphogenesis. The model is based on genetic and biochemical experiments outlined in the text in Ohta *et al.* [8]. Possible phosphorelay components of pathways regulated by CckA, DivL, and PleC have not been identified, as indicated by "?". The localization of DivJ (yellow), CckA (blue), and PleC (green) is summarized from Figs. 3A and 3B.

suggesting that phosphotransfer to CtrA is not stimulated by DivK [43]. Consequently, an unidentified histidine phosphotransferase (Hpt) was proposed to participate in the phosphorelay DivJ → DivK → Hpt → CtrA to catalyze the phosphotransfer from DivK to CtrA [43]. Similar H1 → D1 → H2 → D2 phosphorelay systems have been described in other bacterial systems [46, 47]. If PleC catalyzes the phosphorylation of DivK, as suggested by the genetic and biochemical experiments, then this kinase may regulate motility and stalk formation *via* CtrA by a similar phosphorelay (see Fig. 4).

DivL and CckA Are Essential Cell Cycle HPKs

Although *divJ* mutants are defective in cell division, this gene is not essential for viability [42] (N. Ohta and A. Newton, unpublished results). The PleC kinase functions in the regulation of motility and is also not essential under normal growth conditions [48]. Because both DivK [36] and CtrA [44] are essential proteins, additional protein kinases presumably contribute to CtrA phosphorylation. At least two essential protein kinases appear to fill the bill. One is the tyrosine-containing DivL kinase [35] and the other is the HPK CckA [49]. Evidence that DivL is upstream of CtrA and regulates CtrA activity is the finding that the *sokA301* allele of *ctrA* suppresses the severe cell division phenotype of a cold-sensitive *divL* mutation [35]. Consistent with this observation, a purified cytoplasmic fragment of DivL phosphorylates CtrA in the presence of ATP.

The other essential kinase, CckA, was identified in a screen for conditional mutants that were defective in both cell division and polar morphogenesis [49]. CckA belongs to HPK subfamily 4 and, like other hybrid kinases in *C. crescentus*, it contains an H box motif, a catalytic domain, and a C-terminal receiver module. This kinase is required for CtrA phosphorylation and full activation of CtrA-dependent promoters *in vivo* [49]. Whether CckA phosphorylates CtrA directly or indirectly through DivK or another phosphorelay protein is unknown. The function of the CckA receiver domain has not been established.

Yeast Two-Hybrid Analysis of HPK–RR Interactions

Another approach employed to identify components of these complex signal transduction pathways is the assay of protein–protein interactions using the yeast two-hybrid system [50, 51]. DivK was used as the bait to screen a prey library of *C. crescentus* DNA fragments. The sequences of interacting prey clones identified five proteins, all of which were histidine kinases (N. Ohta and A. Newton, manuscript in preparation). Three of the kinases, DivJ, PleC, and DivL, had been implicated previously in cell cycle regulation [31]. No function has yet been assigned to the two other HPKs, a cytoplasmic kinase and a hybrid kinase. The two new HPKs are, however, similar in sequence to DivJ, PleC, and DivL and belong to HPK subfamily 1. It is also of note that no clones were isolated corresponding either to the proposed phosphotransferase (Hpt; Fig. 4) or to CckA. The latter result is consistent with the idea that CckA does not regulate cell cycle events through a DivK-dependent phosphorelay. However, a number of factors can account for the failure to detect certain protein–protein interaction in a two-hybrid screen [51].

A large number of the clones isolated on the basis of interaction with DivK encode cytoplasmic fragments of either DivJ (73) or PleC (49) and they align to form overlapping arrays. The common amino acid sequences shared by all fragments contain the conserved H box, and the minimal sequences represented in DivJ and PleC appear to define the docking site for DivK. The homologous sequence in the EnvZ kinase forms a four-helix bundle responsible for the dimerization and phosphotransfer to OmpR in *E. coli* [52]. All interacting *divL* clones (8) encode fragments of DivL with the presumptive phosphotransfer domain and the Tyr-500 residue [35]; (N. Ohta and A. Newton, manuscript in preparation). Although DivL phosphorylates CtrA *in vitro* much more efficiently than DivK [35], these two-hybrid results suggest that DivL may also regulate CtrA *via* a DivK-dependent phosphorelay, perhaps during a specific period in cell cycle.

Coordination of Multiple Kinase Activities

To identify signal transduction components that affect cell division regulation by DivL, pseudorevertants of temperature-sensitive *divL* mutations were isolated. Outside suppressors were mapped to *solA* and *solB* (suppressor of *divL*) loci, which correspond to the *CckA* and *divJ* kinase genes, respectively (N. Ohta and A. Newton, unpublished results). Thus, mutations in either DivJ or CckA compensate for defective cell cycle regulation by a mutant DivL protein. Each of the protein kinases, CckA, DivJ, and DivL, is required for normal cell division in *C. crescentus*, and these suppression results indicate that their activities are coordinated in the cell cycle. They also support the conclusion that all three kinases regulate phosphorylation and activation of a common RR, CtrA (Fig. 4). The DivJ and CckA proteins display different and distinctive patterns of subcellular localization (Fig. 3), and one interpretation of suppression data is that the kinase activities are not restricted exclusively to the subcellular compartments in which the HPKs are localized.

A directly related question is how the activities of the sensor kinases are regulated. Possible mechanisms include (i) cell cycle-regulated HPK synthesis; (ii) temporal control of kinase activation in response to specific input signal or cell cycle cues; and (iii) restriction of HPK activity to a single cell compartment. All three mechanisms may apply to the regulation of one or more HPKs.

i. HPK synthesis. DivJ may be the only reported cell cycle kinase whose time of expression is important to its function. Transcription of *divJ* is induced during the G1 to S phase transition at the time of DNA synthesis initiation and stalk formation (Fig. 3C) [34]. The steady-state level of the DivJ protein also increases at this time in the cell cycle [53] (T. Lane and A. Newton, unpublished results). Transcription of *pleC* and *cckA* genes is also cell cycle regulated [12], but levels of the PleC and CckA proteins are constant through the cell cycle (Fig. 3C) [53, 49]. The rate of *divL* transcription is constant in the cell cycle (J. Wu and A. Newton, unpublished results), but the steady-state levels of DivL protein have not been reported.

ii. HPK activity. The DNA synthesis-dependent checkpoint in *C. crescentus* controls cell division initiation [19], and evidence now shows that this checkpoint regulates the level of CtrA phosphorylation [25]. As yet, there is no information on which HPK is regulated or the mechanism(s) by which this or other cell cycle cues influence kinase activity (see earlier discussion).

iii. HPK localization. CtrA was the first two-component protein reported to be localized in *C. crescentus* [14]. Work principally from the Shapiro laboratory, using GFP protein fusions, has demonstrated that DivJ, PleC, and CckA also display distinctive patterns of spatial localization as a function of cell cycle progression (Fig. 3) [54].

Spatial Localization of HPKs Is Cell Cycle Regulated

DivJ

In an analysis of the DivJ–GFP fusion protein, little or no DivJ was present in swarmer cells. DivJ was first detected in the newly formed stalked cell where it accumulates at the base of the stalk after its synthesis during the G1 to S phase transition (Figs. 3B and 3C) [42]. The DivJ protein is stable and remains at the stalked pole throughout the cell cycle before segregating uniquely to the stalked cell at division. These studies also showed that PleC is required for targeting of DivJ to the stalked pole. Identical conclusions about DivJ localization and regulation were reached independently by visualizing

DivJ in thin sections of synchronous *C. crescentus* cells using immunogold electron microscopy (T. Lane and A. Newton, unpublished results).

Conditional *divJ* mutations were originally shown to suppress the Mot$^-$ phenotype of *pleC* mutations, a result suggesting that PleC functions upstream of the DivJ kinase [31]. Epistasis experiments using *divJ* and *pleC* null mutations have extended these results [42]. The finding that PleC regulates DivJ localization has suggested that failure to localize DivJ in *pleC* mutants is responsible for the altered *in vivo* phosphorylation of DivK, as well as the Stk, Mot$^-$ phenotype (see Fig. 2) [42]. As an alternative explanation, PleC may have other targets, which, like DivK, also affect morphogenetic events, including stalk formation and the polar localization of DivJ. A direct test to determine if the localization of DivJ and other HPKs is required for differentiation will be to identify and ablate signal sequences required for polar targeting. If DivJ localization were a requirement, then these mutations should display the developmental phenotype of *pleC* mutants.

PleC

Although the PleC protein is present throughout the cell cycle, this HPK is localized dynamically as a function of cell cycle progression [42]. PleC is concentrated at the flagellated pole of the new swarmer cell, but it disperses during the swarmer-to-stalked cell transition. In late predivisional cells, PleC localizes to the flagellated pole of predivisional cell and then segregates with the daughter swarmer at division (Fig. 3B). The temperature-sensitive period (TSP, Fig. 3C) for motility and stalk formation determined using a conditional *pleC* mutation extends from *ca.* 0.6 to 0.95 of the cell cycle [32]. Thus, the last time that PleC activity is required for gain of motility (0.95) coincides closely with the time of PleC localization to the flagellated pole (Fig. 3B).

CckA

CckA, like the PleC kinase, is present continuously in the cell cycle and changes its subcellular location during cell cycle progression. Interestingly, there is no evidence that the reorganization of CckA localization depends on either protein turnover or synthesis (Fig. 3A) [49]. CckA is dispersed in the newly divided swarmer and stalked cells. The kinase then localizes to the stalked-distal pole of the early predivisional cell after the initiation of DNA synthesis (Fig. 4A). Later, a second, less intense, center of CckA can be detected at the stalked pole. Just before division, CckA disperses within the predivisional cell.

SUMMARY AND PERSPECTIVES

Signal transduction pathways mediated by essential His-Asp proteins regulate the *C. crescentus* cell cycle and couple the developmental program to cell cycle progression. Signal pathways initiated by four sensor protein kinases, DivJ, CckA, DivL, and PleC, regulate the initiation of DNA synthesis, flagellum biosynthesis, cell division, and motility, among other developmental events (Fig. 4). Cues that activate these signal transudation pathways are furnished by the completion of successive steps in the cell division cycle itself. This complex signal transduction network thus determines the order of developmental events and assures the coordination of these events with cell cycle progression [8, 18].

Results of genetic and biochemical experiments [35, 43] argue strongly that the common downstream target regulated by these HPKs is the essential transcription factor CtrA. Another RR, the GGDEF protein PleD, controls the swarmer-to-stalked cell transition by acting as a negative regulator of motility and a positive regulator of flagellum loss and stalk formation (see Fig. 4) [40, 41]. The role of PleC in the activation of flagellar rotation raises the possibility that this kinase also regulates PleD.

Figure 4 is a working model for the organization of sensor kinase-dependent pathways in the cell cycle. With the exception of DivL, whose time of synthesis and function have not been examined, the HPKs are placed in the figure at times corresponding to the cell cycle events they regulate. For example, DivJ is assumed to act early in S phase when it functions to prevent the premature reinitiation of DNA replication [42]. In a similar fashion, PleC is expected to act late in the cell cycle when it regulates motility at a time that overlaps with the end of the TSP for PleC function (Fig. 3C) [32].

There is a striking correspondence between the times of HPK function in the cell cycle and the times at which the respective HPKs are localized topologically (see Fig. 4). Thus, DivJ is localized to the stalked cell pole early in S phase when it functions to control an early cell division cycle event and PleC is localized to flagellated cell pole of the predivisional cell when the cell gains motility. There is, as yet, no direct evidence that polar localization of an HPK is required for its activity in cell cycle regulation.

What mechanism(s) is responsible for the specific targeting of HPKs to either the swarmer or the stalked cell pole? By analogy to protein localization in other systems, targeting presumably depends on signal sequences in the histidine kinases on the one hand, and on topologically localized receptors, or subcellular targets, on the other. Similar questions apply to other polar membrane proteins, e.g., the methylated chemotactic proteins [55], and the differentiated surface structures, e.g., the flagella and pili. Because these membrane proteins and structures are localized at the stalk-distal pole of the predivisional cell (Fig. 1), one can imagine that the same subcellular target serves for polar localization of the regulatory HPKs, as well as the developmental structures. As proposed previously [17], the position of polar localization may be specified by an "organizational center" that is laid down at the new pole as it forms during cell division. This new cell pole would then determine the position of flagellum assembly and protein localization in the next cell cycle.

A related issue is how the hypothetical polar targets are regulated. The specificity of these sites for localizing a given HPK changes as a function of cell cycle progression and results in "kinase swapping." Thus, just before division, CckA disperses within the predivisional cell at a time coinciding with PleC localization to the same flagellated pole (Fig. 3B). Consequently, PleC localized to the flagellated pole may effectively substitute for CckA activity in the swarmer cell compartment and be responsible for CtrA regulation and gain of motility (cf. Figs. 3A and 3B) [42]. Another example of kinase swapping occurs at one cell pole in the new swarmer cell after division (Fig. 3B). The flagellated cell pole, which is recognized by the PleC kinase in the late predivisional cell and the swarmer cell, changes its specificity during the swarmer-to-stalked transition. Coincident with this change, the PleC protein disperses in the cell and the DivJ protein localizes at the base of the newly formed stalk (Fig. 3B) [42]. PleC appears to regulate this change in polar specificity because *pleC* mutants, which do not undergo the swarmer-to-stalk cell transition, also fail to localize the DivJ protein. This result has led to the suggestion that the stalk, or a subcellular component of the stalk, may be the target for the polar accumulation of DivJ [42].

Why are multiple HPKs required in *C. crescentus* for cell cycle regulation? One rationale may be the need to control precisely the expression of a very large number of genes both spatially and temporally in response to cell cycle checkpoints (Fig. 4). Based on a microarray analysis of 79% of the total open reading frames in the *C. crescentus* genome, CtrA controls, directly or indirectly, an estimated 144 genes that are expressed differentially during the cell cycle [12]. A similar argument has been made for regulation of the *B. subtilis* Spo0A protein by a multicomponent phosphorelay in response to three sensor kinases [39, 56]. Spo0A, which regulates sporulation in response to nutrient deprivation, activates or represses the transcription of 520 genes that are not regulated by σ^F [57]. In contrast to *C. crescentus*, whose developmental regulation responds principally to endogenous cell cycle signals [17, 2], the signal transduction pathways for sporulation in *B. subtilis* must integrate multiple nutritional and physiological signals [58]. In addition to these environmental cues, it is of particular interest that phosphotransfer to Spo0A and initiation of sporulation in *B. subtilis* are also controlled by a DNA initiation checkpoint whose regulation is independent of the SOS pathway [59].

Given the recent completion of the *C. crescentus* genome sequence [1], this bacterium is now amenable to the study of general questions of how biological networks are organized and regulated. The detailed organization of signal transduction networks and the regulation of sensor protein kinases in response to cell cycle checkpoints are promising future areas of investigation.

ACKNOWLEDGMENTS

We are grateful to Stephen Sciochetti for critically reading this manuscript. Work from this laboratory was supported by Public Health Service Grants GM22299 and GM58794 from the National Institutes of Health.

REFERENCES

1. Nierman, W. C., Feldblyum, T. V., Laub, M. T., Paulsen, I. T., Nelson, K. E., Eisen, J., Heidelberg, J. F., Alley, M. R. K., Ohta, N., Maddock, J. R., *et al.* (2001). Complete genome sequence of *Caulobacter crescentus*. *Proc. Natl. Acad. Sci. USA* **98**, 4136–4141.
2. Newton, A. (1984). *In* "Microbial Cell Cycle" (P. Nurse and E. Streiblova, eds.), pp. 51–75. CRC Press, Boca Raton, FL.
3. Osley, M. A., and Newton, A. (1974). Chromosome segregation and development in *Caulobacter crescentus*. *J. Mol. Biol.* **90**, 359–370.
4. Marczynski, G. T., Dingwall, A., and Shapiro, L. (1990). Plasmid and chromosomal DNA replication and partitioning during the *Caulobacter crescentus* cell cycle. *J. Mol. Biol.* **212**, 709–722.
5. Degnen, S. T., and Newton, A. (1972). Chromosome replication during development in *Caulobacter crescentus*. *J. Mol. Biol.* **64**, 671–680.
6. Brun, V. Y., and Janakiraman, R. (2000). *In* "Prokaryotic Development" (Y. Brun, and L. Shimkets, eds.), pp. 297–317. ASM Press, Washington, DC.
7. Hung, D., McAdams, H., and Shapiro, L. (2000). *In* "Prokaryotic Development" (Y. V. Brun, and L. Shimkets, eds.). ASM Press, Washington, DC.
8. Ohta, N., Grebe, T. W., and Newton, A. (2000). *In* "Prokaryotic Development" (Y. Brun, and L. Shimkets, eds.), pp. 341–359. ASM Press, Washington, DC.
9. Newton, A. (1972). Role of transcription in the temporal control of development in *Caulobacter crescentus*. *Proc. Natl. Acad. Sci. USA* **69**, 447–451.
10. Wu, J., and Newton, A. (1997). Regulation of the *Caulobacter* flagellar gene hierarchy; not just for motility. *Mol. Microbiol.* **24**, 233–240.
11. Gober, J. W., and England, J. C. (2000). *In* "Prokaryotic Development" (Y. Brun, and L. Shimkets, eds.), pp. 319–339. ASM Press, Washington, DC.
12. Laub, M. T., McAdams, H. H., Feldblyum, T., Fraser, C. M., and Shapiro, L. (2000). Global analysis of the genetic network controlling a bacterial cell cycle. *Science* **290**, 2144–2148.
13. Kelly, A. J., Sackett, M. J., Din, N., Quardokus, E., and Brun, Y. V. (1998). Cell cycle-dependent transcriptional and proteolytic regulation of FtsZ in *Caulobacter*. *Genes Dev.* **12**, 880–893.
14. Domian, I. J., Quon, K. C., and Shapiro, L. (1997). Cell type-specific phosphorylation and proteolysis of a transcriptional regulator controls the G1-to-S transition in a bacterial cell cycle. *Cell* **90**, 415–424.
15. Osley, M. A., and Newton, A. (1977). Mutational analysis of developmental control in

Caulobacter crescentus. *Proc. Natl. Acad. Sci. USA* **74**, 124–128.

16. Ohta, N., Ninfa, A. J., Allaire, A. D., Kulick, L., and Newton, A. (1997). Identification, characterization and chromosomal organization of cell division cycle genes in *Caulobacter crescentus*. *J. Bacteriol.* **179**, 2169–2180.
17. Huguenel, E. D., and Newton, A. (1982). Localization of surface structures during procaryotic differentiation: Role of cell division in *Caulobacter crescentus*. *Differentiation* **21**, 71–78.
18. Ohta, N., and Newton, A. (1996). Signal transduction in the cell cycle regulation of *Caulobacter* differentiation. *Trends Microbiol.* **4**, 326–332.
19. Osley, M. A., and Newton, A. (1980). Temporal control of the cell cycle in *Caulobacter crescentus*: Roles of DNA chain elongation and completion. *J. Mol. Biol.* **138**, 109–128.
20. Osley, M. A., Sheffery, M., and Newton, A. (1977). Regulation of flagellin synthesis in the cell cycle of *Caulobacter*: Dependence on DNA replication. *Cell* **12**, 393–400.
21. Sheffery, M., and Newton, A. (1981). Regulation of periodic protein synthesis in the cell cycle: Control of initiation and termination of flagellar gene expression. *Cell* **24**, 49–57.
22. Terrana, B., and Newton, A. (1976). Requirement of cell division step for stalk formation in *Caulobacter crescentus*. *J. Bacteriol.* **128**, 456–462.
23. Sommer, J. M., and Newton, A. (1988). Sequential regulation of developmental events during polar morphogenesis in *Caulobacter crescentus*: Assembly of pili on swarmer cells requires cell separation. *J. Bacteriol.* **170**, 409–415.
24. Ward, D., and Newton, A. (1997). Requirement of topoisomerase IV *parC* and *parE* genes for cell cycle progression and developmental regulation in *Caulobacter crescentus*. *Mol. Microbiol.* **26**, 897–910.
25. Wortinger, M., Sackett, M. J., and Brun, Y. V. (2000). CtrA mediates a DNA replication checkpoint that prevents cell division in *Caulobacter crescentus*. *EMBO J.* **19**, 4503–4512.
26. Ohta, N., Chen, L. S., Swanson, E., and Newton, A. (1985). Transcriptional regulation of a periodically controlled flagellar gene operon in *Caulobacter crescentus*. *J. Mol. Biol.* **186**, 107–115.

26a. Ohta, N., Masurekar, M., and Newton, A. (1990). Cloning and cell cycle-dependent expression of DNA replication gene *dnaC* from *Caulobacter crescentus*. *J. Bacteriol.* **172**, 7027–7034.

27. Parkinson, J. S., and Kofoid, E. C. (1992). Communication modules in bacterial signaling proteins. *Annu. Rev. Genet.* **26**, 71–112.
28. Koretke, K. K., Lupas, A. N., Warren, P. V., Rosenberg, M., and Brown, J. R. (2000). Evolution of two-component signal transduction. *Mol. Biol. Evol.* **17**, 1956–1970.
29. Grebe, T., and Stock, J. (1999). The histidine protein kinase superfamily. *Adv. Microbial Physiol.* **11**,
30. Stock, J. B., Surette, M. G., Levit, M., and Park, P. (1995). *In* "Two-Component Signal Transduction" (J. A. Hoch and T. J. Silhavy, eds.), pp. 25–51. American Society for Microbiology, Washington, DC.
31. Sommer, J. M., and Newton, A. (1991). Pseudoreversion analysis indicates a direct role of cell division genes in polar morphogenesis and differentiation in *Caulobacter crescentus*. *Genetics* **129**, 623–630.
32. Sommer, J. M., and Newton, A. (1989). Turning off flagellum rotation requires the pleiotropic gene *pleD*: *pleA*, *pleC*, and *pleD* define two morphogenic pathways in *Caulobacter crescentus*. *J. Bacteriol.* **171**, 392–401.
33. Wang, S. P., Sharma, P. L., Schoenlein, P. V., and Ely, B. (1993). A histidine protein kinase is involved in polar organelle development in *Caulobacter crescentus*. *Proc. Natl. Acad. Sci. USA* **90**, 630–634.
34. Ohta, N., Lane, T., Ninfa, E. G., Sommer, J. M., and Newton, A. (1992). A histidine protein kinase homologue required for regulation of bacterial cell division and differentiation. *Proc. Natl. Acad. Sci. USA* **89**, 10297–10301.

35. Wu, J., Ohta, N., Zhao, J.-L., and Newton, A. (1999). A novel bacterial tyrosine kinase essential for cell division and differentiation . *Proc. Natl. Acad. Sci. USA* **96**, 13068–13073.
36. Hecht, G. B., Lane, T., Ohta, N., Sommer, J. N., and Newton, A. (1995). An essential single domain response regulator required for normal cell division and differentiation in *Caulobacter crescentus*. *EMBO J.* **14**, 3915–3924.
37. Stock, A. M., Mottonen, J. M., Stock, J. B., and Schutt, C. E. (1989). Three-dimensional structure of CheY, the response regulator of bacterial chemotaxis. *Nature* **337**, 745–749.
38. Trach, K. A., Chapman, J. W., Piggot, P. J., and Hoch, J. A. (1985). Deduced product of the stage 0 sporulation gene *spo0F* shares homology with the Spo0A, OmpR, and SfrA proteins. *Proc. Natl. Acad. Sci. USA* **82**, 7260–7264.
39. Burbulys, D., Trach, K. A., and Hoch, J. A. (1991). Initiation of sporulation in *B. subtilis* is controlled by a multicomponent phosphorelay. *Cell* **64**, 545–552.
40. Hecht, G. B., and Newton, A. (1995). Identification of a novel response regulator required for the swarmer to stalked cell transition in *Caulobacter crescentus*. *J. Bacteriol.* **177**, 6223–6229.
41. Aldridge, P., and Jenal, U. (1999). Cell cycle-dependent degradation of a flagellar motor component requires a novel type response regulator. *Mol. Microbiol.* **32**, 379–391.
42. Wheeler, R. T., and Shapiro, L. (1999). Differential localization of two histidine kinases controlling bacterial cell differentiation. *Mol. Cell.* **4**, 683–694.
43. Wu, J., Ohta, N., and Newton, A. (1998). An essential, multicomponent signal transduction pathway required for cell cycle regulation in *Caulobacter*. *Proc. Natl. Acad. Sci. USA* **95**, 1443–1448.
44. Quon, K. C., Marczynski, G. T., and Shapiro, L. (1996). Cell cycle control by an essential bacterial two-component signal transduction protein. *Cell* **84**, 83–93.
45. Reisenauer, A., Quon, K., and Shapiro, L. (1999). The CtrA response regulator mediates temporal control of gene expression during the *Caulobacter* cell cycle. *J. Bacteriol.* **181**, 2430–2439.
46. Hoch, J. A. (1995). *In* "Two-Component Signal Transduction" (J. A. Hoch and T. J. Silhavy, eds.), pp. 129–144. ASM Press, Washington, DC.
47. Appleby, J. L., Parkinson, J. S., and Bourret, R. B. (1996). Signal transduction via the multistep phosphorelay: Not necessarily a road less traveled. *Cell* **86**, 845–848.
48. Burton, G., Hecht, G. B., and Newton, A. (1997). Roles of the histidine protein kinase PleC in *Caulobacter crescentus* motility and chemotaxis. *J. Bacteriol.* **179**, 5849–5853.
49. Jacobs, C., Domian, I. J., Maddock, J. R., and Shapiro, L. (1999). Cell cycle-dependent polar localization of an essential bacterial histidine kinase that controls DNA replication and cell division. *Cell* **97**, 111–120.
50. Fields, S., and Song, O. (1989). A novel genetic system to detect protein-protein interactions. *Nature* **340**, 245–246.
51. Brent, R., and Finley, R. L. Jr. (1997). Understanding gene and allele function with two-hybrid methods. *Annu. Rev. Genet.* **31**, 663–704.
52. Tomomori, C., Tanaka, T., Dutta, R., Park, H., Saha, S. K., Zhu, Y., Ishiima, R., Lui, D., Tong, K. I., Kurokawa, H., *et al.* (1999). Solution structure of the homodimeric core domain of *Escherichia coli* histidine kinase EnvZ. *Nature Struct. Biol.* **6**, 729–734.
53. Wheeler, R. T., Gober, J. W., and Shapiro, L. (1998). Protein localization during the *Caulobacter crescentus* cell cycle. *Curr. Opin. Cell Biol.* **1**, 636–642.
54. Shapiro, L., and Losick, R. (2000). Dynamic spatial regulation in the bacterial cell. *Cell* **100**, 89–98.
55. Alley, M. R. K., Maddock, J. R., and Shapiro, L. (1992). Polar localization of a bacterial chemoreceptor. *Genes Dev.* **6**, 825–836.
56. Burkolder, W. F., and Grossman, A. D. (2000). *In* "Prokaryotic Development" (Y. V. Brun and L. J. Shimkets, eds.). ASM Press, Washington, DC.

57. Fawcett, P., Eichenberger, P., Losick, R., and Youngman, P. (2000). The transcriptional profile of early to middle sporulation in *Bucillus subtilis*. *Proc. Natl. Acad. Sci. USA* **97**, 8063–8068.
58. Sonenshein, A. L. (1989). *In* "Regulation of Prokaryotic Development" (I. Smith, R. Slepecky and P. Stelow, eds.), pp. 109–130. American Society for Microbiology, Washington, DC.
59. Burkholder, W. F., Kurtser, I., and Grossman, A. D. (2001). Replication initiation proteins regulate a developmental checkpoint in *Bacillus subtilis*. *Cell* **104**, 269–279.

CHAPTER 19

The Sln1-Ypd1-Ssk1 Multistep Phosphorelay System That Regulates an Osmosensing MAP Kinase Cascade in Yeast

HARUO SAITO
Division of Molecular Cell Signaling, Institute of Medical Science, University of Tokyo, Tokyo 108-8639, Japan

Histidine Kinases in Signal Transduction

When the budding yeast *Saccharomyces cerevisiae* is exposed to a high osmotic environment, the sensor histidine kinase Sln1 activates the stress-responsive high osmolarity glyserol (HOG) mitogen-activated protein (MAP) kinase pathway through the Sln1-Ypd1-Ssk1 multiple phosphorelay reaction. When activated (under low osmotic conditions), the Sln1 histidine kinase autophosphorylates a histidine residue in Sln1. The phosphate is transferred, in stepwise reactions, to an aspartate residue in the Sln1 C-terminal receiver domain, then to a histidine residue in Ypd1, and finally to an aspartate residue in the Ssk1 receiver domain. When phosphorylated, Ssk1 is incapable of activating the Ssk2/Ssk22 MAPKKKs. Under high osmotic conditions, the Sln1 histidine kinase is inhibited. As a consequence, unphosphorylated Ssk1 accumulates, which binds and activates Ssk2/Ssk22 MAPKKKs. These MAPKKKs activate the Pbs2 MAPKK, which then activates the Hog1 MAPK. The activated Hog1 translocates to the nucleus, where it activates various stress-adaptive genes. The budding yeast has another osmosensing mechanism that also activates the HOG MAPK cascade but independent of the Sln1 osmosensor. This alternative mechanism involves the transmembrane anchorage protein Sho1, the small GTPase Cdc42, the PAK-like kinase Ste20, the Ste11 MAPKKK, and the Ste11-binding protein Ste50. The Pbs2 MAPKK serves as a scaffold protein that prevents the potential cross-talk between the osmosensing and the mating pheromone signaling pathways, both of which share the same Ste11 MAPKKK.

INTRODUCTION

Yeast cells in their natural habitats must adapt to extremes of osmotic conditions such as the saturating sugar of drying fruits and the nearly pure water of rain [1]. In budding yeast (*Saccharomyces cerevisiae*), accumulation of the compatible osmolite glycerol is particularly important for hyperosmolarity adaptation [2]. The molecular mechanism by which yeast cells detect extracellular osmolarity changes and regulate, among other things, glycerol synthesis, is termed the high osmolarity glycerol (HOG) response signal transduction pathway.

It is now known that the HOG pathway comprises two independent osmosensing mechanisms. Both osmosensors regulate, independently, a common downstream protein kinase, termed the HOG mitogen-activated protein kinase (MAPK) cascade [3]. One of the upstream osmosensing mechanisms utilizes a sensor histidine kinase Sln1, hence termed the SLN1 branch. The other upstream osmosensing mechanism (SHO1 branch) is named after a key transmembrane protein Sho1, although Sho1 itself is probably not an osmosensor. This chapter briefly surveys the common downstream MAPK kinase signal transducing elements in order to set the stage for a more detailed discussion about the upstream two-component His-Asp phosphorelay osmosensing elements.

THE COMMON DOWNSTREAM PATHWAY

HOG MAPK CASCADE

MAPK cascades are common eukaryotic signaling modules that consist of a MAPK, a MAPK kinase (MAPKK), and a MAPKK kinase (MAPKKK) [4, 5]. A MAPK is activated by a specific MAPK kinase (MAPKK) through phosphorylation of conserved threonine and tyrosine residues in the activation loop of MAPK. A MAPKK is activated by a specific MAPKK kinase (MAPKKK) through phosphorylation of conserved threonine and/or serine residues (Fig. 1). The originally identified MAPK pathway (the mammalian ERK MAPK pathway) is activated by mitogenic factors as its name implies, but

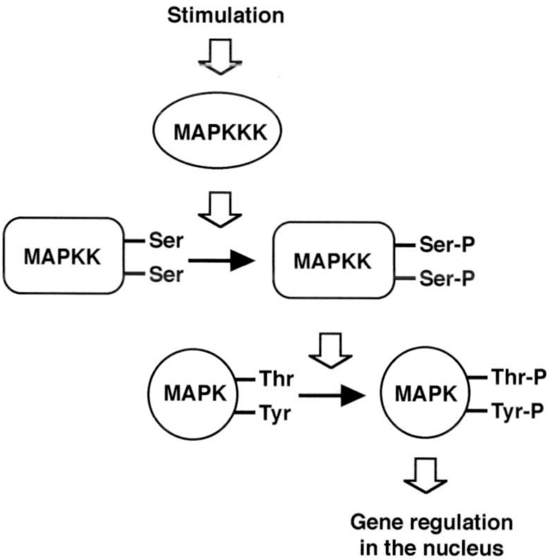

FIGURE 1 Schematic model of a MAP kinase cascade. Various extracellular stimuli are recognized by specific receptors/sensors on the cell surface, and generate intracellular signals that are eventually transmitted to MAPK cascades. A number of distinct MAPK cascades may coexist in a cell. Diverse mechanisms are proposed for MAPKKK activation, including binding of an activator protein, phosphorylation by a kinase, and proteolytic cleavage. It seems possible that some MAPKKKs can be activated by more than one mechanism. Activated MAPKKK phosphorylates and activates a specific MAPKK, which in turn phosphorylates and activates a cognate MAPK. A fraction of activated MAPK is transported to the nucleus to activate transcription factors.

many additional MAPK pathways are now known that are activated by non-mitogenic stimuli, such as cellular stresses. The yeast HOG MAPK pathway is an example of such stress-activated MAPK cascades and is composed of three partially redundant MAPKKKs (Ste11, Ssk2, and Ssk22), a MAPKK (Pbs2), and a MAPK (Hog1) (Fig. 2).

The HOG pathway is essential for yeast cells to adapt to a high osmotic environment. Yeast mutants defective in either the *PBS2* or the *HOG1* gene cannot grow on media containing, for example, 1 *M* sorbitol [6–8]. Because the three MAPKKKs (Ste11, Ssk2, and Ssk22) are partially redundant, disruptions of one or two of these genes, in any combination, do not significantly affect osmosensitivity of the mutant cells. Triple disruption mutants (*ste11Δ ssk2Δ ssk22Δ*) are, however, as sensitive to high osmolarity as *pbs2Δ* or *hog1Δ* mutants [8].

In wild-type yeast cells, hyperosmotic stress induces only a transient activation of the HOG pathway, which lasts for about 20 min. Cells then reestablish a poststress equilibrium in which Hog1 activity is only moderately higher than that of the prestress cells [7]. Downregulation of the HOG

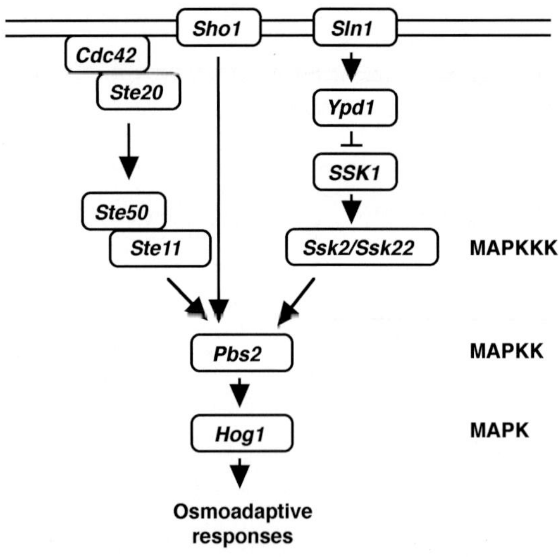

FIGURE 2 The yeast HOG MAPK pathway consists of three partially redundant MAPKKKs (Ste11, Ssk2, and Ssk22), MAPKK (Pbs2), and MAPK (Hog1). The Sln1-Ypd1-Ssk1 multistep phosphorelay regulates two MAPKKKs: Ssk2 and Ssk22. The third MAPKKK, Ste11, is activated by Ste20, a PAK-like protein kinase, and membrane-bound small G-protein Cdc42. Ste50 binds tightly to Ste11 and is necessary for the latter's activity. Sho1 is a membrane protein whose SH3 domain specifically binds Pbs2 MAPKK. Sho1 is necessary for the activation of Pbs2 by Ste11 MAPKKK, but not by Ssk2/Ssk22 MAPKKKs.

pathway is mediated by several protein phosphatases. Because activation of a MAPK requires phosphorylation at both a tyrosine residue and a nearby threonine residue in the characteristic T-X-Y motif, dephosphorylation at either residue can inactivate it. In *S. cerevisiae*, two protein tyrosine phosphatases (Ptp2 and Ptp3) are known to inactivate Hog1 by dephosphorylating the activation phosphotyrosine residue [9, 10]. Also, a protein serine/threonine phosphatase (Ptc1) has been implicated in the negative regulation of Hog1 activity [11, 12].

Whereas Hog1 activation is essential for survival of yeast cells at high osmolarity, prolonged and strong activation of Hog1 is toxic. For example, expression of a constitutively active mutant of the Ssk2 or Pbs2 kinase results in a lethal level of Hog1 activation [10]. Overexpression of the Ptp2 protein tyrosine phosphatase suppresses the lethality caused by the constitutively active Ssk2 or Pbs2 mutant by dephosphorylating Hog1 [10]. Disruption of *PTP2* leads to elevated levels of tyrosine phosphorylated Hog1 following exposure of cells to high osmolarity [9, 10]. Furthermore, disruption of both *PTP2* and another protein tyrosine phosphatase, *PTP3*, results in constitutive phosphorylation at Hog1 Tyr176, even in the absence of increased environmental osmolarity. Disruption of the *PTP3* gene alone, however, has no significant effect, indicating that Ptp3 has only a subsidiary role in the HOG pathway. Ptp3 has been shown to have a more prominent role in the negative regulation of the mating pheromone-responsive MAPK, namely Fus3 [13].

Phosphatase activities of Ptp2 and Ptp3 are induced by a Hog1-mediated mechanism [9, 10]. When the catalytically inactive Hog1$^{K/N}$ mutant protein is expressed in *hog1Δ* cells, it is constitutively tyrosine phosphorylated. In contrast, Hog1$^{K/N}$, expressed together with wild-type Hog1, is tyrosine phosphorylated only when cells are exposed to high osmolarity. Thus, the kinase activity of Hog1 is required for its own tyrosine dephosphorylation. It is not yet clear how Hog1 activates these tyrosine phosphatases.

Paradoxically, even though *ptp2Δ ptp3Δ* mutant cells accumulate tyrosine phosphorylated Hog1, such mutants are viable. This is because serine/threonine phosphatases can also inactivate Hog1 MAPK by dephosphorylating Thr174. Among the many protein serine/threonine phosphatases known in yeast, a type 2C protein phosphatase, Ptc1, seems to be the most important in this respect. Thus, *ptp2Δ ptc1Δ* double mutants are lethal because of Hog1 hyperactivation [11, 12].

Nuclear Translocation of Activated Hog1

The components of the HOG MAPK cascade usually reside in the cytoplasm, but its activation alters gene transcription in the nucleus. Presumably, some component must cross the nuclear envelope. Using GFP fusion of the kinases in the HOG MAPK cascade, it was found that Hog1, but neither Pbs2 nor Ste11, translocates into the nucleus following osmotic stress [14, 15]. When cells carrying Hog1-GFP were exposed to a brief osmotic shock, Hog1-GFP changed its distribution, from being mainly cytoplasmic to accumulating completely within the nucleus. After 30 min in the presence of NaCl, the levels of Hog1-GFP in the cytoplasm were partially restored, indicating that nuclear localization of Hog1 is a transient effect. Interestingly, the time course of the Hog1 nuclear localization correlates with the transient Hog1 phosphorylation that is observed after osmotic stress. When a *pbs2Δ* strain was used for a similar analysis, nuclear translocation of Hog1-GFP did not occur after stress, indicating that phosphorylation and/or activation of Hog1 by Pbs2 is required for Hog1 nuclear translocation.

To distinguish between the possibilities that the protein kinase activity of Hog1 was mediating its translocation and that phosphorylation of the activating residues Thr174/Tyr176 was sufficient, two Hog1 mutants were constructed. $Hog1^{T/A, Y/A}$ is a mutant in which the two phosphorylation sites required for MAPK activation are mutated to Ala. This mutant cannot be phosphorylated after stress and therefore is inactive. The second mutant, $Hog1^{KM}$, can be phosphorylated at Thr174/Tyr176, but the phosphorylated protein is devoid of protein kinase activity due to a mutation of a Lys52 to Met in the active site. The $Hog1^{T/A, Y/A}$ protein failed to localize to the nucleus after osmotic stress. In contrast, $Hog1^{KM}$ behaves identically to wild type Hog1. Thus, phosphorylation of Hog1, but not activation of its kinase activity, is required for Hog1 movement into the nucleus.

The exit of Hog1 from the nucleus correlates with its dephosphorylation. As mentioned earlier, the protein tyrosine phosphatases that dephosphorylate Hog1, namely Ptp2 and Ptp3, are activated by Hog1-mediated phosphorylation [9, 10]. Thus, when a kinase-deficient allele of Hog1 ($Hog1^{KM}$) is expressed in a *hog1*-deficient strain, its level of phosphorylation after stress remains high for a longer period of time due to the lack of phosphatase activity. Interestingly, return of $Hog1^{KM}$-GFP to the cytoplasm in the *hog1Δ* strain, where phosphatases cannot be activated, was much slower than in the wild-type strain. These results suggest that the phosphorylation of Hog1 drives its nuclear accumulation and that Hog1 dephosphorylation mediates its return to the cytoplasm.

Nuclear Events

Following the nuclear translocation of activated Hog1, expression of a distinct set of genes occurs. The *GPD1* and *GPD2* genes encode two isoenzymes for NAD-dependent glycerol 3-phosphate dehydrogenase, which is a key enzyme in glycerol biosynthesis [2, 16, 17, 18]. Accumulation of glycerol counteracts osmotic imbalance. Other induced genes have a more general role in the cellular stress response. Those include the genes that encode cytosolic catalase (*CTT1*), heat shock proteins (*HSP12*, *SHP70*, and *HSP104*), and a DNA damage-induced protein (*DDR2*) [19, 17, 20]. Induction of these genes requires transcription factors, Msn1, Msn2, Msn4, and Hot1 [21–23]. In some cases, however, the induction of gene expression is also mediated by HOG-pathway dependent inactivation of a transcriptional repressor, Sko1 [24]. It is not known, however, whether the activities of any or all of these regulatory proteins are directly regulated by Hog1-mediated phosphorylation. A protein kinase, Rck2, has been found as a strong binder to Hog1 [25]. Rck2 is phosphorylated upon osmotic stress in a Hog1-dependent manner, and Rck2 phosphorylation by Hog1 results in an increase of Rck2 kinase activity. The role of Rck2 in osmoregulation is not yet clear.

THE SLN1 BRANCH

Discovery of the Sln1 Osmosensor

The Sln1 osmosensor was discovered by efforts that were initially unrelated to the osmosensing mechanism. The *SLN1* gene was initially recognized by a mutation that was synthetically lethal with an N-end rule gene, hence the name SLN1 [26]. The relationship between the Sln1 osmosensor and the N-end rule protein degradation pathway is not yet clear. In a separate effort, we were studying the role of yeast tyrosine phosphatases, including Ptp2. Because yeast do not have any tyrosine kinase, it appeared strange that yeast have protein tyrosine phosphatases at all [27–29]. Of course, it was not yet known at the time that those tyrosine phosphatases inactivate MAPKs.

In order to study the physiological function of the Ptp2 tyrosine phosphatase, a screening was conducted for mutants that require expression (or overexpression) of *PTP2* for survival [11, 12]. The isolated mutants had a defect in either one of three genes: *PTC1*, *SLN1*, and *YPD1*. The *PTC1* gene encodes protein serine/threonine phosphatase type 2C, and as explained earlier, *ptc1*Δ *ptp2*Δ double mutants are lethal because of Hog1 hyperactivation. The *ptc1*Δ mutation alone, however, has relatively moderate effects [11, 30]. In contrast, mutations that belong to the other two genes (*SLN1* and *YPD1*) were lethal by themselves, and suppression of their lethality required over-expression of *PTP2*. The lethality of *sln1*Δ and *ypd1*Δ mutations could also be suppressed by the overexpression of *PTC1*.

These findings indicated that the functions of the four genes, *SLN1*, *YPD1*, *PTP2*, and *PTC1*, are intimately interconnected. However, it was obvious that neither Sln1 nor Ypd1 was a substrate of the protein phosphatases (because overexpression of these phosphatases had effects only when Sln1 or Ypd1 was absent). Thus, we hypothesized that a protein X exists that is a substrate of both Ptp2 and Ptc1 phosphatases and that X hyperphosphorylation is toxic to cell growth. We further hypothesized that both Sln1 and Ypd1 negatively regulate the phosphorylation of X. This model would explain, at least formally, the lethality of *sln1Δ* and *ypd1Δ* mutations, as well as their suppression by the overexpression of *PTP2* or *PTC1*. But what is protein X?

In order to identify the hypothetical factor X, we conducted another series of mutant screenings based on the following premise: if X hyperphosphorylation is toxic to cell growth, then either mutational inactivation of X itself or of a protein kinase that phosphorylates X should suppress the lethality of *sln1Δ* and *ypd1Δ* mutations. To test this prediction, extragenic suppressor mutants were isolated from either *sln1* or *ypd1* mutant strains. In both cases, suppressor mutations were isolated in four genes: *SSK1*, *SSK2*, *PBS2*, and *HOG1* [12, 31]. Both Hog1 and Pbs2 had been implicated previously in an osmosensing signal transduction pathway [6, 7]. Furthermore, a rapid, *PBS2*-dependent tyrosine and threonine phosphorylation of Hog1 occurs in response to increases in extracellular osmolarity. Further studies demonstrated that Hog1 is indeed the hypothetical factor X, solving satisfactorily the original question about the substrate of Ptp2 tyrosine phosphatase [12]. However, now there are more puzzling questions: what are the functions of Sln1, Ypd1, Ssk1, and Ssk2 in the HOG pathway?

SSK2 encodes a previously undescribed protein with a C-terminal domain that is similar to MAPKKKs. The long N-terminal noncatalytic domain of about 1200 amino acids has no similarity to any known MAPKKKs. It was later found, however, that a human MAPKKK, MTK1, and its mouse homologue, MEKK4, are similar to SSK2 both in the C-terminal kinase domain and in the N-terminal regulatory domain [32, 33]. To examine if Ssk2 is an upstream activator of the Pbs2-Hog1 pathway, we generated a constitutively activated Ssk2 by eliminating its N-terminal noncatalytic domains. Expression of *SSK2*ΔN indeed induced activation of Hog1 in a *PBS2*-dependent manner, establishing the Ssk2-Pbs2-Hog1 kinase cascade.

Further analyses, however, indicated that Ssk2 is not the sole activator of the Pbs2-Hog1 pathway: unlike *pbs2Δ* or *hog1Δ* mutants that are severely osmosensitive, *ssk2Δ* mutants are osmoresistant and can activate Hog1 in a *PBS2*-dependent manner, indicating the presence of additional MAPKKKs that are functionally redundant to *SSK2*. Initially, an *SSK2*-related gene termed *SSK22* was isolated by cross-hybridization to the *SSK2* probe. Sequence comparison revealed that Ssk2 and Ssk22 are similar both in their

C-terminal kinase domain (69% identity) and in the N-terminal noncatalytic domain (47% identity). That Ssk22 can activate the Pbs2-Hog1 pathway was then proved by using a constitutively active Ssk22 mutant, Ssk22ΔN [31]. However, even *ssk2Δ ssk22Δ* double mutants are osmosensitive and can activate Hog1 in a *PBS2*-dependent manner. Additional genetic screenings revealed that two independent osmosensory mechanisms can activate Pbs2-Hog1: one is mediated by Ssk2/Ssk22 MAPKKKs and the other is mediated by the third MAPKKK, Ste11 [8]. In the remainder of this chapter, I will focus on the first osmosensory mechanism that involves Sln1, Ypd1, Ssk1, and Ssk2/Ssk22 MAPKKKs. The second osmosensory mechanism that activates the Ste11 MAPKKK will be described in a later chapter.

Structures of Sln1, Ypd1, and Ssk1

The osmosensor Sln1 is an example of hybrid histidine kinases, in which both a histidine kinase domain and a receiver domain are encoded covalently in a single protein. The N terminus of Sln1 is topologically similar to the bacterial osmosensor EnvZ: it has an extracellular domain flanked by two transmembrane segments, TM1 and TM2 (Fig. 3). The Sln1 C-terminal cytoplasmic

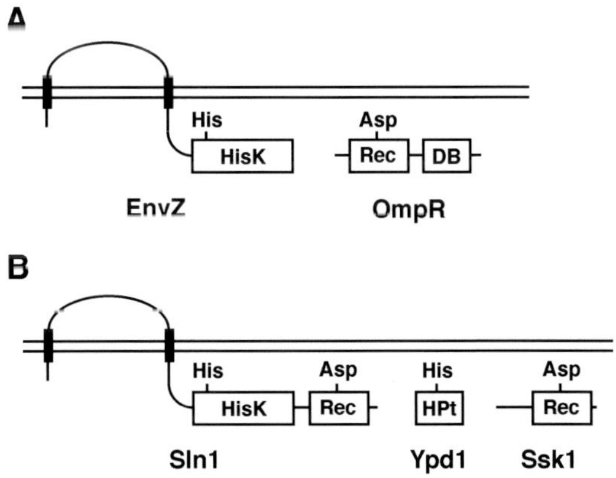

FIGURE 3 Comparison of bacterial and yeast osmoregulatory two-component systems. (A) The *E. coli* osmosensing pathway is composed of two proteins: the sensor histidine kinase EnvZ and the response regulator OmpR. OmpR is a transcription factor that differentially regulates the expression of the two porin genes (OmpC and OmpF) depending on its phosphorylation states. (B) The *S. cerevisiae* osmosensing pathway is composed of three proteins, the sensor histidine kinase Sln1, the intermediate phosphate carrier Ypd1, and the response regulator Ssk1. Ssk1 regulates the activities of the Ssk2/Ssk22 MAPKKKs. Both EnvZ and Sln1 sensor histidine kinases have two transmembrane segments flanking an extracellular domain. HisK, histidine kinase domain; Rec, receiver domain; DB, DNA-binding domain; HPt, histidine-containing phosphotransfer domain; His, histidine phoshorylation site; Asp, aspartate phosphorylation site.

region contains a histidine kinase domain with a histidine phosphorylation site (His576) and a receiver domain with a conserved aspartate phosphorylation site (Asp1144).

It is believed that the Sln1 histidine kinase is active under normal (nonstressful) osmotic conditions; when the environmental osmolarity is increased, its kinase activity is inhibited [34]. The Sln1 osmosensing mechanism is not yet fully understood, but it is probably similar to that of EnvZ. Deletion of the first transmembrane segment (TM1) constitutively activates the Sln1 kinase, whereas removal of both TM1 and the extracellular domain inactivates Sln1 [35]. Replacement of the Sln1 extracellular domain with the dimerization promoting leucine zipper motif restores Sln1 histidine kinase activity (but not the capacity to respond to osmotic changes). Thus, it seems likely that transmembrane segments are essential for osmosensing, whereas the extracellular domain is required for Sln1 dimerization that activates the histidine kinase activity.

Ssk1 also contains a receiver domain in its C terminus with the conserved aspartate phosphorylation site Asp554; however, its N-terminal half is not homologous to any known proteins (except, of course, the Ssk1 homologous in other fungal and yeast species).

Ypd1 is a protein of 167 amino acid without a receiver domain or any other known domain motif. A short sequence around Ypd1 His64, however, has a weak sequence similarity to the sequences around the histidine phosphorylation sites of bacterial chemotactic CheA protein and the C-terminal histidine phosphorylation (HPt) domain of several prokaryotic histidine kinases [36, 37]. Determination of the crystal structures proved that Ypd1 is indeed similar to the HPt domain of the *Escherichia coli* ArcB protein [38–40]. Both Ypd1 and ArcB HPt domain are composed of a four-helix bundle with the phospho-accepting histidine in the middle of one helix.

In Vivo Characterization of the Sln1-Ypd1-Ssk1 Multistep Phosphorelay

The *ypd1*Δ phenotype is very similar to that of *sln1*Δ: both are lethal and their lethality is suppressed either by overexpression of *PTP2* or by either of the *ssk1*Δ, *ssk2*Δ, *pbs2*Δ, of *hog1*Δ mutation. These results suggested that Ypd1 plays a role in the Sln1-Ssk1 two-component system. Furthermore, two-hybrid analyses indicated that Ypd1 interacts with the receiver domains of both Sln1 and Ssk1, whereas there is no direct interaction between Ssk1 and Sln1, suggesting the possibility that Ypd1 functions as an intermediary between Sln1 and Ssk1 [34].

Ypd1 His64 was found essential for the proper function of Ypd1 because $YPD1^{H64Q}$ could not rescue the *ypd1Δ* lethality. Using the *GST-YPD1* fusion gene, the GST-Ypd1 protein isolated from wild-type cells growing in normal media was found to be phosphorylated. This phosphate was alkali resistant and acid labile, suggesting that the phosphorylated amino acid was histidine. In contrast, the GST-$Ypd1^{H64Q}$ mutant protein was not phosphorylated, implying that Ypd1 His64 is the phosphorylation site. Ypd1 phosphorylation is dependent on Sln1 activity, as no phosphorylation was observed in *sln1Δ* mutant cells. Ypd1 is, however, not directly phosphorylated by the histidine kinase activity of Sln1. The Sln1ΔC mutant protein lacks the C-terminal receiver domain but has an intact histidine kinase activity. Nevertheless, the *sln1ΔC* mutant could not support the histidine phosphorylation of Ypd1, indicating that the Sln1 kinase domain alone is not sufficient to phosphorylate Ypd1.

The $Sln1^{D1144N}$ and $Sln1^{H576Q}$ mutant proteins lack, respectively, the aspartate phosphorylation site in the C-terminal receiver domain and the histidine autophosphorylation site. Both these mutants also failed to phosphorylate Ypd1 *in vivo*, indicating that phosphorylation at both Sln1-Asp1144 and Sln1-His576 is required for the phosphorylation of Ypd1-His64.

The phenotypes of *sln1* and *ssk1* mutants suggested that the Sln1p histidine kinase is active under normal osmotic conditions, whereas it is suppressed under hyperosmotic conditions [12]. This prediction was corroborated by *in vivo* phosphorylation experiments that showed that the level of Ypd1 histidine phosphorylation is much lower after osmotic shock than in control cells.

In Vitro Characterization of the Sln1-Ypd1-Ssk1 Multistep Phosphorelay

The origin of the phosphate transferred to Ypd1-His64 and its relationship to the predicted phosphorylation at Ssk1-Asp554 were examined by reconstituting the phosphotransfer reactions *in vitro*. First, we tested whether a phosphate at Sln1-His576 was transferred directly to Sln1-Asp1144. For this purpose, the Sln1-HisK protein that contains the Sln1 histidine kinase domain, but neither the transmembrane segments nor the C-terminal receiver domain, was purified as a GST fusion protein. When purified Sln1-HisK was incubated in the presence of [γ-^{32}P]ATP, ^{32}P was incorporated into the fusion protein by its autohistidine kinase activity. As predicted, the Sln1-$HisK^{K576Q}$ mutant has no autokinase activity.

^{32}P-labeled Sln1-HisK was then mixed with SLN1-Rec, which contains the Sln1 receiver domain, but not the kinase domain. In 15 min, nearly half of the ^{32}P was transferred from ^{32}P-labeled Sln1-HisK to Sln1-Rec, and a corresponding decrease of radioactivity in Sln1-HisK was observed. In contrast, when the Sln1-HisK^{K576Q} or Sln1-Rec^{D1144N} mutant protein was used in place of their wild-type counterpart, no transfer of ^{32}P was observed. Thus, the phosphate residue on Sln1-His576 is transferred to Sln1-Asp1144.

Sln1-His576~P does not transfer its phosphate to Ypd1 *in vitro*, indicating that direct phosphotransfer from Sln1-His576 to Ypd1-His64 does not occur. In contrast, SLN1-Asp1144~P efficiently transferred its phosphate to Ypd1, but not to Ypd1^{H64Q}. Thus, it was concluded that the phosphate group on Ypd1-His64 was derived from Sln1-Asp1144.

Finally, in similar *in vitro* reactions, it was demonstrated that Ssk1-Asp554 was phosphorylated by Ypd1-His64~P. Thus, the reconstituted event can be summarized as follows. First, the autokinase activity of Sln1 phosphorylates Sln1-His576 and then Sln1-His576~P donates its phosphate to Sln1-Asp1144. Sln1-Asp1144~P then transfers the phosphate to Ypd1-His64, which further transfers the phosphate to Ssk1-Asp554. Both mutant analyses and the *in vitro* reconstitution experiment indicated that each phosphotransfer step is obligatory for the next step to occur. For example, there is no short-cut phosphotransfer reaction between Sln1-His576 and Ssk1-Asp554, or no phosphorylation of Ypd1-His64 by the histidine kinase activity of Sln1. Each phosphotransfer reaction, however, seems reversible. For example, the retrograde phosphotransfer from Ssk1-Asp554 to Ypd1-His64 is possible [41].

The Sln1-Ypd1-Ssk1 multistep phosphotransfer reaction (or phosphorelay) may seem an unnecessarily complex scheme to move one phosphate from Sln1 to Ssk1. However, similar multistep phosphorelays are also used in prokaryotic signal pathways, such as the Spo0 sporulation regulatory pathway of *Bacillus subtilis* [37, 42–44]. In the case of the Spo0 pathway, intermediate steps are used to integrate additional signals by modulating the activities of specific aspartate phosphatases [45–47]. It is an interesting possibility that the yeast Sln1-Ypd1-Ssk1 multistep phosphorelay is also regulated by phosphatases.

Activation of Ssk2 MAPKKK by the Ssk1 Response Regulator

The Ssk2-Pbs2-Hog1 kinase cascade is activated under conditions that inactivate Sln1 histidine kinase activity. Mutational inactivation of either the *SLN1* or the *YPD1* gene also activates the Hog1 kinase cascade. In contrast, lack of Ssk1 prevents activation of the Ssk2-Pbs2-Hog1 kinase cascade. These observations are consistent with the model that the unphosphorylated Ssk1 (Ssk1-OH) acts as an Ssk2 activator, whereas phosphorylated Ssk1 (Ssk1~P) is inactive.

Indeed, the Ssk2 kinase is activated by a direct interaction with Ssk1-OH. *In vitro*, the addition of purified Ssk1 activated Ssk2 kinase activity [48]. By two-hybrid analyses, the region of Ssk2 essential for its Ssk1 binding was mapped to Ssk2 residues 294–413. Reciprocally, the region of Ssk1 essential for its Ssk2 binding was mapped to Ssk1 residues 475–670, which coincides with the Ssk1 receiver domain. The N-terminal noncatalytic domains of MAPKKKs, such as Raf-1 and MEKK1, are autoinhibitory for their kinase activity [49]. This is also the case for Ssk2, as an N-terminal domain deletion results in constitutive activation of the kinase [31, 8]. A deletion of the N-terminal 191 amino acids was sufficient to activate the SSK2 kinase domain, indicating that the inhibitory domain is located near its N terminus. Thus, the autoinhibitory region may be close enough to the Ssk1-binding domain (294–413) so that binding of Ssk1 to Ssk2 disrupts autoinhibition and leads to the activation of Ssk2 kinase.

It may seem counterintuitive that the unphosphorylated form of Ssk1 acts as an activator because aspartyl-phosphate on a typical receiver domain protein has a very short half-life, ranging from milliseconds to a few minutes. In order for cells to maintain the levels of dephosphorylated Ssk1 sufficiently low in resting cells, Sln1 will have to consume an enormous amount of ATP only to replenish the rapidly disappearing phosphate on Ssk1. Otherwise, an inadvertent firing of the Ssk2-Pbs2-Hog1 kinase cascade would ensue. A recent finding, however, revealed that cells need not squander their ATP. Aspartyl-phosphate on the purified Ssk1~P turned out to have an unusually long half-life of 42 h (in comparison, phosphate on the Sln1 receiver domain has a more typical half life of 13 min) [41]. The extra stability of Ssk1~P ensures that when the Sln1 histidine kinase is active, there is little unphosphorylated Ssk1. This will, however, create another problem. How is Ssk1~P dephosphorylated rapidly enough when hyperosmotic stress inhibits Sln1 histidine kinase activity? Activation of the Ssk2-Pbs2-Hog1 cascade occurs within a minute of the stress stimulus. A possible solution is the retrograde transfer of phosphate from Ssk1~P to Ypd1 and to the Sln1 receiver domain, which may have a short half-life. Another possibility is participation of an aspartate phosphatase or a histidine phosphatase that dephosphorylates Ssk1~P or Ypd1~P, respectively, in an analogous manner that specific aspartate phosphatases regulate the bacterial Spo0 signal pathway [45–47]. Clearly, there is much more to be learned about the regulation of the Sln1-Ypd1-Ssk1 multistep phosphorelay reaction in osmoregulation.

Sensor Histidine Kinases and Multistep Phosphorelay in Fission Yeast

In the fission yeast *Schizosaccharomyces pombe*, the stress-activated Sty1 MAPK (also called Spc1, Phh1) is activated, not only by hyperosmotic stress, but also by oxidative and heat shock stresses [50–53]. Activation of the Sty1 MAPK is, at least partly, dependent on a two-component system that is very similar to the Sln1-Ypd1-Ssk1 multistep phosphorelay.

The fission yeast genome encodes three histidine kinases (Mak1, Mak2, and Mak3) [54], one Ypd1-like protein (Mpr1; also called Spy1) [53, 55], and two response regulator proteins (Msc4 and Prr1) [56–58]. Mpr1 and Msc4 are likely components of a multistep phosphorelay that regulates the stress-responsive Sty1 MAPK cascade (Fig. 4). The Msc4 response regulator is structurally closely related to the budding yeast Ssk1, and *msc4Δ* mutant cells are defective in activation of the Sty1 MAPK by high osmolarity or oxidative

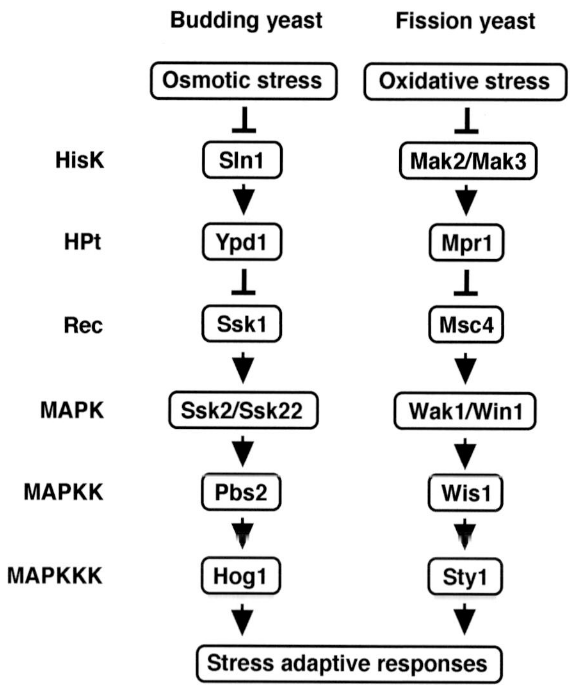

FIGURE 4 Comparison of the stress-responsive MAPK pathways in budding and fission yeast. Budding yeast (*S. cerevisiae*) and fission yeast (*S. pombe*) have very similar MAP kinase cascades that are regulated by two-component signal transducers. The stimulating input signals are, however, different between them. The budding yeast Sln1 histidine kinase responds to osmotic stress, whereas fission yeast Mak1/Mak2 histidine kinases respond to oxidative stress. Accordingly, respective adaptive responses are different. Despite these differences, overall similarity is impressive. Some of the fission yeast genes in this pathway have commonly used synonyms: Mpr1 = Spy1; Wak1 = Wik1; and Sty1 = Spc1 = Phh1. HisK, sensor histidine kinase; HPt, histidine containing phosphotransfer domain; Rec, receiver domain protein that activates MAPKKK.

stress [57, 59]. The Ypd1-like Mpr1 interacts with Msc4 in response to oxidative stress, and in *mpr1Δ* mutant cells, the Sty1 MAPK cascade is activated constitutively [53]. By analogy with the budding yeast SLN1-HOG pathway, it can be predicted that the Wak1 and Win1 MAPKKKs (homologues of Ssk2) are activated by the unphosphorylated response regulator Msc4.

An important difference between the budding yeast and the fission yeast two-component pathways is that the latter responds strongly to oxidative stress, whereas the former does not. This seems to be due to an interesting difference in the structure of their sensor histidine kinases. The three fission yeast histidine kinases (Mak1, 2, and 3) are cytoplasmic proteins and contain one or two so-called PAS domain [54]. The PAS domain has been identified in a variety of bacterial sensors of oxygen or redox [60], and the PAS domain of the bacterial oxygen sensor FixL was shown to bind a heme group, which serves as an oxygen sensor [61]. Thus, it is possible that a bound heme group serves as the oxygen-sensing mechanism in the Mak proteins too, although no heme binding has been demonstrated yet. The absence of a PAS domain in the budding yeast osmosensor Sln1 might explain why oxidative stress does not activate the Hog1 MAPK cascade.

In the absence of either Mak2 or Mak3, no activation of Sty1 MAPK occurs by oxidative stimulation [54]. The requirement for both Mak2 and Mak3 suggests that the two histidine kinases act as a heterodimer. Although Mak1 also contains PAS domains, *mak1* gene deletion does not affect Sty1 activation. Mak1 might be involved in the activation of another response regulator protein, Prr1.

THE SHO1 BRANCH

DISCOVERY OF THE SECOND OSMOSENSING MECHANISM

Although *ssk2Δ ssk22Δ* double mutants cannot transduce any signal from the Sln1-Ypd1-Ssk1 multistep phosphorelay, such mutants can still activate Hog1 upon osmotic shock and, as a consequence, are osmoresistant. This observation led to a hypothesis that a second osmosensing mechanism exists that is independent of the Sln1 osmosensor. In order to find genes that are involved in the hypothetical second pathway, mutant screening was conducted based on the assumption that a mutation in the second pathway, when combined with *ssk2Δ ssk22Δ* double mutation, would prevent Hog1 activation and therefore cause osmosensitivity. In this manner, at least five genes have been found to be involved in the second osmosensory mechanism: a transmembrane protein Sho1; a small GTPase Cdc42; a PAK-like protein kinase Ste20; a MAPKKK Ste11; and a Ste11-binding protein Ste50 (Fig. 5) [8, 31, 62–64].

Using a constitutively active Ste11 mutant, it was demonstrated that Ste11 is another MAPKKK that activates the Pbs2 MAPKK. Thus, three MAPKKKs (Ssk2, Ssk22, and Ste11) can activate the Pbs2-Hog1 kinase cascade. This finding was astonishing because the Ste11 MAPKKK was initially identified as an essential component of the mating pheromone responsive MAP kinase cascade, which is composed of Ste11, the Ste7 MAPKK, and the partially redundant Fus3/Kss1 MAPKs [65]. Mating factors stimulate the Ste11 MAPKKK, which then activates the Ste7-Fus3 cascade. However, mating factors never activate the Pbs2-Hog1 kinase cascade. Osmotic shock also activates the Ste11 MAPKKK, but in this case it only activates the Pbs2-Hog1 kinase cascade, but not the Ste7-Fus3 pathway. In other words, the Ste11 MAPKKK must somehow choose different substrates depending on how it is activated. The mechanism of insulating the two signaling pathways from each other seems to involve the so-called scaffold proteins, Ste5 and Pbs2. Before discussing scaffold proteins, however, it is necessary to discuss briefly the role of the Sho1 protein in osmosensing.

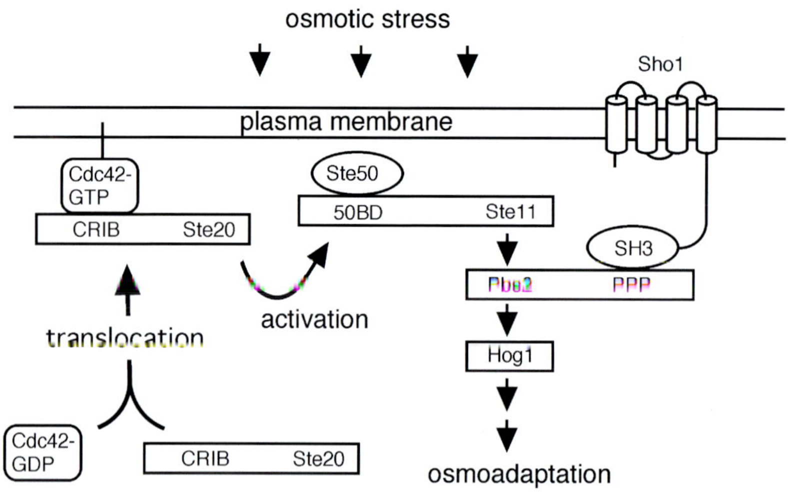

FIGURE 5 Schematic model of the SHO1 branch of the HOG pathway. The membrane-bound protein Sho1 anchors the HOG MAP kinase module to the membrane through its interaction with Pbs2 MAPKK, which in turn interacts with the Ste11 MAPKKK and Hog1 MAPK. In response to osmotic stress, Cdc42-bound, membrane localized Ste20 phosphorylates Ste11, which leads to activation of the Pbs2 MAPKK and the Hog1 MAPK, and eventually to the osmoadaptive response. 50BD, Ste50-binding domain; CRIB, Cdc42/Rac interactive-binding domain; PPP, proline-rich sequence.

Sho1

Sho1 is a membrane protein with four transmembrane segments (TM1–TM4) clustered in its N-terminal region. Both its N and C termini lie in the cytoplasm, exposing two short segments (one between TM1 and TM2 and another between TM3 and TM4) to the extracellular milieu. The C-terminal region contains an SH3 domain, a protein–protein interaction module that binds proline-rich motifs [66, 67]. Its structural features suggested that Sho1 might be an alternative osmosensor in yeast. More recent studies, however, indicated that although Sho1 is essential for the alternative osmosensing pathway, it is probably not an osmosensor [64].

The physiologically relevant target of the Sho1 SH3 domain was identified as Pbs2 MAPKK. A proline-rich sequence (KPLPPLPVA) reminiscent of the consensus SH3-binding sequence was found in the N-terminal noncatalytic domain of Pbs2 and was indeed shown to be the Sho1-binding site [31]. A mutation in the central proline residue in this sequence inhibits the binding of Pbs2 to Sho1 and causes a phenotype that is indistinguishable from the *sho1Δ* mutation, namely the inability to activate Pbs2-Hog1 when the SLN1 pathway is also mutationally inactivated. Thus, interaction between the Sho1 SH3 domain and Pbs2 is essential for Hog1 activation

Complete deletion of Sho1 transmembrane segments (thus its SH3 domain is no longer membrane associated) also abolishes Sho1 activity. Replacement of the Sho1 transmembrane segments with another membrane targeting sequence, however, did not affect Sho1 activity significantly. For example, transmembrane segments from the Ste2 pheromone receptor could functionally substitute Sho1 transmembrane segments. Furthermore, an N-terminal myristoylation site could also functionally replace the Sho1 transmembrane segments. Thus, membrane targeting of the Sho1 SH3 domain is sufficient for its role in the osmotic stress response. In fact, if Pbs2 MAPKK is translocated artificially to the plasma membrane by fusing it to a transmembrane protein, the need for the Sho1 protein in osmosensing could be circumvented, at least partially [64]. Thus, the only essential role of Sho1 is to localize Pbs2 to the plasma membrane. The Sho1 transmembrane segments might be important, however, for the localization of Pbs2 at proper subregions in the plasma membrane so that the efficiency of Hog1 activation is optimized [64, 68].

Scaffold Proteins That Prevent Cross-Talk

It is clear that not all components involved in the SHO1 pathway have been uncovered. In fact, some of the unidentified components might be critically important for osmosensing. With this limitation in mind, the current model of how the SHO1 branch might work and how cross-talk between the osmosensing pathway and the mating pheromone pathway might be prevented is outlined.

The membrane-bound Sho1, through its direct interaction with Pbs2 MAPKK, anchors the Ste11-Pbs2-Hog1 MAP kinase module to selected regions in the plasmamembrane, namely to the bud during the G1 and S phases and to the shmoo tip in cells treated with mating factor [64, 68]. Ste20 binds the small G-protein Cdc42, which is known to localize to similar regions in the plasma membrane when it is in the active (GTP-bound) form [69, 70]. Thus, both Ste20 and Ste11 are localized to the same regions in the plasma membrane, allowing the two proteins to interact with each other. In response to osmotic stress, the membrane-localized Ste20 phosphorylates the membrane-localized Ste11, which then activates Pbs2 MAPKK and the Hog1 MAP kinase in the same complex.

Formation of another multiprotein complex is necessary for the activation of Ste11 by the mating pheromone stimulation. Such a complex contains Ste11, Ste7, Fus3 (or Kss1), and the scaffold protein Ste5 [71–73]. The Ste20 kinase is also implicated in the mating pheromone pathway, but its role is less critical than in the osmoregulatory pathway. More important, the interaction between Ste20 and Cdc42 is dispensable in the mating pathway, suggesting that the membrane localization of Ste20 is not pertinent to the activation of Ste11 in response to the mating pheromone stimulus [70]. Thus, the presence of two forms of protein complexes, both containing Ste11, but localized differently in the cell, may explain why the two signals (mating pheromone and osmotic stress) stimulate only the relevant MAPK cascade.

CONCLUDING REMARKS

Two-component osmosensors homologous to Sln1 are found not only in other yeasts, but even in higher plants, suggesting that a similar mechanism is widely employed by eukaryotic cells [74, 75]. Furthermore, two-component proteins are involved in ethylene and cytokinin responses in higher plants, developmental regulation of ameba (*Dictyostelium*), and pathogenesis of the *Candida* yeasts. Clearly, the two-component system has established itself as an important signaling mechanism in eukaryotes, as well as in prokaryotes. Unlike bacterial counterparts, however, eukaryotic two-component systems are not self-contained signaling modules. Eukaryotic two-component systems usually function with other signaling modules, as the SLN1 pathway functions together with a MAP kinase cascade. It will be very interesting to see how other eukaryotic two-component systems are meshed into eukaryotic signaling network.

REFERENCES

1. Mager, W. H., and Varela, J. C. S. (1993). Osmostress response of the yeast *Saccharomyces* *Mol. Microbiol.* **10**, 253–258.
2. Albertyn, J., Hohmann, S., Thevelein, J. M., and Prior, B. A. (1994). *GPD1*, which encodes glycerol-3-phosphate dehydrogenase, is essential for growth under osmotic stress in *Saccharomyces cerevisiae*, and its expression is regulated by the high osmolarity glycerol response pathway. *Mol. Cell. Biol.* **14**, 4135–4144.
3. Gustin, M. C., Albertyn, J., Alexander, M., and Davenport, K. (1998). MAP kinase pathways in the yeast *Saccharomyces cerevisiae*. *Microbiol. Mol. Biol. Rev.* **62**, 1264–1300.
4. Cobb, M. H., and Goldsmith, E. J. (1995). How MAP kinases are regulated *J. Biol. Chem.* **270**, 14843–14846.
5. Davis, R. J. (2000). Signal transduction by the JNK group of MAP kinases. *Cell* **103**, 239–252.
6. Boguslawski, G. (1992). *PBS2*, a yeast gene encoding a putative protein kinase, interacts with the *RAS2* pathway and affects osmotic sensitivity of *Saccharomyces cerevisiae* *J. Gen. Microbiol.* **138**, 2425–2432.
7. Brewster, J. L., de Valoir, T., Dwyer, N. D., Winter, E., and Gustin, M. C. (1993). An osmosensing signal transduction pathway in yeast. *Science* **259**, 1760–1763.
8. Posas, F., and Saito, H. (1997). Osmotic activation of the HOG MAPK pathway via Ste11p MAPKKK: Scaffold role of Pbs2p MAPKK. *Science* **276**, 1702–1705
9. Jacoby, T., Flanagan, H., Faykin, A., Seto, A. G., Mattison, C., and Ota, I. (1997). Two protein-tyrosine phosphatases inactivate the osmotic stress response pathway in yeast by targeting the mitogen-activated protein kinase, Hog1. *J. Biol. Chem.* **272**, 17749–17755.
10. Wurgler-Murphy, S. M., Maeda, T., Witten, E. A., and Saito, H. (1997). Regulation of the *Saccharomyces cerevisiae HOG1* mitogen-activated protein kinase by the *PTP2* and *PTP3* protein tyrosine phosphatases *Mol. Cell. Biol.* **17**, 1289–1297.
11. Maeda, T., Tsai, A. Y.and M., and Saito, H. (1993). Mutations in a protein tyrosine phosphatase gene (*PTP2*) and a protein serine/threonine phosphatase gene (*PTC1*) cause a synthetic growth defect in *Saccharomyces cerevisiae. Mol. Cell. Biol.* **13**, 5408–5417.
12. Maeda, T., Wurgler-Murphy, S. M., and Saito, H. (1994). A two-component system that regulates an osmosensing MAP kinase cascade in yeast. *Nature* **369**, 242–245.
13. Zhan, X.-L., Deschenes, R. J., and Guan, K.-L. (1997). Differential regulation of *FUS3* MAP kinase by tyrosine-specific phosphatases *PTP2/PTP3* and dual-specificity phosphatase *MSG5* in *Saccharomyces cerevisiae. Genes Dev.* **11**, 1690–1702.
14. Ferrigno, P., Posas, F., Koepp, D., Saito, H., and Silver, P. A. (1998). Regulated nucleo/cytoplasmic exchange of HOG1 MPAK requires the importin b homologs NMD5 and XPO1. *EMBO J.* **17**, 5606–5614.
15. Reiser, V., Ruis, H., and Ammerer, G. (1999). Kinase activity-dependent nuclear export opposes stress-induced nuclear accumulation and retention of Hog1 mitogen-activated protein kinase in the budding yeast *Saccharomyces cerevisiae. Mol. Biol. Cell* **10**, 1147–1161.
16. Marchler, G., Schüller, C., Adam, G., and Ruis, H. (1993). A *Saccharomyces cerevisiae* UAS element controlled by protein kinase A activates transcription in response to a variety of stress conditions. *EMBO J.* **12**, 1997–2003.
17. Hirayama, T., Maeda, T., Saito, H., and Shinozaki, K. (1995). Cloning and characterization of seven cDNAs for hyperosmolarity-responsive (*HOR*) genes of *Saccharomyces cerevisiae. Mol. Gen. Genet.* **249**, 127–138.
18. Ansell, R., Granath, K., Hohmann, S., Thevelein, J., and Adler, L. (1997). The two isoenzymes for yeast NAD-dependent glycerol 3-phosphate dehydrogenase, encoded by *GPD1* and *GPD2*, have distinct roles in osmoadaption and redox regulation. *EMBO J.* **16**, 2179–2187.

19. Schüller, C., Brewster, J. L., Alexander, M. R., Gustin, M. C., and Ruis, H. (1994). The HOG pathway controls osmotic regulation of transcription via the stress response element (STRE) of the *Saccharomyces cerevisiae CTT1* gene. *EMBO J.* **13**, 4382–4389.
20. Treger, J. M., Magee, T. R., and McEntee, K. (1998). Functional analysis of the stress response element and its role in the mutistress response of *Saccharomyces cerevisiae. Biochem. Biophys. Res. Commun.* **243**, 13–19.
21. Martinez-Pastor, M. T., Marchler, G., Schüller, C., Marchler-Bauer, A., Ruis, H., and Estruch, F. (1996). The *Saccharomyces cerevisiae* zinc finger proteins Msn2p and Msn4p are required for transcriptional induction through the stress-response element (STRE). *EMBO J.* **15**, 2227–2235.
22. Schmitt, A. P., and McEntee, K. (1996). Msn2p, a zinc finger DNA-binding protein, is the transcription activator of the multistress response in *Saccharomyces cerevisiae. Proc. Natl. Acad. Sci. USA* **93**, 5777–5782.
23. Rep, M., Reiser, V., Gartner, U., Thevelein, J. M., Hohmann, S., Ammerer, G., and Ruis, H. (1999). Osmotic stress-induced gene expression in *Saccharomyces cerevisiae* requires Msn1p and the novel nuclear factor Hot1p. *Mol. Cell. Biol.* **19**, 5474–5485.
24. Proft, M., and Serrano, R. (1999). Repressors and upstream repressing sequences of the stress-regulated ENA1 gene in *Saccharomyces cerevisiae*: bZIP protein Sko1p confers HOG-dependent osmotic regulation. *Mol. Cell. Biol.* **19**, 537–546.
25. Bilsland-Marchesan, E., Ariño, J., Saito, H., Sunnerhagen, P., and Posas, F. (2000). Rck2 kinase is a substrate for the osmotic stress-activated mitogen-activated protein kinase Hog1. *Mol. Cell. Biol.* **20**, 3887–3895.
26. Ota, I. M., and Varshavsky, A. (1993). A yeast protein similar to bacterial two-component regulators. *Science* **262**, 566–569.
27. Guan, K., Deschenes, R. J., and Dixon, J. E. (1992). Isolation and characterization of a second protein tyrosine phosphatase gene, *PTP2*, from *Saccharomyces cerevisiae. J. Biol. Chem.* **267**, 10024–10030.
28. James, P., Hall, B. D., Whelen, S., and Craig, E. A. (1992). Multiple protein tyrosine phosphatase-encoding genes in yeast *Saccharomyces cerevisiae. Gene* **122**, 101–110.
29. Ota, I., and Varshavsky, A. (1992). A gene encoding a putative tyrosine phosphatase suppresses lethality of an N-end rule-dependent mutant. *Proc. Natl. Acad. Sci. USA* **89**, 2355–2359.
30. Robinson, M. K., van Zyl, W. H., Phizicky, E. M., and Broach, J. R. (1994). *TPD1* of *Saccharomyces cerevisiae* encodes a protein phosphatase 2C-like activity implicated in tRNA splicing and cell separation. *Mol. Cell. Biol.* **14**, 3634–3645.
31. Maeda, T., Takekawa, M., and Saito, H. (1995). Activation of yeast PBS2 MAPKK by MAPKKKs or by binding of an SH3-containing osmosensor. *Science* **269**, 554–558.
32. Gerwins, P., Blank, J. L., and Johnson, G. L. (1997). Cloning of a novel mitogen-activated protein kinase kinase kinase, MEKK4, that selectively regulates the c-Jun amino terminal kinase pathway. *J. Biol. Chem.* **272**, 8288–8295.
33. Takekawa, M., Posas, F., and Saito, H. (1997). A human homolog of the yeast Ssk2/Ssk22 MAP kinase kinase kinases, MTK1, mediates stress-induced activation of the p38 and JNK pathways. *EMBO J.* **16**, 4973–4982.
34. Posas, F., Wurgler-Murphy, S. M., Maeda, T., Witten, E. A., Thai, T. C., and Saito, H. (1996). Yeast HOG1 MAP kinase cascade is regulated by a multistep phosphorelay mechanism in the SLN1-YPD1-SSK1 two-component osmosensor. *Cell* **86**, 865–875.
35. Ostrander, D. B., and Gorman, J.A. (1999). The extracellular domain of the *Saccharomyces cerevisiae* Sln1p membrane osmolarity sensor is necessary for kinase activity. *J. Bacteriol.* **181**, 2527–2534.
36. Ishige, K., Nagasawa, S., Tokishita, S., and Mizuno, T. (1994). A novel device of bacterial

signal transducers. *EMBO J.* **13**, 5195–5202.
37. Uhl, M. A., and Miller, J. F. (1996). Integration of multiple domains in a two-component sensor protein: The *Bordetella pertussis* BvgAS phosphorelay. *EMBO J.* **15**, 1028–1036.
38. Kato, M., Mizuno, T., Shimizu, T., and Hakoshima, T. (1997). Insights into multistep phosphorelay from the crystal structure of the C-terminal HPt domain of ArcB. *Cell* **88**, 717–723.
39. Song, H. K., Lee, J. Y., Lee, M. G., Moon, J., Min, K., Yang, J. K., and Suh, S. W. (1999). Insights into eukaryotic multistep phosphorelay signal transduction revealed by the crystal structure of Ypd1 from *Saccharomyces cerevisiae*. *J. Mol. Biol.* **293**, 753–761.
40. Xu, Q., and West, A. H. (1999). Conservation of structure and function among histidine-containing phoshotransfer (HPt) domains as revealed by the crystal structure of YPD1. *J. Mol. Biol.* **292**, 1039–1050.
41. Janiak-Spens, F., Sparling, J. M., Gurfinkel, M., and West, A. H. (1999). Differential stabilities of phosphorylated response regulator domains reflect functional roles of the yeast osmoregulatory SLN1 and SSk1 proteins. *J. Bacteriol.* **181**, 411–417.
42. Burbulys, D., Trach, K. A., and Hoch, J. A. (1991). Initiation of sporulation in *B. subtilis* is controlled by a multicomponent phosphorelay. *Cell* **64**, 545–552.
43. Tsuzuki, M., Ishige, K., and Mizuno, T. (1995). Phosphotransfer circuitry of the putative multi-signal transducer, ArcB, of *Escherichia coli*: *In vitro* studies with mutants. *Mol. Microbiol.* **18**, 953–962.
44. Hoch, J. A. (2000). Two-component and phoshorelay signal transduction. *Curr. Opin. Microbiol.* **3**, 165–170.
45. Ohlsen, K. L., Grimsley, J. K., and Hoch, J. A. (1994). Deactivation of the sporulation transcription factor Spo0A by the Spo0E protein phosphatase. *Proc. Natl. Acad. Sci. USA* **91**, 1756–1760.
46. Perego, M., Hanstein, C., Welsh, K. M., Djavakhishvili, T., Glaser, P., and Hoch, J. A. (1994). Multiple protein-aspartate phosphatases provide a mechanism for the integration of diverse signals in the control of development in *B. subtilis*. *Cell* **79**, 1047–1055.
47. Perego, M., Glaser, P., and Hoch, J. A. (1996). Aspartyl-phosphate phosphatases deactivate the response regulator components of the sporulation signal transduction system in *Bacillus subtilis*. *Mol. Microbiol.* **19**, 1151–1157.
48. Posas, F., and Saito, H. (1998). Activation of the yeast SSK2 MAP kinase kinase kinase by the SSK1 two-component response regulator. *EMBO J.* **17**, 1385–1394
49. Minden, A., Lin, A., McMahon, M., Lange-Carter, C., Dérijard, B., Davis, R. J., Johnson, G. L., and Karin, M. (1994). Differential activation of ERK and JNK mitogen-activated protein kinases by Raf-1 and MEKK. *Science* **266**, 1719–1723.
50. Kato, T., Okazaki, K., Murakami, H., Stettler, S., Fantes, P. A., and Okayama, H. (1996). Stress signal, mediated by a Hog1-like MAP kinase, controls sexual development in fission yeast. *FEBS Lett.* **378**, 207–212.
51. Samejima, I., Mackie, S., and Fantes, P. A. (1997). Multiple modes of activation of the stress-responsive MAP kinase pathway in fission yeast. *EMBO J.* **16**, 6162–6170.
52. Millar, J. B. A. (1999). Stress-activated MAP kinase (mitogen-activated protein kinase) pathways of budding and fission yeasts. *Biochem. Soc. Symp.* **64**, 49–62.
53. Nguyen, A. N., Lee, A., Place, W., and Shiozaki, K. (2000). Multistep phosphorelay proteins transmit oxidative stress signals to the fission yeast stress-activated protein kinase. *Mol. Biol. Cell* **11**, 1169–1181.
54. Buck, V., Quinn, J., Pino, T. S., Martin, H., Saldanha, J., Makino, K., Morgan, B., and Millar, J. B. A. (2000). Peroxide sensors for the fission yeast stress activated MAP kinase pathway. *Mol. Biol. Cell.*
55. Aoyama, K., Mitsubayashi, Y., Aiba, H., and Mizuno, T. (2000). Spy1, a histidine-containing phosphotransfer signaling protein, regulates the fission yeast cell cycle through the Mcs4

response regulator. *J. Bact.* **182**, 4868–4874.

56. Cottarel, G. (1997). Mcs4, a two-component system response regulator homologue, regulates the *Schizosaccharomyces ombe* cell cycle control. *Genetics* **147**, 1043–1051.
57. Shieh, J.-C., Wilkinson, M. G., Buck, V., Morgan, B. A., Makino, K., and Millar, J. B. A. (1997). The Mcs4 response regulator coordinately controls the stress-activated Wak1-Wis1-Sty1 MAP kinase pathway and fission yeast cell cycle. *Genes Dev.* **11**, 1008–1022.
58. Ohmiya, R., Kato, C., Yamada, H., Aiba, H., and Mizuno, T. (1999). A fission yeast gene (*prr1*$^+$) that encodes a response regulator implicated in oxidative stress response. *J. Biochem.* **125**, 1061–1066.
59. Shiozaki, K., Shiozaki, M., and Russell, P. (1997). Mcs4 mitotic catastrophe suppressor regulates the fission yeast cell cycle through the Wik1-Wis1-Spc1 kinase cascade. *Mol. Biol. Cell* **8**, 409–419.
60. Crews, S. T., and Fan, C. M. (1999). Remembrance of the things PAS: regulation of development by bHLH-PAS proteins. *Curr. Opin. Genet.* **9**, 580–587.
61. Gong, W., Hao, B., Mansy, S. S., Gonzalez, G., Gilles-Gonzalez, M. A., and Chan, M. K. (1998). Structure of a biological oxygen sensor: A new mechanism for heme-driven signal transduction. *Proc. Natl. Acad. Sci. USA* **95**, 15177–15182.
62. O'Rourke, S. M., and Herskowitz, I. (1998). The Hog1 MAPK prevents cross talk between the HOG and pheromone response MAPK pathways in *Saccharomyces cerevisiae*. *Genes Dev.* **12**, 2874–2886.
63. Posas, F., Witten, E. A., and Saito, H. (1998). Requirement of STE50 for osmostress-induced activation of the STE11 mitogen-activated protein kinase kinase kinase in the high-osmolarity glycerol response pathway. *Mol. Cell. Biol.* **18**, 5788–5796.
64. Raitt, D. C., Posas, F., and Saito, H. (2000). Yeast Cdc42 GTPase and Ste20 PAK-like kinase regulate Sho1-dependent activation of the Hog1 MAPK pathway. *EMBO J.* **19**, 4623–4631.
65. Herskowitz, I. (1995). MAP kinase pathways in yeast: For mating and more. *Cell* **80**, 187–197.
66. Cohen, G. B., Ren, R., and Baltimore, D. (1995). Modular binding domains in signal transduction proteins. *Cell* **80**, 237–248.
67. Pawson, T. (1995). Protein modules and signalling networks. *Nature* **373**, 573–580.
68. Reiser, V., Salah, S. M., and Ammerer, G. (2000). Polarized localization of yeast Pbs2 depends on osmostress, the membrane protein Sho1 and Cdc42. *Nature Cell Biol.* **2**, 620–627.
69. Ziman, M., Preuss, D., Mulholland, J., O'Brien, J. M., Botstein, D., and Johnson, D.I . (1993). Subcellular localization of Cdc42p, a *Saccharomyces cerevisiae* GTP-binding protein involved in the control of cell polarity. *Mol. Biol. Cell* **4**, 1307–1316.
70. Leberer, E., Wu, C., Leeuw, T., Fourest-Lieuvin, A., Segall, J. E., and Thomas, D. Y. (1997). Functional characterization of the Cdc42p binding domain of yeast Ste20p protein kinase. *EMBO J.* **16**, 83–97
71. Choi, K. Y., Satterberg, B., Lyons, D. M., and Elion, E. A. (1994). Ste5 tethers multiple protein kinases in the MAP kinase cascade required for mating in *S. cerevisiae*. *Cell* **78**, 499–512.
72. Marcus, S., Polverino, A., Barr, M., and Wigler, M. (1994). Complexes between STE5 and components of the pheromone-responsive mitogen-activated protein kinase module. *Proc. Natl. Acad. Sci. USA* **91**, 7762–7766.
73. Printen, J. A., and Sprague, G. F. (1994). Protein-protein interactions in the yeast pheromone response pathway: Ste5p interacts with all members of the MAP kinase cascade. *Genetics* **138**, 609–619.
74. Nagahashi, S., Mio, T., Ono, N., Yamada-Okabe, T., Arisawa, M., Bussey, H., and Yamada-Okabe, H. (1998). Isolation of *CaSLN1* and *CaNIK1*, the genes for osmosensing histidine kinase homologues, from the pathogenic fungus *Candida albicans*. *Microbiology* **144**, 425–532.

75. Urao, T., Yakubov, B., Satoh, R., Yamaguchi-Shinozaki, K., Seki, M., Hirayama, T., and Shinozaki, K. (1999). A transmembrane hybrid-type histidine kinase in *Arabidopsis* functions as an osmosensor. *Plant Cell* **11**, 1743–1754.

CHAPTER 20

Histidine Kinases of *Dictyostelium*

CHRISTOPHE ANJARD AND WILLIAM F. LOOMIS
Center for Molecular Genetics, Division of Biology,
University of California, San Diego, La Jolla, California 92093

INTRODUCTION

At least 15 members of the histidine kinase family can be recognized in the genomic sequences of *Dictyostelium discoideum*. The predicted products of each of these genes have well-conserved catalytic and receiver domains, although two do not appear to be active kinases, as they lack the histidine in the H motif that is autophosphorylated and one, DokA, appears to function predominantly as a phosphatase. One member of the family, DhkD, is a double histidine kinase with two catalytic and two receiver domains. Alignment of this large family with the Sln1 histidine kinase of yeast extends the sequence profile that characterizes eukaryotic histidine kinases. Biochemical studies of the first member of this family to be discovered, DhkA, have shown that it is a transmembrane receptor kinase that autophosphorylates and relays the phosphate to the receiver aspartate when dimerized. Genetic studies on *dhkA*, *dhkB*, *dhkC*, and *dokA* have indicated their roles in cellular and developmental processes. It is likely that DhkC relays phosphate to the N-terminal receiver domain of the cAMP phosphodiesterase RegA through the H2 domain of RdeA. Phosphorylation of the receiver domain of RegA activates it. DhkA and DhkB appear to inhibit RegA activity by indirectly activating the MAP kinase ERK2. When the carboxy-terminal region of RegA is phosphorylated, phosphodiesterase activity is inhibited. Histidine kinases may also activate the late adenylyl kinase, ACR. Thus, these histidine kinases seem to be focused on regulating cAMP to modulate the activity of the cAMP-dependent protein kinase, PKA.

EUKARYOTIC HISTIDINE KINASES

Phosphorelay from histidine kinases to response regulators controls a myriad of processes in bacteria but appears to be more specialized in eukaryotes. Although multiple histidine kinases have been recognized in plants, *Neurospora* and *Dictyostelium*, only a single member of the family occurs in the yeast, *Saccharomyces cerevisiae*, and none have been found in *Drosophila*, *C. elegans*, or mammals [1–8]. The line that gave rise to animals undoubtedly carried genes encoding histidine kinases, but they all appear to have been lost or converted to other types of protein kinases following divergence from plants and fungi. This chapter focuses on the histidine kinases of the social amoeba, *Dictyostelium discoideum*, with emphasis on those that appear to be involved in cell signaling.

Eukaryotic members of the histidine kinase family can be recognized by a set of shared amino acid motifs that make up the catalytic domain [3, 6]. The histidine that accepts a phosphate from ATP is found in the H1 motif (SHELRTP) where leucine can be replaced by isoleucine, valine, or methionine. About 120 amino acids downstream there is an N motif (Nh_1h_2KFT) where h_1 is A or S and h_2 is usually I, L,V, or F. The ATP-binding domain is characterized by G1 (DS/TGIGI), F (FXXFXQ), and G2 (GS/TGLGL) motifs, where X can be one of many amino acids (Fig. 1).

As in many kinases, an unstable histidine phosphate acts as an intermediate in the reaction, but in members of the histidine kinase family, this intermediate is more stable and can be observed directly in purified material [3, 9]. When histidine kinases form dimers, each subunit phosphorylates the other on a histidine using the γ-phosphoryl group of ATP [10, 11]. Eukaryotic histidine kinases have a receiver domain toward their carboxy termini, which is characterized by two short motifs, D1 and K, the first of which includes an aspartate to which the phosphate is transferred before being relayed to other proteins (Fig. 1). In the *Dictyostelium* DhkA protein, these motifs are FMDCQMP and KPI, respectively [6]. Variations on these motifs are found in all receiver domains.

Phosphorelay to effector proteins in eukaryotes is mediated by small intermediate proteins that carry a histidine in H2 motifs related to **IIDIKGSS**. The amino acids shown in bold are conserved in the products of the *Dictyostelium rdeA* gene and the *Saccharomyces* YPD1 gene, where the initial histidine accepts the phosphate from the aspartyl-phosphate in the receiver domain of histidine kinases. Although the primary sequences of RdeA and Ypd1 outside of the H2 motif are highly divergent, the structures of these relay proteins are so conserved that yeast Ypd1 can replace *Dictyostelium* RdeA in mutant strains [12].

Signals that activate histidine kinases have their effects when phosphate is transferred from the H2 motif to the receiver domain of an effector protein. Several proteins have been found in yeast, *Dictyostelium*, and plants that carry receiver domains and are the apparent targets of phosphorelay from histidine kinases [7, 13]. Some control MAP kinase cascades, others are transcription factors, whereas one is a cAMP phosphodiesterase involved in the control of the cAMP-dependent protein kinase, PKA. The flow of phosphate resulting

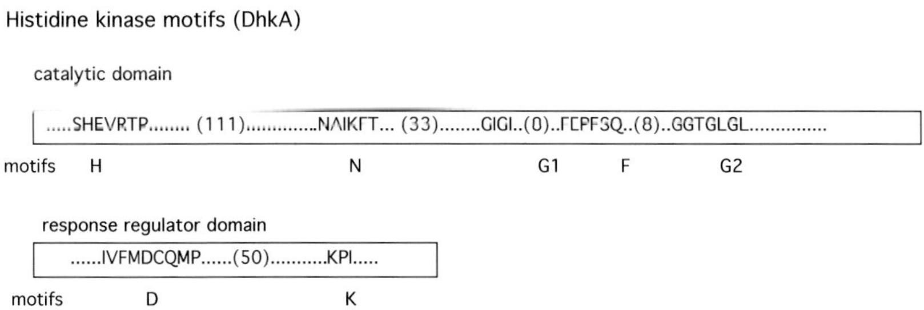

FIGURE 1 Histidine kinase motifs (DhkA). Five short motifs of conserved amino acid sequences found in a region of 175 to 300 amino acids characterize the catalytic domains of all histidine kinases. The sequences and the number of intervening amino acids (given in parentheses) from *Dictyostelium* DhkA are shown. The most highly conserved amino acids are in bold. The H motif carries the phosphorylatable histidine and is followed by the N motif. The ATP binding domain is found 25 to 150 amino acids downstream and is characterized by G1, F, and G2 motifs. All recognized eukaryotic histidine kinases carry a receiver domain toward their carboxy termini that is characterized by a D and a K motif about 50 amino acids apart. The aspartate that receives the phosphate is found in the D motif.

from autophosphorylation of a histidine in the H1 motif of histidine kinases that is relayed to aspartate in the D1 motif of the same molecule, followed by the transfer to histidine in the H2 motif in a small intermediate protein and final relay to aspartate in the D2 motif of effector proteins can be reversed under some conditions such that the histidine kinase acts as a phosphate sink or even a net phosphatase [14–16]. Phosphorelay in these systems is regulated by ligands that activate the histidine kinases.

DICTYOSTELIUM HISTIDINE KINASES

Five *Dictyostelium* genes, *dhkA*, *dhkB*, *dhkC*, *dhkD*, and *dokA*, have been genetically characterized that encode proteins with all of the motifs of eukaryotic histidine kinases [9, 17–19]. Another 8 genes encoding proteins with each of the catalytic domain motifs of histidine kinases can be recognized in the *Dictyostelium* genome by scanning the collection of preliminary sequences made available by the ongoing *Dictyostelium* genome project and the *Dictyostelium* EST database [20] (Table I). Another two genes encode proteins

TABLE I Known and Putative Histidine Kinase Genes in *Dictyostelium*

Locus[a]	Predicted product (amino acids)	cDNAs in EST database	Comments	Accession number
dhkA	2149	SLG564; SSD644; SLB549	Receptor histidine kinase	U42597
dhkB	1969	CFJ365; VFH740	Receptor histidine kinase	AF024654
dhkC	1225	SSM556	Histidine kinase	AF361474
dhkD	1546	SLD672; SLK607	Double histidine kinase	AF361475
dokA	1671	None	Osmotic sensor kinase	X96869
dhkE	1699	None	Potential transmembrane domain	AF362375
dhkF	>1055	None	N terminus not found	AF362368
dhkG	3432	None	Carries a ser/threo kinase domain	AF362369
dhkH	>1147	SSK767	N terminus not found	AF362370
dhkI	1736	None	Complete sequence	AF362371
dhkJ	2062	None	Similar to Nik1	AF362372
dhkK	1213	SSA688	Complete sequence	AF362376
dhkL	>1687	SLC110; SSG478; SLB663; SLF537	N terminus not found	AF362373
dhkM	>1318	SLB260; SSI667; SSF716	N-terminus not found; H motif degenerate	AF362374

[a]Assembled sequences are available on request. Raw data can be accessed at http://www.sdsc.edu/mpr/dicty/.

with the N, G1, F, and G2 motifs but are missing the H motif and are unlikely to be histidine kinases, as they lack the autophosphorylatable histidine. One of these, *acrA*, has been described and shown previously to encode a large protein with a degenerate histidine kinase domain fused to an adenylyl cyclase domain [21]. Sequence coverage available May 2002 is estimated to include over 95% of all the genes, so it is likely that the 15 genes we have recognized account for all of the histidine kinases of *Dictyostelium*.

Although four of the putative histidine kinase sequences assembled from raw genomic reads are incomplete because we were not able to make reliable extensions from the portions encoding the catalytic and receiver domains, it is still apparent that all of the predicted products of the histidine kinase genes are unusually large proteins. This may be the consequence of common heritage from a family of large bacterial genes. Comparisons of the catalytic domains indicate that they are all quite different and are unlikely to have arisen by recent duplications (Fig. 2).

Originally, *dhkD* was thought to be fairly small, encoding a protein of only 710 amino acids [17], but newly available sequences show that the sequence deposited in GenBank had a base deletion that resulted in premature termination. When this error was corrected, it became apparent that *dhkD* encodes a protein more than twice the reported size and that the extension carries both a complete catalytic domain and a receiver domain. Thus, DhkD is a double histidine kinase. We refer to these domains as dhkD1 and dhkD2. The sequence of *dhkE* indicates that its product includes a single putative transmembrane domain that could play a role in signal transduction from the cell surface. While the predicted product of *dhkM* has recognizable N, G1, F, and G2 motifs, there is no evidence for an H domain upstream of the N domain. At present we assume that this member of the family is not a functional histidine kinase. However, DhkM carries a well-conserved receiver domain. EST sequences were found in the collection of cDNAs prepared from the slug stage for 9 out of the 15 histidine kinases, demonstrating that these genes are expressed during development. mRNAs from the other genes appear to be at low abundance during development.

When catalytic domains of the newly recognized hisitidine kinase are aligned with those of the established ones, as well as the yeast histidine kinase, SLN1, it can be seen that not only are the classical motifs conserved but so are certain surrounding sequences (Fig. 2). About 20 amino acids after the H motif there is a leucine-rich region in each of these eukaryotic histidine kinases. A conserved GDXXR motif precedes the N motif by a dozen amino acids in almost all cases. Conserved sequences extending beyond the ATP-binding motifs of these histidine kinases indicate their descent from a common eukaryotic ancestor.

Alignment of the receiver domains of the *Dictyostelium* histidine kinase with that of the yeast SLN1 shows that the classical D and K motifs are well

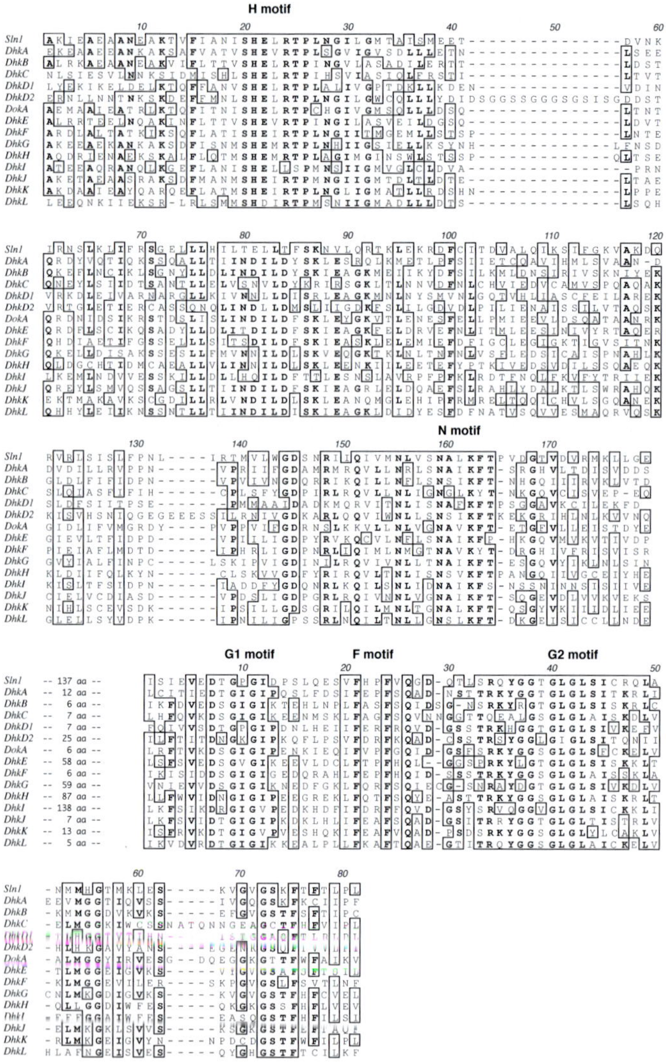

FIGURE 2 Alignment of catalytic domains. Sequences from the only histidine kinase in *Saccharomyces cerevisiae*, Sln1, and 13 histidine kinases of *Dictyostelium discoideum* were aligned using the ClustalW program. Strongly conserved amino acids are shown in bold. One of the *Dictyostelium* genes, *dhkD*, encodes a double histidine kinase, and sequences from the two catalytic domains, DhkD1 and DhkD2, were analyzed separately. A considerable conservation of sequence precedes the H motif. About 25 amino acids downstream there is a conserved region, which is high in leucine. The N motif is surrounded by partially conserved sequences, including the GDXXR motif. A region of variable length separates the N motif from the ATP-binding domain. The G1 and G2 motifs are embedded in larger regions of conserved sequences and are followed by regions with other conserved glycines.

conserved in each of the predicted gene products (Fig. 3). Other well-conserved receiver motifs could also be recognized in this set of eukaryotic histidine kinases: an EDN motif found in the sequence LIV**EDN**HVN of DhkA that precedes the D motif by 46 amino acids and the A motif, ALTA, that precedes the K motif by 23 amino acids. Addition of these motifs to the characterization of receiver domains should increase the reliability of their recognition in previously uncharacterized eukaryotic gene products. The sequences of these receiver domains also line up well with the sequence of the receiver domain of spo0 from *Bacillus subtilis*. Analysis of crystallographic data on spo0 indicates that the phosphoaspartate is found at the beginning of a bend between a β strand and an α helix [22]. The *Dictyostelium* receiver

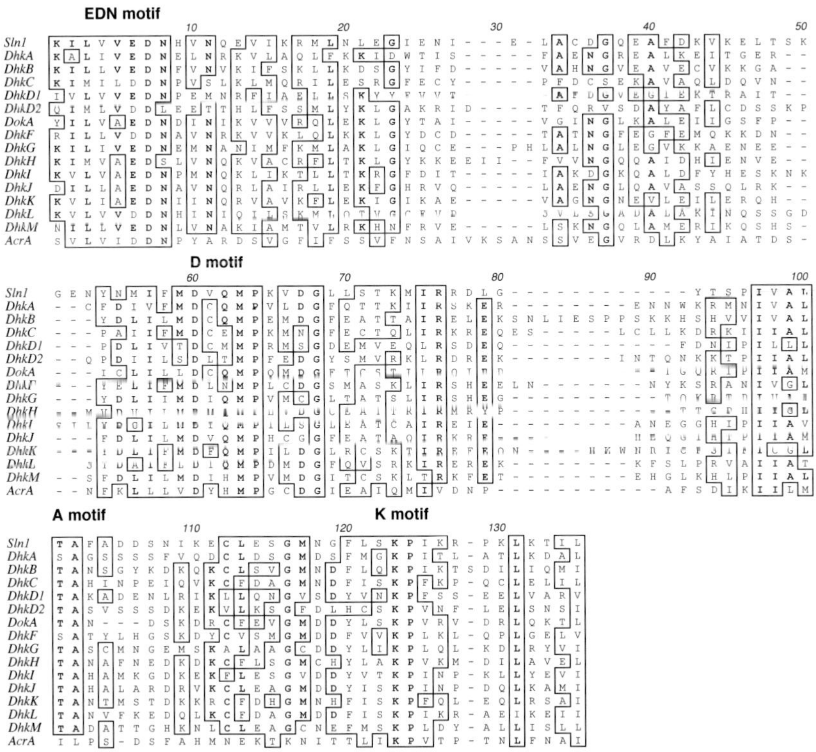

FIGURE 3 Alignment of receiver domains. Receiver domains of the histidine kinases analyzed in Fig. 2 and of AcrA were aligned using the ClustalW program. Strongly conserved amino acids are shown in bold. A motif including EDN precedes the D motif in these eukaryotic proteins. An A motif can also be recognized just upstream of the K motif.

domains may take up similar structures. In the region near the consensus receiver domain of two of the new putative histidine kinases (*dhkL*, *dhkM*), a somewhat more diverged second receiver domain can be recognized. Only molecular genetic studies can tell whether these redundant receivers play a role in phosphorelay.

PHENOTYPIC ANALYSES

Roles of the characterized genes can be partially inferred from the mutant phenotypes of strains in which they are disrupted. Five of the *Dictyostelium* histidine kinases have been subjected to genetic analysis. None of them appears to be essential for growth, as the null mutants all grow well in axenic medium. However, there are developmental consequences to loss of these histidine kinases that indicate the stage at which they may function. Normally, *Dictyostelium* cells express genes that are necessary for synthesizing and responding to cAMP within a few hours of the initiation of development. When the products of these genes accumulate, the cells sense each other, respond chemotactically, and aggregate to form mounds each with up to 10^5 cells [23]. Based on the expression of specific marker genes at the mound stage, cells can be fate mapped to either the stalks or the sori in the final fruiting bodies [24–26]. The cell types sort out within mounds such that the mass of prespore cells is covered by a cap of prestalk cells by 12 h of development. Prestalk cells remain uppermost until culmination at 18 h when they descend through the underlying prespore cells constructing a stalk tube on the way. As prespore cells rise on the elongating stalk, they encapsulate into ellipsoid spores that are surrounded by an impervious spore coat. Progress from one stage to the next is regulated both by intercellular signals and by intracellular responses that include activation of the cAMP dependent protein kinase, PKA [27; 28]. Among other things, PKA leads to the phosphorylation of GBF, a DNA-binding factor, that may control expression of postaggregation genes [29].

DHKA

Random plasmid insertion facilitated by restriction enzyme-mediated integration (REMI) has been used for the near-saturation mutagenesis of genes involved in *Dictyostelium* morphogenesis [30]. Mutants have been recovered that fail to aggregate chemotactically, fail to form migrating slugs, fail to initiate fruiting body formation, or have aberrations in terminal differentiation. One of these mutant strains, AK299, makes exceptionally fragile stalks and almost no spores. Cloning and sequencing the gene disrupted in this strain, *dhkA*, showed that it encodes a "hybrid" histidine kinase with well-conserved motifs in both the catalytic domain and the receiver domain (Fig. 1) [9]. The catalytic domain could be shown to autophosphorylate the histidine in the H1 domain when purified material was incubated with radio-labeled ATP [31]. Site-directed mutations of either the histidine of the H1 or the aspartate of the D1 motif of DhkA that preclude phosphorelay reduce the ability of such constructs to complement the block to sporulation, resulting from deletion of *dhkA*. However, if two constructs, one with a modified H1 but intact D1 and the other with an intact H1 but modified D1, are both introduced into *dhkA*⁻ mutant cells, normal sporulation is recovered [31]. These results are consistent with transfer of the phosphate from the intact H1 site of one subunit to the intact D1 site on the other subunit in a dimer.

DhkA is not expressed while cells are still growing or during the first 8 h of development, but is expressed at high levels thereafter in both prespore and prestalk cells. Disruption of *dhkA* results in five-fold overexpression of a prestalk-specific marker gene, *ecmA* [9]. Two types of prestalk cells, PST-A and PST-0, were initially distinguished by the levels at which they express *ecmA*: PST-A cells are found at the anterior of slugs and express *ecmA* at high levels, whereas PST-0 cells are found just behind them and express *ecmA* at a lower level [32]. Expression of *ecmA* is controlled by distinct *cis*-acting elements in PST-0 cells that are repressed in PST-A cells. The overexpression of *ecmA* seen in *dhkA*⁻ cells results from loss of this repression such that the PST-0 elements drive transcription not only in PST-0 cells but also in PST-A cells [9]. While overexpression of *ecmA* itself is unlikely to account for the long, fragile stalks seen in *dhkA*⁻ null strains, other coordinately expressed PST-0 specific genes may also be affected and result in abnormal stalks. The consequences to loss of DhkA cannot be overcome by developing mutant cells together with wild-type cells in chimeric aggregates, indicating that DhkA functions in a cell autonomous manner. Thus, it is unlikely to be responsible for the production or release of signals that regulate cell-type-specific differentiation but may play a role in responding to such signals.

Expression of prespore specific marker genes such as *cotB* is normal in *dhkA*⁻ null stains [9]. However, very few viable spores are made. Spore formation is not rescued by developing mutant cells together with wild-type cells, indicating that the block to sporulation is also a cell autonomous phenoptype. However, activation of PKA by addition of the cell-permeable analog 8-bromo-cAMP to culminants of *dhkA*⁻ strains induces encapsulation such that within 2 h as many spores are made as in wild-type strains [9]. These results suggest that DhkA may function in a pathway that acts upstream of PKA activation to induce the rapid encapsulation of prespore cells, as well as the repression of PST-0-specific genes in prestalk cells.

DHKB

Using degenerate primers based on the H1 and G2 motifs, several other *Dictyostelium* histidine kinases were amplified and sequenced [17]. They contained all of the motifs associated with the catalytic domain as well as the receiver domain. One of them, *dhkB*, was found to be expressed in growing cells and throughout development. Null mutants in *dhkB* were generated by homologous recombination and were found to go through all the morphological stages leading to fruiting bodies but to form defective spores [17]. Normally, spore dormancy is maintained by the germination inhibitor, disadenine. *dhkB*⁻ null spores fail to remain dormant in the sori and germinate within a few hours. Overexpression of the gene encoding the catalytic subunit of PKA, *pkaC*, in *dhkB*⁻ cells results in spores that remain dormant for days. Treatment with 8-Br-cAMP also extends dormancy dramatically [17]. It appears that DhkB is required to maintain PKA activity in dormant spores, possibly by keeping the level of cAMP high.

DHKC

Another of the histidine kinase genes amplified by polymerase chain reaction (PCR), *dhkC*, was found to be expressed in developing cells but not in growing cells; its mRNA can first be observed at 4 h of development and remains high thereafter [18]. Disruption of *dhkC* results in strains that grow well but aggregate an hour early and begin to form fruiting bodies 4 h earlier than wild-type strains. Expression of early marker genes such as *dscA* and *csaA* is accelerated during aggregation, indicating that *dhkC*⁻ cells are rapid developers. Other strains that are rapid developers have defects in *pkaR* that encodes the regulatory subunit of PKA, *rdeA* that encodes a small protein with an H2 motif, and *regA* that encodes the cytoplasmic cAMP phosphodiesterase, which has a receiver domain in the N-terminal portion of the protein [9, 12, 33–35]. Cells transformed with a construct that overexpresses the central region of *dhkC* encoding only the catalytic domain form tipped mounds 2 h later than wild-type strains and stay as slugs for prolonged periods of time. Removal of the N-terminal input domain renders many histidine kinases constitutive [36]. The dominant gain-of-function phenotype resulting from constitutive DhkC function could be overcome by mutating *regA*, indicating that this phosphodiesterase acts downstream [18]. The addition of 8-Br-cAMP to activate PKA also leads to rapid culmination of these overexpressing cells. The similarity of the phenotypes seen in *dhkC*, *rdeA*, *regA*, and *pkaR* mutants suggests that activation of DhkC leads to phosphorelay via the H2 of RdeA to the receiver domain of RegA. Phosphotransfer from RdeA to RegA has been demonstrated *in vitro* and shown to result in increased phosphodiesterase activity [37, 38]. Stimulation of RegA activity would be expected to lower the concentration of internal cAMP, thereby reducing PKA activity and slowing progression through the stages [27]. Loss of any of these genes would be expected to result in rapid development, which is what is observed. It is not known how DhkC activity is regulated to control the rate of early development, but it is likely to involve an internal signal, as DhkC is not expected to be membrane localized [18].

DHKD

Sequences available from the *Dictyostelium* genome project have allowed us to correct and extend the sequence of the *dhkD* gene [17]. This gene was thought to encode a small protein with a single histidine kinase catalytic domain and a receiver domain. The extended sequence encodes a large protein (176 kDa) with the unusual feature of carrying two catalytic domains and two receiver domains. It physiological function remains to be determined.

DOKA

PCR amplification yielded another gene, *dokA*, that encodes a histidine kinase [19]. The receiver domain of this protein could be shown to accept phosphate from radiolabeled acetyl-phosphate. The gene is expressed at a low level in growing cells and at a high level after 12 h of development. Disruption of *dokA* results in strains that are hypersensitive to high osmotic conditions; when incubated in 400 m*M* sorbitol, they have a half-life of only about an hour, whereas wild-type cells are unaffected [19]. Within a few minutes after the addition of 400 m*M* sorbitol to wild-type cells, the concentration of cAMP increases to 6 pmol/5 $\times 10^7$ cells, whereas the concentration increases to only 2 pmol/5 $\times 10^7$ cells in *dokA*$^-$ null cells. As mentioned earlier, deletion of the N-terminal input domain of histidine kinases often results in constitutive activity and it was found that expression of a construct encoding just the catalytic and receiver domains of DokA gives a dominant gain-of-function phenotype [16]. Such cells aggregate 2 h earlier than wild-type cells and are able to do so even in the presence of the nonhydrolyzable cAMP analog cAMP-S that blocks the aggregation of wild-type cells [16]. The concentration of cAMP was found to be three times higher in these cells than in wild-type cells. However, this effect does not appear to depend on the histidine kinase activity of DokA. Expression of a construct encoding a truncated DokA in which the histidine in the H1 motif is replaced by glutamine so that it cannot accept a phosphate also results in rapid development and resistance to cAMP-S. Expressing the catalytic domain alone has no effect on the cells, whereas expressing the receiver domain alone gives the dominant gain-of-function phenotype. Cells overexpressing the receiver domain have cAMP levels four times higher than wild-type cells. Replacing the aspartate in the D1 motif with an alanine such that the receiver domain could not accept a phosphate resulted in inactive protein. Cells expressing this mutated form had normal levels of cAMP. Ott *et al.* [16] suggested that DokA acts as a phosphatase that blocks activation of the cAMP phosphodiesterase RegA when its receiver domain interacts with RdeA and hydrolyzes the phosphate in the H2 motif. It is not yet clear when the catalytic domain of DokA might act as a kinase or how it is activated to confer protection to high osmotic conditions.

DOUBLE MUTANTS

When independent signal transduction pathways converge on an effector, the consequences of modifying components in one or the other pathway are often found to be additive in double mutants. Both DhkA and DhkB play roles in encapsulation and dormancy of spores, but mutant strains lacking one or the other still form a few percentage of spores. However, double mutants lacking both of these histidine kinases make 10-fold less viable spores than either of the single mutants [31]. The addition of 8-Br-cAMP to culminants of the double mutant induced them to sporulate efficiently. It appears that the signal transduction pathways emanating from both DhkA and DhkB affect PKA activity in an additive manner.

Further evidence that DhkA leads to activation of PKA comes from suppressor studies on the phenotype of *dhkA*⁻ null mutants. Inactivation of either *regA* or *pkaR* in a *dhkA*⁻ background increases the efficiency of sporulation more than 4-fold. Likewise, overexpression of *pkaC* increases the efficiency of sporulation at least 10-fold, most likely because there is insufficient PKA-R to regulate the extra PKA-C [31, 39]. Sporulation in such cells would then be expected to be independent of high levels of cAMP normally required to keep PKA-R from inhibiting the activity of PKA-C.

STRUCTURE AND FUNCTION OF DhkA

The primary amino acid sequence of the N terminal region of DhkA includes two potential membrane-spanning domains separated by a 310 amino acid loop [9]. The loop could be shown to be extracellular by inserting a MYC_6 epitope at position 900 and demonstrating that it is sensitive to proteases added to intact cells expressing the construct [31]. A MYC_6 epitope inserted at position 2025 near the carboxy terminus of DhkA was not degraded by extracellular proteases unless the cells were broken. Insertion of the MYC_6 epitope into the extracellular loop severely compromised the ability of the transgene to complement a *dhkA*⁻ mutation [31]. However, the addition of monoclonal antibodies to the MYC_6 epitope stimulates encapsulation several-fold, much as cross-linking of surface antigens with antibodies can activate mammalian lymphocytes [31]. The response of $dhkA^{900MYC6}$ cells to anti-MYC antibodies indicates that the extracellular loop is involved in the activation of DhkA.

A deletion of the 5′ end of *dhkA* that removes the transmembrane and extracellular domains results in partially constitutive DhkA (Wang, Soderbom, and Loomis, unpublished results). This truncated version gives a dominant gain-of-function phenotype in wild-type cells such that 25% of the cells encapsulate after development in low-density monolayers. Under these conditions, cell signaling is ineffective and wild-type cells fail to encapsulate.

During normal development, prespore cells start to encapsulate when the sorus nears the top of the elongating stalk. Using cells carrying a construct in which bacterial β-galactosidase is driven by the regulatory region of a spore-specific gene, *spiA*, it is possible to visualize the wave of expression as it passes over the sorus [10]. Encapsulation starts at the top where the prespore cells are juxtaposed to the prestalk cells and sweeps down over the sorus during the next hour or so. It appears that prestalk cells signal prespore cells when fruiting body formation is nearing completion. Once a prespore cell has encapsulated, it can no longer move and so would be penalized for premature sporulation. Prestalk cells release a small peptide called SDF-2 that stimulates encapsulation of prespore cells [39]. Secretion of this peptide requires the prestalk-specific gene *tagC*, which encodes a member of the ATP-dependent transporters with an attached protease domain [41]. When added to monolayer cultures at nanomolar concentrations, SDF-2 induces encapsulation within 30 min [39]. However, it has no effect on cells lacking DhkA. Moreover, $dhkA^{900MYC6}$ cells require a 100-fold higher concentration of SDF-2 before encapsulation is stimulated [31]. These results indicate that the extracellular loop of DhkA binds SDF-2 and activates internal kinase activity.

Because mutations in either *regA* or *pkaR* have been found to suppress the block to sporulation seen in *dhkA*⁻ mutants, activation of DhkA by SDF-2 appears to result in inhibition of the phosphodiesterase encoded by *regA* such that cAMP can accumulate and bind to PKA-R, resulting in increased PKA activity. Studies carried out during the aggregation stage indicate that the MAP kinase encoded by *erkB* can phosphorylate threonine moieties found in the carboxy-terminal region of RegA and inhibit its *in vivo* activity [6, 42, 43] (Lu and Kuspa, in preparation). ERK2 also plays a role during culmination, as shifting an *erkB*^{ts} mutant to the nonpermissive temperature at the mound stage blocks encapsulation. Overexpression of *pkaC* bypasses the need for ERK2 [42]. As discussed in the following chapters, both yeast and *Arabidopsis* histidine kinases regulate MAP kinase cascades, suggesting that this may be a conserved strategy. Clark *et al.* [44] have reported that a Raf-like protein kinase, Ctr1, binds directly to the histidine kinase ETR1. Because Raf-like kinases are known to activate MAP kinase cascades, the pathway connecting the receptor histidine kinase and MAP kinase activation may be relatively short and direct. The *Dictyostelium* gene, *splA*, that encodes a close homologue of Ctr1 has been shown to be essential for efficient sporulation [45]. Evidence indicates that SplA acts upstream of PKA in that encapsulation can be induced by 8-Br-cAMP or overexpression of *pkaC* (Chae and Loomis, unpublished results). Present data are consistent with the histidine kinase DhkA activating ERK2 through SplA such that the MAP kinase phosphorylates and inhibits RegA. The resulting decrease in phosphodiesterase activity leads to an increase in cAMP and PKA activity that triggers rapid exocytosis of prespore vesicles and encapsulation (Fig. 3).

THE LATE ADENYLYL CYCLASE ACR

Two different genes encoding adenylyl cyclases are expressed during development. *acaA* encodes the enzyme that is involved in synthesizing cAMP for relay of the chemotactic signal, whereas *acrA* encodes the enzyme that is responsible for synthesizing most of the internal cAMP late in development [21, 46]. There is a region toward the N terminus of ACR that has all of the motifs of a histidine kinase catalytic domain except the H1 motif. Clearly it cannot have histidine kinase activity because the critical histidine in H1 has been lost. Further along the molecule the receiver domain has well-conserved D and K motifs and may easily still be functional. Phosphorylation of the aspartate in the D2 motif may activate this adenylyl cyclase, leading to a higher rate of cAMP synthesis. However, the specific activity of ACR during the culmination of strains lacking DhkA, DhkB, or DhkC is not significantly different from that in wild-type cells [21]. Either there is redundancy such that RdeA is phosphorylated by multiple histidine kinases and can donate to several receiver domains or one or more of the histidine kinases that have yet to be fully characterized (Table I) is responsible for the activation of ACR. Adenylyl cyclases with attached histidine kinase catalytic and receiver domains are found in a variety of cyanobacteria [47, 48]. *Dictyostelium* may have inherited such a gene and subsequently lost the histidine kinase function while retaining the receiver domain. Because most adenylyl cyclases act as dimers, it is possible that the remnants of the histidine kinase catalytic domain may function in the dimerization of ACR.

Mutants lacking ACR have a cell autonomous block to sporulation that can be partially overcome by the addition of 8-Br-cAMP [21, 49]. Sporulation is also recovered in *acrA*$^-$ null strains when they overexpress a G-protein-independent form of ACA showing that the histidine kinase domains of ACR play no essential role during terminal differentiation. However, ACR plays a critical role in generating the cAMP that activates PKA late in development. While the signal transduction pathways initiated by DhkA and DhkB appear to converge on RegA, inhibiting this cAMP phosphodiesterase would make little difference if there were not an adenylyl cyclase to synthesize cAMP (Fig. 4).

SUMMARY AND PERSPECTIVES

Convergence of signal transduction pathways can integrate diverse signals from the environment [50]. Multiple histidine kinases may relay through RdeA because no other gene in the available genomic or EST databases has the conserved Hpt domain. Moreover, it has been shown that even the bacterial enzyme CheA can phosphorylate RdeA *in vitro* [38]. RdeA itself may not provide much specificity and relay phosphate to a variety of receiver domains, as the yeast homologue YPD1 is able to fulfill its function when RdeA is miss-

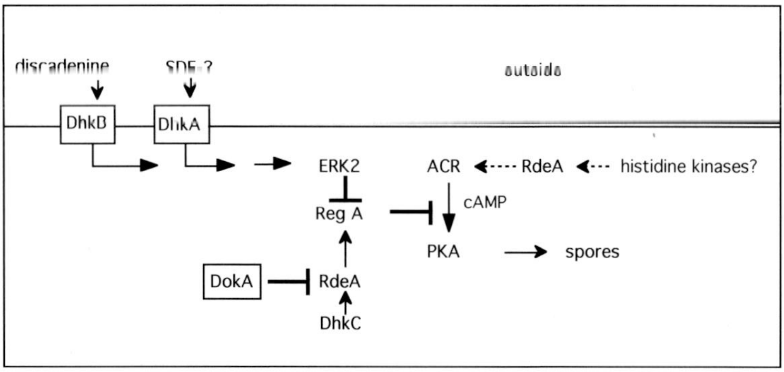

FIGURE 4 Proposed roles of histidine kinases in a culmination network. When the membrane associated histidine kinase DhkA is activated by SDF-2, it stimulates the MAP kinase cascade that leads to ERK2 activity. Discadenine activates DhkB to stimulate the MAP kinase cascade. Phosphorelay from DhkC via the Hpt protein RdeA activates RegA, whereas ERK2 inhibits it. DokA may inhibit phosphorelay by dephosphorylating RdeA under conditions of high osmotic pressure. When cAMP accumulates to levels that activate PKA, prespore cells are induced to encapsulate.

ing [12]. Interconnected pathways and signal transduction networks can be compared to the flow of water through wetlands. Dozens of shallow channels carry water from where it enters a swampy area to the outlet. Blocking one or two channels will not stop the flow. Thus cells lacking one of the histidine kinases may be attenuated in a particular signaling pathway but be able to partially compensate with the function of other histidine kinases. It will be an interesting challenge to follow the course of these signal transduction pathways to their various outcomes.

ACKNOWLEDGMENTS

We are grateful to the laboratories who have provided sequences as part of the *Dictyostelium* genome project: The Baylor Sequencing Center, Houston, Texas (A. Kuspa and R. Gibbs), where sequencing is supported by the NIH; the Institute of Biochemistry, Cologne, together with the Institute of Molecular Biotechnology, Jena, Germany (G. Glöckner , A. Rosenthal, L. Eichinger, and A. Noegel), where sequencing is supported by the Deutsche Forschungsgemeinschaft (Nos. 113/10-1 and 10-2); and the EUDICT consortium supported by The European Union (M.-A. Rajandream, D. Lawson, and B. Barrell). We thank Stephan Schuster, Charles Singleton, and Peter Thomason for discussions of recent results. This work was supported by a grant from the NSF (No. 9728463).

REFERENCES

1. Chang, C., Kwok, S. F., Bleecker, A. B., and Meyerowitz, E. M. (1993). Arabidopsis ethylene-response gene ETR1: Similarity of product to 2-component regulators. *Science* **262**, 245–249.
2. Kakimoto, T. (1996). CKI1, a histidine kinase homolog implicated in cytokinin signal transduction. *Science* **274**, 982–985.
3. Alex, L., and Simon, M. (1994). Protein histidine kinases and signal transduction in prokaryotes and eukaryotes. *Trends Genet.* **10**, 133–138.
4. Alex, L., Borkovich, K., and Simon, M. I. (1996). Hyphal development in *Neurospora crassa*. Involvement of a two-component histidine kinase. *Proc. Natl. Acad. Sci. USA* **93**, 3416–3421.
5. Loomis, W. F., Shaulsky, G., and Wang, N. (1997). Histidine kinases in signal transduction pathways of eukaryotes. *J. Cell Sci* **110**, 1141–1145.
6. Loomis, W. F., Kuspa, A., and Shaulsky, G. (1998). Two-component signal transduction systems in eukaryotic microorganisms. *Curr. Opin. Microbiol.* **1**, 643–648.
7. Thomason, P., and Kay, R. (2000). Eukaryotic signal transduction via histidine-aspartate phosphorelay. *J. Cell Sci.* **113**, 3143–3150.
8. Posas, F., Wurgler-Murphy, S. M., Maeda, T., Witten, E., A., Thai, T. C., and Saito, H. (1996). Yeast HOG1 MAP kinase cascade is regulated by a multistep phosphorelay mechanism in the SLN1 YPD1 SSK1 "two component" osmosensor. *Cell* **86**, 865–875.
9. Wang, N., Shaulsky, G., Escalante, R., and Loomis, W. F. (1996). A two-component histidine kinase gene that functions in *Dictyostelium* development. *EMBO J.* **15**, 3890–3898.
10. Surette, M., Levit, M., Liu, Y., Lukat, G., Ninfa, E., Ninfa, A., and Stock, J. (1996). Dimerization is required for the activity of the protein histidine kinase CheA that mediates signal transduction in bacterial chemotaxis. *J. Biol. Chem.* **271**, 939–945.

11. Bilwes, A. M., Alex, L. A., Crane, B. R., and Simon, M. I. (1999). Structure of CheA, a signal-transducing histidine kinase. *Cell* **96**, 131–141.
12. Chang, W. T., Thomason, P. A., Gross, J. D., and Newell, P. C. (1998). Evidence that the RdeA protein is a component of a multistep phosphorelay modulating rate of development in *Dictyostelium. EMBO J.* **17**, 2809–2816.
13. Shaulsky, G., Escalante, R., and Loomis, W. F. (1996). Developmental signal transduction pathways uncovered by genetic suppressors. *Proc. Natl. Acad. Sci. USA* **93**, 15260–15265.
14. Perego, M., and Hoch, J. A. (1996). Protein aspartate phosphatases control the output of two-component signal transduction systems. *Trends Genet.* **12**, 97–101.
15. Keener, J., and Kustu, S. (1988). Protein kinase and phosphoprotein phosphatase activities of nitrogen regulatory protein NTRB and NTRC of entric bacteria: Roles of the conserved amino-terminal domain of NTRC. *Proc. Natl. Acad. Sci. USA* **80**, 3599–3603.
16. Ott, A., Oehme, F., Keller, F., and Schuster, S. (2000). Osmotic stress response in *Dictyostelium* is mediated by cAMP. *EMBO J.* **19**, 5782–5792.
17. Zinda, M. J., and Singleton, C. K. (1998). The hybrid histidine kinase dhkB regulates spore germination in *Dictyostelium discoideum. Dev. Biol.* **196**, 171–183.
18. Singleton, C. K., Zinda, M. J., Mykytka, B., and Yang, P. (1998). The histidine kinase dhkC regulates the choice between migrating slugs and terminal differentiation in *Dictyostelium discoideum. Dev. Biol.* **203**, 345–357.
19. Schuster, S. C., Noegel, A. A., Oehme, F., Gerisch, G., and Simon, M. I. (1996). The hybrid histidine kinase DokA is part of the osmotic response system of *Dictyostelium. EMBO J.* **15**, 3880–3889.
20. Morio, T., Urushihara, H., Saito, T., Ugawa, Y., Mizuno, H., Yoshida, M., Yoshino, R., Mitra, B. N., Pi, M., Sato, T., Takemoto, K., Yasukawa, H., Williams, J., Maeda, M., Takeuchi, I., Ochiai, H., and Tanaka, Y. (1998). The *Dictyostelium* developmental cDNA project: Generation and analysis of expressed sequence tags from the first-finger stage of development. *DNA Res.* **5**, 335–340.
21. Soderbom, F., Anjard, C., Iranfar, N., Fuller, D., and Loomis, W. F. (1999). An adenylyl cyclase that functions during late development of *Dictyostelium. Development* **126**, 5463–5471.
22. Varughese, K. I., Madhusudan, Zhou, X. Z., Whiteley, J. M., and Hoch, J. A. (1998). Formation of a novel four-helix bundle and molecular recognition sites by dimerization of a response regulator phosphotransferase. *Mol. Cell* **2**, 485–493.
23. Devreotes, P. N. (1982). Chemotaxis. *In* "The development of *Dictyostelium discoideum*" (W. F. Loomis, ed.), pp. 117–168. Academic Press, New York.
24. Williams, J. G., Duffy, K. T., Lane, D. P., Mcrobbie, S. J., Harwood, A. J., Traynor, D., Kay, R. R., and Jermyn, K. A. (1989). Origins of the prestalk-prespore pattern in *Dictyostelium* development. *Cell* **59**, 1157–1163.
25. Fosnaugh, K. L., and Loomis, W. F. (1993). Enhancer regions responsible for temporal and cell-type-specific expression of a spore coat gene in *Dictyostelium. Dev. Biol.* **157**, 38–48.
26. Firtel, R. A. (1995). Integration of signaling information in controlling cell-fate decisions in *Dictyostelium. Genes Dev.* **9**, 1427–1444.
27. Loomis, W. F. (1998). Role of PKA in the timing of developmental events in *Dictyostelium* cells. *Microbiol. Mol. Biol. Rev.* **62**, 684.
28. Aubry, L., and Firtel, R. (1999). Integration of signaling networks that regulate *Dictyostelium* differentiation. *Annu. Rev. Cell Dev. Biol.* **15**, 469–517.
29. Hopper, N. A., Sanders, G. M., Fosnaugh, K. L., Williams, J. G., and Loomis, W. F. (1995). Protein kinase A is a positive regulator of spore coat gene transcription in *Dictyostelium. Differentiation* **58**, 183–188.
30. Kuspa, A., and Loomis, W. F. (1992). Tagging developmental genes in *Dictyostelium* by

restriction enzyme-mediated integration of plasmid DNA. *Proc. Natl. Acad. Sci. USA* **89**, 8803–8807.
31. Wang, N., Soderbom, F., Anjard, C., Shaulsky, G., and Loomis, W. F. (1999). SDF-2 induction of terminal differentiation in *Dictyostelium discoideum* is mediated by the membrane-spanning sensor kinase DhkA. *Mol. Cell. Biol.* **19**, 4750–4756.
32. Early, A. E., Gaskell, M. J., Traynor, D., and Williams, J. G. (1993). Two distinct populations of prestalk cells within the tip of the migratory *Dictyostelium* slug with differing fates at culmination. *Development* **118**, 353–362.
33. Abe, K., and Yanagisawa, K. (1983). A new class of rapid developing mutants in *Dictyostelium discoideum*: Implications for cyclic AMP metabolism and cell differentiation. *Dev. Biol.* **95**, 200–210.
34. Simon, M. N., Pelegrini, O., Veron, M., and Kay, R. R. (1992). Mutation of protein kinase-A causes heterochronic development of *Dictyostelium*. *Nature* **356**, 171–172.
35. Shaulsky, G., Fuller, D., and Loomis, W. F. (1998). A cAMP-phosphodiesterase controls PKA-dependent differentiation. *Development* **125**, 691–699.
36. Parkinson, J. S. (1993). Signal transduction schemes of bacteria. *Cell* **73**, 857–871.
37. Thomason, P. A., Traynor, D., Cavet, G., Chang, W. T., Harwood, A. J., and Kay, R. R. (1998). An intersection of the cAMP/PKA and two-component signal transduction systems in *Dictyostelium*. *EMBO J.* **17**, 2838–2845.
38. Thomason, P. A., Traynor, D., Stock, J. B., and Kay, R. R. (1999). The RdeA-RegA system, a eukaryotic phospho-relay controlling cAMP breakdown. *J. Biol. Chem.* **274**, 27379–27384.
39. Anjard, C., Zeng, C., Loomis, W. F., and Nellen, W. (1998). Signal transduction pathways leading to spore differentiation in *Dictyostelium discoideum*. *Dev. Biol.* **193**, 146–155.
40. Richardson, D. L., Loomis, W. F., and Kimmel, A. R. (1994). Progression of an inductive signal activates sporulation in *Dictyostelium discoideum*. *Development* **120**, 2891–2900.
41. Shaulsky, G., Kuspa, A., and Loomis, W. F. (1995). A multidrug resistance transporter serine protease gene is required for prestalk specialization in *Dictyostelium*. *Genes Devel.* **9**, 1111–1122.
42. Aubry, L., Maeda, M., Insall, R., Devreotes, P. N., and Firtel, R. A. (1997). The *Dictyostelium* mitogen-activated protein kinase ERK2 is regulated by ras and cAMP-dependent protein kinase (PKA) and mediates PKA function. *J. Biol. Chem.* **272**, 3883–3886.
43. Laub, M. T., and Loomis, W. F. (1998). A molecular network that produces spontaneous oscillations in excitable cells of *Dictyostelium*. *Mol. Biol. Cell* **9**, 3521–3532.
44. Clark, K., Larsen, P., Wang, X., and Chang, C. (1998). Association of the *Arabidopsis* CTR1 Raf-like kinase with ETR1 and ERS ethylene receptors. *Proc. Natl. Acad. Sci.* **95**, 5401–5406.
45. Nuckolls, G. H., Osherov, N., Loomis, W. F., and Spudich, J. A. (1996). The *Dictyostelium* dual-specificity kinase splA is essential for spore differentiation. *Development* **122**, 3295–3305.
46. Pitt, G. S., Milona, N., Borleis, J., Lin, K. C., Reed, R. R., and Devreotes, P. N. (1992). Structurally distinct and stage-specific adenylyl cyclase genes play different roles in *Dictyostelium* development. *Cell* **69**, 305–315.
47. Katayama, M., and Ohmori, M. (1997). Isolation and characterization of multiple adenylate cyclase genes from the cyanobacterium *Anabena* sp. strain PCC 7120. *J. Bacteriol.* **179**, 3588–3593.
48. Kasahara, M., Yashiro, K., Sakamoto, T., and Ohmori, M. (1997). The *Spirulina platensis* adenyllate cyclase gene, cyaC, encodes a novel signal transduction protein. *Plant Cell Physiol.* **38**, 828–836.
49. Anjard, C., Soderbom, F., and Loomis, W. F. (2001). Requirements for the adenylyl cyclases in development of *Dictyostelium*. *Development* **128**, 3649–3654.
50. Hoch, J. (1993). The phosphorelay signal transduction pathway in the initiation of *Bacillus subtilis* sporulation. *J. Cell Biochem.* **51**, 55–61.

CHAPTER 21

Ethylene Perception in *Arabidopsis* by the ETR1 Receptor Family: Evaluating a Possible Role for Two-Component Signaling in Plant Ethylene Responses

RONAN C. O'MALLEY AND ANTHONY B. BLEECKER
Department of Botany, University of Wisconsin, Madison, Wisconsin 53706

The simple two-carbon gas ethylene is an important plant hormone that has dramatic yet specific effects on growth and development at every stage of a plant's life. Genetic and biochemical evidence has demonstrated that *Arabidopsis* detects the hormone through a family of ethylene receptors consisting of ETR1 and its four isoforms. These ethylene receptors are composed of a hydrophobic N-terminal domain that has a high affinity for ethylene, followed by a cytoplasmic GAF-like domain and a C-terminal domain that is related to bacterial two component signal transducers. Although ETR1 shows histidine kinase activity *in vitro*, it is the only receptor isoform that contains all the necessary conserved residues and domains typically required for two component signaling, which raises the question of whether this is the primary mechanism of signaling. The ETR1 and ERS1 histidine kinase domain and ETR1 response regulator domains do interact with CTR1, a Raf-like kinase required for the suppression of ethylene responses, suggesting some role for these domains in signaling even if it is not based on kinase activity. A similar interaction between a response regulator and MAPKKK in the osmotic response pathway in yeast and evidence concerning two-component signaling in plant cytokinin responses are discussed to help evaluate a possible role for ETR1 two-component signaling in plant ethylene responses. © 2003, Elsevier Science (USA).

INTRODUCTION

In order to tailor physiological responses to prevailing conditions, organisms depend on the ability to interpret relevant environmental information at the cellular level. To accomplish this, cells utilize arrays of sensor proteins, each sensitive to specific environmental stimuli. Once a sensor perceives its stimulus, it generally transmits an intracellular signal through a network of downstream proteins to ultimately alter gene expression and cell biochemistry. In an overwhelming number of cases, these sensors are receptor-linked kinases, and the intracellular signal consists of sequential, reversible phosphorylation of a series of kinases in the response pathway. Kinases, large families in both prokaryotes and eukaryotes, are classified according to the amino acid residue that they phosphorylate and include serine, threonine, tyrosine, and histidine kinases (HK). In animals, essentially all signaling pathway kinases are serine, threonine, or tyrosine kinases, whereas prokaryotes appear to depend almost entirely on histidine kinases. It was thought that this exclusivity suggests that serine, threonine, and tyrosine kinases were eukaryotic, whereas histidine kinases were prokaryotic, but this presumption was toppled in 1993 with the discovery of histidine kinases in both plant and fungi kingdoms [1, 2].

ETR1, the first plant histidine kinase discovered, was identified in *Arabidopsis* in a screen for plants with aberrant ethylene responses [3]. Ethylene, a volatile plant hormone, plays a well recognized role in fruit ripening, synchronizes important developmental transitions during germination, senescence, and organ abscission, and has a pivotal role in whole plant responses to predation, wounding, and mechanical and environmental stresses

[4]. In 1993, ETR1 was isolated when it was found that a point mutation in this gene conferred ethylene insensitivity in *Arabidopsis* [1]. Four additional ETR1 homologues, ETR2, ERS1, ERS2, and EIN4, have also been isolated, and a similar point mutation in each homologue produces dominant ethylene insensitivity [5–7]. Sequence analysis of the carboxy terminus of *ETR1* family members shows strong homology to histidine kinases, whereas binding studies with radiolabeled ethylene show that the amino terminus of ETR1 and ERS1 have submicromolar affinity for ethylene [5–7]. Binding studies have also shown that the point mutations responsible for ethylene insensitivity abolish ethylene binding, providing compelling evidence that ETR1 and its homologous histidine kinases are ethylene receptors [8].

Histidine kinases are commonly known as two-component systems in reference to the two distinct signal transducers involved in response activation: the HK, and the receiver domain (RD), an aspartate phosphoreciever. The activity of the HK is generally regulated by an amino-terminal signal sensor domain, which is sensitive to a specific chemical or physical stimulus. When the receptor domain interacts with its signal, HK kinase activity is either activated or inhibited, which alters the phosphorylation state of a specific histidine residue also located in the HK. This phosphate is then transferred to an aspartate residue in the RD. Phosphorylation of the RD alters its molecular interactions with either an attached domain or downstream protein to affect gene regulation and/or cell biochemistry. In about 70% of prokaryote two-component systems, the HK and RD are located on separate proteins, whereas the other 30% are what are known as hybrid kinases, in which the sensor, HK, and RD are located in the same protein [10]. ETR1 and several of its homologues, as well as all the other eukaryotic histidine kinases identified to date, are hybrid kinases [11].

Identified at around the time that *ETR1* was cloned, SLN1 from *Saccharomyces cerevesiae* has been shown to be an osmosensing histidine hybrid kinase receptor that signals in a two-component fashion, providing evidence for a functional role for these transducers in eukaryotes [2]. Since 1993, other two-component systems have been identified in the yeasts *Saccharomyces pombe* and *Candida albicans*, in the slime mold *Dictyostelium discoideum*, and in *Neurospora crassa* [12–16]. Although completion of the *Arabidopsis* genome has revealed that histidine kinases are a small family in plants as compared to serine, threonine, and tyrosine kinases, strong evidence indicates that CRE1, an *Arabidopsis* histidine kinase, is a receptor for the plant hormone cytokinin [17]. A role for this common prokaryotic transducer in eukaryotic signaling, particularly in plant hormone response pathways, is generating a great deal of interest and study. This chapter assesses a role for ETR1 family histidine kinase signaling in the ethylene response pathway by presenting the current biochemical and genetic evidence and examining these data in the context of other prokaryotic and eukaryotic, particularly plant, two-component regulatory systems.

ETR1 FAMILY GENE STRUCTURE AND BIOCHEMISTRY

Sequence analysis has allowed for the identification of four distinct domains in the *ETR1* gene family: an N-terminal membrane-associated ethylene-binding domain, a GAF-like linker domain, a histidine kinase, and a receiver domain [1, 8, 18]. Biochemical work supports some of these assignments, demonstrating ethylene-binding activity in the sensor domain and histidine kinase activity in the HK [8, 19]. However, primary sequence analysis also shows that ETR1 is the only complete histidine kinase, as the other receptor isoforms lack features critical for histidine kinase signaling, which, at least on first analysis, appears inconsistent with two-component signal transduction [5, 20]. This section describes each domain of the ethylene receptors in terms of both gene structure and biochemical analysis to provide a descriptive picture of the ethylene receptor family.

ETHYLENE SENSOR DOMAIN

In the majority of cases in which biochemical information has been collected for two-component systems, it is typically kinase activity and its relationship to a physiological response that have been best characterized. In fact, in some cases, the precise stimuli responsible for activating the two-component system have not even been identified. For the ETR1 family, the opposite situation is true, as the best-characterized domain in these receptors is the ethylene-sensing region. Furthermore, dozens of ethylene agonists and antagonists have been characterized, providing a detailed picture of the physiochemical requirements of this system [21–23]. As a result, it seems appropriate to present a detailed discussion of the ethylene binding domain in order to provide some insight into the types of sensor-–ligand interactions capable of affecting histidine kinase activity.

Radiolabeled ethylene-binding experiments in *S. cerevesiae* expressing the amino terminal 128 amino acids of ETR1 [ETR1(1–128)] have shown that this region is necessary and sufficient for ethylene binding [8, 21]. These first 128 amino acids of ETR are predicted to encode three transmembrane α helices (Fig. 1A) [25]. The other four ethylene receptors share strong homology to these three α helices, whereas ERS2, ETR2, and EIN4 are also predicted to have additional N-terminal transmembrane α helices [6, 7]. Because ethyl-

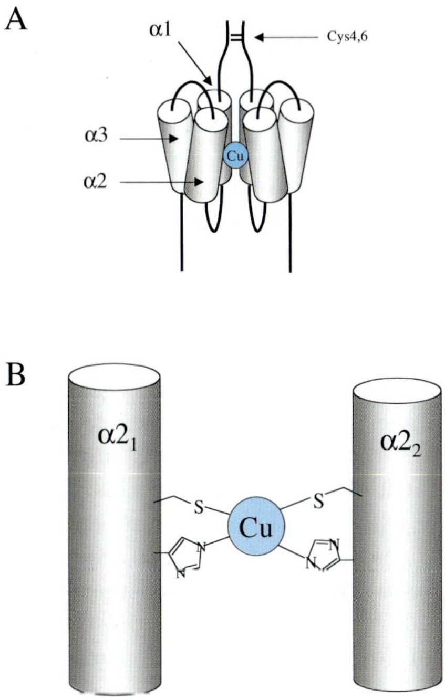

FIGURE 1 A model of the ETR1 ethylene sensor domain. (A) Yeast-expressed ETR1(1–128) has been shown to form membrane-associated dimers with disulfide linakages between the two Cys^4 and the two Cys^6. Hydropathy analysis predicts three possible transmembrane α helices in this region. Biochemical evidence indicates that each ETR1 dimer binds one copper ion. (B) Based on structural predictions and binding studies with various mutants, the Cys^{65} and His^{69} on the second transmembrane helix have been proposed as possible copper ligands.

ene has greater solubility in a lipid bilayer than in water, it is reasonable that a transmembrane region rather than an extracellular domain is responsible for binding [4].

ETR1 and ERS1 have both been shown to form membrane-associated homodimers in plant extracts and when expressed in *S. cerevesiae* [9, 25]. These dimers are disulfide linked, as high concentrations of DTT are required to dissociate them [25]. The ethylene-binding region, ETR1(1–128), is responsible for dimerization, as it forms a membrane-associated dimer when expressed in yeast, whereas yeast-expressed ETR1[128–738) runs as a monomer in PAGE [25]. Cysteines at positions 4 and 6 are responsible for disulfide bond formation, for only when both are substituted with serines does the ETR1 run as a monomer in nonreducing PAGE (Fig. 1A) [25]. Evidence from other two component systems indicates that histidine kinases are only active as dimers, as it is the kinase region of one monomer that phosphorylates the histidine phosphoreceiver in the other monomer [10]. As is the case for ETR1, the sensor domain is typically important for dimer formation [10].

Binding experiments with a purified ETR1(1–128)-GST have demonstrated that ETR1 requires a copper cofactor for ethylene binding [24]. Maximal ethylene binding occurs when ETR1 and Cu are in a 2:1 ratio, suggesting that each ETR1 dimer binds one copper (Fig. 1A) [24]. The original dominant insensitive mutant (C65Y) binds neither copper nor ethylene [8, 24]. Two additional insensitive ETR1 mutants (I62F and H69A) have also been shown to lack ethylene binding [26]. Based on an α helical structure, ^{65}Cys and ^{69}His are located on the same face of the second transmembrane α helix (Fig. 1B) [24]. As these two amino acids are common metal ligands, our current model contends that an ETR1 dimer binds a single copper ion with each monomer providing a ^{65}Cys and a ^{69}His [24]. Ethylene is proposed to bind to the protein through a p interaction with the copper cofactor [24]. Such a π–metal interaction is consistent with the physiochemical requirements for olefin agonists. Indeed, in 1967, based on trends in the activity of olefin agonists, Burg and Burg [21] proposed this as a likely interaction.

The activity of a class of olefinic ethylene antagonists is consistent with the involvement of a π–metal interaction in ethylene perception [21, 24]. These olefinic antagonists include such structurally diverse compounds as cyclopropene, *trans*-cyclooctene, and *cis*-butene, which do share a common feature, a highly strained double bond [27]. In each case, this strain energy can be relieved through the formation of an olefin–metal complex, which allows for partial rehybridization of the sp2 vinylic carbon (ideal bond angle of 120°) closer to an sp3 carbon (ideal bond angle 109°) [28]. In the case of cylclopropene compounds, where the best overlap would be achieved with a 60° angle, the less obtuse angle allows for better overlap of the orbitals in the ring [28]. This partial rehybridization can also relieve the sterically unfavorable interactions in the ethylene antagonists *trans*-cyclooctene and *cis*-butene by allowing partial rotation around the double bond. This release of strain energy can result in a very stable interaction, explaining why these compounds are so effective in competing with ethylene for the copper-binding site [27]. [^{14}C]-Ethylene competition assays with the ethylene antagonist 1-methylcyclopropyl (MCP) show that MCP competes effectively for yeast-expressed ETR1 and ERS1 with an extremely low K_I (10 nl/liter) [9]. This K_I is, however, 10-fold higher than the EC_{50} of MCP action (1 nl/liter) [9]. If the other isoforms show similar K_I, it would imply that inhibition occurs at 10% receptor occupancy.

An ethylene–copper cofactor interaction is also consistent with the ethylene antagonist properties of silver(I) compounds [28]. The proposed mechanism of Ag(I) inhibition of ethylene responses is through the displacement of the copper cofactor from the ethylene receptor [28, 29]. It was originally proposed that copper displacement by silver would disrupt ethylene association, but this is not the case, as ETR1 loaded with silver has a strong affinity for ethylene [24]. One model consistent with this result is that the ETR1 loaded with silver cannot undergo a conformational change required for the activation of ethylene responses.

Cu(I) has been proposed to be the likely form of the native metal cofactor [24]. This conclusion is based in part on the evidence that Ag(I) can bind to ethylene receptors [24]. Also, ethylene forms much more stable complexes with Cu(I) than Cu(II) because the greater electron density on Cu(I) allows for better back bonding to the π-antibonding orbital of ethylene [24]. Cu(I) has 10 outershell valance electrons, and with two electrons from each of the four ligands (two histidines, two cystines), the bonding orbitals of copper would be completely occupied with 18 electrons, a noble gas configuration. In *in vitro* ethylene-binding experiments, ETR1 is reconstituted with $CuSO_4$, a Cu(II) ion, which seems inconsistent with a Cu(I) cofactor. However, considering the ease with which copper can be reduced in water and the reductive potential in a cell lysate, it is reasonable that some Cu(I) would be present for loading.

Further support for a Cu(I) cofactor comes from evidence concerning the in vivo biogenesis of the ethylene receptors. RAN1, an *Arabidopsis* homologue to a Cu(I) chaperone found in humans and yeast, was discovered in a screen for plants with an aberrant ethylene antagonist response [30]. It has since been demonstrated that plants with a RAN1 null mutation show a constitutive ethylene response, a result consistent with a nonfunctioning ethylene receptor [31]. This evidence suggests that RAN1 is required for loading the ETR1 family with Cu(I) to produce active receptors [31].

One unanswered question concerns how the ethylene associates with the ETR1–Cu(I) complex, as no empty d-orbital would be available to accept the ethylene π electrons. One speculative possibility is that the copper–receptor interaction is not static and that ethylene can substitute for one of the amino acid ligands. Interpolation of an ethylene molecule could lock the receptor into a new conformation, altering the HK activity. Another possibility could be that ethylene binding may cause another ligand to dissociate. There is physiochemical evidence to support this because the potency of various alkene agonists correlates to their *trans* effect, a measure of the amount an alkene destabilizes the ligand bound to the opposite side of the metal center [21, 27].

FixL, a hemoglobin-based oxygen sensor from *Rhizobia*, provides an interesting comparison for ETR1 as it is a histidine kinase regulated by a metal-based gas sensor domain [32]. FixL has been studied extensively, and crystallographic structural data with and without ligands have led to a model that oxygen binding changes the spin state of the iron from high spin to low spin [32]. This chemical change at the iron center is proposed to then affect a conformational change in the protein as the result of a rearrangment of the porphoryin ring coordination geometry [33]. This conformational change on oxygen binding inactivates the FixL histidine kinase to initiate a physiological effect [32]. Interestingly, although FixL and ETR1 sensor regions are different at both the level of primary sequence and their metal cofactor, they do have some similarities in terms of ligand specificity, as both ethylene responses and FixL activity are affected by carbon monoxide and cyanide-based compounds [33]. As these compounds generally have strong affinity for metals, this similarity does not necessarily imply that ETR1 would work by the same mechanism as FixL. However, FixL does show how a ligand–metal interaction can affect protein conformation by altering the coordination chemistry of a metal cofactor, perhaps providing some insight into the ETR1 ethylene–copper signaling mechanism.

GAF-LIKE DOMAIN

The linker region between the hydrophobic sensor domain and the HK of ETR1 has no assigned function as of this time, although the region does have some similarity to GAF domains [34]. GAF domains have been found in over 90 proteins with a variety of functions and are characterized by similarities in predicted structure and amino acid usage more than sequence homology [34]. A GAF-like domain that has some sequence homology to the ethylene receptor linker is the chromophore-binding domain in plant and cyanobacterial phytochromes. However, a cystine essential for chromophore binding is not conserved in the linker domain, suggesting that this is probably not the role of the linker domain. The largest class of GAF containing proteins are phosphodiesterases, in which the GAF domain plays a role in cyclic GMP and cyclic AMP binding [34]. While it is an intriguing possibility that ethylene receptors may be sensitive to cyclic nucleotides in addition to ethylene, cyclic nucleotide binding assays with ETR1 have been unsuccessful (B. Binder, unpublished data). The sequence similarity of the linker domain to these GMP-binding domains is not especially high, and it may be that because the GAF-like sequence encodes a structural motif that has been recruited for a range of applications over evolution, homology alone is unlikely to indicate a function.

HISTIDINE KINASE-COUPLED RECEPTOR

The typical HK contains 240 conserved amino acids known as the catalytic histidine kinase core. X-ray crystallography and nuclear magnetic resonance solution studies show the HK to be composed of two distinct regions: the conserved histidine phosphoacceptor domain and the catalytic kinase domain [35, 36]. The conserved phosphoacceptor domain is a 40 amino acid long stretch, which folds into two α helices with the histidine phosphate-accepting residue pointing outward in an accessible location [36]. The remaining 200 amino acids compose the catalytic kinase domain, which itself contains four regions of higher conservation: N, F, G1, and G2 boxes [37]. G1 and G2 boxes are glycine-rich stretches of sequence that resemble nucleotide-binding motifs found in other proteins [38]. Biochemical evidence indicates that G1 and G2 boxes are critical for histidine phosphorylation [39]. The function of N and F boxes has not been determined, but their conservation and positions in the crystal structure relative to the nucleotide-binding site suggest some catalytic role.

Of the five ethylene receptors present in *Arabidopsis*, only ETR1 and ERS1 contain all the required elements for a functional HK [18]. The three other ethylene receptors are each deficient in one or more of the required signaling components (Fig. 2). ETR2 and ERS2 lack the conserved histidine phospho-

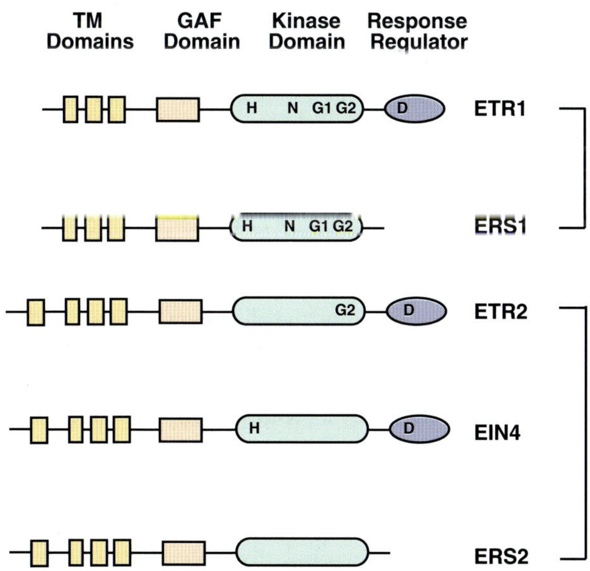

FIGURE 2 Structural features of the ETR1 gene family. The histidine kinase regulatory domains of ETR1 and ERS1 contain all the hallmark features of a functional kinase. ERS2, ETR2, and EIN4 are each deficient in some of these features. The receiver domains of ETR1, ETR2, and EIN4 show conservation of the phosphate accepting aspartate. ERS1 and ERS2 do not have receiver domains.

acceptor [6, 7]. EIN4, ERS2, and ETR2 are missing conserved residues required for a functional ATPase, which is critical for kinase activity in bacterial two-component systems [6, 7]. All of these receptors do appear to be functional, however, as a dominant insensitive allele has been isolated for each of the isoforms [5, 7]. The incomplete kinase region does raise the question as to whether signaling occurs by a two-component phosphorelay mechanism.

RECEIVER DOMAIN

The canonical receiver domain consists of a 125 amino acid region, with a conserved aspartate required as the phosphoacceptor [10]. Phosphotransfer from the HK to the RD is the result of a nucleophilic attack by the aspartate on the histidine-bound phosphate [10]. *In vitro* studies show that the aspartate will accept a phosphate from a number of small organic molecules with high energy phosphate bonds, which suggests that the HK may simply provide an easily hydrolyzable phosphate and that the RD is not involved in catalyzing the reaction [10].

Just as several of the receptors appear to have incomplete HKs, the entire receiver domain is missing from ERS1 and ERS2 (Fig. 2) [5, 7]. The remaining three receptors, ETR1, ETR2, and EIN4, do have RDs that appear to be functional, as they contain the conserved aspartate phosphoreceiver [5, 7]. Indeed, X-ray crystollography of the ETR1 RD shows that it has strong structural homology to the characterized bacterial RDs (CheY, CheB, PhoB, and NarL), even though the sequence homology is not especially high [40].

Based on the mechanism of cross-phosphorylation, only the ETR1 dimer would be capable of signaling in a two-component fashion. This raises some question as to whether this is the actual mechanism of transduction. Given that all five receptors appear to be expressed in all tissues of *Arabidopsis*, one possibility that could allow some of the degenerate kinases to participate in signaling is heterodimers between the various isoforms [7, 20]. In this scenario, receptor isoforms without kinase activity could still be phosphorylated at their histidine phosphoacceptor by a receptor with an active ATPase.

KINASE ACTIVITY IN THE CYTOSOLIC PORTION OF ETR1

The chemistry of phosphate bonds to histidine and aspartate residues is distinct from that of serine, threonine, and tyrosine, substrates for the majority of eukaryote kinases. One of the most obvious differences is in the stability of the bond to hydrolysis. The general trend is that phosphorylated serine and threonine are the most stable, followed by tyrosine, histidine, and finally aspartate, which has a half-life of around 5 min at neutral pH and only seconds at acidic or basic pH [41]. The kinetics of the phosphohistidine hydrolysis is also pH dependent, as it is quite stable at neutral or basic conditions, but unstable in acid [42]. This pH-dependent hydrolysis rate has been a useful tool for determining whether a protein-bound phosphate is attached at a histidine, and therefore whether an observed kinase activity is histidine specific.

In vitro studies show that the cytosolic portion of ETR1, which lacks the ethylene binding region, can accept a phosphate from γ^{32}P-ATP [19]. Analysis of pH stability indicates that the phosphate is bound to a histidine, as the radioactivity can be washed off with acidic buffer, but not with basic or neutral buffer [19]. A point mutation at the conserved histidine receiver (H401S) abolishes the activity, which indicates that histidine kinase activity is specific [19]. Mutations in G1 and G2 boxes of the HK also abolish activity, indicating that phosphorylation requires a functional kinase [19]. Phosphorylation requires either magnesium or manganese, metal cofactors commonly required in kinase activity [19]. Manganese imbues stronger activity, suggesting that this less common cofactor is likely to be the physiologically significant metal [19]. The intrinsic kinase activity in this truncated ETR1 supports the possibility of a two-component phosphorelay signaling mechanism in the ethylene pathway. However, there still remains a question as to the regulatory role of the ethylene sensor domain on kinase activity. It will also be vital to establish the relationship among ethylene binding, kinase activity, and ethylene responses in order to clarify the role of two-component signaling in ethylene perception.

MUTATIONAL ANALYSIS OF THE ETHYLENE PATHWAY

While it is the central role of ethylene in regulating fruit ripening and flower abscission that makes it of great agricultural importance, it has been a set of ethylene responses in etiolated seedlings that have been exploited most successfully to gain information about the ethylene signal transduction chain. These effects, known collectively as the triple response, consist of inhibition of root and shoot elongation, radial enlargement of the hypocotyl, and an exaggeration of the apical hook of etiolated seedlings grown in the presence of ethylene [4]. These dramatic ethylene effects are highly specific and can be detected clearly in young *Arabidopsis* seedlings, which facilitate the management of the large population required for a successful screen. These features have made the triple response an ideal choice for identifying plants defective in ethylene perception and response.

One of the first mutants isolated by this method was completely insensitive to ethylene [3]. This mutant permitted cloning of the *ETR1* gene [1]. The hormone insensitive allele, *etr1-1*, contains a change in a single base pair (C65Y), which has been shown to abolish ethylene binding [1]. Subsequent mutant screens led to the isolation of *ETR2* and *EIN4*, two of the four *ETR1* isoforms [6, 43]. Although it has not been established that these isoforms bind ethylene, the substituted amino acid in these mutated isoforms is located in the ethylene-binding domain. No mutations of the native genes have been isolated for the remaining two isoforms, *ERS1* and *ERS2*. These two isoforms do appear to be functional *in vivo*, however, as a point mutation in genomic clones of these genes homologous to the etr1-1 mutation produces ethylene-insensitive offspring when transferred into wild-type *Arabidopsis* [5, 7]. In all of these cases, the insensitive mutations are dominant, such that a single mutant allele in any of these receptors produces a plant completely insensitive to ethylene [1, 5, 7].

No loss-of-function null mutations for ETR1 family members turned up in the original screens, but were subsequently isolated using a variety of techniques [20]. Plants homozygous for null mutations in one or two of the receptors show no detectable change in their response to ethylene [20]. However, loss-of-function mutants in three or four receptors generally result in constitutive ethylene responses in air grown seedlings [20]. These data suggest that ethylene receptors actively suppress ethylene responses in the absence of ethylene, and ethylene releases this suppression.

A null mutant for a different gene, CTR1, a MAPKKK-like kinase, also produces a constitutive ethylene response phenotype in air grown seedlings [44]. This indicates that the CTR1 gene product is also involved in suppressing ethylene responses. Epistatic analysis reveals that CTR1 acts downstream of the ETR1 family, as etr1-1 does not rescue CTR1 null mutants from a constitutive triple response [44]. Yeast two-hybrid screens and immunoprecipitation have been used to demonstrate an interaction between ETR1 and ERS1 and the noncatalytic region of CTR1 [45]. Together, these observations are consistent with these proteins interacting directly in the transduction chain.

Based on genetic data, a plausible proposal is that the relationship between the receptor family and CTR1 represents an inverse agonist mechanism for ethylene action [20]. An inverse agonist mechanism describes a situation in which a response is actively repressed in the absence of an agonist. Agonist binding induces the response by deactivating the inhibition. According to this model, unoccupied receptors activate CTR1 mediated suppression of the ethylene response pathways (Fig. 3A). Upon ethylene exposure, the hormone binds to the ethylene receptors and deactivates them. This releases CTR1 suppression of the ethylene response pathway (Fig. 3B). In the CTR1 null mutant, this suppression never exists. The plant shows constitutive ethylene

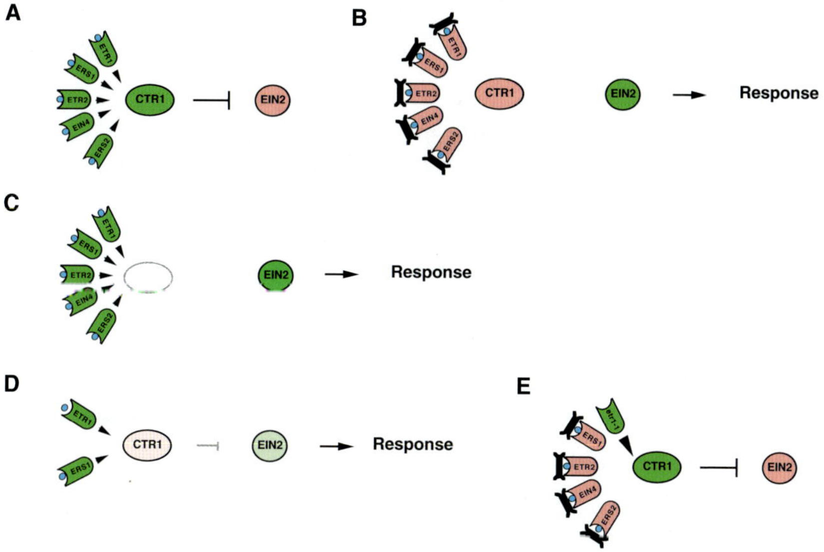

FIGURE 3 Inverse agonist model of ethylene action. (A)At low ethylene concentrations, unoccupied receptors activate CTR1 which suppresses ethylene responses. (B) When the concentration of the hormone increases, ethylene binds to the receptors, deactivates this signaling, and the CTR1 suppression of ethylene responses is lifted. (C) In CTR1 null mutants, ethylene responses are always on, regardless of the activity of the unoccupied receptors. (D) In triple receptor null plants, the reduced receptor number results in insufficient activation of CTR1 to maintain the suppression of ethylene responses. (E) The dominant-insensitive allele etr1-1 cannot bind ethylene and maintains activation of CTR1 even when the other receptors are deactivated by ethylene.

responses throughout its life (Fig. 3C). The constitutive responses in the receptor triple nulls are the result of insufficient ethylene receptors signaling to activate CTR1 suppression (Fig. 3D). In the case of dominant-insensitive mutants, the model suggests that because ethylene cannot deactivate the mutant receptor, it continues to activate CTR1 suppression. This holds true even when wild-type receptors are deactivated (Fig. 3E). Such a mechanism is consistent with the dominant nature of these alleles.

In addition to the ETR1 family and CTR1, screens based on the triple response have led to the cloning of several additional components of the ethylene response pathway (Fig. 4). Downstream of CTR1 is EIN2, which shows sequence homology to N-ramp metal transporters and a novel carboxy-terminal sequence [45]. A homozygous null mutation in EIN2 produces ethylene insensitive plants even in CTR1 null backgrounds, and overexpression of the carboxy-terminal sequence in *Arabadopsis* produces constitutive ethylene responses [45, 46]. This suggests that ethylene responses are mediated by the

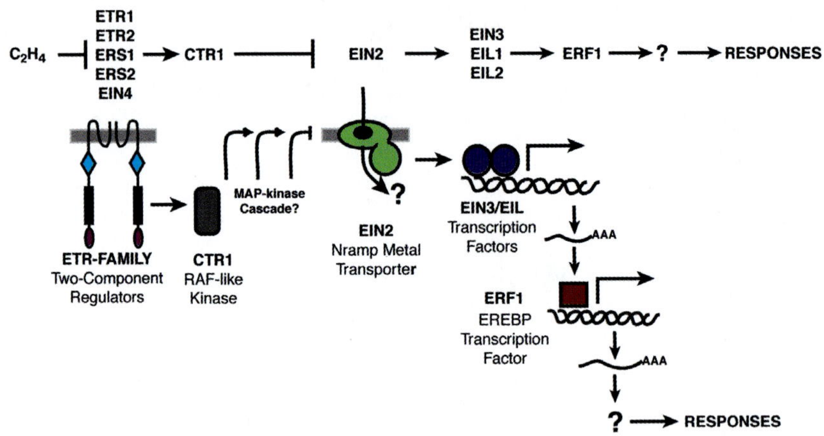

FIGURE 4 A model of the ethylene signal transduction pathway. Epistatic analysis has identified the ETR1 family of two-component related proteins as the first step in the ethylene response pathway. Biochemical evidence indicates that members of the ethylene receptor family interact directly with CTR1, a RAF-like MAPKKK. Together, receptor–CTR1 complexes appear to be responsible for negatively regulating EIN2, a membrane-localized protein related to a superfamily of metal transporters. The cytoplasmic portion of EIN2 has been shown to be involved in the activation of EIN3, a member of a family of nuclear-localized transcription factors. Activated EIN3 upregulates a second transcription factor, ERF1. Ectopic expression of ERF1 has been shown to activate a subset of ethylene responses.

carboxy terminus of EIN2, which is normally repressed by the N-ramp domain, but are activated by ethylene through upstream components [45]. How this regulation occurs is not yet understood, nor is it known how EIN2 activates the downstream transcription factor, EIN3 [47]. Overexpression of EIN3 in *Arabadopsis* results in constitutive ethylene responses even in an EIN2 null background [47]. EIN3 binds as a homodimer to the promoter region of a second transcription factor, ERF1, a member of the plant EREBP family [48]. At this point the pathway appears to show some bifurcation, as overexpression of ERF1 in *Arabadopsis* produces constitutive activation of some but not all ethylene responses [48].

TWO-COMPONENT SIGNALING THROUGH MAPK KINASES IN *SACCHAROMYCES CEREVISIAE* AND *ARABIDOPSIS*

Genetic evidence indicates that ethylene receptors are histidine like-kinases that act in concert with the CTR1 MAPKKK-like protein to regulate ethylene responses. Evidence of an interaction between CTR1-ETR1 and CTR1-ERS1 suggests that these proteins are likely to form a signaling complex *in vivo* [45]. Despite this, there is still no unequivocal demonstration of a role, if any, for the two-component regulation of CTR1 in the ethylene response pathway. The yeast osmosensing pathway does, however, provide a clear example of MAPKKK regulation by a two-component system that may provide some hints as to how regulation occurs in the ethylene response pathway.

In this yeast pathway, a receptor-coupled histidine hybrid kinase, Sln1, transfers a phosphate to a cytosolic RD, Ssk1, via Ypd1, a histidine-containing phosphostransfer protein (HPt) [49, 50]. At normal osmolarity the Sln1 HK is active, resulting in a high concentration of P-Ssk1, which actively inhibits two MAPKKKs; Ssk2 and Ssk22 [50]. At high osmolarity, Sln1 is deactivated, causing the P-Ssk1 concentration to drop and release the inhibition of Ssk2 and Ssk22. These MAPKKKs can then activate the rest of the MAPK kinase cascade, activating cellular responses to high osmolarity [50]. Thus, the Ssk1/Ssk2-Ssk22 constitutes a clear demonstration that an RD phosphorylation state can specifically regulate a MAPK cascade. Interestingly, there is a second yeast osmosensor, Sho1, that is functionally redundant for Sln1, but does not signal through the two-component pathway [49]. Instead, Sho1 is capable of directly activating the MAPKK, Pbs1 [49]. Because either Sln1 or Sho1 activity is required to activate a high osmolarity response, Dsln1Dsho1 yeast die on high osmolarity medium [49].

In addition to providing the first functional demonstration of a eukaryotic two-component signaling system, the yeast Sln1 pathway has provided the platform for experiments to demonstrate a role for two *Arabidopsis* histidine kinases in the regulation of osmotic (ATHK1) and cytokinin (CRE1) responses. ATHK1, originally isolated as a highly expressed cDNA in dehydrated plants, has high sequence similarity to Sln1, including the putative N-terminal osmosensor domain [51]. A functional role for ATHK1 as an *Arabidopsis* osmosensor is further bolstered by experiments in which ATHK1 is shown to successfully rescue Δsln1Δsho1 yeast from dying on low osmolarity medium, indicating that this HK is signaling actively through the yeast pathway [51]. Furthermore, ATHK1, like Sln1, appears to function as an osmosensor, as it can activate the MAPK kinase components of the osmotic response pathway [51].

A second *Arabidopsis* histidine kinase, CRE1, has been discovered in a screen for mutants impaired in cytokinin responses in plant callus [17]. To establish a connection between cytokinin binding and histidine kinase activity, the authors expressed the *CRE1* gene in sln1Δ yeast. While sln1Δ*CRE1* yeast do not grow on low osmolarity media, they will grow when media contain any of four structurally distinct cytokinins. Furthermore, while *trans*-zeatin, a potent cytokinin, rescues the yeast, *cis*-zeatin, structurally very similar but a much less active cytokinin, does not rescue. Additionally, mutations at either the conserved histidine or aspartate of the HK and RD, respectively, abolished this rescue. These data provide strong evidence that cytokinin is required for CRE1 kinase activity and that CRE1 is a cytokinin receptor [17].

A role for two-component regulation in *Arabidopsis* for both osmosensing and activating cytokinin responses does provide strong precedence for ethylene responses being mediated in a similar fashion. It is important to note, however, that unlike CRE1 and ATHK1, ETR1 cannot rescue sln1Δsho1Δ yeast [51]. This suggests that either ETR1 cannot transfer a phosphate to Yph1, the yeast HPt, or ETR1 is not an active histidine kinase. Sequence analysis is consistent with a possible incompatibility between ETR1 and Yph1, as ATHK1 and Sln1 are found to be much more similar to each other than either one is to ETR1 [51].

In addition to the example of a RD/MAPKKK interaction in yeast, structural homology supports a possible role for an ETR1 family two-component activation of CTR1. CTR1 is most homologous to the MAPKKK RAF, with a 41% amino acid identity in the kinase domain [44]. In the typical RAF MAPK pathway, the G protein RAS is an important activator of RAF activity. While ETR1 RD shows low sequence homology to RAS, a comparison of the crystal structure of the RD of ETR1 and that of RAS does reveal a high degree of structural similarity [40, 45]. In conjunction with data showing an interaction between CTR1 and ETR1 RDs, this structural homology does lend credence to the direct activation of CTR1 by the ethylene receptor's receiver domains. However, evidence that ETR1 interacts directly with the RAF-like CTR1 clearly distinguishes this system from the phosphorelay mechanism used by Sln1. This leaves open the possibility that ethylene receptors have evolved a unique mechanism for activating downstream effectors.

REFERENCES

1. Chang, C., Kwok, S. F., Bleecker, A. B., and Meyerowitz, E. M. (1993). *Arabidopsis* ethylene-response gene *ETR1*: Similarity of product to two-component regulators. *Science* **262**, 539–544.
2. Ota, I. M., and Varshavsky, A. A. (1993). Yeast protein similar to bacterial 2 component regulators. *Science* **262**, 566–569.
3. Bleecker, A. B., Estelle, M. A., Somerville, C., and Kende, H. (1988). Insensitivity to ethylene conferred by a dominant mutation in *Arabidopsis thaliana*. *Science* **241**, 1086–1089.
4. Abeles, F. B., Morgan, P. W., and Saltveit, M. E. Jr. (1992). "Ethylene in Plant Biology" 2nd Ed. Academic Press, San Diego.
5. Hua, J., Chang, C., Sun, Q., and Meyerowitz, E. M. (1995). Ethylene insensitivity conferred by *Arabidopsis ERS* gene. *Science* **269**, 1712–1714.
6. Sakai, H., Hua, J., Chen, Q. G., Chang, C., Medrano, L.J., Bleecker, A. B., and E. M. Meyerowitz, E. M. (1998). *ETR2* is an *ETR1*-like gene controlling ethylene signal transduc-

tion. *Proc. Natl. Acad. Sci. USA* **95**, 5812–5817.

7. Hua, J., Sakai, H., Nourizadeh, S., Chen, Q. G., Bleecker, A. B., Ecker, J. R., and Meyerowitz, E. M. (1998). *EIN4* and *ERS2* are members of the putative ethylene receptor gene family in *Arabidopsis*. *Plant Cell* **10**, 1321–1332.
8. Schaller, G. E., and Bleecker, A. B. (1995). Ethylene-binding sites generated in yeast expressing the *Arabidopsis ETR1* gene. *Science* **270**, 1809–1811.
9. Hall, A. E. , Findell, J. L., Schaller, G. E., Sisler, E. C., and Bleecker, A. B. (2000). Ethylene perception by the ERS1 protein in *Arabidopsis*. *Plant Physiol.* **123**, 1449–1457.
10. Hoch, J. A., and Silhavy, T. J. (1995). "Two-Component Signal Transduction." ASM Press, Washington, DC.
11. Pao, G. M., and Saier, M. H. (1997). Nonplastid eukaryotic receiver domains have a monophyletic origin and evolved from their bacterial precursors in parallel with their cognate sensor kinases. *J Mol. Evol.* **44**, 605–613.
12. Shieh, J. C., Wilkinson, M.G., Buck, V., Morgan, B. A., Makino, K., and Millar, J. B. A. (1997). The Mcs4 receiver domain coordinately controls the stress activated Wak1-Wis1-Sty1 MAP kinase pathway and fission yeast cell cycle. *Genes Dev.* **11**, 1008–1022.
13. Alex, L. A., Borkovich, K. A., and Simon, M. I. (1996). Hyphal development in *Neurospora crassa*: Involvement of a two-component histidine kinase. *Proc. Nat. Acad. Sci. USA* **93**, 3416–3421.
14. Corner, B. E., and Magee, P. T. (1997). *Candida* pathogenesis: Unravelling the threads of infection. *Curr. Biol.* 7, R691–R694.
15. Schuster, S. S., Noegel, A. A., Oehme, F., Gerisch, G., and Simon, M. (1996). The hybrid histidine kinase DokA is part of the osmotic response system of *Dictyostelium*. *EMBO J.* **15**, 3880–3889.
16. Shaulsky, G., Escalante, R., and Loomis, W.F. (1996). Developmental signal transduction pathways uncovered by genetic suppressors. *Proc. Natl. Acad. Sci. USA* 93, 15260–15265.
17. Inoue, T., Higuchi, M., Hashimoto, Y., Seki, M., Kobayashi, M., Kato, T., Tabata, S., Shinozaki, K., and Kakimoto, T. (2001). Identification of CRE1 as a cytokinin receptor for *Arabidopsis*. *Nature* **409**, 1060–1063.
18. Bleecker, A. B., Esch, J. J., Hall, A. E., Rodriguez, F. I., and Binder, B. M. (1998). The ethylene-receptor family from *Arabidopsis*: Structure ad function. *Phil. Trans. R. Soc. B* **353**, 1405–1412.
19. Gamble, R. L., Coonfield, M. L., and Schaller, G. E. (1998). Histidine kinase activity of the ETR1 ethylene receptor from *Arabidopsis*. *Proc. Natl. Acad. Sci. USA* 95, 7825–7829.
20. Hua, J., and Meyerowitz, E. M. (1998). Ethylene responses are negatively regulated by a receptor gene family in *Arabidopsis thaliana*. *Cell* **94**, 261–271.
21. Burg, S. P., and Burg, E. A. (1967). Molecular requirements for the biological activity of ethylene. Plant Physiol. **42**, 144–152.
22. Sisler, E. C. (1977). Ethylene activity of some π-acceptor compounds. *Tobacco Sci.* **21**, 43–45.
23. Sisler, E. C., Serek, M., Dupill, E., and Goren, R. (1999). Inhibition of ethylene responses by 1-methylcyclopropene and 3-methylcyclopropene. *Plant Growth Regul.* **27**, 105–111.
24. Rodriguez, F. I., Esch, J. J., Hall, A. E., Binder, B. M., Schaller, G. E., and Bleecker, A. B. (1999). A copper cofactor for the ETR1 receptor from *Arabidopsis*. *Science* **283**, 996–998.
25. Schaller, G. E., Ladd, A. N., Lanahan, M. B., Spanbauer, J. M., and Bleecker, A. B. (1995). The ethylene response mediator ETR1 from *Arabidopsis* forms a disulfide-linked dimer. *J. Biol. Chem.* **270**, 12526–12530.
26. Hall, A. E., Chen, Q. G.,Findell, J. L., Schaller, G. E., and Bleecker, A. B. (1999). The relationship between ethylene binding and dominant insensitivity conferred by mutant forms of the ETR1 ethylene receptor. *Plant Physiol.* **121**, 291–299.

27. Sisler, E. C., and Yang, S. F. (1984). Anti-ethylene effects of cis-2-butene and cyclic olefins. *Phytochemistry* **23**, 2765–2768.
28. Beyer, E. M. Jr. (1976). A potent inhibitor of ethylene action in plants [Silver nitrate tested on peas, Cattleya orchid, and cotton]. *Plant Physiol.* **58**(3), 268–271.
28. Crabtree, R. H. (1988). "The Organometallic Chemistry of the Transition Metals." J. Wiley, New York.
29. Veen, H., and van de Geijn, S. C. (1978). Mobility and ionic form of silver as related to longevity of cut carnations [*Dianthus caryophyllus*]. *Planta* **140**(1), 93–96.
30. Hirayama, T., Kieber, J. J., Hirayama, N., Kogan, M., Guzman, P., Nourizadeh, S., Alonso, J. M., Dailey, W. P., Dancis, A., and Ecker, J. R. (1999). Responsive-to-antagonist1, a Menkes/Wilson desease-related copper transporter, is required for ethylene signaling in *Arabidopsis*. *Cell* **97**(3), 383–393.
31. Woeste, K. E., and Kieber, J. J. (2000). A strong loss-of-function mutation in RAN1 results in constitutive activation of the ethylene response pathway as well as a rosette-lethal phenotype. *Plant Cell* **12**, 443–455.
32. Gilles-Gonzalez, M. A., Gonzalez, G., and Perutz, M. F. (1995). Kinase activity of oxygen sensor FixL depends on the spin state of its heme iron. *Biochemistry* **34**, 232–236.
33. Gong, W., Hao, B., and Chan, M. K. (2000). New mechanistic insights from structural studies of the oxygen-sensing domain of *Bradyrhizobium japonicum* FixL. *Biochemistry* **39**, 3955–3962.
34. Aravind, L., and Ponting, C. P., (1997). The GAF domain: An evolutionary link between diverse phototransducing proteins. *Trends Biochem. Sci.* **22**, 458–459.
35. Bilwes, A. M., Alex, L. A., Crane, B. R., and Simon, M. I. (1999). Structure of CheA, a signal-transducing histidine kinase. *Cell* **96**, 131–141.
36. Tanaka, T., Saha, S. K., Tomomori, C., Ishima, R., Liu, D., Tong, K. I., Park, H., Dutta, R., Qin, L., Swindells, M. B., Yamazaki, T., Ono, A. M., Kainosho, M., Inouye, M., and Ikura, M. (1998). NMR structure of the histidine kinase domain of the *E. coli* osmosensor EnvZ. *Nature* **396**, 88–92.
37. Park, H., Saha, S. K., and Inouye, M. (1998). Two-domain reconstitution of a functional protein histidine kinase. *Proc. Natl. Acad. Sci. USA* **95**, 6728–6732.
38. Alex, L. A., and Simon, M. I. (1994). Protein histidine kinases and signal transduction in prokaryotes and eukaryotes. *Trends Genet.* **10**, 133–138.
39. Yang, Y., and Inouye, M. (1991). Intermolecular complementation between two defective mutant signal-transducing receptors of *Escherichia coli*. *Proc. Natl. Acad. Sci. USA* **88**, 11057–11061.
40. Muller-Dieckmann H., Grantz, A. A., and Kim, S. (1999). The structure of the signal receiver domain of the *Arabidopsis thaliana* ethylene receptor ETR1. *Structure* **7**, 1547–1556.
41. Pirrung, M. C. (1999). Histidine kinases and two component signal transduction systems. *Chem. Biol.* **6**, R167–R175.
42. Hultquist, D. E., Moyer, R. W., and Boyer, P. D. (1966). The preperation and characterization of 1-phosphohistidine and 3-phosphohistidine. *Biochemistry* **5**, 322–331.
43. Roman, G., Lubrasky, B., Kieber J. J., Rothenberg, M., and Ecker, J. R. (1995). Genetic analysis of ethylene signal transduction in *Arabidopsis thaliana*: Five novel mutant loci integrated into a stress-response pathway. *Genetics* **139**, 1393–1409.
44. Kieber, J. J., Rothenberg, M., Roman, G., Feldmann, K. A., and Ecker, J. R. (1993). *CTR1*, a negative regulator of the ethylene response pathway in *Arabidopsis*, encodes a member of the Raf family of protein kinases. *Cell* **72**, 427–441.
45. Alonso, J. M., Hirayama, T., Roman, G., Nourizadeh, S., and Ecker, J. R. (1999). EIN2, a bifunctional transducer of ethylene and stress responses in *Arabidopsis*. *Science* **284**, 2148–2152.

46. Clark, K. L., Larsen, P. B., Wang, X., and Chang, C. (1998). Association of the *Arabidopsis* CTR1 Raf-like kinase with the ETR1 and ERS ethylene receptors. *Proc. Natl. Acad. Sci. USA* **95**, 5401–5406.
47. Chao, Q., Rothenberg, M., Solano, R., Roman, G., Terzaghi, W., and Ecker, J. R. (1997). Activation of the ethylene gas response pathway in *Arabidopsis* by the nuclear protein ETHYLENE-INSENSITIVE3 and related proteins. *Cell* **89**, 1122–1144.
48. Guzman, P.,. and Ecker, J. R. (1990). Exploiting the triple response of *Arabidopsis* to identify ethylene-related mutants. *Plant Cell* **2**, 512–523.
49. Solano, R., Stepanova, A., Chao, Q. M., and Ecker, J. R. (1998). Nuclear events in ethylene signaling: A transduction cascade mediated ETHYLENE-INSENSITIVE3 and ETHYLENE-RESPONSE-FACTOR1. *Genes Dev.* **12**, 3703–3714.
50. Maeda, T., Takekawa, M., and Saito, H. (1995). Activation of yeast PBS2 MAPKK by MAPKKKs or by binding of an SH3-containing osmosensor. *Science* **269**, 554–558.
51. Posas, F., Wurgler-Murphy, S. M., Maeda, T., Witten, E. A., Thai, T. C., and Saito, H., (1996). Yeast HOG1 MAP kinase cascade is regulated by a multistep phosphorelay mechansim in the SLN1-YPD1-SSK1 "two-component" osmosensor. *Cell* **86**, 865–875.
52. Urao, T., Yukubaov, B., Satoh, R., Yamaguchi-Shinozaki, K., Seki, M., Hirayama, T., and Shinozaki, K. (1999). A transmembrane hybrid-type histidine kinase in *Arabidopsis* functions as an osmosensor. *Plant Cell* **11**, 1743–1754.

CHAPTER 22

Pathogenicity and Histidine Kinases: Approaches Toward the Development of a New Generation of Antibiotics

J. HUBBARD,* M. K. R. BURNHAM,† AND J. P. THROUP†

*Computational and Structural Sciences, GlaxoSmithKline, Harlow, United Kingdom and
†Antimicrobials and Host Defence, GlaxoSmithKline, Collegeville, Pennsylvania 19426

The role of two-component signal transduction (TCST) systems as regulators of both pathogenicity and growth in gram-negative and gram-positive bacteria is reviewed. A comprehensive study of the entire TCST complement of both *Streptococcus pneumoniae* and *Staphylococcus aureus* and the effect of defined

TCST mutations on viability and pathogencity is also described. It is apparent that TCSTs interact to control both pathogenicity and bacterial cell growth/survival in an unexpectedly complex manner. Nevertheless, the requirement of selected TCSTs for growth either *in vivo* or *in vitro* by pathogenic bacteria has raised the possibility that such systems may form attractive antimicrobial drug targets. High throughput screening of compound libraries used to identify inhibitors of the enzymatic components of these systems gave hits that were unattractive for further development because of their chemical properties or that had low activity. As this has been the common experience of similar screens by other pharmaceutical companies, the use of structural-based screens to find routes to more useful types of inhibitors is suggested.

INTRODUCTION

ROLE IN PATHOGENICITY

In its simplest terms, bacterial pathogenicity can be viewed as the ability of a bacterium to grow and divide within a host. Thus any factor, which contributes to the survival of the pathogen in this environment, can be considered a virulence factor. However, bacterial growth *in vivo* (i.e., during infection) is a multifactorial dynamic process in which the invading pathogen encounters a diverse array of environmental conditions. To maximize growth, the infecting organism must sense each of these environmental parameters and then coordinate an appropriate adaptive response. To this end, many pathogenic bacteria have developed a series of sophisticated signal transduction systems, which are used to sense specific stimuli, integrate this information, and then elicit an appropriate response.

Over the past decade as the study of bacterial virulence has progressed, it has emerged that two-component signal transduction (TCST) systems (or His-Asp phosphorelays) are employed by many different bacterial pathogens during infection (Table I). These systems can be divided arbitrarily into two types: those that control the production of factors, which promote bacterial growth *in vivo*, and those that mediate resistance to antimicrobial agents, which are typically administered to infected individuals. The number of TCST loci harbored by individual pathogens varies, ranging from 30 for *Escherichia coli* [1] to 4 for *Helicobacter pylori* [2]. It has been suggested that the number of TCST systems possessed by a single pathogen reflects the lifestyle of the organism in question. *E. coli* can be found in a variety of environments and must therefore be able to adapt accordingly; in contrast, *H. pylori* leads a highly specialized lifestyle is found almost exclusively within the mucus lining of the stomach, as conditions which this organism encounters are rela-

TABLE I Two-Component Signal Transduction Systems Involved in the Regulation of Bacterial Pathogenicity

Gene pair	Organism	Function	Reference
Two-component systems implicated in bacterial virulence			
AgrA/C	*Staphylococcus aureus*	Global virulence regulator	Novick *et al.* (1995)
AlgR2/1	*Pseudomonas* spp.	Alginate capsule synthesis	Deretic *et al.* (1991)
BvgS/A	*Bordatella* spp.	Virulence factor production (hemolysin, toxin, hemagglutinin)	Akerley *et al.* (1992)
RcsB/C	*Escherichia coli*	Capsule	Jayaratne *et al.* (1993)
CheA/Y	*Helicobacter pylori*	Motility and colonization of gastric mucosa	Foynes *et al.*, 2000)
CpxR/A	*Shigella sonnei*	Regulation of virulence regulator	Nakayama *et al.* (1998)
CsrS/R	*Streptococcus pyogenes*	Multiple virulence factors	Levin and Wessels (1998)
FsrC/A	*Enterococcus faecalis*	Global virulence regulator	Qin *et al.* (2000)
GacA/S	*Pseudomonas* spp.	Exoprotease and multiple virulence factors	Tan *et al.* (1999)
EnvZ/	*Salmonella* spp.	Survival in macrophages	Lindgren *et al.* (1996)
OmpR	*Shigella* spp.	Virulence	Bernardini *et al.* (1993)
	Yersinia spp.	Virulence	Dorrell *et al.* (1998)
PhoQ/P	*Salmonella* spp.	Survival in macrophages	Miller *et al.* (1989)
	Yersinia spp.	Survival in macrophages	Oyston *et al.* (2000)
PmrB/A	*Salmonella typhimurium*	Resistance to host peptides	Kox *et al.* (2000)
ToxS/R	*Vibrio cholerae*	Toxin and hemolysin synthesis	Ottemann *et al.* (1992)
IpA/B/D	*Shigella felxneri*	Host cell invasion	Yao and Palchchaudhuri (1992)
Mga	Group A streptococci	M-protein virulence	McIver (1995)
Spv	*Salmonella*	Virulence plasmid	Grob *et al.* (1997)
SaeS/R	*S. aureus*	Exoprotein synthesis	Giraudo *et al.* (1998)
LisR/K	*Listeria monocytogenes*	Virulence	Cotter *et al.* (1999)
Two-component systems implicated in bacterial resistance to antimicrobial agents			
VanR/S	*Enterococcus faecalis*	Vancomycin resistance	Arthur *et al.* (1992)
RprX/Y	*Bacteroides fragilis*	Porin mediated tetracycline resistance	Rasmussen, *et al.* (1993)
ArlR/S	*Staphylococcus aureus*	NorA expression	Fournier *et al.* 2000)
CiaH/R	*Streptococcus pneumoniae*	Cefotaxime resistance	Giammarinaro *et al.* (1999)
PhoP/Q	*Pseudomonas aeruginosa*	Aminoglycoside resistance	Macfarlane *et al.* (2000)
LrgA/B	*Staphylococcus aureus*	Penicillin tolerance	Groicher *et al.* (2000)
VncR/S	*Streptococcus pneumoniae*	Vancomycin tolerance	Novak *et al.* (1999)

tively constant, it has no need for the complex regulatory systems harbored by its more ubiquitous cousins.

Like their nonvirulent counterparts, TCST systems associated with pathogenicity are drawn from four major families (Fig. 1); the majority of these fall

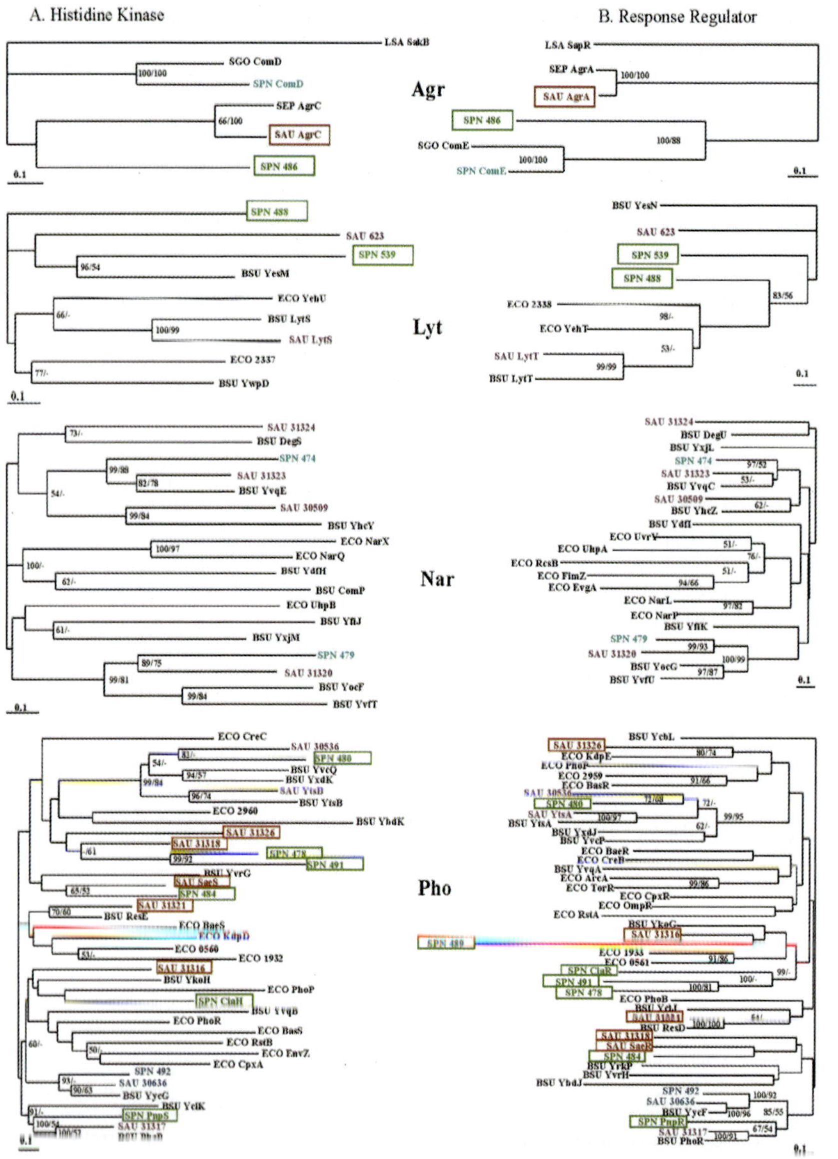

FIGURE 1 Phylogenetic trees using neighbor joining of histidine kinases and response regulators from key bacterial species, including *E. coli* (ECO), *S. aureus* (SAU), *Bacillus subtilis* (BSU), and *S. pneumoniae* (SPN). In each case, either the designated gene name or the ORF identification number is given. Those TCST systems known to be involved in the regulation of pathogenicity are boxed. Histidine kinase (HK) and response regulator (RR) proteins have been assigned to the Agr, Lyt, Nar, or Pho subfamilies on the basis of homology.

into the PhoPR subfamily presumably reflecting the wide distribution of these systems in both pathogenic and nonpathogenic bacteria.

Several so-called quorum-sensing systems have also been identified from gram-positive bacteria [3–5] and have been linked to the regulation of bacterial pathogenicity. Thus far the only example of a hybrid kinase known to control bacterial pathogenicity is the BvgAS TCST pair from *Bordetella pertussis* [2]. However, examples of hybrid kinases linked to the regulation of pathogenicity in eukaryotes are beginning to emerge [6].

Virulence Gene Regulation in *Salmonella enterica* Serovar *Typhimurium*

Of the TCST systems that have been linked to bacterial pathogenicity, the PhoPQ system of *Salmonella enterica* is arguably the best studied, as it is one of the few systems for which the stimulus (Mg^{2+}) is known. In *Salmonella* the PhoPQ system activates the expression of a series of essential virulence attributes, including entry into phagocytic cells and resistance to antimicrobial peptides, in response to magnesium deprivation [7]. It has been postulated that the low magnesium environment within macrophages activates the membrane bound histidine kinase PhoQ, which in turn phosphorylates the cognate response regulator, resulting in the activation of *pag* (PhoP-activated genes) or the repression of *prg* (PoP-repressed genes) transcription [7]. Conversely, increased levels of Mg^{2+} and Ca^{2+} can repress signaling by PhoQ [8]. Until recently, the precise mechanism by which changes in the extracellular concentration of Mg^{2+} or Ca^{2+} modulate PhoQ activity remained unknown. Studies by Castelli *et al.* [9] have revealed that the interaction of magnesium with the periplasmic domain of PhoQ promotes a conformational change in the protein that results in the activation of a hitherto unknown phosphatase activity harbored by the sensor protein. PhoQ then dephosphorylates the response regulator PhoP, thereby modulating the expression of PhoP-dependent loci. Hence activation of the phosphatase activity, as opposed to the downregulation of the autokinase activity of the PhoQ protein, is the means by which the PhoPQ TCST system is controlled (Fig. 2).

Interestingly, searches for members of the PhoPQ regulon identified a second two-component system, PmrAB, which is apparently induced during infection [10] and is required for the resistance to antimicrobial proteins of human neutrophil agents [11]. However, transcription of PmrA-activated genes can also be induced by mild acid [12] and high iron [13]. These researchers have revealed that a small protein PmrD, which is upregulated by PhoPQ, mediates the activation of the PmrAB system at the post-transcriptional level. However, transcription of PmrAB-activated genes is independent of PmrD when cells are grown in high iron. Because the action of PmrD is apparently phosphorylation dependent, Kox *et al.* [13] proposed that the activator may either promote phosphorylation of the PmrA histidine kinase and/or inhibit its dephosphorylation. Presumably, this complex cascade allows *S. enterica* to monitor multiple environmental signals and therefore accurately ascertain when conditions warrant the deployment of this organism's full pathogenic potential (Fig. 2).

Regulation of Pathogenicity in Gram-Positive Pathogens

A number of TCST systems belonging to the Pho subfamily have now been shown to regulate pathogenicity in gram-negative bacteria [14, 15]. The best studied virulence-associated TCST system from gram positive bacteria is the AgrAC system from *Staphylococcus aureus*. AgrAC belongs to an emerging class of TCST systems that respond to cell density or, more specifically, the accumulation of a small diffusible signaling peptide. In *S. aureus*, surface proteins such as adhesins and protein A are produced early in the growth

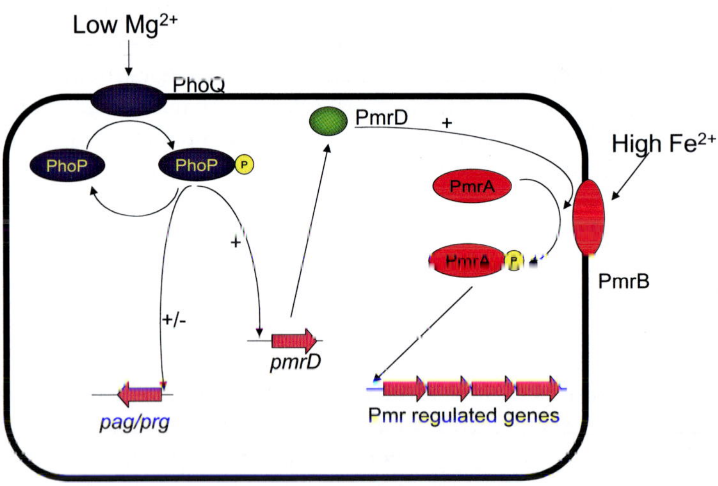

FIGURE 2 Model for the activation of PhoPQ and Pmr regulons by the PhoPQ system. Under conditions containing low magnesium, the phosphatase activity of PhoQ is repressed, resulting in accumulation of the phosphorylated response regulator PhoP. PhoP then promotes transcription of the *pmrD* gene in addition to other phoPQ-controlled loci. The PmrD protein then activates the PmrAB system at the posttranscription level.

phase and their expression is downregulated as growth continues. In contrast, most secreted proteins, such as toxins, hemolysins, proteases, and TSST-1, are expressed preferentially at higher cell densities in the postexponential phase of growth [16]. In this way, colonization and the establishment of infection are promoted at low cell densities, then as the number of invading bacteria increases extracellular proteins are released that promote tissue breakdown and the spread of infection to new sites. The cell density dependent regulation of these factors is controlled by the global regulatory locus Agr [17]. The *agr* gene cluster is divided into two divergent transcription units termed RNAII and RNAIII driven by the P2 and P3 promoters, respectively, which are active only during the midexponential phase of growth [18]. The RNAIII transcript is the effector of the *agr* locus, which serves to control the expression of over a dozen genes encoding potential virulence factors by an as yet undetermined mechanism [19]. The P2 operon contains four genes: *agrA*, *B*, *C*, and *D*. Genes *agrA* and *agrC* encode a histidine kinase and response regulator, respectively [18], which belong to the Agr subfamily of TCST systems (Fig. 1). *agrD* and *agrB* are responsible for the production of a small peptide autoinducer: a thiolactone-containing octapepetide [20]. It is thought that the peptide binds covalently to the third extracellular loop of the membrane-bound histidine kinase AgrC, which ultimately results in autophosphorylation at a conserved histidine residue [21]. This in turn leads to activation of the cognate response regulator (AgrA), which, in conjunction with a second transcription factor (SarA), serves to upregulate the transcription of both RNAIII and *agrABCD*, increasing the synthesis of the signaling peptide thereby creating a positive feedback loop (Figs. 3A and 3B) [22].

Studies have identified a second signaling molecule termed RAP (RNA activating protein) [23], which appears to upregulate RNAIII expression. An additional small molecular weight peptide termed RIP (RNA-inhibiting peptide) is capable of downregulating *agr*-induced gene expression [23]. Interestingly, both proteins are thought to mediate their effects via a 21-kDa protein called TRAP (target of RAP). Rap activates TRAP via the phosphorylation of a histidine residue within the TRAP protein [24], ultimately resulting in the upregulation of RNAIII synthesis. Conversely, both RIP and AgrD thiolactone peptides inhibit TRAP phosphorylation. Balaban *et al.* [24] proposed that the autoinduction of virulence occurs in a two-step process (Fig. 3). As cell growth continues, the autoinducer RAP accumulates and induces the phosphorylation of its target molecule TRAP, resulting in the induction of *agrABCD* expression. Once Agr is activated, the AgrD peptide accumulates quickly, leading to activation of the AgrAC signal transduction cascade and the subsequent upregulation of RNAIII synthesis. The build up of the thiolactone and RIP in turn stimulates the desphosphorylation of TRAP, limiting the signal transduction cascade. Presumably by harnessing these complex

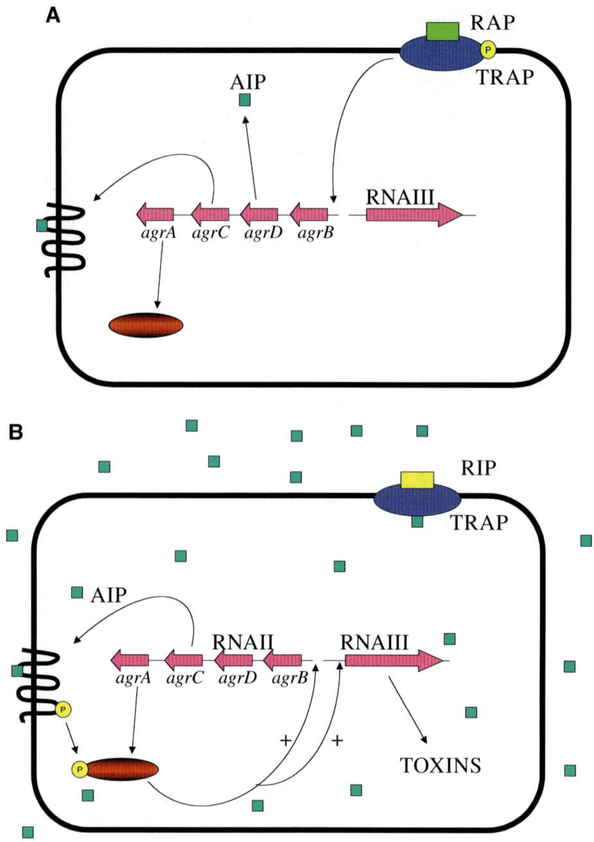

FIGURE 3 Model for the regulation of *S. aureus* virulence protein synthesis via both TRAP and Agr. As the cell density of *S. aureus* increases, the autoinducer RAP accumulates and induces the phosphorylation of its target molecule TRAP, which in turn activates the P2 promoter, resulting in the synthesis of AIP its receptor AgrC and the associated response regulator AgrA (A). The accumulation of AIP results in phosphorylation of AgrA, which then acts to upregulate the expression of the *agrA-D* operon and the regulatory effector RNAIII. Production of RNAIII in conjunction with additional regulators such as *sar* and *sae* leads to the expression of exoproteins and the suppression of surface proteins. AIP in conjunction with RIP then act to dephosphorylate TRAP, thereby limiting the Agr signal cascade (B).

signaling mechanisms *S. aureus* is able to display the appropriate response in accordance with the prevailing environmental conditions. Indeed, it is now becoming clear that the regulation of pathogenicity in *S. aureus* is dependent on several other environmental parameters, including oxygen tension, ion concentration, and osmotic pressure [25], indicating that additional regulatory determinants serve to control pathogenicity in this organism. In support of this hypothesis, our own analysis of the complete TCST gene complement of *S. aureus* has identified three additional loci, which, when inactivated, lead to a repression of pathogenicity in a murine pyelonephritis model of infection (J. P. Throup *et al.*, unpublished results). Two previously unidentified loci, 31321 and 31316, apparently play a pivotal role in the regulation of staphylococcal pathogenicity, as disruption of either locus leads to a three log attenuation of growth *in vivo*. While the function of the 31316 locus remains unexplored, analysis of the 31321 locus revealed that both the histidine kinase and the response regulator shared considerable homology with the *resDE* pair of *B. subtilis* (J. P. Throup *et al.*, unpublished results). Subsequent studies revealed that this locus is required for the growth of *S. aureus* under oxygen-limiting conditions and may be involved in the regulation of certain virulence determinants [26].

The disruption of a third TCST system termed SaeSR [27] (J. P. Throup *et al.*, unpublished results) led to a two log attenuation in virulence when tested in the pyelonephritis model. Additional studies have demonstrated that this locus is also involved in the regulation of exoprotein production in *S. aureus*, indicating that at least three individual TCST loci interact to regulate this particular group of pathogenic determinants. Interestingly, when the entire set of *S. aureus* TCST mutants was examined in a subcutaneous biofilm model, only the *saeSR* deletion exhibited any degree of attenuation (J. P. Throup *et al.*, unpublished results). Thus it would appear that different TCST loci are required for the manifestation of different clinical diseases.

A parallel analysis of the entire TCST gene complement of *S. pneumoniae* [5] identified 13 TCST systems and one additional response regulator of which eight are apparently required for infection, as the deletion of each locus led to a dramatic attenuation of growth in a mouse respiratory tract model of infection. Time course infection studies of selected TCST systems indicated that at least some of these regulators are not required throughout infection. Surprisingly, when analyzed in either the otitis media model or the bacteremia mouse model of infection (J. P. Throup *et al.*, unpublished results) [28], none of the TCST mutants examined displayed any degree of attenuation. The differing levels of attenuation may result from the fundamental differences between demands placed on bacteria in each model of infection. In the respiratory tract infection model, bacteria are introduced intranasally, where they must adapt and overcome a series of host immune responses before initiating infection. In contrast, both the bacteremia model and the otitis media model involve delivery of the bacterial inoculum directly into the site of infection, removing many of the host defenses and therefore the requirement for adaptation.

ARE HISTIDINE KINASES GOOD ANTIBACTERIAL TARGETS ?

Ideally, an antibacterial target is essential to the viability of a wide spectrum of pathogens most likely to be encountered in the clinic and community, and to avoid toxicity problems with antibiotic compounds developed using it, has no close homologues in mammals. Although there are mammalian protein sequences, which possess some of the conserved signature sequences of the histidine kinases [29], the considerable sequencing efforts of the last decade have revealed none that are convincing homologues in the vertebrates. There are few reports of individual histidine kinases being essential in pathogenic bacteria. A temperature-sensitive lethal mutation was characterized in the YycG kinase of *Staphylococcus aureus*, and it was shown subsequently that the chromosomal copy of the gene could only be disrupted in the presence of an additional wild-type copy [30]. When our laboratory carried out a comprehensive program of disruption of each of 10 putative histidine kinases, including the YycG homologue identified by genomic analysis in *S. pneumoniae* [5], none was found to be essential. A similar program disrupting 10 putative histidine kinase genes other than *yycG* in *S. aureus* gave no indication of individual essentiality (unpublished results). Mutation of all the predicted TCST systems in the other important pathogens *Haemophilus influenzae* [31] and *Helicobacter pylori* [32] did not find any histidine kinases necessary for growth *in vitro* either.

Histidine kinases share considerable identity at the amino acid [33] and structural levels [29, 34]. Consequently, it is reasonable to expect to find compounds that inhibit the enzymatic activity of several histidine kinases belonging to different TCST systems in a single bacterial species. Although no single histidine kinase appears to be essential in the majority of pathogens *in vitro*, such compounds may be antibacterial. The feasibility of this approach should be tested by the attempted genetic deletion of more than one type of histidine kinase in a bacterium, but there are no reports of this to date.

HIGH THROUGHPUT SCREENS FOR INHIBITORS OF HISTIDINE KINASES AS ANTIBACTERIALS

Several high-throughput screens (HTS) of compound libraries of pharmaceutical companies to find inhibitors of histidine kinase autophosphorylation have been carried out in the last decade, but the compounds discovered have all been unattractive to further development due to their hydrophobic nature or detergent-like properties [35]. Interestingly, although whole cell screens or assays configured with both a histidine kinase and a cognate response regulator have been used on several occasions, nearly all the TCST inhibitors found target autophosphorylation of the histidine kinase. Using a coupled *E. coli* CheA/CheY enzymatic assay to screen compound libraries for inhibitors of either autophosphorylation of CheA with ATP or subsequent phosphorylation of CheY, our experience was essentially the same. We identified 36 compounds that gave greater than 50% inhibition of activity at less than 100 μ*M* concentration of which 18 were detergents, polysulfonic acids, or polyphenolic sulfonic acids. The other hits, which were all active against CheA rather than CheY, fell into three classes (Fig. 4). There were 13 piperazinylmethyl biphenyls, e.g., SB-382553. This class had inhibition plots with high slopes indicative of cooperative detergent-like binding, a conclusion supported by high activity in a liposome lysis assay. The second class was composed of four acylated dipeptides, e.g., SB-371763, which we considered insufficiently potent for progression. The third class represented by SB-346611 gave an I_{50} of approx 8 μ*M* against both CheA/Y and EnvZ/OmpR but was not antibacterially active. Synthesis of a series of compounds related to SB-346611 did not lead to any structures of greater potency or activity against whole cells.

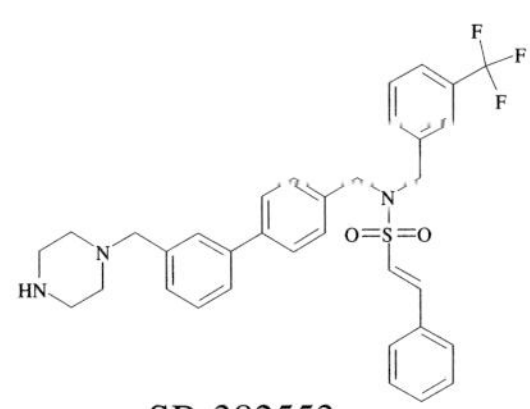

SB-382553

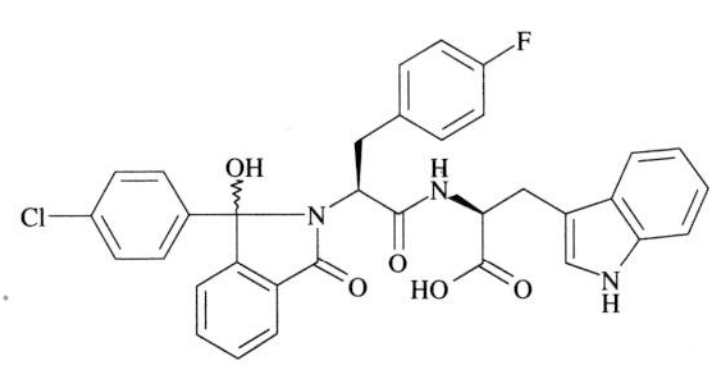

SB-371763

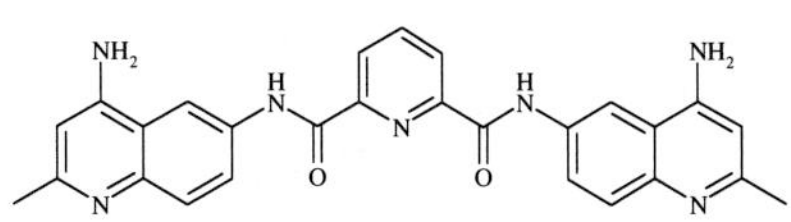

SB-346611

FIGURE 4 Representatives of the three classes of CheA inhibitors detected in a high-throughput screen of compound libraries.

ALTERNATIVES TO HIGH THROUGHPUT SCREENS: POSSIBILITIES FOR STRUCTURE-BASED SCREENING FOR IDENTIFICATION OF HISTIDINE KINASE INHIBITORS

Improvements in both screening technology and availability of recombinant protein mean that it is now possible to screen large compound libraries, and hence find new chemical starting points for lead generation. Using compounds derived from this approach is ideal from a patent viewpoint as totally distinct chemical entities may be identified and pursued. However, as was the case with histidine kinases [35], HTS technologies are not always successful in the generation of leads that are suitable for development into a novel antibiotic. An alternative to HTS is the rational design of inhibitors whereby computational methods are used to predict which compounds will bind to the active site. There are successful examples of rational design methods reviewed in the literature [36, 37] and a recent example of an inhibitor designed using mechanistic and structural data to a eukaryotic protein kinase [38], but the approaches have not yet been described in the literature for bacterial HK.

This section briefly outlines some of these novel methods, including structural based screening and so-called evolutionary approaches. Although there are differences between these methods, they can all be considered to be hybrids between HTS and rational design. We then review the possibilities and limitations of applying them to bacterial histidine kinases as target proteins.

STRUCTURE-BASED SCREENING

The first group of alternative novel methods we would like to consider is "structural-based screening". There are at least three approaches to structural-based screening (SBS) that appear to be employed currently in the pharmaceutical industry. These are X-ray crystallography and nuclear magnetic resonance (NMR) alone or in combination with, for example, surface plasmon resonance techniques for the screen. Computational approaches may be used both to design the compound libraries and to use structural data to build models of how hits from HTS are binding as part of the lead optimization process.

What do we mean by structural based screening, why should it work in situations when HTS fails, and could it be used to identify inhibitors for HK? The first stage in SBS uses a biophysical approach as the screen to identify ligands that are bound in a specific place in the three-dimensional structure of the protein (normally the active site) rather than using the inhibitory effects of a ligand in a bioassay. The second stage uses information on how the hits are orientated in the active site together with computational approaches to design more potent ligands. The information collected suggests how to link two different ligands that bind simultaneously to different pockets on the protein active site or to combine nonoverlapping features of different ligands that bind to the same site into one molecule. The principle in both is similar in terms of increasing both the entropy and the enthalpy of ligand binding and therefore increasing the binding energy.

One reason why SBS may identify novel compounds that HTS fails to identify is that theoretically the compound library may be bigger or more diverse in chemical space. Although they may contain many thousands of compounds and natural products, the libraries used in HTS that most pharmaceutical companies possess occupy areas of chemical space that are skewed to varying degrees so that they may not contain all compounds that could be made that may be leads. The "virtual libraries" employed in Fesik's SAR by NMR approach [39] are constructed by targeting independently two physically close but different sites on the protein with different small molecules. These are linked together in one molecule so that the library of potential linked molecules contains many millions of compounds as the number of linked compounds is the N^2, where N = number of compounds screened for each site. The chemical diversity is increased as any chemical entity identified for site 1 could be combined with site 2. Thus such methods could potentially identify new chemistry outside the HTS library and offer an alternative or supportive method of identification of hits not encompassed by existing HTS collections.

HK and TCST have been screened by HTS by several pharmaceutical companies. So far this method has failed to produce anything other than highly hydrophobic compounds that are unattractive for drug development. HK is involved in a high level of hydrophobic interactions, which may drive the identification of such ligands in HTS as suggested by the studies of Stephenson [40] on the mechanism of action of inhibitors of bacterial two-component signal transduction systems. In such cases, SBS may provide a viable alternative, as compounds causing inhibition by these mechanism are easily discounted. If relevant parts of the protein can be identified, SBS can target different activities of the protein simultaneously. HK has at least five activities that could potentially be targeted.

Histidine Kinases as a Target for SBS

HK are a very interesting target class to consider for structural based design. Both class 1 and class 2 HK are too large to be used as integral proteins for SBS by NMR (which so far has been applied to proteins with an upper limit of approximately 40 kDa). The class 1 family is membrane bound and shows considerable structural and functional diversity. However, the core units involved in the signal transduction pathway throughout the family are highly conserved. Several isolated domains and combinations of domains of HK have now been described [41–44], e.g., the A and B domains of EnvZ, which are subdomains of the cytoplasmic region of a typical class 1 HK. Both of these proteins are of suitable size and properties for NMR . The kinase domain of a class 2 HK CheA has also been characterized by crystallography. A structural understanding of the protein/protein interfaces that are important in the two-component signal transduction pathway involving HK can be arrived at from X-ray and NMR studies of protein complexes.

Possible Sites on the Protein Structure to Target

The mechanism of TCST reveals that there are three enzymatic activities of HK, autokinase, phosphotransferase to RR and RR-P phosphatase, that could be targeted alongside several protein/protein recognition sites (dimerization, recognition between the RR and HK proteins).

Thus HK as a family of proteins to target has many practical features that make it amenable to the novel structural-based screening methods that are being developed at the current time.

The most obvious enzymatic site for targeting is the conserved catalytic domain that binds ATP. This domain is distinct in sequence and folds from Ser, Thr, and Tyr kinases of eukaryotes [42,44]. Both X-ray and NMR data are available on the structure of the ATP-binding site. The example we describe shows how rational design, possibly combined with the SBS approach, would work on the ATP-binding domain on EnvZ (EnvZ- B domain). The kinase domain of EnvZ (residues 223–450) can be divided into domain A (residues 223–289), which contains His-243, the site of autophosphorylation, and domain B (residues 290–450), which binds ATP [41]. The B domain, which is a small monomeric protein suitable for SBS, can be overexpressed easily, in both native and isotopically labeled forms. Structurally the domain can be described as an α/β sandwich fold, consisting of five β sheets and three α helices. Also of structural significance because of the contacts it makes with AMP-PNP is the "central loop" (residues 385–409), which is structurally not well defined due to its flexibility. The three-dimensional structure of the B–domain in complex with the nonhydrolyzable analogue of ATP, AMP-PNP was determined by NMR [42]. NMR data show where AMP-PNP is localized, making contacts with αhelix3, the central loop, and further contacts to strands F and G. In this way, the ATP-binding site is inferred and is shown to consist of the very highly conserved sequences (N, D/F, and G boxes), which are found in other family members.

Structural-based screening by NMR uses changes in chemical shifts in HSQC spectra to identify where ligands are interacting with a protein. As seen in Fig. 5, almost all the amino acids in EnvZ give rise to a resonance in an HSQC spectrum. The position of a resonance is dependent on its unique environment. Figure 5 also shows the HSQC spectrum of the apo form of the B domain of EnvZ. The HSQC-based NMR approach relies on the identification of resonances from amino acids that occur in distinct separate pockets in the folded protein. Observing chemical shift changes that occur when a series of compounds is added to the protein then targets these. These chemical shifts occur as the environment around the amino acid giving rise to the NMR signal is changed by the close proximity of a ligand on binding.

Figure 5 shows the dramatic changes in the spectrum on binding of the ATP analog AMP-PNP. Resonances can be divided into roughly three groups: those that show little or no change, those that show medium chemical shift changes (less than 0.2 ppm), and those that show large chemical shift changes and broadening. These later changes imply binding in this area of the protein. For example, as shown in Fig. 6 using residues A379, G375, and A348, which show large changes, these residues are in close proximity to the ligand in the NMR structure of the AMP-PMP–Mg^{2+} complex. It is possible to

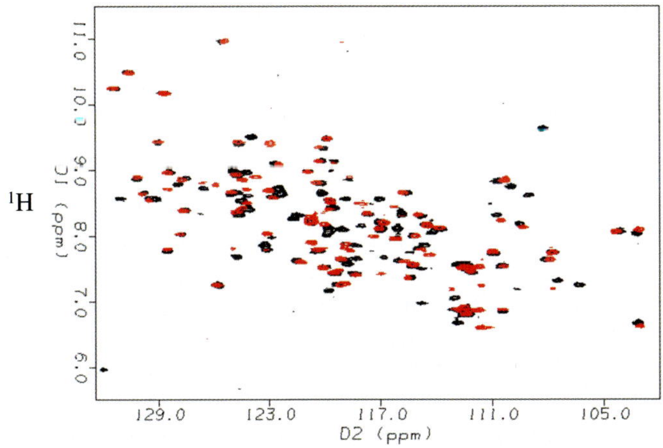

FIGURE 5 Overlay of ^{15}N-edited HSQC spectra of apo-EnvZ(B):(290–450) (black) and EnvZ(B):(290–450)/AMP-PNP/Mg^{2+} (red). Sample conditions: 1.0–1.5 m*M* in 20 m*M* sodium phosphate, 50 m*M* KCl, 0.5 m*M* 4-(2-aminoethyl)-benzenesulfonyl fluoride, 50 μ*M* sodium azide, pH 7.0, with or without 5 m*M* $MgCl_2$ and 5 m*M* AMP-PNP.

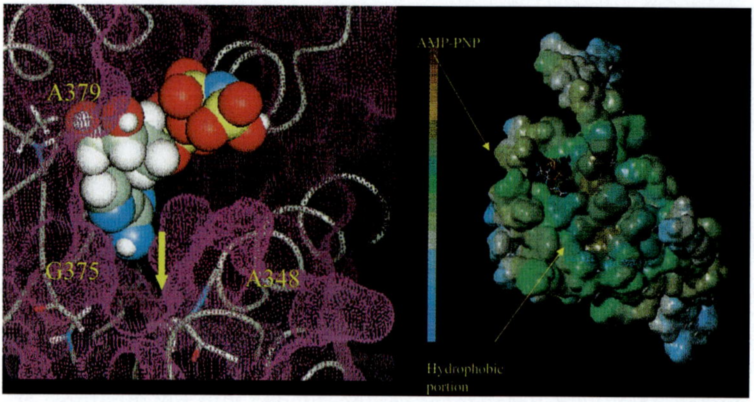

FIGURE 6 Models of EnvZ(B):(290–450) with bound AMP-PNP showing (left) examples of amino acids that exhibit large chemical shift changes on ligand binding. These are potentially good resonances to monitor in identifying ligands binding in the active site in SBS. The right-hand side also highlights a potential remote hydrophobic binding site (see text for details).

interpret these spectral changes in terms of different but very close in space sites that could be targeted separately by different molecules to obtain novel ATP competitive ligands. This was done using either smaller analog of AMP-PNP, e.g., adenine or adenosine, or observing further changes when adding Mg^{2+} to the AMP-PNP complex of the EnvZ B domain.

As shown in Fig. 7, when Mg^{2+} is added to the AMP-PNP–EnvZ(B) complex, a subset of changes in the spectrum is seen relative to those observed when AMP-PNP/Mg^{2+} is added simultaneously. These changes occur in resonances from closely localized residues (in the three-dimensional structure), and do not occur when the adenine pocket is occupied by the presence either of adenine only or of AMP-PNP without Mg^{2+}. These changes thus identify a second binding site responsible for the PO_4–Mg^{2+} interaction of the ATP.

By comparison of chemical shift changes of these ligands together with knowledge of the protein structure, molecules that bind specifically to the active site can be determined. For example, SKF-26413 showed significant chemical shift changes (Fig. 8) similar to AMP-PNP, and apparently binds to the same site.

Not all chemical shift changes were as straightforward to interpret, however. Chemical shift changes at V304 and L305 are observed for the ligand SKF-26413 and the AMP-PNP analogues (Fig. 8). Because these residues lie at a small hydrophobic cavity, a second binding site could be a possibility. It was possible to use NMR to determine the K_d for the interactions of these molecules. Calculations of the K_d for these ligands using residues at the active

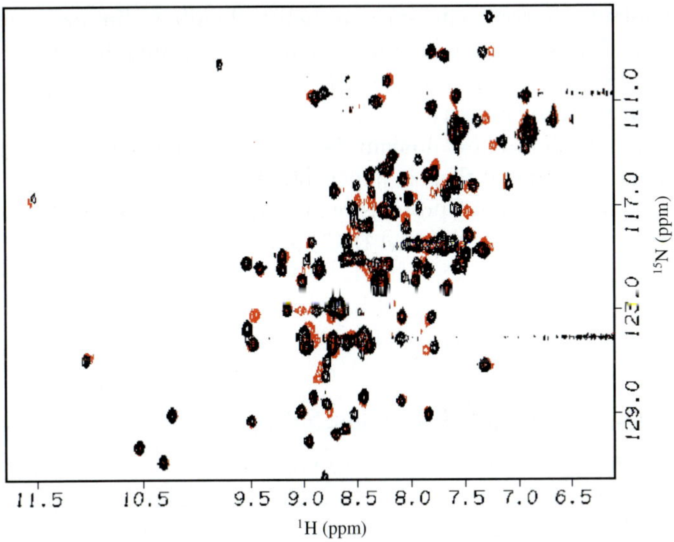

FIGURE 7 Overlay of ^{15}N-edited HSQC spectra of EnvZ(B):(290–450)/AMP-PNP (black) and EnvZ(B):(290–450)/AMP-PNP/Mg^{2+} (red) showing the effects of addition of $MgCl_2$.

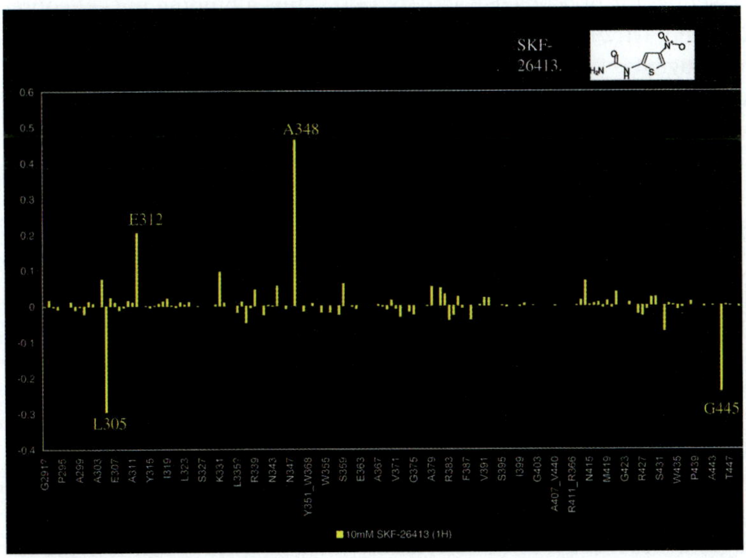

FIGURE 8 ^{1}H chemical shift changes for EnvZ(B):(290–450) on binding of SKF-26413.

site and putative second site showed similar binding affinities, suggesting only one binding site. A possible role would be a communication role. Thus changes in the active site could be relayed to signal the binding of ATP to domain B through to domain A.

This example shows how domain dissection in the HK EnvZ provides the ATP-binding B domain that is amenable to structural based screening. However, a large number of proteins bind ATP. The ATP site of HK resembles two ATPases very closely, bacterial DNA gyrase B and Hsp90, indicating that these proteins belong to a superfamily of ATPases unlike other kinases. Thus it remains to be seen if structural-based screening will introduce sufficient selectivity against other proteins kinases.

Targeting Other Activities of HK Using SBS

The autokinase activity of HK could be targeted by combining residues from the ATP-binding site together with the region around the invariant phosphorylated histidine. A monomeric histidine kinase containing two A domains and one B domain (mass 33 kDa) has been derived from EnvZ [41] that could be amenable to SBS. Each domain type would contribute one binding site rather than both binding sites arising from one domain.

Many HK act as phosphatases to RR. The cell may regulate RR activity by modulating these opposing roles of the HK. Residue Thr 247 of EnvZ domain A is essential for phosphatase activity [45]. Thus a binding pocket containing His243 and Thr247 may make domain A alone a target for developing antibacterial agents.

Alternatively, protein recognition regions could be targeted, including the dimerization and response regulator interface. It is interesting to note that most HK inhibitors discovered through HTS act at the dimer interface [35]. Another possibly more tractable site is the recognition interface between the phosphotransfer domain of HK and the downstream RR. Here a relatively small region of approximately nine highly conserved residues appears to be the key to binding throughout the family [46].

Significant mobility exists in the ATP binding site of HK, and conformational changes appear to occur on ATP binding. Thus the structure of the apo protein (yet to be determined)may be required for drug design. Mobility may also make it difficult to determine the interactions critical in autophosphorylation. This information would be required to target simultaneously the A and B domains. SBS methods appear to have been applied successfully to single members of a protein family. Commercially viable HK inhibitors would need to target a wide range of diverse family members. Although this problem is shared by any method of drug development, this adds extra challenges in the development of antibiotics.

Evolutionary Approaches

An alternative method for antibiotic development that has been successful in the development of many drugs, including antiinfectives, is the so-called evolutionary approach. Compounds with known interactions with the target protein (and possibly biological activity) are improved, e.g., certain side effects are removed. Starting compounds may be either existing drugs (me-too drugs) or natural products. A structural version of this approach that has been used both in our laboratories and the Abbott Laboratories [47] involves obtaining fragments of known drugs and determining the mode of binding using HSQC-based NMR. Alternatives to parts of the drug molecule that one wishes to modify could then be found before reconstruction of the full lead compound.

There is close similarity in the global fold of the catalytic domain of HK to the ATPase domain of DNA gyrase and heat shock protein 90 (HSP90) [48] for which there are known potent inhibitors and structural information on how these inhibitors bind. We investigated the possibility of binding to HK of two of these inhibitors, novobiocin and geldamycin. We were able to demonstrate the inhibition of TCST in our in-house assay with novobiocin, an inhibitor of DNA gyrase but not geldamycin, due to compound interference in the assay. Geldamycin was also without effects on HSQC spectra of the B domain of EnvZ. NMR was used to probe the mechanism of binding of novobiocin using the complete molecule and fragments of novobiocin representing either the coumarin or sugar moieties. The fragments alone had very little effects on the NMR spectrum, whereas novobiocin caused line broadening consistent with protein aggregation. Structural studies on GyrB have shown that the sugar ring overlaps the binding site for the adenine ring of ATP, obscuring substrate binding [49]. Thus it appears that binding of novobiocin in EnvZ is different enough to that in DNA gyrase so that ATP binding is unaffected. The effects of novobiocin are similar to most of the inhibitors identified by HTS in they are predominantly due to nonspecific hydrophobic interactions rather than substrate competition.

Other Approaches

A number of peptides have been discovered using a phage-display combinatorial peptide library that inhibit the response regulator protein VanR. Further studies on these identified 12-mer peptides that are thought to act as minimalist analog of kinase in the phosphotransfer site [50]. Roychoudhury *et al.* [51] have also found a series of hexapeptides that are inhibitors of CheA autophosphorylation. Structure and rational design approaches have been described to try to make nonpeptide, smaller molecules for drug leads from peptides. Nevertheless, the design of peptidomimetic ligands with biological activities *in vivo* is still a major challenge.

Future of These Approaches and HTS for Inhibitors

There are clearly a number of relatively unexplored avenues for finding inhibitors of HK using SBS and rational design approaches. These approaches may have advantages with hydrophobic proteins such as HK, as most compounds found through HTS seem to act as nonspecific hydrophobic agents changing protein aggregation states [35]. Developments to improve the efficiency and speed of structural methods for drug design are likely to come by further combination with computational approaches [52], e.g., via *in silico* screening prior to the structural screens. Another "virtual NMR screening" approach that could be integrated with SBS has been described by Marrone *et al.* [53].

One aspect of SBS not investigated yet is that the screen is against a relatively static conformation of the protein, and potential inhibitors do not have access to the many conformations it may adopt during turnover. So far this is not documented to cause serious consequences in the application of the methods we have described. In rational design approaches, conformational flexibility is regarded as an added difficulty. As shown by both NMR and X-ray studies, HK has significant flexibility in the ATP-binding domain. Methods for including dynamic fluctuations of a protein in rational-design approaches are being developed [54] and could be particularly relevant to HK.

REFERENCES

1. Mizuno, T., (1997). Compilation of all genes encoding two-component phosphotransfer signal transducers in the genome of *Escherichia coli*. *DNA Res.* **4**, 161–168.
2. Beier, D., Schwarz, B., Fuchs, T. M. and Gross, R. (1995). *In vivo* characterization of the unorthodox BvgS two-component sensor protein of *Bordetella pertussis*. *J. Mol. Biol.* **248**, 596–610.
3. Ji, G., Beavis, R. C., and Novick, R. P. (1995). Cell density control of staphylococcal virulence mediated by an octapeptide pheromone. *Proc. Natl. Acad. Sci. USA* **92**, 12055–12059.
4. Van Wamel, W. J., van Rossum, G., Verhoef, J., Vandenbroucke-Grauls, C. M., and Fluit, A. C. (1998). Cloning and characterization of an accessory gene regulator (*agr*)-like locus from *Staphylococcus epidermidis*. *FEMS Microbiol. Lett.* **163**, 1–9.
5. Throup, J. P., Koretke, K. K., Bryant, A. P., Ingraham, K. A., Chalker, A. F., Ge, Y, Marra, A., Wallis, N. G., Brown, J. R., Holmes, D. J., Rosenberg, M., and Burnham, M.K.R. (2000). A genomic analysis of two-component signal transduction in *Streptococcus pneumoniae*. *Mol.*

Microbiol. **35**, 566–576.

6. Calera, J. A., and Calderone, R. (1999). Histidine kinase, two-component signal transduction proteins of *Candida albicans* and the pathogenesis of candidosis. *Mycoses* **42**, s49–s53.
7. Groisman, E. A. (1998). The ins and outs of virulence gene expression: Mg^{2+} as a regulatory signal. *Bioessays* **20**, 96–101.
8. Garcia Vescovi, E., Soncini, F. C, and Groisman, E. A. Mg^{2+} as an extracellular signal: Environmental regulation of *Salmonella* virulence. *Cell* **84**, 165–74.
9. Castelli, M. E., Garcia Vescovi, E., and Soncini, F. C. (2000). The phosphatase activity is the target for Mg^{2+} regulation of the sensor protein PhoQ in *Salmonella*. *J. Biol. Chem.* **275**, 22948–22954
10. Heithoff, D. M., Conner, C. P., Hanna, P. C., Julio, S. M., Hentschel, U., and Mahan, M. J. (1997). Bacterial infection as assessed by *in vivo* gene expression. *Proc. Natl. Acad. Sci. USA* **94**, 934–339.
11. Roland, K. L., Martin, L. E., Esther, C. R., and Spitznagel, J. K. (1993). Spontaneous *pmrA* mutants of *Salmonella typhimurium* LT2 define a new two-component regulatory system with a possible role in virulence. *J. Bacteriol.* **175**, 4154–4164.
12. Soncini, F. C., and Groisman, E. A. (1996). Two-component regulatory systems can interact to process multiple environmental signals. *J. Bacteriol.* **178**, 6796– 6801.
13. Kox, L. F. F., Wosten, M. M. S. M., and Groisman, E. A. (2000). A small protein that mediates the activation of two-component system by another two-component system *EMBO J.* **19**, 1861–1872.
14. Bernardini, M. L., Sanna, M. G., Fontaine, A., and Sansonetti, P. J. (1993). OmpC is involved in invasion of epithelial cells by *Shigella flexneri*. *Infect. Immun.* **61**, 3625–3635.
15. Dorrell, N., Li, S. R., Everest, P. H., Dougan, G., and Wren, B. W. (1998). Construction and characterisation of a *Yersinia enterocolitica* O:8 *ompR* mutant. *FEMS Microbiol. Lett.* **165**, 145–151.
16. Novick, R. P., and Muir, T. W. (1999). Virulence gene regulation by peptides in staphylococci and other Gram-positive bacteria. *Curr. Opin. Microbiol.* **2**, 40–45.
17. Recsei, P., Kreiswirth, B., O'Reilly, M., Schlievert, P., Gruss, A., and Novick, R. P. (1986). Regulation of exoprotein gene expression in *Staphylococcus aureus* by agr. *Mol. Gen. Genet.* **202**, 58–61.
18. Novick, R. P., Projan, S. J., Kornblum, J., Ross, H. F., Ji, G., Kreiswirth, B., Vandenesch, F., and Moghazeh, S. (1995). The *agr* P2 operon: An autocatalytic sensory transduction system in *Staphylococcus aureus*. *Mol. Gen. Genet.* **248**, 446–458.
19. Novick, R. P., Ross, H. F., Projan, S. J., Kornblum, J., Kreiswirth, B., and Moghazeh, S. (1993). Synthesis of staphylococcal virulence factors is controlled by a regulatory RNA molecule. *EMBO J.* **12**, 3967–3975.
20. Mayville, P., Ji, G., Beavis, R., Yang, H., Goger, M., Novick, R. P, and Muir, T. W. Structure-activity analysis of synthetic autoinducing thiolactone peptides from *Staphylococcus aureus* responsible for virulence. *Proc. Natl. Acad. Sci. USA* **96**, 1218–1223.
21. Lina, G., Jarraud, S., Ji, G,, Greenland, T., Pedraza, A., Etienne, J., Novick, R. P., and Vandenesch, F. (1998). Transmembrane topology and histidine protein kinase activity of AgrC, the *agr* signal receptor in *Staphylococcus aureus*. *Mol. Microbiol.* **28**, 655–662.
22. Cheung, A. L, Koomey, J. M., Butler, C. A., Projan, S. J., and Fischetti, V. A. (1992). Regulation of exoprotein expression in *Staphylococcus aureus* by a locus (*sar*) distinct from *agr*. *Proc. Natl. Acad. Sci. USA* **89**, 6462–6466.
23. Balaban, N., and Novick, R. P. (1995). Autocrine regulation of toxin synthesis by *Staphylococcus aureus*. *Proc. Natl. Acad. Sci. USA* **92**, 1619–1623.
24. Balaban, N., Goldkorn, T., Gov, Y., Hirshberg, M., Koyfman, N., Matthews, H. R., Nhan, R. T., Singh, B., and Uziel, O. (2000). Regulation of *S. aureus* pathogenesis via TRAP. *J. Biol. Chem.*

25. Chan, P. F., and Foster, S. J. (1998). The role of environmental factors in the regulation of virulence-determinant expression in *Staphylococcus aureus* 8325–4. *Microbiology* **144**, 2469–2479.
26. Yarwood, J. M. and Schlievert, P. M. (2000). Identification and characterization of a two component system with a role in anaerobic respiration of *Staphylococcus aureus*. Abstracts of The American Society of General Microbiology 100th General Meeting Los Angeles, May 21–25, 2000.
27. Giraudo, A. T, Calzolari, A., Cataldi, A. A., Bogni, C., and Nagel, R. (1999). The *sae* locus of *Staphylococcus aureus* encodes a two-component regulatory system. *FEMS Microbiol. Lett.* **177**, 15–22.
28. Lange, R., Wagner, C., de Saizieu, A., Flint, N., Molnos, J., Stieger, M., Caspers, P., Kamber, M., Keck, W., and Amrein, K. E. (1999). Domain organization and molecular characterization of 13 two-component systems identified by genome sequencing of *Streptococcus pneumoniae*. *Gene* **237**, 223–234.
29. Grebe, T. W., and Stock, J. B. (1999). The histidine protein kinase superfamily. *Adv. Microb. Physiol.* **41**, 139–227.
30. Martin, P. K., Li, T., Sun, D., Biek, D. P., and Schmid, M. B. (1999). Role in cell permeability of an essential two-component system in *Staphylococcus aureus*. *J. Bacteriol.* **181**, 3666–3673.
31. Gwinn, M.L., Yi, D., Smith, H.O., and Tomb, J.F. (1996). Role of the two-component signal transduction and the phosphoenolpyruvate: Carbohydrate phosphotransferase systems in competence development of *Haemophilus influenzae* Rd. *J. Bacteriol.* **178**, 6366–6368.
32. Beier, D., and Frank, R. (2000). Molecular characterization of two-component systems of *Helicobacter pylori*. *J. Bacteriol.* **182**, 2068–2076.
33. Hoch, J. A., and Silhavy, T. J. (1995). "Two-Component Signal Transduction." ASM Press, Washington, DC.
34. Dutta, R., Qin, L., and Inouye, M. (1999). Histidine kinases: Diversity of domain organization. *Mol. Microbiol.* **34**, 633–640.
35. Macielag, M. J., and Goldschmidt, R. (2000). Inhibitors of bacterial two-component signalling systems. *Expert Opin Invest. Drugs* **9**, 2351–2369.
36. Lipkowitz, K.B., and Boyd, D.B. (eds). (1997). "Reviews in Computational Chemistry," Vol. 11. Wiley-VCH, New York.
37. Van de Waterbeemd, H., Testa, B., and Folkers, G. (eds) (1997). "Computer-Assisted Lead Finding and Optimization." Wiley-VCH, Basel, Switzerland.
38. Parang, K., Till, J. H., Ablooglu A. J., Kohanski, R. A, Hubbard, S. R., and Cole, P. A. (2001). Mechanism-based design of a protein kinase inhibitor. *Nature Struct. Biol.* **8**, 37–41.
39. Shuker, S. B., Hajduk, P. J., Meadows, R. P., and Fesik, S. W. (1996). Discovering high-affinity ligands for proteins: SAR by NMR. *Science* **274**, 1531–1534.
40. Stephenson, K., Yamaguchi, Y., and Hoch, J.A. (2000). The mechanism of action of inhibitors of bacterial two-component signal transduction systems." *J. Biol. Chem.* **275**, 38900–38904.
41. Qin, L., Dutta, R., Kurokawa, H., Ikura, M., and Inouye M. (2000). A monomeric histidine kinase derived from EnvZ, an *Escherichia coli* osmosensor. *Mol. Microbiol.* **36**, 24–32.
42. Tanaka, T., Saha, S. K., Tomomori, C., Ishima, R., Liu, D., Tong, K. I., Park, H., Dutta, R., Qin, L., Swindells, M. B., Yamazaki, T., Ono, A. M., Inouye, M., and Ikura, M. (1999). NMR structure of the histidine kinase domain of the *E. coli* osmosensor EnvZ. *Nature* **396**, 88–92.
43. Tomomori, C., Tanaka, T., Dutta, R., Park, H. Y., Saha, S. K., Zhu, Y., Ishima, R., Liu, D. J., Tong, K. I., Kurokawa, H., Qian. H., Inouye, M., and Ikura, M.(1999). Solution structure of the homodimeric core domain of *Escherichia coli* histidine kinase EnvZ. *Nature Struct. Biol.* **6**, 729–734.
44. Bilwes, A. M., Alex, L. A., Crane, B. R., Melvin I, and Simon, M. I. (1999). Structure of CheA, a signal-transducing histidine kinase. *Cell* **96**, 131–141.

45. Dutta, R., Yoshida, T., and Inouye, M. (2000). The critical role of the conserved Thr^{247} residue in the functioning of the osmosensor EnvZ, a histidine kinase/phosphatase, in *Escherichia coli*. *J. Biol. Chem.* **275**, 38645–38653.
46. Zapf, J., Sen, U., Madhusudan, Hoch, J. A., and Varughese, K. I. (2000). A transient interaction between two phosphorelay proteins trapped in a crystal lattice reveals the mechanism of molecular recognition and phosphotransfer in signal transduction. *Structure Fold. Des.* **8**, 851–862.
47. Hajduk, P. J., Gomtsyan, A., Didomenico, S., Cowart, M., Bayburt, E.K.., Solomon, L., Severin, J., Smith R., Walter, K., Holzman, T. F., Stewart, A., McGaraughty, S., Jarvis, M. F., Kowaluk, E. A. and Fesik, S. W. (2000). Design of adenosine kinase inhibitors from the NMR-based screening of fragments. *J.Med.Chem.* **43**, 4781–4786.
48. Robinsom V. L., Buckler, D. R., and Stock, A. M. (2000). A tale of two components: A novel kinase and a regulatory switch. *Nature Struct. Biol.* **70**, 626–633.
49. Heddle, J. G., Barnard, F. M., Wentzell, L. M., and Maxwell, A. (2000). The interactions of drugs with DNA gyrase: A model for the molecular basis of quinolone action. *Nucleosides Nucleotides Nucleic Acids* **19**, 1249–1264.
50. Ulijasz, A. T., Kay, B. K., and Weisblum, B. (2000). Peptide analogues of the VanS catalytic center inhibit VanR binding to its cognate promotor. *Biochemistry* **39**, 11417–11424.
51. Roychoudhury, S., Blondelle, S. E., Collins, S. M., Davis, M. C., McKeever, H. D., Houghten, R. A., and Parker., C. N. (1998). Use of combinatorial library screening to identify inhibitors of a bacterial two-component signal transduction kinase. *Mol. Diversity* **4**, 173–182.
52. Kobinyi, H. (1998). Combinatorial and computational approaches in structure-based drug design. *Curr. Opin. Drug Disc. Dev.* **1**, 16–27.
53. Marrone, T. J., Luty, B. A, and Rose, P. W. (2000). Discovering high-affinity ligands from the computationaly predicted structures and affinities of small molecules bound to a target: A virtual screening approach. *Perspect. Drug Disc. Design* **20**, 209–230.
54. Carlson, H. A., Masukawa, K. M., and McCammon, J. A. (1999). Method for including the dynamic fluctuations of a protein in computer-aided drug design. *J. Phys. Chem. A* **103**, 10213–10219.

CHAPTER 23

Molecular Evolution of Histidine Kinases

KRISTIN K. KORETKE, CRAIG VOLKER, MICHAEL L. BOWER, AND ANDREI N. LUPAS

GlaxoSmithKline, Collegeville, Pennsylvania 19426

Two-component signal transduction (TCST) systems are the principal means for coordinating responses to environmental changes in bacteria. Phylogenetic analyses show that (i) these systems probably originated in bacteria and radiated into archaea and eukaryotes by lateral gene transfer and that (ii) proteins forming these systems have coevolved extensively, with only occasional recruitment of components from other TCST systems. Three kinase superclusters, containing 14 clades can currently be defined from phylogeny and sequence patterns. Some of these clades are largely species specific, suggesting extensive differentiation of TCST systems during speciation. Three of the clades contain (or are formed exclusively of) serine kinases, which apparently originated independently at different times in the evolution of histidine kinases. One clade, containing mitochondrial pyruvate dehydrogenase kinases, is found consistently at the root of the histidine kinase tree and may represent the ancestral state of the system. The observation that the receiver domain of response regulators is related to a superfamily of phosphatases suggests a model for the evolution of TCST in which the original function of the response regulator was that of a posttranslational antagonist to a protokinase with serine-phosphorylating activity. © 2003, Elsevier Science (USA).

INTRODUCTION

As described elsewhere in this book, two-component signal transduction (TCST) systems form the central signalling machinery in bacteria. To a lesser extent, they are found in plants, fungi, slime molds, and some archaea. However, they have not been detected in any metazoan so far. They are named for their two main components, histidine kinase and response regulator, which transduce a sensory input (typically extracellular) into a cellular response (typically transcriptional regulation). Transduction of the signal involves transfer of a high-energy phosphate group from ATP via a histidine residue in the kinase to an aspartate residue in the receiver domain of the response regulator. Although most systems use a linear phosphorelay from one kinase to one response regulator, some use more complicated paths, involving a branching of the signal (chemotaxis) or multiple phosphorylated components (sporulation, adaptation to anaerobic conditions) (Fig. 1A).

Histidine kinases are an extremely diverse group of proteins, whose common element is the kinase domain. Their name stems from the ability of the kinase domain to transfer the γ-phosphate group of ATP to a histidine residue in a heterologous domain. They are closely related to several other groups of kinases, notably bacterial anti-sigma F factors, plant phytochromes, and two mitochondrial enzymes — pyruvate dehydrogenase kinase and branched chain α-ketoacid dehydrogenase kinase (which will be referred to jointly as PDKs). These have serine phosphorylating activity and do not require either a histidine-carrying domain or a response regulator for their function.

This chapter addresses the domain composition of histidine kinases, the evolutionary origin of the kinase domain, the diversification of TCST systems, and the relationship of histidine kinases to their serine-phosphorylating relatives.

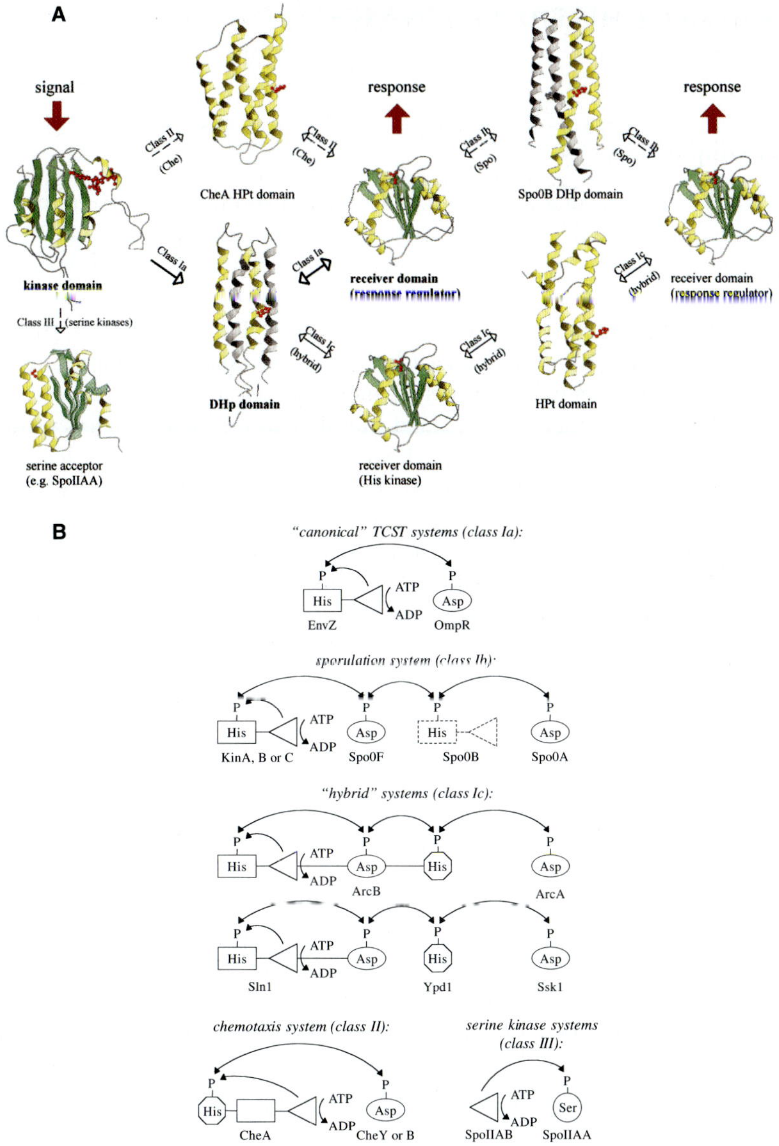

FIGURE 1 Phosphorelay paths in TCST systems and related kinases. The classification represents an extension of the one made by Bilwes *et al.* [20] and is described in the text. (See also color plate.) (A) Structure gallery of domains involved in phosphate transfer. The bound nucleotide and phosphate acceptor residues are shown in red. (B) Schematic overview showing the order in domain arrangements underlying the diversity of histidine kinases. The kinase domain is shown as a triangle, the DHp domain as a rectangle, the HPt domain as an octagon, the receiver domain as an oval, and serine acceptors as a circle.

DOMAINS OF HISTIDINE KINASES

Overview

Sensory Input Domains

Histidine kinases are modular proteins formed of multiple, diverse domains. The sometimes bewildering diversity hides an underlying regularity in domain arrangement (Fig. 1B). The N-terminal part typically contains one or more sensory domains, which often occur in conjunction with transmembrane segments, as most kinases act as receptors for extracellular signals. Although some sensory domains clearly belong to conserved families, found in a wide range of signaling proteins, most appear to be functionally and structurally unique. Indeed, as a rule of thumb, extracellular domains usually appear unique, whereas the recognizably conserved domains are intracellular. Of course, many extracellular domains of histidine kinases may prove to be related to each other or to other protein families once their structures have been determined. However, in that case, their sequences have diverged beyond our current ability to establish relatedness.

The most frequent sensory domains encountered in histidine kinases are PAS (or a C-terminally extended version, PAS/PAC) [1, 2] and GAF [3, 4]. These two domains belong to the same structural fold, termed "profilin-like" in the SCOP database (http://scop.mrc-lmb.cam.ac.uk/scop) [5], and may have common ancestry. They are both versatile ligand-binding domains involved in transducing light, redox, cyclic, nucleotide, and possibly other signals, depending on the bound ligand. About one-third of all histidine kinases listed in the SMART database (http://smart.embl-heidelberg.de/) [6] contain PAS and about one-fifth contain GAF; of the latter, about one-half contain both GAF and PAS. Other conserved sensory domains include Cache [7], cyclic monophosphate nucleotide binding domain (PBPb) [8], Forkhead associated domain (FHA) [9] (all intracellular); and periplasmic binding protein domain (PBPb) [10] and PDZ [11] (both extracellular). Each is found in only a small number of kinases. Based on the spectrum of sensory domains in proteins observed today, it would seem that none was ancestrally associated with TCST or invented within its context and that all were coopted from other systems.

Core Signaling Pathway

Following the sensory input domains, many kinases contain a small conserved region, termed the 'linker region' [12, 13] or HAMP domain [14], sometimes in multiple, sequential copies. Its function is not yet understood, but mutant phenotypes show that it plays a critical role in signal transduction. C-terminal to the sensory and linker domains are found the two domains required for phosphotransfer: the H box or DHp (dimerization and histidine phosphotransfer) domain [15] followed by the kinase domain. These two domains are often referred to jointly as the "transmitter" [16]. Many histidine kinases end with the transmitter; one class, however, contains a C-terminal "receiver" domain (homologous to that of response regulators), generally followed by a second type of His-acceptor domain called HPt (histidine phosphotransfer). These domains extend the phosphorelay chain and provide additional regulatory input. A variant of the extended phosphorelay chain, which substitutes a transmitter for the HPt domain, is found in the sporulation system and may also occur in other systems, such as TodST [17] and PhcSRQ [18]. In the sporulation system, the second transmitter (Spo0B) is catalytically inactive as a kinase and highly degenerate, but recognizable as a transmitter from its structure [19].

A variant of this domain organization is found in the chemotaxis kinase CheA. As seen from its structure [20], it contains a transmitter domain consisting of consecutive DHp and kinase domains; however, the DHp domain lacks the catalytic histidine and is used exclusively for dimerization. Transfer of the ATP γ-phosphate is made instead to an N-terminal HPt domain [21], which is used for phosphorelay to CheY and CheB.

Kinase Classification

The position in the sequence of the histidine acceptor domain relative to the kinase domain has led Bilwes *et al.* [20] to divide histidine kinases into two classes: class I containing all kinases except CheA and its homologues, which are grouped separately as class II. The same division can be obtained using the structural nature of the histidine acceptor domain as a classification criterium (namely whether the initial phosphoryl transfer is made to a DHp or to an HPt domain). This seems to us a more robust criterium. We further propose that class I is usefully divided into three subclasses based on the further path of phosphate transfer (Fig. 1B). Histidine kinases that transfer directly to a response regulator form subclass a; this class contains most TCST systems. Kinases that transfer via a receiver and a second transmitter form subclass b; currently, this class consists primarily of the sporulation system, in which the second transmitter is degenerate. Finally, kinases that transfer via a receiver and an HPt domain form subclass c; this class contains most eukaryotic systems and a large number of the prokaryotic 'hybrid' kinases. At this point, phosphoryl transfer in class II appears to follow a path equivalent to that in class Ia. However, a substantial subset of CheA-like kinases have a C-terminal receiver domain (*Myxococcus* FrzE; *Helicobacter* CheF; several *Synechocystis* CheA homologues) and may use an extended phosphorelay chain equivalent to subclasses Ib or Ic. Finally, serine-phosphorylating kinases transfer directly to a serine residue in an acceptor protein; we propose to group these in class III.

Other Domains

In addition to the sensory and phosphate-transfer domains just listed, histidine kinases may contain several other conserved domains, all with a very limited distribution. CheA and related chemotaxis kinases usually contain a C-terminal domain homologous to CheW [20]; this domain can be used as a marker domain for the chemotaxis kinases, as it is not found elsewhere in the histidine kinase family. Also, a small number of kinases (almost all eukaryotic) contain an N-terminal Ser/Thr/Tyr kinase domain or a C-terminal adenylate cyclase domain. The large number of sequences becoming available from genome studies will undoubtedly uncover further histidine kinases with unusual domains.

The following sections discuss the domains involved in the core signaling pathway.

The Linker (HAMP) Domain

The "linker region" or HAMP domain is conserved broadly in histidine kinases as well as in chemoreceptors, bacterial nucleotidyl cyclases, and phosphatases [13–15]. In the SMART database, about a quarter of the histidine kinases contain the HAMP domain and about half of all HAMP domains are found in histidine kinases. Of course, as in all such statements based on sequence similarity detection, it should be kept in mind that many instances of a domain may go unrecognized due to sequence divergence. In the case of the HAMP domain, such a situation appears particularly likely, as the domain is very short (about 45 residues), does not contain invariant residues, and appears to be formed of two amphipathic helices, connected by a long loop [13–15, 22]. Amphipathic helices are repetitive structures whose properties are dominated by the pattern of hydrophilic and hydrophobic residues, resulting in the rapid divergence of homologous sequences and the rapid convergence of heterologous sequences to a sequence identity of about 20%. Some kinases contain multiple copies of the HAMP domain in tandem, with 12 being the largest number identified (in *Dictyostelium* DhkJ), implying that it represents an autonomously folding unit. The length of the loop relative to the helices and the predicted coiled-coil nature of the structure [23] suggest that the HAMP domain forms a parallel four-helical coiled coil, resembling the helix-loop-helix domains of eukaryotic transcription regulators (Fig. 2). Although this is different from the structures proposed so far [14, 22], it is best compatible with the observed sequence properties of HAMP domains.

The HAMP domain typically occurs at the C-terminal end of the last transmembrane segment in receptor-type histidine kinases and chemoreceptors. Mutation and cysteine disulfide scanning studies show that it plays a critical role in signal transduction and that it may be involved in transmembrane localization, but that it does not affect dimerization [13, 22]. The HAMP domain is also found in kinases lacking transmembrane regions, suggesting that it can exert its function in a soluble context. Current models for HAMP activity interpret experimental results in terms of conformational changes propagated between the periplasmic domain and the transmitter [13, 14, 22].

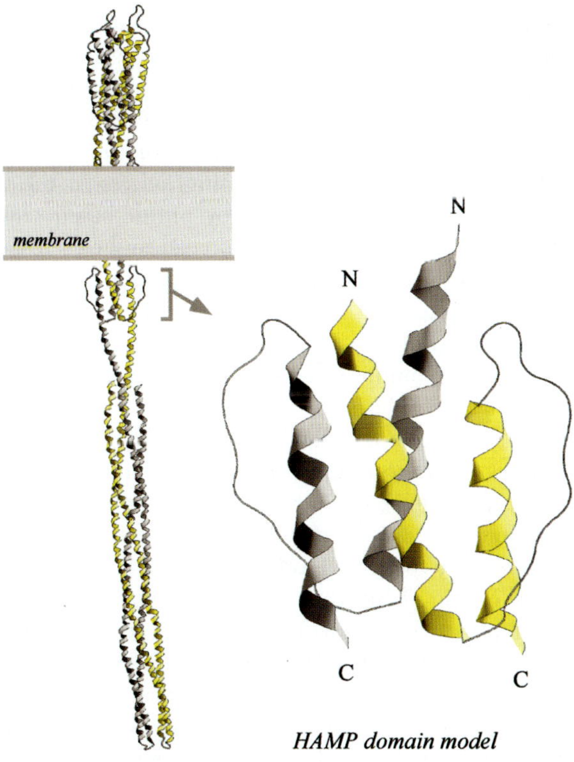

FIGURE 2 Model structure of the HAMP domain shown in the context of a bacterial chemotaxis receptor. The receptor model was based on structures of the Tar extracellular domain (1VLS) and the Tsr intracellular domain (1QU7). The HAMP domain was modeled on the structure of the HLH domain from the eukaryotic transcription factor Max (1AN2).

DHp and HPt Domains

Two basic types of histidine acceptor domains are involved in histidine-aspartate phosphotransfer reactions (Fig. 1): DHp and HPt. All class I kinases contain active-site histidines within the DHp dimerization domain. This domain consists of a left-handed four-helix bundle, formed by one α hairpin from each subunit [20, 24]; in the SCOP database, the fold is classified as ROP like. The acceptor histidine is found on the N-terminal helix of each hairpin, embedded within a region of conserved but not invariant residues.

A different type of DHp domain is found in the Spo0B protein of *Bacillus* species, which is involved in the bidirectional transfer of phosphoryl groups between Spo0F and Spo0A. Here, the domain also consists of a four-helix bundle, formed by one α hairpin from each subunit, with the acceptor histidine located on the N-terminal helix [19]. However, this bundle is right handed. The similarity between the two types of DHp domains may be the result of convergence; historically, Spo0B has been considered a variant of the HPt domain because of its position in the phosphorelay chain [25, 26]. However, several considerations led us to conclude that Spo0B is a degenerate transmitter and that the two types of DHp domains are descended from a common ancestor: (i) The histidine-containing helices of the two domains have sequence similarity that is detectable by profile-based sequence search methods such as PSI-Blast. Such a relationship cannot be established between either DHp variant and the analogous acceptor domain, HPt. (ii) The domain structure of Spo0B is that of a transmitter and the C-terminal domain of Spo0B has the same fold as the kinase domain, except that it has lost the residues forming the ATP-binding site ([20]; see also the SCOP classification and Fig. 3). (iii) Helical bundles can rearrange easily into alternate structures that are energetically almost equivalent. For example, the ROP protein, which has the same fold as the canonical DHp domain, can be caused to rearrange from a left-handed to a right-handed bundle by a single point mutation [27]. Thus, divergence by point mutation can account for the differences between the two types of structures observed today.

The second basic type of histidine acceptor domain is HPt. This domain serves as a phosphorylated intermediate in hybrid kinases (class Ic) and as the primary phosphate acceptor in chemotaxis kinases (class II). The structure of the HPt domain consists of a monomeric, left-handed four helix bundle [21, 28, 29] and is elaborated by an additional N terminal (Ypd1, ArcB) or C-terminal (CheA) helix. HPt and DHp are similar in overall size and secondary structure, with the active histidines solvent exposed and in similar local environments near the center of the bundle. However, the structural similarity is superficial and the histidine-containing helices have conserved

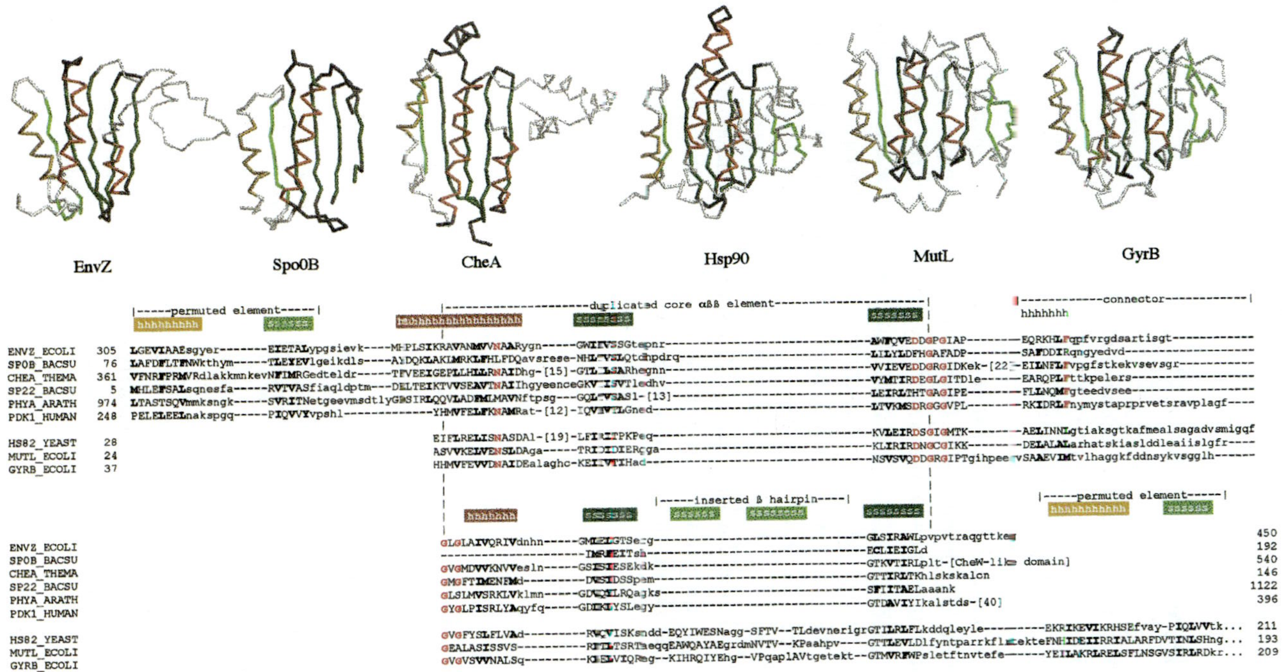

FIGURE 3 Gallery of structures belonging to the Hsp90/topoII/histidine kinase superfamily. The repeated αββ element is shown in dark colors, and the circularly permuted and inserted elements are shown in light colors. The connector was left gray. The position of these elements in the sequences is shown beneath the structures. The alignment reflects the structural superposition of the domains using STAMP [46]. For proteins of unknown structure, the alignment was made using conserved patterns of hydrophobic residues and the predicted secondary structure, obtained with JPred[2] (http://jura.ebi.ac.uk:8888/) [47]).

sequence patterns that are different. Thus, DHp and HPt most likely originated through convergent evolution.

The Kinase Domain

The structure of the ATP-binding domain of histidine kinases, which was determined for EnvZ (class Ia) and CheA (class II), belongs to a superfamily that includes the ATP-binding domains of Hsp90, MutL, and type II topoisomerases (Fig. 3) [20, 30]. The conserved structural core consists of an antiparallel, four-stranded β sheet flanked on one side by three α helices, which surround the ATP-binding site. In addition, an αβ element, which is in an equivalent structural position, is circularly permuted in the sequence of histidine kinases relative to other proteins with this fold. The structural similarity is mirrored in a set of conserved sequence motifs, primarily associated with nucleotide binding, which strongly imply descent from a common ancestor.

Phylogenetic analysis of the sequences in this superfamily by distance methods indicates that all kinases, namely PDKs, histidine kinases, anti-sigma F factors, and phytochromes, arose from a single ancestral protokinase (Fig. 4) [31]. In addition, the serine-phosphorylating kinases are not monophyletic, indicating that they evolved independently. The deepest branchpoint

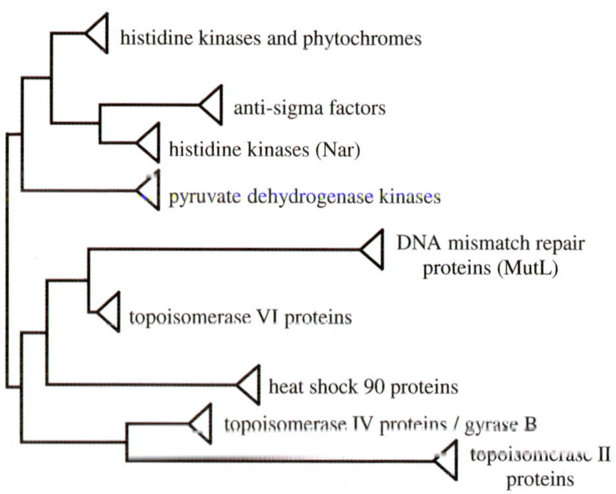

FIGURE 4 Neighbor-joining phylogenetic tree of the Hsp90/topoII/histidine kinase superfamily. The phylogeny was constructed using the program NEIGHBOR of the PHYLIP 3.57c package (http://evolution.genetics.washington.edu/phylip.html; [48]). From Koretke *et al.* [31].

in the kinase clade is formed by PDKs; it is therefore conceivable that the protokinase had Ser- rather than His-phosphorylating activity, even though most members of the family now phosphorylate histidine. This phylogenetic analysis must be viewed with some caution due to the great divergence of the sequences in the dataset. Nevertheless, monophyly of kinases is supported by structural features, specifically by the absence of a β hairpin located between strands 3 and 4 of the conserved domain and by the circular permutation mentioned earlier. Both structural features distinguish histidine kinases from Hsp90, MutL, and topoisomerases, and sequence analysis shows that they share these features with anti-sigma F factors, phytochromes, and PDKs (Fig. 3). (It should be mentioned here that the sequences specifying these features were not included in the phylogenetic calculations and therefore provide an independent verification of the result.)

Further analysis of the histidine kinase structure showed that the conserved domain consists of two consecutive αββ elements, whose structures are superimposable with less than a 2 Å root mean square deviation in backbone atoms (Fig. 5). Surprisingly, a similar arrangement of secondary

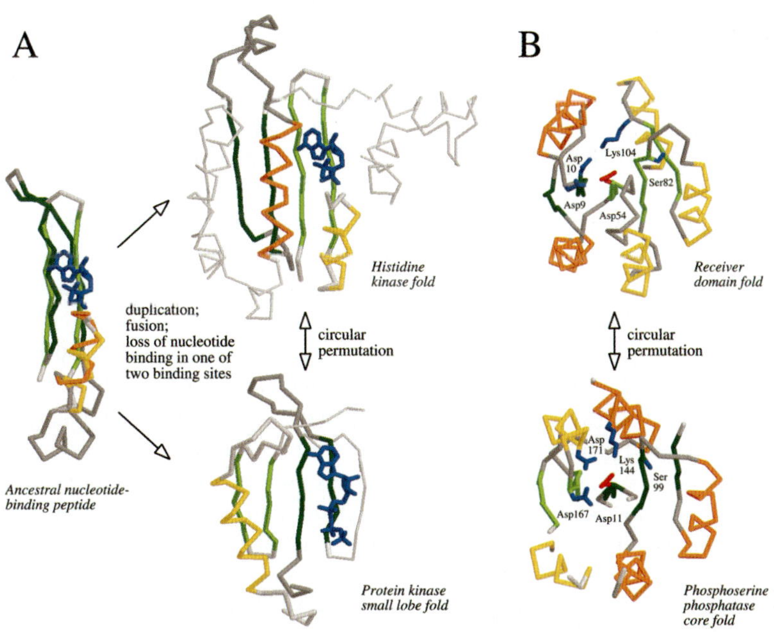

FIGURE 5 Circular permutation events that may relate (A) the histidine kinase and protein kinase small lobe folds and (B) the receiver domain and HAD superfamily folds. (A) A superposition of the two αββ domains that may have given rise to the histidine kinase fold. N-terminal segments are shown in dark colors and C-terminal segments in light colors.

structure elements, in circularly permuted form, is observable in the small (ATP-binding) lobe of eukaryotic protein kinases [31]. Several structural considerations suggest that this similarity might be more than a superficial analogy: (i) In histidine kinases, the N terminus of one αββ element is in close proximity to the C terminus of the other, a prerequisite of circular permutation. (ii) In both folds, the nucleotide is bound in a similar location and with a similar orientation relative to the protein, and important nucleotide interactions are made by two glycine-containing loops, which originate from the same parts of the structure. (iii) The structurally similar region covers virtually the entire conserved core of both folds. (iv) Although the protein kinase small lobe is part of the so-called ATP grasp fold (jointly with the peptide-binding large lobe), a recent structural analysis by Grishin [32] has concluded that only the large lobe is homologous among the members of this fold, with the small lobe having been recruited among structurally similar but unrelated proteins. Thus, it seems possible that both histidine and protein kinases evolved from an ancestral nucleotide-binding domain, which may initially have consisted of a single αββ element (Fig. 5).

The Receiver Domain

Receiver domains are characteristic of response regulators, but also occur quite frequently in histidine kinases; about a quarter of the histidine kinases listed in the SMART database contain this domain. Kinases with receiver domains are generally referred to as "hybrid." Most belong to class Ic and contain the receiver domain C-terminal to the kinase domain, often followed by an HPt domain. A subset of these kinases contain two receiver domains in tandem (*Dictyostelium* DhkJ, DhkL; *Synechocystis* slr1759, slr2098), although in many cases, the first of the two domains appears to be degenerated (*E. coli* BarA, RcsC; *Vibrio cholerae* LuxN). Other hybrid kinases include a substantial subset of CheA-like kinases, which also contain the receiver domain at their C-terminal end (*Myxococcus* FrzE, *Helicobacter* CheF), some soluble kinases of bacteria and archaea, which contain an N-terminal receiver domain (*Myxococcus* AsgA, AsgD; all archaeal hybrid kinases), and a small number of "double" kinases, which contain a receiver domain between two transmitters (*Pseudomonas putida* TobS, TodS; *Dictyostelium* DhkD).

The receiver domain consists of a doubly wound, parallel, five-stranded β sheet with strand order 21345 [33]; the SCOP database classifies it as flavodoxin-like. The critical residues are two aspartates located at the C-terminal end of β1, which form part of the Mg^{2+}-binding site, an aspartate at the C-terminal end of β3, which accepts the phosphate, and a lysine in a loop connecting β5 to α5, which coordinates the acceptor aspartate. The fold shows striking similarity to that of P-loop NTPases, such as Ras, which has led to speculation that these folds might be homologous [31, 34]. Ridder and Dijkstra [35], however, make a convincing argument that the receiver domain represents a circularly permuted form of the HAD superfamily [36], which includes P-type ATPases, phosphatases, epoxide hydrolases and haloacid dehalogenases. Conservation of the Mg^{2+}-binding site, the acceptor aspartate, and the lysine in structurally equivalent positions, as well as the overall similarity of the folds, strongly suggests homology between the receiver domain and the HAD superfamily (Fig. 5).

EVOLUTION OF HISTIDINE KINASES

PHYLOGENETIC SPECTRUM

Homologues of histidine kinases and response regulators are found throughout the living world in archaea, bacteria, and eukaryotes. In most cases, they are identified by sequence similarity to known TCST proteins, without confirming experimental evidence; occasionally, residues known to be critical for phosphotransfer are missing or domains are truncated, suggesting that some annotations may change as experimental biology catches up with genomics. Because of this uncertainty, global tallies of kinases and response regulators show variations depending on the criteria used for inclusion. Numbers in the following paragraphs are from Koretke *et al.* [31] unless specified otherwise.

In bacteria, the large number of genomes determined over the past years has shown that TCST systems are truly universal, being represented in organisms from every major branch. As the sole exception so far, mycoplasms (*M. genitalium* and *M. pneumoniae*) have been found to lack TCST proteins, but this is clearly the result of gene loss rather than an ancestral feature, as TCST proteins are otherwise richly represented in low GC gram-positive organisms. Although universal in distribution, TCST proteins vary sustantially in number among organisms. The largest number in a single genome appears to be 148, corresponding to the 72 (putative) histidine kinases and 76 (putative) response regulators of *Pseudomonas aeruginosa* listed in the SMART database. The smallest, as mentioned, is none in mycoplasms. The variations are not correlated with phylogeny, and large differences are observed between related bacterial species. Thus, among γ-proteobacteria, the number of TCST proteins ranges from 148 in *Pseudomonas* and 66 in *E. coli* to 9 in *Haemophilus influenzae*, and among low GC gram-positive bacteria from 70 in *Bacillus subtilis* to none in mycoplasms. Large variations are also observed between organisms from the same environmental niche, e.g., between the intracellular pathogens *Mycoplasma* and *Rickettsia*. The only rule that seems to apply with some consistency is that more TCST proteins occur in free-living species (*Escherichia*, *Bacillus*, *Synechocystis*) than in pathogenic ones (*Haemophilus*, *Rickettsia*, *Mycoplasma*).

Among archaea, representation of TCST proteins is much patchier. Some organisms, such as *Methanobacterium thermoautotrophicum* and *Archaeoglobus fulgidus*, contain substantial numbers of histidine kinases (16 and 14, respectively) and response regulators (10 and 11, respectively), and in the recently determined genome of *Halobacterium* NRC-1, 13 histidine kinases and 6 response regulators were identified. Other archaea only contain a single TCST system, namely chemotaxis (*Pyrococcus horikoshii, P. abysii*), and most archaea seem entirely devoid of TCST proteins (*P. furiosus*, *Methanococcus jannaschii*, *Aeropyrum pernix*, *Thermoplasma acidophilum*, *T. volcanii*, *Pyrobaculum islandicum*).

Among eukaryotes, only fungi, slime molds, and plants appear to contain TCST proteins. The complete genome of the yeast *Saccharomyces cerevisiae* contains 1 histidine kinase and 3 response regulators, and the genome of the mustard weed *Arabidopsis thaliana* contains 11 histidine kinases and 24 response regulators (of which 8 are classified as degenerate or pseudoresponse regulators) [37]. No histidine kinases have been found in most protists, such as *Plasmodium*, *Trypanosoma*, or *Giardia*, or in the complete (or nearly complete) genomes of the metazoans *Caenorhabditis elegans*, *Drosophila melanogaster*, mouse, and human.

MAJOR CLADES OF HISTIDINE KINASES

Phylogenetic studies [31, 38, 39], as well as classifications based on sequence motifs [40, 41], have subdivided histidine kinases into several major clades. The divisions arrived at in the various studies are largely congruent (Table I) and presumably reflect an underlying biological reality. As discussed in this section, several lines of evidence point to their existence. First, however, it may be useful to make some remarks on factors that are critical to the success of phylogenetic analysis in such a large and diverse group as the histidine kinases. Although much is made of the differences in flavor between methods (distance, parsimony, maximum likelihood), the primary determinant of success in any phylogenetic reconstruction is the quality of the underlying alignment. Because molecular phylogeny is based on the assumption that aligned residues are derived from the same position of an ancestral sequence, only positions in the alignment that are likely to be equivalent should be considered. Gaps and unalignable regions artificially increase the apparent evolutionary distance between sequences, and, because errors in alignment are more likely to occur between dissimilar sequences, result in a systematic bias.

TABLE I Histidine Kinase Clades Derived from Pylogenetic Analysis and Sequence Patterns, and Their Representation in *E. coli* and *B. subtilis*

Koretke *et al.* [31]; Throup *et al.* [39]	Grebe and Stock [40]	Kim and Forst [41]	*E. coli*	*B. subtilis*
Supercluster I				
Hybrid	Hpk1	Type I	BarA, ArcB, RcsC, TorS, EvgS	—
SAM	Hpk3h	Type I	—	—
Syn	Hpk1	Type I	KdpD	—
Ntr	Hpk4	Type I	AtoS, NtrB, HydH	KinA, KinB, KinC, YkrQ, YkvD
Pho	Hpk1,2,3	Type I	YgiY, BasS, YbcZ, YedV, PhoR, CreC, RstB, EnvZ, CpxA, YfhK, BaeS, PhoQ,	YdhK, YhqB, YycG, ResE, YclK, PhoR, YrkQ, YvrG, YkoH, YvcQ, YxdK, YtsB, SpaK, YcbA?
Supercluster II				
Arf	Hpk6	Type II	—	—
Cit	Hpk5	Type II	YidH, CitA	CitS, YufL, YdhF
Che	Hpk9	Type V	CheA	CheA
Lyt	Hpk8	Type IV	b2380, YehU	LytS, YwpD, YesM YccG
Agr	Hpk10	—	—	—
Supercluster III				
Nar	Hpk7	Type III	UhpB, NarX, NarQ	ComP, YdfH, YfiJ, YhcY, YvqE, DegS, YocF, YvfI, YxjM
Anti-s	—	—	—	SpoIIAB
Mth	Hpk11	Type III	—	—
Outgroup				
PDK	—	—	—	

In extreme cases, the inclusion of erroneously aligned or nonequivalent regions yields apparently infinite distances (i.e., a complete lack of detectable relatedness). Such errors happen most frequently in automated alignment programs and lead to meaningless phylogenies. A second important factor is the ratio between the number of aligned residues and the number of sequences in the alignment [referred to technically as operation taxonomic units (OTUs)]. In general, ratios smaller than 1 lead to poor resolution in the basal nodes of a tree. Both factors play critical roles in the accuracy of TCST phylogenies.

In an analysis of TCST proteins from 20 species, phylogenyetic reconstruction using distance and neighbor-joining methods yielded 13 major kinase clades, arranged in three superclusters (Fig. 6) [31]; further analysis of

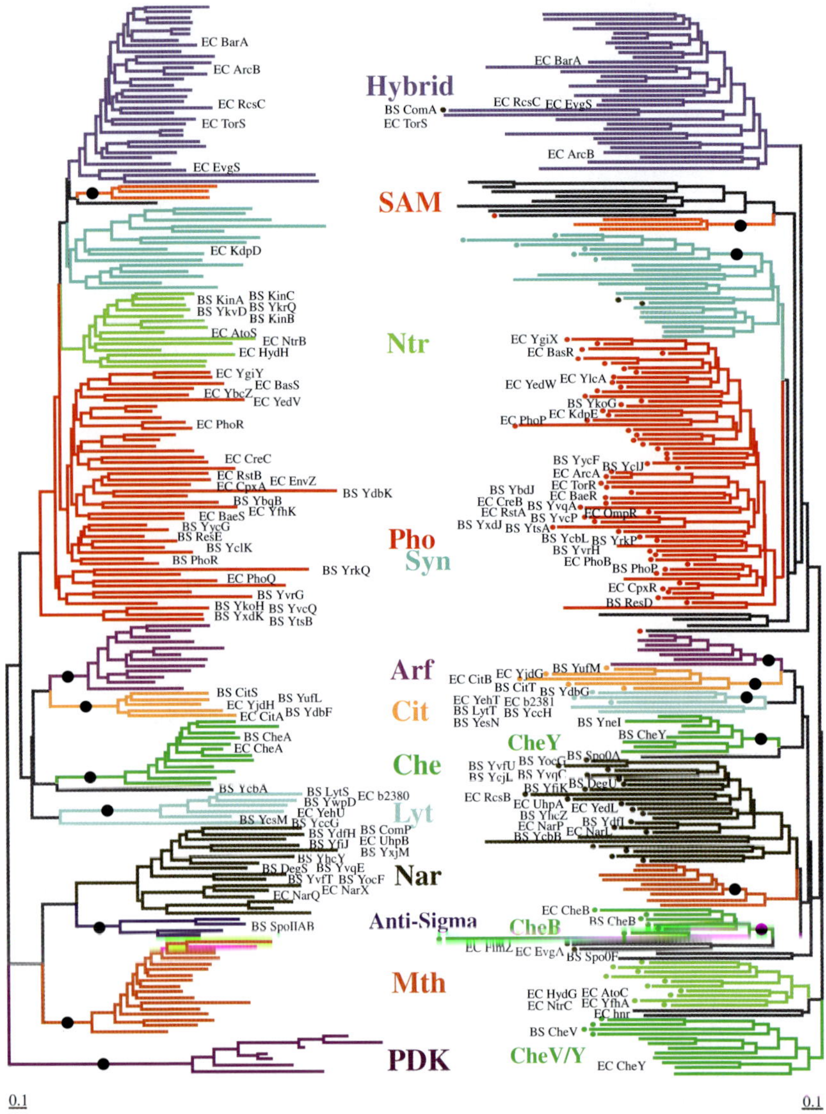

FIGURE 6 Neighbor-joining phylogenetic trees of histidine kinases (left) and response regulators (right). Multiple sequence alignments only included amino acid positions alignable in all sequences (133 in histidine kinases and 107 in response regulators). The neighbor-joining phylogeny was constructed using NEIGHBOR [48]. The scale bar represents 0.1 expected amino acid replacements per site as estimated by PROTDIST [48] using the Dayhoff PAM substitution matrix. Nodes that occurred in more than 50% of 500 random bootstrap replicates trees are marked with black dots. Reproduced in modified form from Koretke *et al.* [31].

Staphylococcus and *Streptococcus* species yielded one additional clade in supercluster II (Agr) [39]. Despite the low ratio between the number of aligned residues and OTUs, bootstrap support was obtained for 7 of these clades (marked by black dots in Fig. 6). Further support was obtained by maximum parsimony, which confirmed the main clades derived in the neighbor-joining tree. As already seen in the phylogenetic reconstruction of the kinase/Hsp90/topoII superfamily (Fig. 4), PDKs formed the outgroup to the other kinases and anti-sigma F factors were grouped with the Nar clade. The clades and superclusters closely parallel the divisions arrived at through sequence pattern analyses by Grebe and Stock (subfamilies HPK1–11) [40] and Kim and Forst (types I–V) [41] (Table I).

The smallest supercluster (supercluster III; type III) is formed of three clades: Anti-sigma, containing anti-sigma F factors; Nar (Hpk7), containing regulators of anaerobic respiration (NarQ/NarP and NarX/NarL), sugar phosphate uptake (UhpB/UhpA), and degradative enzyme expression (DegS/DegU); and Mth (Hpk 11), containing almost exclusively proteins of the archaeon *M. thermoautotrophicum*.

The second supercluster, in which practically every clade is bootstrap supported, contains five clades: Agr (Hpk10), containing virulence regulators of *Staphylococcus* (AgrC/AgrA) and transformation competence regulators of *Streptococcus* (ComD/ComE); Lyt (Hpk8; type IV), named for the LytS and LytT proteins, involved in *N*-acetylmuramoyl-L-alanine amidase biosynthesis; Che (Hpk9; type V), containing exclusively chemotaxis kinases; Cit (Hpk5), named for CitA and CitB, involved in the expression of citrate-specific fermentation genes; and Arf (Hpk6), formed entirely of proteins from the archaeon *A. fulgidus*. Cit and Arf are grouped together as type II by Kim and Forst [41] and also form a group in phylogenetic analysis (Fig.6).

The final supercluster (supercluster I; type I) contains by far the largest clades and correspondingly has the lowest resolution. Indeed, pattern analysis only reproduces two of its five clades: SAM (Hpk3h), containing proteins from *Synechocystis*, *Archaeoglobus*, and *Methanobacterium*; and Ntr (Hpk4), containing systems that regulate nitrogen assimilation (NtrB/NtrC and NtrY/NtrX), acetoacetate metabolism (AtoS/AtoC), and hydrogenase activity (HydH/HydG). The three other clades (Pho, Syn, and Hybrid) are amalgamated into subfamilies Hpk1–3; however, as will be seen in the next section, their existence and identity are strongly supported by the coevolution of response regulators. Pho contains TCST systems involved in phosphate regulation (PhoR/PhoB), virulence (PhoQ/PhoP), osmoregulation (EnvZ/OmpR), and anaerobic nitrite reduction (ResD/ResE); Syn is formed almost exclusively of proteins from *Synechocystis*, with three kinases from *Escherichia* (KdpD), *Rickettsia*, and *Mycobacterium*; and Hybrid contains all eukaryotic kinases, in agreement with the analysis of Pao and Saier [38], as well as many bacterial kinases, particularly from *E. coli* and *Synechocystis*. Approximately two-thirds of the kinases in the Hybrid clade, including all eukaryotic kinases except phytochrome, are hybrid kinases and most bacterial hybrid kinases fall into this clade [31].

Several interesting observations follow from these phylogenetic clades: Some are largely organism specific, suggesting extensive diversification of systems during speciation. No clade contains proteins from both archaea and eukaryotes, despite their presumed evolutionary relatedness. Bacterial phylogeny does not correlate well with the observed clustering, suggesting extensive lateral transfer. Despite the considerable size of some clades, none contain representatives from each bacterial species, indicating that none of the TCST systems observable today have a "universal" function conserved across bacteria. Although the emergence of further (mainly species-specific) clades must be expected in future analyses, it seems likely that all the large, broadly represented clades have been defined at this point.

Coevolution of Histidine Kinases and Response Regulators

A priori, two competing models for the evolution of TCST systems appear plausible; the recruitment model and the coevolution model. The recruitment model suggests that novel TCST systems evolve through gene duplication of one component, which then co-opts components from heterologous systems to yield a new specificity. The coevolution model suggests that novel TCST systems evolve by global duplication of all their components and subsequent differentiation.

A molecular phylogeny of receiver domains, performed in parallel with the kinase phylogeny by Koretke *et al.* [31], resulted in a set of clades that closely mirrored the kinase clades (Fig. 6). These clades agree well with the results of Pao and Saier [42], their classes 1 through 5 corresponding — in order — to the clades Ntr, Pho, Nar, CheB, and Hybrid in Fig. 6. The astonishing parallelism of the kinase and receiver phylogenies extended beyond individual clades to the formation of superclusters (indeed, only Ntr was found outside its supercluster, as defined by the kinase phylogeny). In addition, pairs of histidine kinases and response regulators that are known to interact were overwhelmingly found in cognate clades, as were most of the histidine kinases and response regulators found in chromosomal vicinity. All these observations point to coevolution being the dominant mechanism by which novel TCST systems arise.

Coevolution extends to the carboxyl-terminal domains of response regulators, whose activity (generally DNA binding) is regulated by the N-terminal receiver domains. In the study of Koretke *et al.* [31], almost all response regulators with homologous carboxyl-terminal domains fell within the same clade (marked by dots in Fig. 6). As the sequences of these domains were not included in the phylogenetic calculations, they also provide an independent confirmation of the receiver domain phylogeny.

Hybrid kinases also provide clear evidence for a recruitment mechanism at work; indeed, response regulator recruitment may be the rule in this clade. For example, four of the five hybrid kinases of *E. coli* (ArcB, TorS, RcsC, and EvgS) are known to signal through response regulators found in noncognate clades: ArcB and TorS through the Pho-type regulators ArcA and TorR, and RcsC and EvgS through the Nar-type regulators RcsB and EvgA. The fifth hybrid kinase, BarA, is thought to signal through a noncognate regulator as well (via OmpR, found in the Pho clade) [43]. It may be that recruitment will be found to play an important role in other TCST systems with complex phosphorelay paths, e.g., in class Ib. In addition, the phylogenetic placement of the CheY, CheB, and CheV/Y clades (in which only CheY has the location anticipated from the kinase tree) may also indicate recruitment (Fig. 6).

Polyphyletic Origin of Eukaryotic and Archaeal Systems

The conventional view of the universal tree is that archaea and eukaryotes are sister groups rooted in the bacteria and that all three urkingdoms of life are separate, monophyletic groups [44, 45]. However, the histidine kinase phylogeny clearly deviates from this canonical view: (i) archaea and eukaryotes are not sister groups. Eukaryotic histidine kinases and response regulators cluster with bacterial TCST proteins in a monophyletic group (Hybrid) that is not closely related to any of the archaeal clades. (ii) Archaea do not form a monophyletic group. Most of their TCST proteins form species-specific clades that are separate and most closely related to bacterial clades in different superclusters. (iii) Although TCST proteins do not occur universally in any of the urkingdoms, their representation is much more limited in archaea and eukaryotes. Despite the large number of sequenced bacterial genomes, only mycoplasms have been found to lack TCST systems, and these are obligate pathogens known to have reduced their gene complement greatly. Among the archaea, however, 11 sequenced genomes have already uncovered 6 that lack TCST proteins entirely and 2 that contain only a single system (Che). Among eukaryotes, TCST proteins are limited to fungi, slime molds, and plants. Thus far, no representatives have been found in animals or protists (these only contain PDKs, which were clearly acquired with the mitochondrial endosymbiont).

These observations suggest that TCST systems originated in bacteria after their separation from the last common ancestor and radiated into archaea and eukaryotes through independent lateral gene transfer events. Indeed, the species-specific clades observed in archaea point to multiple, independent transfer events within this kingdom. They also show that novel functional requirements may arise during the evolution of a species, which remain unfilled by endogenous genes, allowing acquired genes to establish themselves and diversify rapidly.

Serine versus Histidine Phosphorylation

Three branches of serine-phosphorylating kinases are embedded phylogenetically within the histidine kinase family: bacterial anti-sigma F factors, plant phytochromes, and mitochondrial PDKs. All three phosphorylate their substrate proteins directly and are simplified greatly relative to TCSTs, lacking both histidine acceptor domains and response regulators. Phylogenetic and structural analyses indicate that they originated from one ancestral protokinase. However, as shown earlier, they are themselves polyphyletic, i.e., they arose independently at different times in the evolution of histidine kinases.

The presence, in one phylogenetic clade, of proteins with both serine and histidine-phosphorylating activity raises the question of precedence. Phytochromes and anti-sigma F factors are most likely derived from ancestors with histidine phosphorylating activity, but pyruvate dehydrogenase kinase forms the deepest branchpoint within the kinase clade, and the absence of histidine acceptor and receiver domains may represent a primitive feature. The evolution of the same simplified state three times independently is compatible with the notion that this process represents reversion to an ancestral state. We therefore propose the following — highly speculative — scenario for the evolution of TCST systems (Fig. 7): Initially the protokinase may have had serine-phosphorylating activity and modulated a target enzyme in response to a regulatory signal (much like PDKs do today). In time, the system acquired a protophosphatase of the HAD superfamily, which dephosphorylated the target enzyme, itself forming a transient phospho-aspartate intermediate. Circular permutation altered the range of conformational changes available to the phosphatase in its phosphorylated form; upon fusion with a DNA-binding domain, this altered protein proved capable of regulating transcription by conformational change, yielding the first response regulator. Furthermore, the tendency of prokaryotes for grouping proteins with related functions into one operon may well have caused the phosphatase to fuse with the very DNA-binding protein responsible for the transcriptional regulation of the enzyme whose activity was regulated posttranslationally by the phos-

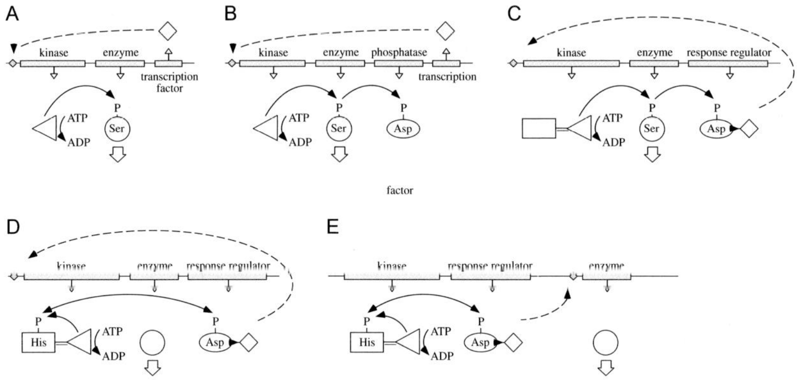

FIGURE 7 Hypothetical sequence of events by which TCST systems may have evolved. Geometric symbols are as in Fig. 1B; the diamond represents the transcription factor. Genes encoding the individual components are shown as gray boxes. (A) Ancestral system consisting of a metabolic enzyme and a serine kinase, which regulates the activity of the enzyme by posttranslational modification. The operon is under the control of a transcription factor. (B) The system has recruited a phosphatase, which counteracts the activity of the kinase. (C) After circular permutation and fusion with the transcription factor, the phosphatase acquires the ability to regulate transcription of the operon. The kinase acquires a dimerization domain. (D) By a point mutation to histidine, the dimerization domain becomes the primary target of phosphorylation by the kinase, and transfer is made directly to the phosphatase. Regulation of enzyme activity now occurs exclusively at the transcriptional level. (E) The first TCST system. The enzyme is now encoded elsewhere on the chromosome and the system is poised to spread by operon duplication and differentiation.

phatase. In time, the regulatory output of the phosphatase became the primary output of the system. At this point, direct transfer of the phosphoryl group from the kinase via a histidine residue (in a domain previously acquired for dimerization purposes) closed the circle, uncoupling transcriptional from posttranslational regulation. This system represented the last common ancestor of TCST systems, with all further domains (HAMP, HPt, PAS, GAF) being acquired subsequently and always only in individual subgroups of the TCST family. This scenario provides a plausible explanation for the ancestral position of PDKs and may represent a useful framework in which to consider the sequential evolution of a complex system (TCST) from a much simpler precursor.

CONCLUSION

As outlined in this chapter, histidine kinases probably originated in bacteria and radiated from there into the other domains of life. In most cases, they coevolved with their cognate response regulators, although in some cases, they subsequently recruited additional, heterologous receiver domains. Lateral transfer between bacteria appears to have played an important part in histidine kinase evolution, and their occasional clustering in species-specific clades suggests extensive functional diversification during speciation.

Histidine kinases arose from a protokinase with an ancient ATPase fold, which may itself have originated from the duplication of a nucleotide-binding peptide and which may be related by circular permutation to the small lobe of protein kinases. In the process of differentiation, histidine kinases evolved branches with serine-phosphorylating activity at least three times independently. The basal position of one such branch (the PDKs) suggests a scenario in which an original, simple system for posttranslational regulation gradually evolved into the complex system for transcriptional regulation that most TCST systems represent today. Of course, with data currently available, such scenarios are deeply rooted in speculation, but the rapid progress of structural and functional genomics, as well as the development of new bioinformatics tools, holds hope for a better understanding of the origins of this fascinating enzyme family in the near future.

REFERENCES

1. Zhulin, I. B., Taylor, B. L., and Dixon, R. (1997). PAS domain S-boxes in Archaea, Bacteria and sensors for oxygen and redox. *Trends Biochem. Sci.* **22**, 331–333.
2. Ponting, C. P., and Aravind, L. (1997). PAS: A multifunctional domain family comes to light. *Curr. Biol.* 7, R674–R677.
3. Aravind, L., and Ponting, C. P. (1997). The GAF domain: An evolutionary link between diverse phototransducing proteins. *Trends Biochem. Sci.* **22**, 458–459.
4. Ho, Y. S., Burden, L. M., and Hurley, J. H. (2000). Structure of the GAF domain, a ubiquitous signalling motif and a new class of cyclic GMP receptor. *EMBO J.* **19**, 5288–5299.
5. Lo Conte, L., Ailey, B., Hubbard, T. J., Brenner, S. E., Murzin, A. G., and Chothia, C. (2000). SCOP: A structural classification of proteins database. *Nucleic Acids Res.* **28**, 257–259.
6. Schultz, J., Copley, R. R., Doerks, T., Ponting, C. P., and Bork, P. (2000). SMART: A web-based tool for the study of genetically mobile domains. *Nucleic Acids Res.* **28**, 231–234.
7. Anantharaman, V., and Aravind, L. (2000). Cache: A signaling domain common to animal Ca^{2+}-channel subunits and a class of prokaryotic chemotaxis receptors. *Trends Biochem Sci.* **25**, 535–537.
8. McKay, D. B, and Steitz, T. A. (1981). Structure of catabolite gene activator protein at 2.9 A resolution suggests binding to left-handed B-DNA. *Nature (London)* **290**, 744–749.
9. Durocher, D., Henckel, J., Fersht, A. R., and Jackson, S. P. (1999). The FHA domain is a modular phosphopeptide recognition motif. *Mol. Cell.* **4**, 387–394.
10. Sack, J. S., Saper, M. A., and Quiocho, F. A. (1989). Periplasmic binding protein structure and function: Refined X-ray structures of the leucine/isoleucine/valine-binding protein and its complex with leucine. *J. Mol. Biol.* **206**, 171–191.
11. Ponting, C. P. (1997). Evidence for PDZ domains in bacteria, yeast, and plants. *Protein Sci.* **6**, 464–468.

12. Park, H., and Inouye, M. (1997). Mutational analysis of the linker region of EnvZ, an osmosensor in *Escherichia coli. J. Bacteriol.* **179**, 4382–4390.
13. Williams, S. B., and Stewart, V. (1999). Functional similarities among two-component sensors and methyl-acceptingchemotaxis proteins suggest a role for linker region amphipathic helices intransmembrane signal transduction. *Mol. Microbiol.* **33**, 1093–1102.
14. Aravind, L., and Ponting, C. P. (1999). The cytoplasmic helical linker domain of receptor histidine kinase andmethyl-accepting proteins is common to many prokaryotic signalling proteins. *FEMS Microbiol. Lett.* **176**, 111–116.
15. Dutta, R., Qin, L., and Inouye, M. (1999). Histidine kinases: Diversity of domain organization. *Mol. Microbiol.* **34**, 633–640.
16. Parkinson, J. S., and Kofoid, E.C. (1992). Communication modules in bacterial signaling proteins. *Annu. Rev. Genet.* **26**, 71–112.
17. Lau, P. C. K., Wang, Y., Patel, A., Labbé, D., Bergeron, H., Brousseau, R., Konishi, Y., and Rawlings, M. (1997). A bacterial basic region leucine zipper histidine kinase regulating toluene degradation. *Proc. Natl. Acad. Sci. USA* **94**, 1453–1458.
18. Clough, S. J., Lee, K. E., Schell, M. A., and Denny, T. P. (1997). A two-component system in *Ralstonia (Pseudomonas) solanacearum* modulates production of PhcA-regulated virulence factors in response to 3-hydroxypalmiticacid methyl ester. *J. Bacteriol.* **179**, 3639–3648.
19. Varughese, K. I., Madhusudan, Zhou, X. Z., Whiteley, J. M., and Hoch, J.A. (1998). Formation of a novel four-helix bundle and molecular recognition sites by dimerization of a response regulator phosphotransferase. *Mol. Cell.* **2**, 485–493.
20. Bilwes, A. M., Alex, L.A., Crane, B. R., and Simon M. I. (1999). Structure of CheA, a signal-transducing histidine kinase. *Cell.* **96**, 131–141.
21. Mourey, L., Da Re, S., Pedelacq, J. D., Tolstykh, T., Faurie, C., Guillet, V., Stock, J. B., and Samama, J. P.(2001). Crystal structure of the CheA histidine phosphotransfer domain that mediates response regulator phosphorylation in bacterial chemotaxis. *J. Biol. Chem.*
22. Butler, S. L., and Falke, J. J. (1998). Cysteine and disulfide scanning reveals two amphiphilic helices in the linker region of the aspartate chemoreceptor. *Biochemistry* **37**, 10746–10756.
23. Singh, M., Berger, B., Kim, P. S., Berger, J. M., and Cochran, A. G. (1998). Computational learning reveals coiled coil-like motifs in histidine kinase linker domains. *Proc. Natl. Acad. Sci. USA* **95**, 2738–2743.
24. Tomomori, C., Tanaka, T., Dutta, R., Park, H., Saha, S. K., Zhu, Y., Ishima, R., Liu, D., Tong, K. I., Kurokawa, H., Qian, H., Inouye, M., and Ikura, M. (1999). Solution structure of the homodimeric core domain of *Escherichia coli* histidine kinase EnvZ. *Nature Struct. Biol.* **6**, 729–734.
25. Appleby, J. L., Parkinson, J. S., and Bourret, R. B. (1996). Signal transduction via the multi-step phosphorelay: Not necessarily a road less traveled. *Cell* **86**, 845–848.
26. Perraud, A. L., Weiss, V., and Gross, R. (1999). Signalling pathways in two-component phosphorelay systems. *Trends Microbiol.* **7**, 115–120.
27. Glykos, N. M., Cesareni, G., and Kokkinidis, M. (1999). Protein plasticity to the extreme: Changing the topology of a 4-alpha-helical bundle with a single amino acid substitution. *Structure Fold. Des.* **7**, 597–603.
28. Xu, Q., and West, A. (1999). Conservation of structure and function among histidine-containing phosphotransfer (Hpt) domains as revealed by crystal structure YPD1. *J. Mol. Biol.* **292**, 1039–1050.
29. Kato, M., Mizuno, T., Shimizu, T., and Hakoshima, T. (1997). Insights into multistep phosphorelay from the crystal structure of the C-terminal HPt domain of ArcB. *Cell* **88**, 717–723.
30. Tanaka, T., Saha, S. K., Tomomori, C., Ishima, R., Liu, D., Tong, K. I., Park, H., Dutta, R., Qin, L., Swindells, M. B., Yamazaki, T., Ono, A. M., Kainosho, M., Inouye, M., and Ikura, M. (1998). NMR structure of the histidine kinase domain of the *E. coli* osmosensor EnvZ. *Nature (London)* **396**, 88–92.

31. Koretke, K. K., Lupas, A. N., Warren, P. V., Rosenberg, M., and Brown, J. R. (2000). Evolution of two-component signal transduction. *Mol. Biol. Evol.* **17**, 1956–1970.
32. Grishin, N.V. (1999). Phosphatidylinositol phosphate kinase: A link between protein kinase and glutathione synthase folds. *J. Mol. Biol.* **291**, 239–247.
33. Stock, A.M., Martinez-Hackert, E., Rasmussen, B.F., West, A.H., Stock, J.B., Ringe, D., and Petsko, G.A. (1993). Structure of the Mg(2+)-bound form of CheY and mechanism of phosphoryl transfer in bacterial chemotaxis. *Biochemistry* **32**, 13375–13380.
34. Artymiuk, P.J., Rice, D.W., Mitchell, E.M., and Willett, P. (1990). Structural resemblance between the families of bacterial signal-transduction proteins and of G proteins revealed by graph-theoretical techniques. *Protein Eng.* **4**, 39–43.
35. Ridder, I. S., and Dijkstra, B. W. (1999). Identification of the Mg(2+)-binding site in the P-type ATPase and phosphatase members of the HAD (haloacid dehalogenase) superfamily by structural similarity to the response regulator protein CheY. *Biochem. J.* **339**, 223–226.
36. Koonin, E. V, and Tatusov, R. L. (1994). Computer analysis of bacterial haloacid dehalogenases defines a large superfamily of hydrolases with diverse specificity: Application of an iterative approach to database search. *J. Mol. Biol.* **244**, 125–132.
37. Arabidopsis Genome Initiative. (2000). Analysis of the genome sequence of the flowering plant *Arabidopsis thaliana*. *Nature (London)* **408**, 796–815.
38. Pao, G. M., and Saier, M. H., Jr. (1997). Nonplastid eukaryotic response regulators have a monophyletic origin and evolved from their bacterial precursors in parallel with their cognate sensor kinases. *J. Mol. Evol.* **44**, 605–613.
39. Throup, J. P., Koretke, K. K., Bryant, A. P., Ingraham, K. A., Chalker, A. F., Ge, Y., Marra, A.,Wallis, N. G., Brown, J. R., Holmes, D. J., Rosenberg, M., and Burnham, M. K. (2000). A genomic analysis of two-component signal transduction in *Streptococcus pneumoniae*. *Mol. Microbiol.* **35**, 566–576.
40. Grebe, T. W., and Stock, J. B. (1999). The histidine protein kinase superfamily. *Adv. Microb. Physiol.* **41**, 139–227.
41. Kim, Dj., and Forst, S. (2001). Genomic analysis of the histidine kinase family in bacteria and archaea. *Microbiology* **147**, 1197–1212.
42. Pao, G. M., and Saier, M.H. Jr. (1995). Response regulators of bacterial signal transduction systems: Selective domain shuffling during evolution. *J. Mol. Evol.* **40**, 136–154.
43. Nagasawa, S., Tokishita, S., Aiba, H., and Mizuno, T. (1992). A novel sensor-regulator protein that belongs to the homologous family of signal-transduction proteins involved in adaptive responses in *Escherichia coli*. *Mol. Microbiol.* **6**, 799–807.
44. Brown, J. R., and Doolittle, W. F. (1997). Archaea and the prokaryote-to-eukaryote transition. *Microbiol. Mol. Biol. Rev.* **61**, 456–502.
45. Woese, C. R., Kandler, O., and Wheelis, M.L. (1990). Towards a natural system of organisms. Proposal for the domains Archaea, Bacteria and Eucarya. *Proc. Natl. Acad. Sci. USA* **87**, 4576–4579.
46. Russell, R. B., and Barton, G.J. (1992). Multiple protein sequence alignment from tertiary structure comparison: Assignment of global and residue confidence levels. *Proteins* **14**, 309–323.
47. Cuff, J. A., and Barton, G. J. (2000). Application of multiple sequence alignment profiles to improve protein secondary structure prediction. *Proteins* **40**, 502–511.
48. Felsenstein, J. (1993). PHYLIP (Phylogeny Inference Package) version 3.5c. Department of Genetics, University of Washington, Seattle.

INDEX